# Advances in postharvest management of cereals and grains

It is widely recognised that agriculture is a significant contributor to global warming and climate change. Agriculture needs to reduce its environmental impact and adapt to current climate change whilst still feeding a growing population, i.e. become more 'climate-smart'. Burleigh Dodds Science Publishing is playing its part in achieving this by bringing together key research on making the production of the world's most important crops and livestock products more sustainable.

Based on extensive research, our publications specifically target the challenge of climate-smart agriculture. In this way we are using 'smart publishing' to help achieve climate-smart agriculture.

Burleigh Dodds Science Publishing is an independent and innovative publisher delivering high quality customer-focused agricultural science content in both print and online formats for the academic and research communities. Our aim is to build a foundation of knowledge on which researchers can build to meet the challenge of climate-smart agriculture.

For more information about Burleigh Dodds Science Publishing simply call us on +44 (0) 1223 839365, email info@bdspublishing.com or alternatively please visit our website at www.bdspublishing.com.

**Related titles:**

Preventing food losses and waste to achieve food security and sustainability
Print (ISBN 978-1-78676-300-6); Online (ISBN 978-1-78676-302-0, 978-1-78676-303-7)

Achieving sustainable cultivation of barley
Print (ISBN 978-1-78676-308-2); Online (ISBN 978-1-78676-310-5, 978-1-78676-311-2)

Integrated disease management of wheat and barley
Print (ISBN 978-1-78676-216-0); Online (ISBN 978-1-78676-219-1, 978-1-78676-218-4)

Achieving sustainable cultivation of wheat Volume 2: Cultivation techniques
Print (ISBN 978-1-78676-020-3); Online (ISBN 978-1-78676-022-7, 978-1-78676-023-4)

Achieving sustainable cultivation of maize Volume 2: Cultivation techniques, pest and disease control
Print (ISBN 978-1-78676-012-8); Online (ISBN 978-1-78676-014-2, 978-1-78676-015-9)

Chapters are available individually from our online bookshop:
https://shop.bdspublishing.com

climate•SMART•publishing
IN AGRICULTURAL SCIENCE

BURLEIGH DODDS SERIES IN AGRICULTURAL SCIENCE
NUMBER 88

# Advances in postharvest management of cereals and grains

Edited by Professor Dirk E. Maier, Iowa State University, USA

burleigh dodds
SCIENCE PUBLISHING

Published by Burleigh Dodds Science Publishing Limited
82 High Street, Sawston, Cambridge CB22 3HJ, UK
www.bdspublishing.com

Burleigh Dodds Science Publishing, 1518 Walnut Street, Suite 900, Philadelphia, PA 19102-3406, USA

First published 2020 by Burleigh Dodds Science Publishing Limited


Library of Congress Control Number: 2020936950

British Library Cataloguing in Publication Data
A catalogue record for this book is available from the British Library

ISBN 978-1-78676-352-5 (Print)
ISBN 978-1-78676-355-6 (PDF)
ISBN 978-1-78676-354-9 (ePub)
ISSN 2059-6936 (print)
ISSN 2059-6944 (online)

DOI 10.19103/AS.2020.0072

Typeset by Deanta Global Publishing Services, Dublin, Ireland

# Contents


Series list x
Acknowledgements xvi
Introduction xvii

**Part 1 Postharvest losses and their causes**

1 Post-harvest losses of cereals and other grains: opportunity among issues and challenges 3
*Steven T. Sonka, University of Maryland, USA*

1 Introduction 3
2 Is zero loss a desirable goal - what does economics tell us? 7
3 What is the 'hidden tax' resulting from excessive post-harvest loss (PHL)? 10
4 Can we move from measuring loss to managing to reduce PHL? 13
5 Can we overcome the PHL paradox? 15
6 Examples demonstrating the potential for sustainable reduction of PHL 19
7 Summary 23
8 Where to look for further information 26
9 References 27

2 Advances in understanding fungal contamination in cereals 31
*Kizito Nishimwe, Iowa State University, USA and University of Rwanda, Rwanda; Julie Aiza L. Mandap, Iowa State University, USA and University of The Philippines Los Baños, The Philippines; and Gary P. Munkvold, Iowa State University, USA*

1 Introduction 31
2 Fungi affecting cereal grains 34
3 Fungal contamination in maize 42
4 Fungal contamination in rice 48
5 Fungal contamination in wheat 52
6 Conclusion 54
7 References 55

3 Advances in detection and management of phosphine resistance in stored grain pests 67
*David Schlipalius and Paul Ebert, University of Queensland, Australia*

1 Introduction 67
2 The biochemical and genetic basis of resistance 80
3 Alternative fumigants 82
4 Conclusion and future trends 85
5 References 85

4 Advances in understanding rodent pests affecting cereal grains 93
*Peter R. Brown, CSIRO Health and Biosecurity, Australia; Grant R. Singleton, International Rice Research Institute, The Philippines; Steven R. Belmain, Natural Resources Institute (NRI) - University of Greenwich, UK; Nyo Me Htwe, Plant Protection Division - Myanma Agriculture Service, Myanmar; Loth Mulungu, Sokoine University of Agriculture, Tanzania; Mashaka Mdangi, Ministry of Agriculture, Tanzania; and Regino Cavia, University of Buenos Aires, Argentina*

1 Introduction 93
2 Impact of rodents on stored grain 99
3 Rodent pest species and their biology 102
4 Types of rodent damage 105
5 Strategies to reduce rodent damage 106
6 Case studies 110
7 Conclusion 113
8 Future trends in research 114
9 Where to look for further information 117
10 References 117

**Part 2 Storage technologies**

5 Advances in bulk storage of cereals and grains 125
*Michelle A. Friedmann and Dirk E. Maier, Iowa State University, USA*

1 Introduction 125
2 Storage structure types: upright silo structures 127
3 Horizontal building structures 137
4 On-the-ground piles 144
5 Maintaining and preserving quality and quantity: preventative strategies 150
6 Maintaining and preserving quality and quantity: intervention strategies 154
7 Best management practices: sanitation, loading, aeration and monitoring (SLAM) 161
8 Best management practices: dry, inspect, clean and examine (DICE) 162

9 Conclusion 163
10 Abbreviations 164
11 Where to look for further information 166
12 References 166

6 Developments in the use of hermetic bags for grain storage 171
*Dieudonne Baributsa, Purdue University, USA; and Ma Cristine Concepcion Ignacio, Iowa State University, USA and University of the Philippines Los Baños, The Philippines*

1 Introduction 171
2 Development of hermetic storage bags 172
3 Hermetic bags for grain storage 175
4 Dissemination of hermetic bags to reduce postharvest losses among smallholder farmers 178
5 Challenges in the use of hermetic storage bag technology 181
6 Case study: Purdue Improved Crop Storage (PICS) bag 185
7 Summary and future trends 191
8 Where to look for further information 192
9 References 193

7 Advances in insect pest management in postharvest storage of cereals: detection and monitoring 199
*Paul Fields, Agriculture and Agri-Food Canada, Canada; Fuji Jian, University of Manitoba, Canada; and Dianxuan Wang, Henan University of Technology, China*

1 Introduction 199
2 How to sample 200
3 Sampling theory 215
4 Future trends in research 221
5 Where to look for further information 222
6 Acknowledgements 222
7 References 222

8 Advances in insect pest management in postharvest storage of cereals: use of controlled atmosphere and temperature control 231
*Shlomo Navarro and Hagit Navarro, Green Storage Ltd., Israel*

1 Introduction 231
2 Controlled and modified atmospheres 232
3 Temperature manipulation 245
4 Conclusions and future trends 257
5 Acknowledgement 260
6 References 260

9 Biologically based control strategies for managing stored-product insect pests 267
*Benjamin Fürstenau and Garnet Marlen Kroos, Julius Kühn-Institut (JKI), Germany*

1 Introduction 267
2 Semiochemicals and their use in stored-product protection 275
3 Biological control of stored-product pests 285
4 Acknowledgements 296
5 Where to look for further information 297
6 References 298

10 Advances in insect pest management in postharvest storage of cereals: novel techniques 319
*Peter Follett, USDA-ARS, USA; Georgios Akepsimaidis and Nicolas Meneses, Bühler AG, Switzerland; and Matthew Murdoch and Heidi Kotilainen, Bühler UK Limited, UK*

1 Introduction 319
2 Irradiation for control of stored product pests 320
3 Sources of ionizing radiation 324
4 Case study 1: low-energy electron beam treatment 328
5 Case study 2: high-energy X-ray treatment 332
6 Conclusion 334
7 Future trends 334
8 Where to look for further information 335
9 References 335

11 Advances in post-harvest detection and control of fungal contamination of cereals 339
*Naresh Magan, Esther Garcia-Cela, Carol Verheecke-Vaessen and Angel Medina, Cranfield University, UK*

1 Introduction 339
2 Factors affecting the activity of spoilage moulds and mycotoxin contamination 341
3 Volatiles as an early indicator of spoilage of stored cereals 342
4 Carbon dioxide monitoring for real-time identification of the risk of spoilage moulds and mycotoxins in stored cereals 344
5 Control strategies to minimise spoilage and mycotoxin contamination of stored grain: monitoring of relative humidity (RH) and temperature and the use of molecular methods 345
6 Control strategies: hermetic storage of grain and use of natural compounds 348

7 Control strategies: biological control and irradiation techniques 350
8 Control technologies: cold plasma technology and ozone fumigation 351
9 Conclusions 354
10 Where to look for further information 355
11 References 356

12 Advances in techniques for monitoring the quality of stored cereal grains 363
*Roger G. Aby and Dirk E. Maier, Iowa State University, USA*

1 Introduction 363
2 Methods and techniques used to monitor stored grain 364
3 Conclusion 382
4 Future trends 383
5 Where to look for further information 384
6 References 384

13 Supporting smallholder farmers in developing countries to improve postharvest management of staple grains: the role of loss reduction technologies 389
*Brighton M. Mvumi, University of Zimbabwe, Zimbabwe; and Tanya Stathers, Natural Resources Institute (NRI) - University of Greenwich, UK*

1 Introduction 389
2 Improvement strategies 394
3 Institutional, policy and market-related issues for PHM transformation 416
4 Creating sustainable markets: case study models 424
5 Conclusion and future trends 430
6 Where to look for further information 435
7 References 437

Index 445

# Series list

| Title | Series number |
|---|---|
| **Achieving sustainable cultivation of maize - Vol 1**<br>From improved varieties to local applications<br>Edited by: Dr Dave Watson, CGIAR Maize Research Program Manager, CIMMYT, Mexico | 001 |
| **Achieving sustainable cultivation of maize - Vol 2**<br>Cultivation techniques, pest and disease control<br>Edited by: Dr Dave Watson, CGIAR Maize Research Program Manager, CIMMYT, Mexico | 002 |
| **Achieving sustainable cultivation of rice - Vol 1**<br>Breeding for higher yield and quality<br>Edited by: Prof. Takuji Sasaki, Tokyo University of Agriculture, Japan | 003 |
| **Achieving sustainable cultivation of rice - Vol 2**<br>Cultivation, pest and disease management<br>Edited by: Prof. Takuji Sasaki, Tokyo University of Agriculture, Japan | 004 |
| **Achieving sustainable cultivation of wheat - Vol 1**<br>Breeding, quality traits, pests and diseases<br>Edited by: Prof. Peter Langridge, The University of Adelaide, Australia | 005 |
| **Achieving sustainable cultivation of wheat - Vol 2**<br>Cultivation techniques<br>Edited by: Prof. Peter Langridge, The University of Adelaide, Australia | 006 |
| **Achieving sustainable cultivation of tomatoes**<br>Edited by: Dr Autar Mattoo, USDA-ARS, USA & Prof. Avtar Handa, Purdue University, USA | 007 |
| **Achieving sustainable production of milk - Vol 1**<br>Milk composition, genetics and breeding<br>Edited by: Dr Nico van Belzen, International Dairy Federation (IDF), Belgium | 008 |
| **Achieving sustainable production of milk - Vol 2**<br>Safety, quality and sustainability<br>Edited by: Dr Nico van Belzen, International Dairy Federation (IDF), Belgium | 009 |
| **Achieving sustainable production of milk - Vol 3**<br>Dairy herd management and welfare<br>Edited by: Prof. John Webster, University of Bristol, UK | 010 |
| **Ensuring safety and quality in the production of beef - Vol 1**<br>Safety<br>Edited by: Prof. Gary Acuff, Texas A&M University, USA & Prof. James Dickson, Iowa State University, USA | 011 |
| **Ensuring safety and quality in the production of beef - Vol 2**<br>Quality<br>Edited by: Prof. Michael Dikeman, Kansas State University, USA | 012 |
| **Achieving sustainable production of poultry meat - Vol 1**<br>Safety, quality and sustainability<br>Edited by: Prof. Steven C. Ricke, University of Arkansas, USA | 013 |
| **Achieving sustainable production of poultry meat - Vol 2**<br>Breeding and nutrition<br>Edited by: Prof. Todd Applegate, University of Georgia, USA | 014 |

**Achieving sustainable production of poultry meat - Vol 3** 015
Health and welfare
Edited by: Prof. Todd Applegate, University of Georgia, USA

**Achieving sustainable production of eggs - Vol 1** 016
Safety and quality
Edited by: Prof. Julie Roberts, University of New England, Australia

**Achieving sustainable production of eggs - Vol 2** 017
Animal welfare and sustainability
Edited by: Prof. Julie Roberts, University of New England, Australia

**Achieving sustainable cultivation of apples** 018
Edited by: Dr Kate Evans, Washington State University, USA

**Integrated disease management of wheat and barley** 019
Edited by: Prof. Richard Oliver, Curtin University, Australia

**Achieving sustainable cultivation of cassava - Vol 1** 020
Cultivation techniques
Edited by: Dr Clair Hershey, formerly International Center for Tropical Agriculture (CIAT), Colombia

**Achieving sustainable cultivation of cassava - Vol 2** 021
Genetics, breeding, pests and diseases
Edited by: Dr Clair Hershey, formerly International Center for Tropical Agriculture (CIAT), Colombia

**Achieving sustainable production of sheep** 022
Edited by: Prof. Johan Greyling, University of the Free State, South Africa

**Achieving sustainable production of pig meat - Vol 1** 023
Safety, quality and sustainability
Edited by: Prof. Alan Mathew, Purdue University, USA

**Achieving sustainable production of pig meat - Vol 2** 024
Animal breeding and nutrition
Edited by: Prof. Julian Wiseman, University of Nottingham, UK

**Achieving sustainable production of pig meat - Vol 3** 025
Animal health and welfare
Edited by: Prof. Julian Wiseman, University of Nottingham, UK

**Achieving sustainable cultivation of potatoes - Vol 1** 026
Breeding improved varieties
Edited by: Prof. Gefu Wang-Pruski, Dalhousie University, Canada

**Achieving sustainable cultivation of oil palm - Vol 1** 027
Introduction, breeding and cultivation techniques
Edited by: Prof. Alain Rival, Center for International Cooperation in Agricultural Research for Development (CIRAD), France

**Achieving sustainable cultivation of oil palm - Vol 2** 028
Diseases, pests, quality and sustainability
Edited by: Prof. Alain Rival, Center for International Cooperation in Agricultural Research for Development (CIRAD), France

**Achieving sustainable cultivation of soybeans - Vol 1** 029
Breeding and cultivation techniques
Edited by: Prof. Henry T. Nguyen, University of Missouri, USA

**Achieving sustainable cultivation of soybeans - Vol 2** 030
Diseases, pests, food and non-food uses
Edited by: Prof. Henry T. Nguyen, University of Missouri, USA

**Achieving sustainable cultivation of sorghum - Vol 1** 031
Genetics, breeding and production techniques
Edited by: Prof. William Rooney, Texas A&M University, USA

**Achieving sustainable cultivation of sorghum - Vol 2** 032
Sorghum utilization around the world
Edited by: Prof. William Rooney, Texas A&M University, USA

**Achieving sustainable cultivation of potatoes - Vol 2** 033
Production, storage and crop protection
Edited by: Dr Stuart Wale, Potato Dynamics Ltd, UK

**Achieving sustainable cultivation of mangoes** 034
Edited by: Prof. Víctor Galán Saúco, Instituto Canario de Investigaciones Agrarias (ICIA), Spain & Dr Ping Lu, Charles Darwin University, Australia

**Achieving sustainable cultivation of grain legumes - Vol 1** 035
Advances in breeding and cultivation techniques
Edited by: Dr Shoba Sivasankar et al., formerly International Crops Research Institute for the Semi-Arid Tropics (ICRISAT), India

**Achieving sustainable cultivation of grain legumes - Vol 2** 036
Improving cultivation of particular grain legumes
Edited by: Dr Shoba Sivasankar et al., formerly International Crops Research Institute for the Semi-Arid Tropics (ICRISAT), India

**Achieving sustainable cultivation of sugarcane - Vol 1** 037
Cultivation techniques, quality and sustainability
Edited by: Prof. Philippe Rott, University of Florida, USA

**Achieving sustainable cultivation of sugarcane - Vol 2** 038
Breeding, pests and diseases
Edited by: Prof. Philippe Rott, University of Florida, USA

**Achieving sustainable cultivation of coffee** 039
Edited by: Dr Philippe Lashermes, Institut de Recherche pour le Développement (IRD), France

**Achieving sustainable cultivation of bananas - Vol 1** 040
Cultivation techniques
Edited by: Prof. Gert H. J. Kema, Wageningen University and Research, The Netherlands & Prof. André Drenth, University of Queensland, Australia

**Global Tea Science** 041
Current status and future needs
Edited by: Dr V. S. Sharma, formerly UPASI Tea Research Institute, India & Dr M. T. Kumudini Gunasekare, Coordinating Secretariat for Science Technology and Innovation (COSTI), Sri Lanka

**Integrated weed management** 042
Edited by: Emeritus Prof. Rob Zimdahl, Colorado State University, USA

**Achieving sustainable cultivation of cocoa** 043
Edited by: Prof. Pathmanathan Umaharan, Cocoa Research Centre – The University of the West Indies, Trinidad and Tobago

**Robotics and automation for improving agriculture** 044
Edited by: Prof. John Billingsley, University of Southern Queensland, Australia

**Water management for sustainable agriculture** 045
Edited by: Prof. Theib Oweis, ICARDA, Jordan

**Improving organic animal farming** 046
Edited by: Dr Mette Vaarst, Aarhus University, Denmark & Dr Stephen Roderick, Duchy College, UK

**Improving organic crop cultivation** 047
Edited by: Prof. Ulrich Köpke, University of Bonn, Germany

**Managing soil health for sustainable agriculture - Vol 1** 048
Fundamentals
Edited by: Dr Don Reicosky, Soil Scientist Emeritus USDA-ARS and University of Minnesota, USA

**Managing soil health for sustainable agriculture - Vol 2** 049
Monitoring and management
Edited by: Dr Don Reicosky, Soil Scientist Emeritus USDA-ARS and University of Minnesota, USA

**Rice insect pests and their management** 050
E. A. Heinrichs, Francis E. Nwilene, Michael J. Stout, Buyung A. R. Hadi & Thais Freitas

**Improving grassland and pasture management in temperate agriculture** 051
Edited by: Prof. Athole Marshall & Dr Rosemary Collins, IBERS, Aberystwyth University, UK

**Precision agriculture for sustainability** 052
Edited by: Dr John Stafford, Silsoe Solutions, UK

**Achieving sustainable cultivation of temperate zone tree fruit and berries - Vol 1** 053
Physiology, genetics and cultivation
Edited by: Prof. Gregory A. Lang, Michigan State University, USA

**Achieving sustainable cultivation of temperate zone tree fruit and berries - Vol 2** 054
Case studies
Edited by: Prof. Gregory A. Lang, Michigan State University, USA

**Agroforestry for sustainable agriculture** 055
Edited by: Prof. María Rosa Mosquera-Losada, Universidade de Santiago de Compostela, Spain & Dr Ravi Prabhu, World Agroforestry Centre (ICRAF), Kenya

**Achieving sustainable cultivation of tree nuts** 056
Edited by: Prof. Ümit Serdar, Ondokuz Mayis University, Turkey & Emeritus Prof. Dennis Fulbright, Michigan State University, USA

**Assessing the environmental impact of agriculture** 057
Edited by: Prof. Bo P. Weidema, Aalborg University, Denmark

**Critical issues in plant health: 50 years of research in African agriculture** 058
Edited by: Dr Peter Neuenschwander and Dr Manuele Tamò, IITA, Benin

**Achieving sustainable cultivation of vegetables** 059
Edited by: Emeritus Prof. George Hochmuth, University of Florida, USA

**Advances in breeding techniques for cereal crops** 060
Edited by: Prof. Frank Ordon, Julius Kuhn Institute (JKI), Germany & Prof. Wolfgang Friedt, Justus-Liebig University of Giessen, Germany

**Advances in Conservation Agriculture - Vol 1** 061
Systems and Science
Edited by: Prof. Amir Kassam, University of Reading, UK and Moderator, Global Conservation Agriculture Community of Practice (CA-CoP), FAO, Rome, Italy

**Advances in Conservation Agriculture - Vol 2** 062
Practice and Benefits
Edited by: Prof. Amir Kassam, University of Reading, UK and Moderator, Global Conservation Agriculture Community of Practice (CA-CoP), FAO, Rome, Italy

**Achieving sustainable greenhouse cultivation** 063
Edited by: Prof. Leo Marcelis & Dr Ep Heuvelink, Wageningen University, The Netherlands

**Achieving carbon-negative bioenergy systems from plant materials** 064
Edited by: Dr Chris Saffron, Michigan State University, USA

**Achieving sustainable cultivation of tropical fruits** 065
Edited by: Prof. Elhadi M. Yahia, Universidad Autónoma de Querétaro, Mexico

**Advances in postharvest management of horticultural produce** 066
Edited by: Prof. Chris Watkins, Cornell University, USA

**Pesticides and agriculture** 067
Profit, politics and policy
Dave Watson

**Integrated management of diseases and insect pests of tree fruit** 068
Edited by: Prof. Xiangming Xu and Dr Michelle Fountain, NIAB-EMR, UK

**Integrated management of insect pests: Current and future developments** 069
Edited by: Emeritus Prof. Marcos Kogan, Oregon State University, USA & Emeritus Prof. E. A. Heinrichs, University of Nebraska-Lincoln, USA

**Preventing food losses and waste to achieve food security and sustainability** 070
Edited by: Prof. Elhadi M. Yahia, Universidad Autónoma de Querétaro, Mexico

**Achieving sustainable management of boreal and temperate forests** 071
Edited by: Dr John Stanturf, Estonian University of Life Sciences , Estonia

**Advances in breeding of dairy cattle** 072
Edited by: Prof. Julius van der Werf, University of New England, Australia & Prof. Jennie Pryce, Agriculture Victoria and La Trobe University, Australia

**Improving gut health in poultry** 073
Edited by: Prof. Steven C. Ricke, University of Arkansas, USA

**Achieving sustainable cultivation of barley** 074
Edited by: Prof. Glen Fox, University of California-Davis, USA and The University of Queensland, Australia & Prof. Chengdao Li, Murdoch University, Australia

**Advances in crop modelling for a sustainable agriculture** 075
Edited by: Emeritus Prof. Kenneth Boote, University of Florida, USA

**Achieving sustainable crop nutrition** 076
Edited by: Prof. Zed Rengel, University of Western Australia, Australia

**Achieving sustainable urban agriculture** 077
Edited by: Prof. Johannes S. C. Wiskerke, Wageningen University, The Netherlands

**Climate change and agriculture** 078
Edited by Dr Delphine Deryng, NewClimate Institute/Integrative Research Institute on Transformations of Human-Environment Systems (IRI THESys), Humboldt-Universität zu Berlin, Germany

**Advances in poultry genetics and genomics** 079
Edited by: Prof. Samuel E. Aggrey, University of Georgia, USA; Prof. Huaijun Zhou, University of California-Davis, USA; Dr Michèle Tixier-Boichard, INRAE, France; and Prof. Douglas D. Rhoads, University of Arkansas, USA

**Achieving sustainable management of tropical forests** 080
Edited by: Prof. Jürgen Blaser, Bern University of Life Sciences, Switzerland; and Dr Patrick D. Hardcastle, Forestry Development Specialist, UK

**Improving the nutritional and nutraceutical properties of wheat and other cereals** 081
Edited by: Prof. Trust Beta, University of Manitoba, Canada

**Achieving sustainable cultivation of ornamental plants** 082
Edited by: Emeritus Prof. Michael Reid, University of California-Davis, USA

**Improving rumen function** 083
Edited by: Dr C. S. McSweeney, CSIRO, Australia; and Prof. R. I. Mackie, University of Illinois, USA

**Biostimulants for sustainable crop production** 084
Edited by Youssef Rouphael, Patrick du Jardin, Patrick Brown, Stefania De Pascale and Giuseppe Colla

**Improving data management and decision support systems in agriculture** 085
Edited by: Dr Leisa Armstrong, Edith Cowan University, Australia

**Achieving sustainable cultivation of bananas - Volume 2** 086
Germplasm and genetic improvement
Edited by: Prof. Gert Kema, Wageningen University, The Netherlands; and Prof. Andrè Drenth, The University of Queensland, Australia

**Reconciling agricultural production with biodiversity conservation** 087
Edited by: Prof. Paolo Bàrberi and Dr Anna-Camilla Moonen, Scuola Superiore Sant'Anna, Pisa, Italy

**Advances in postharvest management of cereals and grains** 088
Edited by: Prof. Dirk Maier, Iowa State University, USA

**Biopesticides for sustainable agriculture** 089
Edited by: Prof. Nick Birch, formerly The James Hutton Institute, UK; and Prof. Travis Glare, Lincoln University, New Zealand

**Understanding and improving crop root function** 090
Edited by: Emeritus Prof. Peter Gregory, University of Reading, UK

**Understanding the behaviour and improving the welfare of chickens** 091
Edited by: Prof. Christine Nicol, Royal Veterinary College - University of London, UK

**Advances in measuring soil health for sustainable agriculture** 092
Edited by: Prof. Wilfred Otten, Cranfield University, UK

**Supporting smallholders in achieving sustainable agriculture** 093
Edited by: Dr Dominik Klauser and Dr Michael Robinson, Syngenta Foundation for Sustainable Agriculture (SFSA), Switzerland

**Advances in horticultural soilless culture** 094
Edited by: Prof. Nazim Gruda, University of Bonn, Germany

**Reducing greenhouse gas emissions from livestock production** 095
Edited by: Dr Richard Baines, Royal Agricultural University, UK

**Understanding the behaviour and improving the welfare of pigs** 096
Edited by: Emerita Prof. Sandra Edwards, University of Newcastle, UK

**Genome editing for precision crop breeding** 097
Edited by: Dr Matthew Willman, Cornell University, USA

# Acknowledgements

We wish to acknowledge the following for their help in reviewing particular chapters:

- Chapter 7: Emeritus Professor David Hagstrom, Kansas State University, USA
- Chapter 9: Professor Agatino Russo, University of Catania, Italy

# Introduction

Post-harvest losses of cereals and other grains, whether from spoilage microorganisms or insect pests, remain a significant issue in both the developed and developing world. Challenges include restrictions on chemicals for decontamination and increasing levels of insect resistance. This collection reviews the wealth of research addressing those challenges. Part 1 assesses the causes of postharvest losses from fungal contamination, insect and rodent pests. Chapters in Part 2 review advances in bulk and other storage technologies, including developments in detection as well as management of insect pests using techniques such as controlled atmospheres, biocontrol techniques and irradiation. The book also reviews advances in the detection and control of fungal contamination together with ways of monitoring the quality of stored cereal grains.

## Part 1 Postharvest losses and their causes

Chapter 1 sets the scene for the book by looking at post-harvest loss from a managerial perspective. The chapter first reviews quantitative, qualitative, economic, societal and environmental aspects of post-harvest loss. Since complete elimination of PHL, given the likely costs, is not economically feasible, the chapter discusses defining appropriate targets and goals. It emphasises the need to assess the economic effects as well as physical losses from PHL. The chapter then reviews the reasons why so many interventions in this area fail (the post-harvest paradox). As the chapter points out, sustained success is achieved when an organically, self-sustaining market-based supply chain for the technology is established by effective collaboration e.g. through public-private partnerships. An effective supply chain provides and enhances the technology – generating value for the farmer, market chain participants and consumers. The chapter shows how this market framework can be established in practice through three case studies: deploying metal silos in Central America (the Postcosecha Program), the use of triple layer bags in Africa (the PICS Project) and reducing PHL maize losses in Tanzania (the YieldWise Initiative).

More than 150 species of filamentous fungi and yeasts have been isolated from cereal grains. Fungi are responsible for the highest fraction of loss to grain in storage and are regarded as the most important cause of spoilage in stored grain and cereal food products. Fungi cause cereal losses by reducing grain yields, by spoiling cereals and rendering them unfit for human and animal consumption, or by production of secondary metabolites like mycotoxins, which are harmful to humans and animals. Chapter 2 focuses on fungal ecology and storage fungal contamination in three major types of cereals (maize, wheat, and rice) which constitute more than 90% of total world cereal production. It reviews what we know about the ecophysiology of species belonging to *Aspergillus, Penicillium* and *Eurotium* as well as mechanisms of infection, factors affecting proliferation of molds and ecological succession of molds in storage.

As Chapter 3 points out, phosphine gas has become the world's primary fumigant for disinfestation of stored grain. Development and increase of strong resistance to phosphine is a threat to the continued use of phosphine and there are limited alternatives currently available. Understanding how phosphine resistance arises is critical in managing resistance problems and extending the useful life of this fumigant. The most recent advances have been in our understanding of the two genes that contribute to phosphine resistance in all pest species that have been examined in countries around the world. Resistance gene identification has led to the development of a monitoring tool based on the DNA sequence. Large surveys to test for resistance genes coupled with population genetics has provided us with information on how resistance arises and how it spreads, as well as in helping to identify alternative fumigants.

Post-harvest losses by rodents are estimated to around 5-10%, which is similar to the level of post-harvest losses caused by insects. Chapter 4 reviews advances in understanding rodent pests affecting cereal grains. It assesses the limited number of studies measuring post-harvest rodent damage and losses. The main pests are the widespread commensal rodent species. Rodents cause direct losses by consuming grain, but they also spoil grain (contamination with hair, urine and faeces), and expose grains to further mould or insect damage. There is a risk of transfer of diseases and zoonoses to humans and livestock. Rodents damage infrastructure such as the warehouses and stores themselves, but also cables and water pipes. As the chapter shows, there is relatively little information about the ecology of the main rodent pest species in grain stores. This remains an important research topic to improve management recommendations about the timing and location of control strategies.

## Part 2 Storage technologies

Chapter 5 provides a comprehensive and wide-ranging review of key research and best practice on the design and construction of silos, horizontal building structures and on-the-ground piles as methods of bulk grain storage. The chapter also discusses the range of preventative strategies for preserving grain quality (such as sealing and aeration) as well as intervention techniques such as gas fumigation, modified atmosphere, ozone and cold plasma treatments. Finally, it summarises key best management practices such as SLAM (Sanitation, Loading, Aeration and Monitoring) and DICE (Dry Grain, Inspect, Clean and Examine) management techniques.

Ch 6 discusses developments in the use of hermetic bags in grain storage. Smallholder farmers' interest in hermetic bags has been driven by the need to reduce grain storage losses due to insect pests as well as limitations in current storage methods. Hermetic bags commercially available for grain storage include single-, double- and triple-layer plastic bags. For the last 12 years, more than 20 million hermetic bags have been sold to smallholder farmers and other users in Africa (primarily) and Asia. The rapid increase in the use of hermetic bags is due to the scalability of both awareness activities, including training to create demand, and supply chain initiatives to improve availability through the private sector. Ongoing challenges to increase adoption of hermetic bags include limited awareness and availability of the technologies in rural areas, variability in the quality of the hermetic bags, and lack of standards to ensure effectiveness across different products. The chapter includes a case study on the use of the Purdue Improved Crop Storage (PICS) bag.

Chapter 7 surveys advances in detection and monitoring in managing insect pests in postharvest storage of cereals. Stored-product insects damaging grain are difficult to locate, making sampling and detection challenging. The chapter outlines the advantages and disadvantages of key methods for detecting insects: imaging, NIR, visible light, microwaves, X-rays, sound and DNA techniques. It discusses factors affecting trapping including species, grain temperature, trap placement, trap type and insect population density The chapter also discusses challenges and advances in absolute or relative sampling with respect to sample number, sample unit and sample size, economic thresholds, sequential sampling and predictive models.

As Chapter 8 shows, oxygen-depleted controlled or modified atmospheres (CA/MA) generated by a variety of different methods and alteration of temperature of stored cereals have been used successfully to replace fumigants for insect control. CA or MA have been used in specially constructed rigid structures for the preservation of cereal grains. Plastic structures suitable for long-term, large-scale storage or for intermediate storage of grain in bags or in bulk have also been developed. The chapter also reviews the use of temperature in insect control. Aeration e.g. is a widely used method to modify the grain bulk microclimate to render it unfavorable for the development of damaging organisms in the grain. Chilled aeration with refrigerated air can be used when ambient air may not be sufficient to cool grain sufficiently to manage pests and maintain grain quality.

Chapter 9 reviews biologically-based control strategies for managing stored-product insect pests. Several biologically-based strategies to protect stored plant products from infestation are currently used but they still need to be further optimized for a more effective implementation in practice. This includes the use of semiochemicals for the monitoring and control of pest populations and the release of natural enemies of stored-product pests (i.e.

parasitoids) as biological control agents. This chapter provides an overview of current knowledge about these two alternative control strategies and discusses how to increase their efficacy or improve control of insects infesting stored products by combining different approaches.

As Chapter 10 indicates, irradiation is a viable alternative to fumigation as a pre- or post-shipment quarantine treatment for control of stored product insects and pathogens in cereals and grains. Electron beam (e-beam) and x-ray are machine sources of ionizing radiation that avoid the problems associated with the transportation of radioactive isotopes. The chapter assesses the advantages and disadvantages of these techniques. Electron beam has low penetration but high efficiency. X-ray has excellent penetration properties but is less energy-efficient than electron beam techniques. As the chapter points out, e-beam is ideal for high throughput treatment, whereas x-ray systems are ideal for bulk or pallet-scale treatment of cereals and grains. The chapter includes a case study showing the use of low energy electron beam (LEEB) and x-rays to disinfest rice. The chapter shows that irradiation may be helpful in controlling fumigant resistant insect populations (e.g. to phosphine) and could help manage resistance by preventing the spread of resistant insects in exported cereals and grains.

Chapter 11 discusses advances in postharvest detection and control of fungal contamination of cereals, based on understanding the role of abiotic factors such as water availability and temperature and their effects on the ecology of spoilage and mycotoxigenic fungi. The chapter reviews the use of volatile organic compounds (VOCs) as an early indicator of spoilage. The potential of measuring $CO_2$ for the early detection of fungal spoilage moulds in stored grain is examined in combination with predictive modelling of boundary conditions for growth and toxin production as tools for the development of DSS systems. Control measures including abiotic factors, use of natural preservatives, biocontrol agents, and physical methods such as irradiation, cold plasma treatments, hermetic bag systems and fumigation with ozone ($O_3$) are considered. Finally, the potential of developing real time approaches for better post-harvest management of stored cereals and other commodities are discussed.

Building on the previous chapter, Chapter 12 reviews advances in sensor technologies for monitoring the quality of stored cereal grains. State-of-the-art monitoring systems have sensors that can early detect the onset of spoilage before "hotspots" develop and alert the user of their location through real-time data that can be delivered directly to cloud-enabled devices. The chapter reviews the range of types and applications of sensor for monitoring temperature, relative humidity and moisture content, carbon dioxide and oxygen, phosphine and other gases, as well as volatile organic compounds.

Building on themes in Chapter 1, the final chapter in the book, Chapter 13, discusses ways of better supporting smallholder farmers in developing countries to improve postharvest management (PHM) of staple grains. The chapter analyses PHM practices, available technologies and delivery models, and examines factors causing non- or poor adoption of the technologies in developing countries. Missing parts of the PHM jigsaw puzzle are identified and possible practical solutions offered. The chapter proposes strategies to support smallholders in building sustainable and robust PHM interventions to enhance technology uptake for effective reduction of postharvest loss in quantity and quality.

# Part 1

## Postharvest losses and their causes

# Chapter 1

## Post-harvest losses of cereals and other grains: opportunity among issues and challenges

*Steven T. Sonka, University of Maryland, USA*

1 Introduction
2 Is zero loss a desirable goal – what does economics tell us?
3 What is the 'hidden tax' resulting from excessive post-harvest loss (PHL)?
4 Can we move from measuring loss to managing to reduce PHL?
5 Can we overcome the PHL paradox?
6 Examples demonstrating the potential for sustainable reduction of PHL
7 Summary
8 Where to look for further information
9 References

### 1 Introduction

Managing post-harvest loss (PHL) is a challenge endemic to agricultural and food systems. Time and resources are devoted to producing an attractive field of grain or an orchard of trees brimming with fruit. But to be available to consumers, that produce must be harvested, stored, transported, and processed within systems that may operate over several months and at considerable distances. Without appropriate care and attention, loss can occur at distressingly high levels.

In developing countries, PHL routinely exists at excessive levels, resulting in reduced well-being for small-holder farmers (SHFs), higher prices for consumers, diminished food security, and unnecessary environmental degradation. Natural and human resources employed to produce agricultural products, which are then not consumed, enlarge the environmental footprint of agriculture. In developed nations, food waste (loss at the retail and consumer level) comprises the dominant component of total food waste and loss. This chapter, however, is focused on loss at the producer level and supply chain level in developing nations.

http://dx.doi.org/10.19103/AS.2020.0072.01

A 2011 FAO report estimated that one-third of food production is lost or wasted, not reaching consumers (Gustavsson et al., 2011). In developing countries, activities in the post-harvest and processing stages account for more than 40% of the loss. In recent years, recognition of the extent of PHL and its negative impacts has fueled interest in the topic as a societal problem. Indeed, simply typing the phrase amount of PHL into a web browser retrieves nearly 900 000 hits with eye-popping numbers, such as:

- 'At $1 trillion, it's an opportunity cost so large that it nearly equals the entire gross domestic product of sub-Saharan Africa. By most official statistics that's the value of food that gets lost or wasted every year in the production process - roughly 30 percent of the amount that gets harvested' (Mendoza, 2016).
- Estimates range from 5 to 30% or more, and in sub-Saharan Africa alone, the value of PHL overall is thought to be around $4 billion (£2.5 billion) a year (Van Vark, 2012).
- In sub-Saharan Africa, up to 50% of crop production may be lost before produce reaches the consumer. Losses result from poor storage and handling practices and a lack of infrastructure. With an estimated value of US$4 billion, these losses threaten the food security and livelihoods of millions (Aphlis, 2019).

## *1.1 PHL - not a new societal concern*

Excessive PHL and its associated negative impacts are not new topics of global interest. In 1975, the then US secretary of state, Henry Kissinger, made an eloquent plea detailing the need for global efforts to reduce PHL. Following his exhortation, the UN General Assembly adopted a resolution, calling for at least a 50% reduction within ten years (Bourne, 1977). However, more than 40 years later, many developing countries are still struggling to find sustainable solutions to mitigate PHL of cereals and other grains.

When making presentations on the PHL topic, I find that audience members are surprised that 'way back' in 1975 the United Nations considered PHL to be an issue worthy of attention. I then ask:

- Was the UN resolution successful? Did we, globally, reduce PHL by 50%?
- The first, and most common, response is, 'We don't know!'
- While, empirically, that answer is correct, my belief is that PHL likely was reduced significantly - as a global average.

The support for that assertion is based upon the massive reduction of PHL in the agricultural systems of developed nations that has occurred over the last 4 decades. In the 20 years or so before the UN resolution, I was growing up on a small farm in Iowa, in the Central United States. For much of that time, the practices and technologies employed at harvest and during storage resulted in losses that were considerably higher than those which exist today in that community. Indeed, the maize storage facilities typically used on our farm in the 1950s are similar to those I personally observed on small farms in Kenya in 2019.

There are many reasons why PHL levels are substantially reduced in places such as Iowa while remaining stubbornly high in places such as Kenya. The second part of this chapter's title provides a somewhat simplistic explanation: in agricultural systems where losses have been reduced, opportunity drove decision makers to take action to resolve the issues and challenges associated with reducing PHL.

## *1.2 PHL is multidimensional*

In this chapter, I hope to focus attention on a few key factors associated with that opportunity perspective. One of these factors is that PHL is multidimensional. However, PHL is often addressed as a singular phenomenon with simple causes and remedies.

Five key dimensions are noted below:

- Quantitative loss refers to the magnitude of physical loss of grain after harvest.
    - Generally, this type of loss is easily observable as, for example, grain laying along the road after spillage during transport.
    - Reported estimates of loss, especially those aggregated at regional or national levels, often include only quantitative loss.
    - Physical losses that occur during harvest are often included within estimates of PHL.
- Qualitative loss occurs when key attributes of the grain have been diminished even though the amount of grain, physically, has not been reduced.
    - In some instances, the status of the key quality attributes is observable. For example, unthreshed rice laying on the ground can become discolored and have cracked kernels. Such grain is discounted in the market.
    - In other instances, such as aflatoxin contamination or poor nutritional value, the diminished quality attribute can't be identified by visual observation. Often, however, buyers and processors, farther down the supply chain, are equipped to detect such quality issues.

- Economic loss refers to the reductions in returns that can occur throughout the supply chain of interest – from the SHF to the consumer.
  - Economic losses can result from both quantitative and qualitative sources of loss.
  - As noted in Section 2 of the chapter, the extent of economic loss also is determined by the amount and value of the resources needed to reduce loss. If an action to reduce loss requires more resources than are saved by reducing the loss, then taking that action actually could increase economic loss.
  - Conversely, excessive PHL levels exist when, with available technologies and practices, actions can be taken where the value of the crop saved exceeds the cost of saving it. These settings offer economic opportunity.
  - Grain that is of diminished quality (or when it is in excess) often is not completely discarded. For example, the highest economic value for wheat may be to be used in baking. However, the use of lower quality wheat as livestock feed does generate some value. In those instances, the economic loss is the differential in the value between the wheat used in baking and the value earned as livestock feed.
  - As illustrated in Section 3 of the chapter, SHFs often are subjected to an opportunity loss when the lack of storage capabilities requires that they sell their crop at harvest. In such instances, farmers do not have the opportunity to earn returns from storage.
- Societal loss has important distributional implications, particularly in developing countries.
  - SHFs often bear a considerable portion of the adverse economic consequences of excessive levels of PHL.
  - Low-income consumers, whether in rural or urban settings, also can be adversely affected by higher food costs or restricted supplies resulting from excessive post-harvest.
- Environmental loss refers to the bundle of natural resources used to produce those portions of the crop which are not effectively employed because of excessive PHL.

### *1.3 Moving forward*

It is important to stress that PHL reduction is a multifaceted, systemic issue. Referring back to my Iowa farm experiences, and in similar locales, the PHL reduction required the contributions of many perspectives operating in an interrelated fashion, fit to the prevailing cultural and social context. Similarly, multiple perspectives will be needed to make significant future progress in developing nations.

The perspective employed in this chapter is managerial in nature, with a heavy reliance on the discipline of economics. Again, let me stress that this is only one of several important perspectives which will be required to make substantial progress. Implementation of effective market-based systems has been shown to foster sustained PHL-reducing results. However, as detailed later, catalytic efforts from social and government sectors likely are necessary to facilitate the needed change within agricultural sectors in many developing countries.

Employing that managerial perspective, the following four interrelated questions frame the chapter's remaining sections:

- Is zero loss a desirable goal – what does economics tell us?
- What is the 'hidden tax' from excessive PHL?
- Can we move from measuring loss to managing to reduce PHL?
- Can we overcome the PHL paradox – where despite infield demonstrations and pilot projects which consistently document significant benefits from employing appropriate PHL-reducing technologies, scaled adoption of those technologies doesn't follow?

The summary section which concludes the chapter includes a discussion of research needs and opportunities.

## 2 Is zero loss a desirable goal – what does economics tell us?

Instances where excessive PHL exists often are quite visible. In India, torn and deteriorating jute bags with rice spilling from them and in Brazil, highways with a continual stream of soybeans laying along the road are oft-pictured examples. The technical and physical solutions similarly are visible and can have immediate effect. More durable bags and improved handling practices can reduce rice spillage, and better transport practices can limit loss of soybeans along the highway. Some observers of these scenes assert that 'we need to stop such wasteful behavior. We need to have a goal of ZERO post-harvest loss.'

While physical loss and technical solutions are key components of efforts to prevent and/or reduce PHL and food waste, they operate within larger economic, managerial and social contexts. Despite that reality, the economics of PHL and, more importantly, the economics of reducing PHL are not correspondingly well known. Reducing loss almost always increases gross revenues. However, relatively few comparisons are available which focus on the economic benefits. This deficiency is important because the decision maker, whether it be an SHF, private sector manager, or official in a government or NGO agency, has to make choices. Typically, they are faced with choosing among several alternatives, each with potential benefits. Although economic viability doesn't have to be the primary determinant, comparing financial benefits and costs is an important factor in choosing among alternatives.

Review of PHL literature, however, reveals that economic analysis of expected net benefits and/or actual net benefits achieved tends not to be reported. Having said that, it is important to emphasize that careful economic analysis of PHL reduction is difficult, mainly because of the lack of useful empirical data.

This section of the chapter employs a simple economic framework to explore two decision phenomena associated with PHL reduction:

- The first examines the goal asserted by some that PHL should be eliminated completely from food supply chains.
- The second conceptually demonstrates how implementation of appropriate technology can effectively reduce PHL and provide economic benefits.

Figure 1 employs a classic supply/demand framework to explore the goal asserted by some that PHL should be eliminated completely from food supply chains. That chart depicts a simplistic graph of supply and demand, where:

- The vertical axis indicates economic value.
- The horizontal axis indicates the level of reduced loss, where movements to the right indicate more loss reduction.

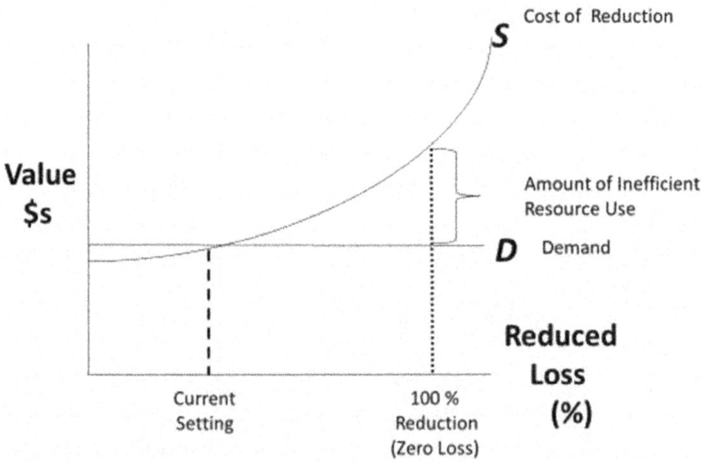


**Figure 1** Economic interpretation of the goal to eliminate post-harvest loss (PHL).

- The horizontal solid line, **D**, represents the demand for loss reduction from the perspective of the individual decision maker.
- The upward sloping solid line, **S**, is the supply curve for loss reduction, that is, the cost of reducing increasing amounts of loss. The upward slope of the line indicates that the per unit cost of loss reduction increases as more reduction is undertaken. In other words, there are diminishing marginal returns.
- The line marked **D** intersects with the line marked **S** at the point which represents the current setting.
- Further to the right on the horizontal axis, a dotted vertical line extends to intersect the supply curve, **S**. This hypothetical point is noted to indicate where achievement of 100% reduction of PHL might occur. That point corresponds to zero PHL.

The costs associated with achieving that level of loss reduction exceed the value of achieving that goal. The bracket labeled amount of inefficient resource use illustrates the amount of that inefficiency. Although expressed in economic terms here, the likely result of achieving zero PHL would be less sustainable food systems relative to the environmental and social dimensions of sustainability as well.

The prior economic illustration focused on the goal of zero PHL. The same general framework can be used to show the potential for gain from the reduction of PHL – which is the opportunity that likely exists in many food chains in the developing world. This second message is shown in Fig. 2.

The graphic in Fig. 2 illustrates the prior point that effective PHL reduction can provide important benefits. Here:

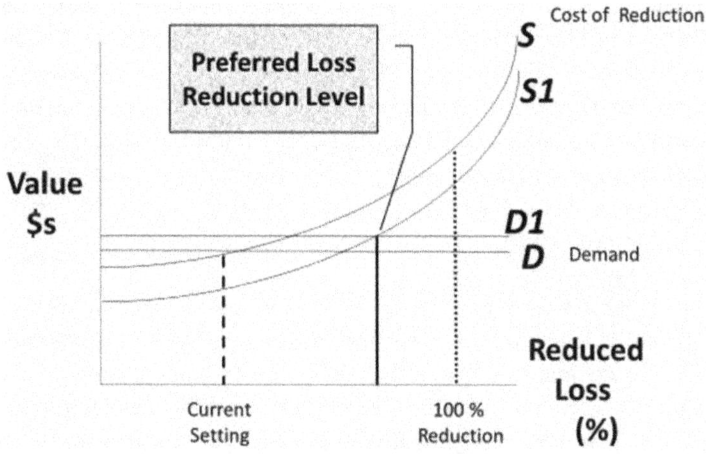


**Figure 2** Illustration of economically sustainable PHL reduction.

- The line D1 indicates that higher quality products, likely to result from reduced loss, have more value to buyers.
- The line S1 reflects the cost-reducing impact of successful interventions to reduce PHL.
- The intersection of lines D1 and S1 represents the preferred level of loss reduction in economic terms.

The message of Fig. 2 is that successful interventions to reduce PHL can indeed provide economic benefits. Again, because such efforts reduce the impact of food chains on the natural environment, they are likely to contribute to the goal of developing more sustainable food chains. Further, reducing PHL often requires development of products and application of services in rural communities, where farmers reside and agricultural supply chains originate. Therefore, it is reasonable to expect that reduction of PHL can contribute to the triple bottom line associated with sustainability.

## 3 What is the 'hidden tax' resulting from excessive post-harvest loss (PHL)?

As mentioned previously, much of what I know about agricultural and food systems was learned growing up on a small Iowa farm. My parents devoted considerable attention to producing our crops and livestock, including implementing practices to reduce PHLs. However, considerable attention also was focused on the decisions of how much to sell and when. These decisions significantly affected the family's well-being.

Interest in reducing PHL of cereals and other grains tends to focus on the amounts of loss in physical terms (kilograms or bushels). A less recognized dimension, however, is that the inability to effectively store produce also acts as a 'hidden tax', especially on SHFs, which extends well beyond the direct physical losses.

In this section, we explore this 'hidden tax' with empirical data from Africa and Central America. A few years ago, I participated in a workshop of agribusiness executives interested in PHL reduction in Africa. We reviewed findings from a World Food Programme project assessing PHL among 400 SHFs in Uganda and Burkina Faso (Costa, 2014). Results from traditional storage methods are presented in Fig. 3.

The results are dramatic. Storage employing traditional practices results in more than 20% loss within 30 days and most of the stored grain is lost within 90 days. The key implication is straightforward - traditional storage losses are unacceptably large. As one African executive in the group commented, 'in reality, no one stores their grain for 60 or 90 days - because of the type of losses shown here'.

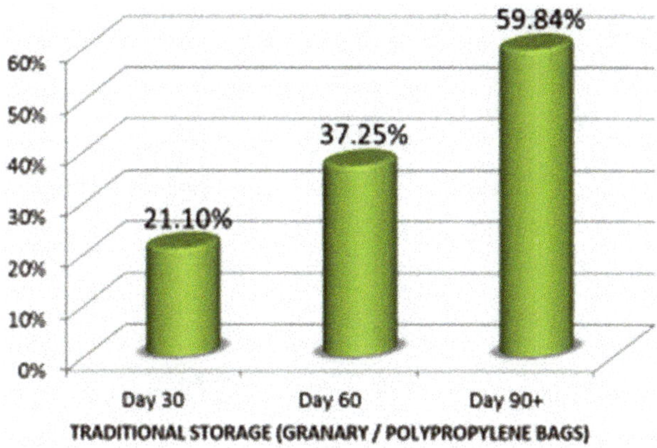


**Figure 3** Average recorded losses (maize) in Uganda, Dec 2013–Apr 2014 (Costa, 2014).

The economically devastating result for SHFs is that their product often is sold at harvest when prices are lower and then purchased for consumption later when prices are higher. Selling at harvest also means that SHFs don't participate in the marketing process. Instead they are trapped in a sell low/buy high cycle.

Although scant overall, some empirical evidence of the negative financial impact of this cycle is available. Evidence from the Purdue Improved Cowpea Storage (PICS) project reveals the nature of this source of loss. Commodity prices typically are low at harvest and then rise until the next crop's harvest. Agricultural economists have quantified this relationship in eight African countries (Chapoto and Jayne, 2010). On average, prices in the high-priced month were 184% higher than those in the low-priced month (ratios for specific countries are in Fig. 4).

While these price differences suggest that there are potential economic benefits from storage, there also are additional costs. Fortunately, we can look to additional work (Jones et al., 2011; Lowenberg-Deboer, 2014) as to the net benefits of SHF storage. Table 1 documents that effective storage (in this case, use of PICS bags) has a positive impact on the income of SHFs.

The Postcosecha Program in Central America provides additional evidence of the hidden tax effect. Over the last 20 years, the Swiss Agency for Development and Cooperation (SDC) successfully fostered the adoption of metal silos for grain storage in four countries. With approximately 670 000 metal silos adopted, 415 000 rural households were helped, preserving about 380 000 tons of grain annually (Fischler et al., 2011).

Post-program impact assessment indicated that maize farmers using metal silos received a higher price than those selling directly after harvest. In 2009, silo users received an average price of $32.44/100 kg while non-silo users received an average of $27.78/100 kg (Fig. 5). Generally, metal silo users gained 23%

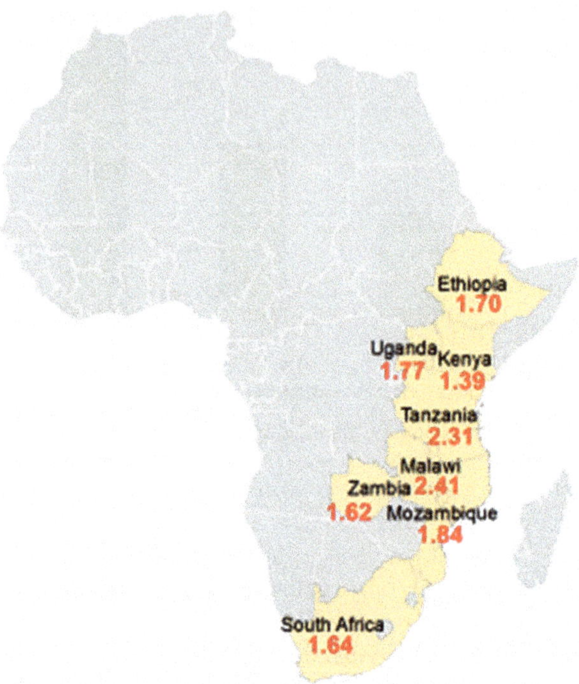


**Figure 4** Economic assessment of PICS bags in selected African countries (Chapoto and Jayne, 2010).

**Table 1** Adjusted real crop income on 100 kg of maize in Tanzania (Tanzania shilling)

| **Method of storage and timing of sale** | Sold at harvest | Sold in February, traditional storage, no insecticide | Sold in February, stored in PICS bags |
|---|---|---|---|
| **Adjusted real crop income** | 7651 | 8836–12 501 | 18 749 |

more than nonusers. For a medium-sized farm in this region, the benefit-cost ratio (BCR) was 2.6 and the internal rate of return (IRR) was 55% – with partial subsidies. Without subsidies, the BCR was estimated to be 1.4 and the IRR 22%.

These data illustrate the relationship between post-harvest storage and enhanced well-being for farmers. Additionally, important systemic benefits arise when SHFs are engaged in the marketing process. In my boyhood farming community, I saw how marketing of stored commodities expanded the farmer's decision-making perspectives. When farmers are routinely engaged in storage and marketing, the effect extends to the community. The skills and capabilities needed for effective marketing can foster additional economic activities in the community. Conversely, excessive PHL, because of the lack of storage and marketing capabilities, acts as a hidden tax on the community as well.

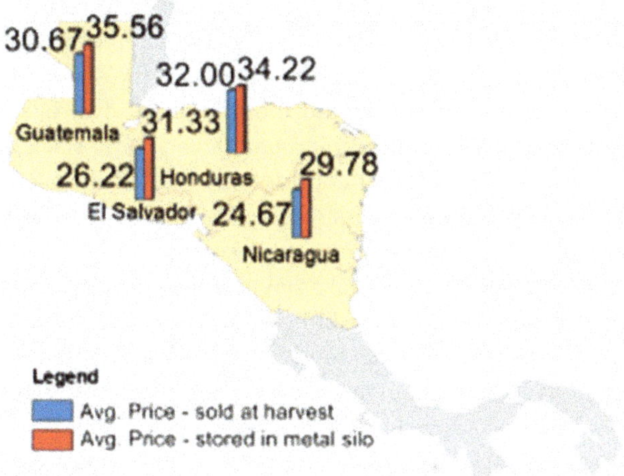


**Figure 5** Postcosecha Program countries and average prices of maize ($/100 kg) in 2009 (Fischler et al., 2011).

## 4 Can we move from measuring loss to managing to reduce PHL?

A widely held precept of modern management, attributed to both Peter Drucker and W. Edwards Deming, can be summarized as, 'You Can't Manage What You Don't Measure!' Numerous systems to improve performance, such as total quality management, six sigma, and just-in-time supply, rely upon sophisticated, continual measurement of ongoing operations to drive progress. Although not normally mentioned, these activities are economically beneficial only if the cost of measurement is less than the benefits achieved from employing the resulting data.

Reflecting on my small farm experience in Iowa, my parents made many decisions to improve efficiency, often resulting in reduced PHL. While we continually sought to understand loss and its sources, the actual measurement of loss from ongoing operations was too expensive in terms of the time and material required. For example, after maize harvest, we could have attempted to count the ears of maize remaining on the ground. However, there were more productive uses of the time and resources that would have been required to count that PHL.

Within some components of food supply chains, systems exist to monitor and measure ongoing operations (e.g. within more sophisticated food processing plants). In those settings, expanding the focus to assess PHLs may be economically feasible. However, on small-holder farms and in those parts of the food chain near to the farm, ongoing measurement systems tend not to exist.

Because of the high cost of physical measurement, surveys of farmers based upon recollection of loss often are used to attempt to quantify loss. A conversation I had with a small-holder rice farmer in the state of Bihar in India a few years ago demonstrated to me the danger of overreliance on recall surveys. The farmer and I were standing near a very large stack of bundles of unthreshed rice. He reported that considerable loss would occur because of rodents and spoilage before the grain could be threshed. I asked, 'How much of your grain will be lost?' He replied, 'I don't know'.

The farmer didn't know the total amount of rice that was contained in the stacks of grain. Therefore, how could he know how much was lost? Often in such survey settings, respondents tend to provide a number that they think the interviewer wants to hear. In this case, I appreciated his honest and direct response that he didn't know the quantitative answer.

It turns out, however, that assessing levels of PHL is not the only setting where continual, comprehensive measurement is not feasible. For example, estimates of the extent and nature of poverty are important factors affecting policy in developing countries. However, continually and comprehensively quantifying poverty across populations would be a time- and resource-intensive undertaking. Attempting to do so would be quite costly.

Because policy makers need to understand the changing nature of poverty, methods have been developed which provide useful and cost-effective means to track poverty. In general terms, a three-step process is employed:

1 Careful research is done to identify valid and quantifiable indicators of poverty. An example indicator could be the number of meals the family consumes in a week.
2 Large-scale and representative survey research is conducted. Such surveys focus on obtaining valid estimates for the indicator variables across the population of interest (e.g. quantifying the distribution of families in terms of the number of meals consumed weekly).
3 The sources of information in (1) and (2) are then combined to provide credible estimates of poverty. An important attribute of this approach is that estimates can be enhanced either as additional survey information is obtained or as research provides improved poverty indicators.

The framework just described seems to be applicable as we attempt to understand PHL. Again, a three-step approach could:

A Conduct careful research relating levels of loss to the alternative post-harvest practices employed. For example, hand threshing of rice remains a common practice in some parts of India. Mechanical threshers and combines are alternative practices that could be employed. (In this context, the post-harvest practices are analogous to the poverty indicators noted previously in (1) above).

B Large-scale surveys, focused on quantifying the extent to which practices are actually employed by farmers, can be conducted. Because these surveys would focus on what farmers actually do, recall-based responses are likely to be provided with minimal error.

C In a fashion similar to (3) above, the information obtained in (A) and (B) can be combined to provide credible estimates of losses.

To this point, the discussion has focused on estimating the extent of physical loss. However, it is important to note that physical losses are only one component of the economic effects of excessive levels of PHL. Two key economic parameters associated with PHL are quality and the timing of sale of the farmer's product (the hidden tax effect discussed previously).

The quality of the farmer's output often is highly related to the practices employed on the farm and in the transport of the product to the market. For example, hand-cut rice often is left for several days to dry in the field. The combined effect of heat in the day and dew overnight can reduce the desirability of the rice, therefore resulting in economic discounts at the market. Research efforts, as noted in (A) above, can assess the relationships between practices and the quality to provide data needed to estimate economic losses because of quality deficiencies.

Further, if SHFs don't have capabilities to adequately store or condition their crops, they often suffer severe economic penalties by having to sell at harvest when market prices are typically at their lowest. For staple crops, this penalty is multiplied when the farm family purchases the same product several months after harvest when prices are high.

Lack of access to appropriate post-harvest technologies and practices leads to substantial economic losses from both the effect of reduced quality and inappropriate market timing. In tracking the gains from reducing PHL, the effects of those parameters the effects of these economic factors must be assessed as they can outweigh the value of the purely physical losses.

## 5 Can we overcome the PHL paradox?

Consider the rice farmer in the state of Bihar in India. At harvest, it's often easy to see when it has been a good production year for rice. Awaiting threshing, bundles of rice are organized into stacks twice as tall as the farmer with the length and width of a small building. The farmer is pleased that yields had been good. However, the farmer knows that total success is yet to be achieved.

Threshing would be done by hand, with men and women beating the bundles on the ground to separate the rice from the straw. Over the several weeks needed to thresh the crop, losses due to rodents and weather-induced deterioration would be large using the practices and technologies available. Much of the potential success from growing an abundant crop does not result in income for the farmer or food supplies for the farmer's family or consumers.

## *5.1 The paradox of PHL reduction*

In developing countries, PHL routinely exists at excessive levels resulting in reduced well-being of SHFs, diminished food security, and unnecessary environmental degradation. It is often noted that almost one-third of the food produced globally does not reach consumers because of PHL. While not new, the problem is stubbornly persistent; even though:

- Technologies are available which can effectively reduce PHL.
- The technology is unfamiliar where the PHL levels are high, appropriate technologies generally are 'low-tech' in terms of their application, and
- Pilot projects routinely document that significant economic benefits would result from adoption of loss-reducing technologies.

This disappointing result has been called the PHL paradox (Sonka et al., 2018). Interestingly the paradox typically is framed as farmers 'failing to adopt' the technology of interest. That framing illustrates an incomplete view of success for interventions to reduce PHL. A view where the objective of pilot project interventions is framed as demonstrating that PHL levels can be reduced but in a specific, limited setting where external resources artificially reduce user costs.

Instead, effectively managing PHL could be framed as a market development opportunity, not as a technical problem. Although introduction of a technology novel to the area often is a necessary and difficult step, demonstrating that the technology works in a pilot setting generally is insufficient. Sustained success is achieved when an organically, self-sustaining market-based supply chain for the technology is established. That supply chain provides and enhances the technology - generating value for the farmer, market chain participants and consumers.

Let's consider two figures to illustrate this idea. Figure 6 stereotypically depicts a pilot effort focused on demonstrating that the intervention technologies will work - reduce PHL - in the field.

In Fig. 6, the horizontal bar refers to the supply chain of an agricultural commodity, let's say maize. That chain is shown as extending from production to the eventual end user market. As an example, let's say that the intervention is focused on improving storage at the SHF level. This could be done through the use of hermetic storage bags, small metal silos in the farm household, or community storage systems. All these approaches can work.

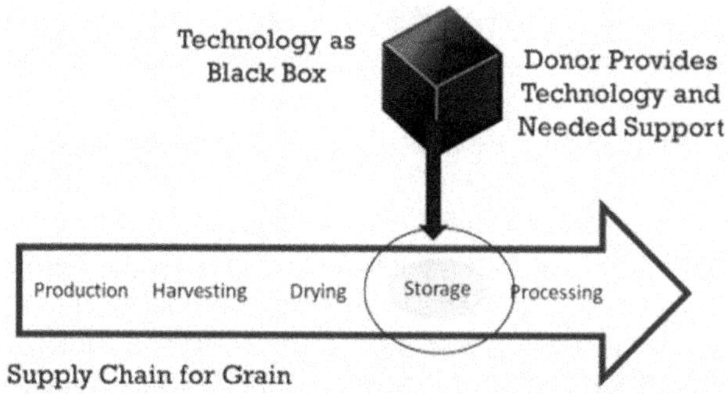


**Figure 6** Intervention to demonstrate that a technology will 'work' - its use will reduce PHL.

Again, stereotypically, the technologies needed are provided by the donor (or donors) as a package to be tested and adopted. This is depicted in Fig. 6 as a black box. In addition to technology, the project likely will provide education, support and service within the context of that black box. Reducing PHL in the field is difficult and requires considerable dedicated effort. Often that dimension of success is not attained. However, even when that aspect of success is achieved, in only a few cases has sustained, scaled adoption followed.

In Fig. 7, the 'black box' of Fig. 6 is replaced with a vertical graphic depicting components of a supply chain for the technology of interest. Now there are two supply chains. One for the agricultural commodity, maize, and one for the technology which enables improved storage - bags, silos or community storage.

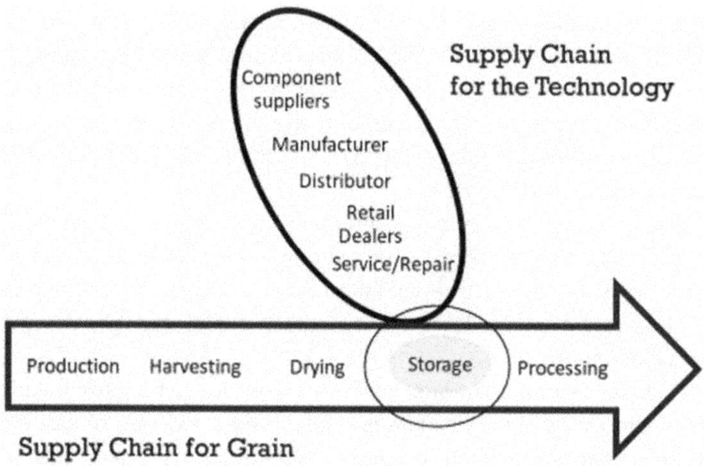


**Figure 7** Structure of a technology supply chain to support sustained adoption.

Significantly, the components of the technology supply chain extend beyond the physical elements of production and delivery. Included are key marketplace mechanisms such as service, user support and product improvement and evolution over time. These mechanisms are vital, if scaled adoption is to occur.

To sharpen the distinction, let's compare the two pilot efforts. One successfully demonstrates that PHL can be reduced. The second establishes a market-driven supply chain for the technology necessary to reduce PHL. A key question is, what happens after each project successfully ends? For the first, the black box of technology and service disappears. For the second, actors along the vertical supply chain have a natural incentive to expand the extent of adoption and to improve the technologies needed for PHL reduction.

## 5.2 Why doesn't the market handle this?

Earlier in this section, one dimension of the PHL paradox was framed as SHFs failing to adopt economically attractive technologies. And pilot efforts which don't target effective supply chain implementation as the end result are components of the dilemma. But that statement leads to a few follow-up inquiries:

- Why should the pilot effort have to do more than demonstrate technical success?
- Don't the market economies and the profit motive exist to exploit opportunities where innovations have been shown to make money?
- Isn't that what entrepreneurs do?

Those questions are reasonable and relevant - in settings where the institutions that facilitate exchange and market dynamics are strong. However, in many developing country settings, especially in rural areas, such institutions tend not to be strong. For example, formal credit mechanisms may not exist or may be difficult to access. And enforcement of contracts tends to be uncertain, restricting the entrepreneur's ability to establish alliance relationships needed to access supplies and services.

Academics employ the term, 'institutional void', to identify such deficiencies (Shah et al., 2017). Void may be too strong a term in many situations. It is common, however, for those capabilities to be inadequate to facilitate a strong, vibrant ecosystem which fosters the establishment and growth of formal business innovation.

The presence of institutional voids provides at least a partial explanation for the existence of the PHL paradox. Even with evidence of opportunities which can contribute to economic and societal objectives, inadequacies in the institutions surrounding the relevant market potential impede implementation.

There are a few instances where sustained success has been achieved in terms of PHL reduction beyond the pilot effort. Those instances have the common feature that a market-driven supply chain for the needed technology was established as part of the development initiative.

In these instances, effective collaboration among the social, private and public sectors was recognized as essential in achieving positive outcomes. A key dimension is the catalytic role played by the social sector in circumventing key institutional voids that otherwise would have inhibited the establishment of a sustaining supply chain. This is more complex than simply providing technology as a 'black box'. Sustaining success was only achievable when the capabilities of an effective market-driven supply chain were attained.

Although the rice farmer in Bihar suffers significant losses while his rice awaits threshing, he is using the 'best practices' commonly available. The use of improved technologies likely could reduce PHL and economically enhance the farmer's well-being. Collaborative efforts of the private, social and public sectors can foster development of the supply chains needed to make improved technologies routinely available. Then the farmer will be able to achieve success which better transforms the hard work invested to achieve good yields into enhanced family well-being.

## 6 Examples demonstrating the potential for sustainable reduction of PHL

The previous section has focused on redefining success for interventions aspiring to reduce PHL in developing country agriculture. An important first step for such an effort is to demonstrate that application of appropriate technologies can reduce PHL in the target setting. From an economic perspective, this step is necessary to document that potential demand exists. However, that step, by itself, is often insufficient. A necessary step is that an effective system of supply for the technology be available. It is important to note that a market-based supply chain can accomplish more than simply physical distribution. The marketing, support and development functions of a well-functioning supply chain provide the impetus to foster a sustained and scaled adoption of the technology.

The remainder of this section will briefly describe three interventions which provide empirical evidence in support of these concepts. Because of space constraints, these illustrations are necessarily brief. However, references providing more detailed information are provided.

### *6.1 Postcosecha Program: metal silos in Central America*

Over a 20-year period, the SDC fostered and facilitated the successful adoption of approximately 670 000 metal silos to SHFs in Honduras, Guatemala, Nicaragua, and El Salvador. Metal silos helped 415 000 rural households preserve about 380 000 tons of grain each year, accounting for approximately 21% of annual grain production in the four countries (Fischler et al., 2011). Metal silos were adopted as an outcome of the Postcosecha Program, aiming to advance food security and reduce rural poverty. With success in the four countries, the program was further extended to other countries in Central America and Africa by SDC or jointly with other organizations (Food and Agriculture Organization, 2008; SDC, 2012).

In addition to the extensive replication, a careful post-program review (Fischler et al., 2011) showed that the production and adoption of metal silos continued after the program was withdrawn in 2003. Based on the three sources of benefit – reduced grain loss, better selling price, and less need to buy high-priced grain for family consumption – an approximate cost-benefit analysis was conducted. The estimated BCRs to the farmer participants ranged from 2.3 to 3.5 and the IRR ranged from 47% to 95%.

The Postcosecha Program developed and supported a market environment to facilitate the adoption of the metal silo technology. SDC fostered the demand for metal silos and, in so doing, actively engaged stakeholders from both public and private sectors. These public-private partnerships promoted the advantages of metal silos and initially helped farmers acquire metal silos with financial support. Direct economic benefits to farmers drove the adoption of metal silos. On average, SHFs who used metal silos received a higher price and they gained 23% more revenue for selling grain than non-silo users (Fischler et al., 2011).

The Program also assisted in establishing a system to supply metal silos. The SDC trained more than 2000 tinsmiths in the techniques of metal silo production, distribution, and marketing (Fischler et al., 2011). The training courses taught tinsmiths the techniques of metal silo production as well as the concepts necessary to run a business. Trained tinsmiths were able to start their own metal silo businesses. In response to an effective demand, the future supply of metal silos was thus secured. With the established market environment, sufficient demand and effective supply led to sustained technology transfer.

## *6.2 PICS project: triple layer bags in Africa*

The PICS project is an excellent example of the importance of the supply chain in a development project. From the selection of actors to build the supply chain, to the timing of communication and marketing activities, to the external partnerships created, this example illustrates the importance of addressing each link in the chain, as well as adapting to the local context.

The PICS project originated at Purdue University in Indiana, in the United States, to protect stored cowpea from pest damage without the use of insecticides. The technology was developed in response to the USAID-supported Bean/Cowpea Collaborative Research Support Program (CRSP) having identified cowpea storage as the key constraint to greater productivity in the 1980s. Professor Larry Murdock developed several nonchemical storage technologies, after working with SHFs and discovering that a main detriment to cowpea producers came from not being able to receive better prices immediately after harvest and that storage losses were too high. Adoption surveys were administered, and the triple layer storage bag was shown to be the most likely technology to be accepted.

PICS bags have been generally well-received in Africa. Within a five-year period, over 2 million bags were sold across 10 countries, 1.7 million households were reached, and the regional IRR was about 29%. This was in part because of the bags' cost-effectiveness, high returns on storage, and the perceived benefits to income and health.

Developing the supply chain was a major factor in the success of the project. Utilization of business consultants to develop the chain; coordination of communication, marketing, and extension activities; and in-country manufacturing with strong actors were crucial. When communication and technical activities were not coordinated, take-up was not as high as was expected, proving the need for effective market planning. The awareness of local context was important, as activities had to be tailored to fit local needs. For example, in areas where it was not acceptable for women to join men at meetings, extension activities were created for women-only meetings or with women associations, to better reach all audiences.

Throughout its life, the project trained farmers in hermetic storage methods and sold bags across ten countries in West and Central Africa with funding from the Bill & Melinda Gates Foundation and the assistance of many partners. Over the project's lifetime, PICS has grown from being a project merely about bags for better cowpea storage to an international implementation project, targeting additional grains and crops (Lowenberg-DeBoer, 2014).

### *6.3 YieldWise Initiative - maize in Tanzania*

In 2016, the Rockefeller Foundation launched YieldWise Food Loss - focused on reducing food loss of fruits, vegetables, and staple crops in Kenya, Nigeria, and Tanzania - countries where up to half of all food grown is lost (Rockefeller Foundation, 2018). This discussion will focus on the YieldWise experience to date for maize in Tanzania.

The YieldWise Initiative incorporates the following four action pillars:

- Access to technologies: promoting the adoption of appropriate loss-reducing technologies.

- Access to finance: collaborating with financial institutions to develop credit products that can be accessed by farmers and farmer-based organizations.
- Aggregation and training: training farmers and other supply chain actors in post-harvest management and facilitating development of local aggregation centers.
- Access to markets: stimulating demand by engaging actors across the diverse ecosystem of buyers.

Across these pillars, as appropriate, the YieldWise Initiative engages both the private sector as key partners and government entities for collaboration. Early results are encouraging, indicating a reduction in loss between 20% and 30%. More farmers are being connected to various market channels, which have provided assured markets for their produce. This has also contributed to a high uptake and utilization of loss-reducing technologies in the maize value chain in Tanzania (Rockefeller Foundation, 2018).

A food staple of major importance for Tanzania's economy, maize is cultivated by most farmers to ensure both household food sufficiency and for commercial purposes (Ipsos, 2018). Illustrating this importance, maize occupies 45% of the country's cultivated land. As a result, Tanzania is East Africa's largest maize producer.

Starting in 2016, the YieldWise Initiative introduced SHFs to a series of PHL-reducing technologies. By the end of 2016, more than 25 000 SHFs, had been trained in the use of at least one of these technologies (WRI, 2019). They include:

- Mechanized dehusking.
- Mechanized threshing.
- Tarpaulin use (as opposed to grain laying on dirt).
- Improved storage practices.

Prior to implementation of YieldWise, the use of the target technologies was quite limited. By mid-2017, when data for the report was collected, a strong increase in technology application was noted. For example;

- The use of mechanical dehusking almost doubled, reaching an adoption rate of 24% among beneficiary farmers.
- The use of threshing technologies increased significantly with about 43% of farmers using the technology.
- From the start of the YieldWise program to mid-2017, the utilization of tarpaulins for sun drying increased to 88% among beneficiary farmers.
    - Increased use of tarpaulins was not limited to the farm level, with gains in use being reported throughout the value chain.

- Use of superior storage technologies (hermetic bags and silos) had not increased as dramatically as originally hoped, reaching only 10% among beneficiary farmers. That level, however, was twice that reported by control group farmers.

The YieldWise initial accomplishments are impressive. Over the project's duration, post-harvest losses were reported as substantially reduced. The beneficiary farmer group reported losses of 19.4%. In contrast, losses in the control farmer group exceeded 39.4% (Ipsos, 2018).

## 7 Summary

The challenge of managing PHL is a feature of all agricultural and food systems. A bountiful field of grain at harvest doesn't guarantee success. Rather that produce typically needs to be harvested, stored, transported, and processed within systems that may operate over several months and at considerable distances. Without appropriate care and attention, loss can occur at distressingly high levels. Particularly in developing countries, PHL routinely exists at excessive levels resulting in reduced well-being for SHFs, higher prices for consumers, diminished food security, and unnecessary environmental degradation.

This chapter adopts a decision-maker perspective to explore issues and challenges relating to PHL management in the developing agricultural setting. In this context, decision makers include small-holder and commercial farmers, managers in supply chains, government officers, and representatives of the social sector. The chapter is framed from an opportunity perspective - what do we need to know to enable the actors in the sector to enhance its performance and provide improved well-being to sector participants and to consumers?

A key question for any manager contemplating action is, what is the appropriate goal or target that we should be striving to achieve? The chapter's first section addressed two facets relevant to that question. The first is the slogan that is often used supposedly to motivate interest in PHL reduction - that we should strive to achieve zero loss! By employing a simple economic framework that section demonstrated that zero loss is unlikely to be an economic or environmentally appropriate goal. In simple terms, SHFs and other decision makers in the sector realize that eliminating all loss almost always comes at a high cost and requires excess resource use.

The second facet illustrates how the use of more effective PHL-reducing practices can bring value to decision makers and to the sector. Often ineffective practices result in quality losses as well as a diminished amount of useful grain. In those instances, the use of more effective practices can increase the demand (and the price received) for the agricultural product. Improved practices also shift the cost curve associated with achieving PHL reductions. The combination of increased demand and lower cost represents the opportunity for economic as well as associated societal gain in the sector.

Research focused on reducing PHL has suffered from inadequate funding relative to other aspects of food production and marketing. Additional work to provide improved practices, including the incorporation of advanced information technology, is certainly warranted. However, the implementation of those improved practices should be driven by appropriate targets and goals. Economic analysis which reveals those targets and goals is urgently needed to inform private and public decision makers.

PHL occurs and is present in ways which are often very visible. Storage bags with grain spilling from holes in the bag or rotten kernels of grain are easy to see. Therefore, the focus of attention often centers on the amount of physical loss. However, from an economic perspective, physical loss may be overshadowed by economic losses throughout the supply chain. This is particularly the case for SHFs who, because of inappropriate storage opportunities, are forced to sell their grain at harvest when prices are lower and then must purchase grain for consumption when prices are higher. In the second section of the chapter, this dilemma is identified as a 'hidden tax' from excessive PHL.

If we are to have effective goals for PHL reduction, we need to understand the impact of the hidden tax and appreciate the opportunity provided by reducing the effects of that tax. Research, which documents, explains and quantifies the benefits of reducing that hidden tax, would aid decision making. In addition to the storage aspects of this issue, such research also should document and measure gains from any quality enhancements arising from the use of more effective PHL-reducing practices.

The preceding paragraphs in this summary focused on our current inability to quantify appropriate goals and targets for PHL reduction. Implicitly, useful, quantified measures of loss are required to rectify this deficiency. The third section of the chapter addresses the challenges that result because effective measurement of loss is not routinely conducted, especially at upstream stages of developing country supply chains. The cause for this deficiency is that the cost of doing actual measurement outweighs the operational benefits of conducting the measurement. In practice, where estimates of loss exist, they tend to be based upon recall from individual actors in the sector. This recall typically occurs sometime after the loss occurs and is based upon observation of unmeasured physical amounts. It is reasonable to expect that such estimates can lack reliability.

The opportunity relating to this deficiency is to conduct research and then develop and employ tools which can provide useful guidance to decision makers not reliant only on recall estimates of loss. Part of the research needed should focus on developing quantitative linkages between the use of improved PHL practices and expected reduction of loss. For example, what are the losses from use of metal silos or hermetic storage to store maize versus the losses from the traditional practice of storing maize in open sheds? Such technical research must document the additional costs from use of improved practices as well. The second research need should focus on the development of robust investment tools which decision makers, at all relevant stages of the supply chain, can employ to estimate potential net gains from reducing PHL.

The chapter's final section focused on overcoming the 'PHL paradox.' That paradox describes the setting when sustained, scaled adoption doesn't result even though pilot projects often document the significant benefits from employing appropriate PHL-reducing technologies. Two elements of this challenge were addressed. The first relates to interventions, typically led by donor institutions, which fund the demonstrations and pilot projects for promising technologies. However, such efforts tend to provide the technologies and resolve any implementation issues. Once the project ends, however, supply chains for the technology often aren't available, increasing the likelihood that scaled adoption doesn't follow. Therefore, to achieve broader success, such interventions need to foster supply chain development for the technology as an endogenous aspect of the effort.

The second aspect of the PHL paradox explores the effect of relatively weak market institutions in many settings where PHL is excessively high. In countries with stronger institutions, market forces and entrepreneurial action drive the use of a technology - if strong evidence exists to document the potential for economic gain. However, where such institutions are relatively weak, there is a need for catalyzing activities to overcome these weaknesses. Institutions in the social or public sector provide the needed facilitating resources to establish viable technology supply chains.

The role and mechanisms for market development, therefore, need to be much better understood and vehicles to overcome the institutional deficiencies need to be routinely available. Research, both conceptual and empirical, will empower institutional actors and private sector actors to achieve sustained success when conducting interventions to reduce PHL.

There are many reasons why PHL levels stubbornly persist in developing countries even though technologies exist which can significantly reduce such losses. Typically, excessive PHL is considered to be a problem which needs to be fixed. An alternative perspective is advanced in this chapter. That perspective considers excessive PHL to be an opportunity for economic and societal gain. The challenge is to empower decision makers to pursue those opportunities by taking actions which resolve the issues and challenges associated with reducing PHL.

A multifaceted, multidimensional approach is needed to exploit this opportunity. Managerial perspectives, solidly based on economic analysis, should be a component within that approach.

## 8 Where to look for further information

A few years ago, I had the honor of helping to establish the ADM Institute for the Prevention of Postharvest Loss at the University of Illinois (https://postharvestinstitute.illinois.edu/). Several of the Institute staff were quite proficient in obtaining information from Internet sources. Taking advantage of that resource, considerable effort was devoted to an extensive review of literature focused on:

- the economics of post-harvest loss reduction, and
- identifying loss-reduction interventions that had achieved sustained, scaled success.

Unfortunately, the results of that effort were disappointing and, I believe that a similar effort done today would experience similar frustrations.

In those relatively few instances where economic estimates were found, the analysis tended to concentrate on documenting that the intervention 'worked' in the context of the pilot intervention being assessed. The resulting benefit/cost ratio invariably was found to be positive. This, of course, is good. However, those findings aren't necessarily good predictors of sustained, scaled adoption (Shah et al., 2017).

Sections 6.1 and 6.2 of this chapter describe two exemplars in terms of quantifying the economic opportunities associated with successful interventions to reduce PHL. Section 6.1 refers to the POSTCOSECHA program (Fischler et al., 2011) in Central America. Section 6.2 focuses on the on-going PICS intervention in Africa (Lowenberg-DeBoer, 2014). Both instances are distinctive in terms of achieving long-term scaled adoption. Careful analysis documented that positive benefit/cost ratios were achieved without significant cost subsidies. Further they documented the development of unsubsidized markets to provide and support the needed technologies.

Both interventions also are distinctive because the interventions were relatively long-term in nature. The intervention described in Section 6.3 is much more typical because it reports on an intervention's result after only a few years of operation. Unfortunately, the norm today is that PHL interventions tend to be conducted as projects with relatively short (3–5 years) length. While those timeframes are not ideal, useful economic information can be derived from their results by focusing on the following considerations:

- Document multiple dimensions of the loss reduction achieved. Actual measurement of post-harvest loss (and any reduction) is a relatively costly activity. Therefore, the norm for quantifying loss is to rely upon recall-based estimates that are likely to be imprecise at best. Because of the cost factor, recall-based estimates will continue to be used. Generally, project reports also identify the extent of additional use of improved technologies and practices. A well-structured project should have identified the loss-reduction levels that can be expected through use of those technologies. Generally those relationships are not quantitatively linked with the extent of project technology used. However, the resulting estimate may well be as, or more, informative than reliance on recall-based estimates (Sonka, forthcoming 2020).

- Extend the economic analysis beyond volume measures of loss (weight, number of pieces) to incorporate the following two dimensions.
  - Inadequate post-harvest practices often have significant detrimental impacts on the quality of product as it moves through the supply chain and arrives at the market. The economic aspects of that decline in quality often are not quantified, even though it can be substantial.
  - SHFs in developing countries tend to be in the unfortunate position of having to sell at harvest when prices generally are low. Better post-harvest practices can provide improved market access, thereby enhancing the farmer's market position. Documenting the economic benefits of these improved practices is critically important.
- Document factors that can be linked to the potential for sustained impact. Identifying market linkages formed within the project can be an effective means of documenting that potential. In addition to documenting the use of an improved technology or practice, it is important to identify the extent to which that use was facilitated by market transactions and the extent to which associated markets linkages have become the norm. For example, recent field research focused on the maize intervention in Tanzania (Section 6.3) has documented that market relationships now are employed by SHFs to acquire tarps and hermetic storage bags. Further, maize processors report that, because of the intervention, they 'now know who to call to get, clean dry grain' (Sonka, 2020). These developments provide evidence indicating continuation of the practices introduced in the intervention.

Being able to identify and access literature that reports on the economics of PHL reduction remains an unresolved challenge. Hopefully entities such as the newly formed Consortium for Innovation in Post-Harvest Loss and Food Waste Reduction (https://sites.google.com/iastate.edu/phlfwreduction/) can provide support to reduce the difficulty of identifying and accessing such literature in the future.

## 9 References

Aphlis (2019). https://www.aphlis.net/en/page/7/about-aphlis#/ (accessed on March 04, 2019).

Bourne, M. C. (1977). Post-harvest food losses - the neglected dimension in increasing the world food supply. Cornell International Agriculture Mimeograph, 53. Retrieved from http://pdf.usaid.gov/pdf_docs/PNABJ082.pdf (accessed on March 30, 2019).

Chapoto, A. and Jayne, T. S. (2010). Maize price instability in Eastern and Southern Africa: the impact of trade barriers and market interventions. Prepared for the COMESA Policy Seminar on "Variation in Staple Food Prices: Causes, Consequence, and Policy Options", Maputo Mozambique, January 25–26, 2010, under the Comesa-MSU-IFPRI African Agricultural Markets Project (AAMP).

Costa, S. J. (2014). Reducing food losses in sub-Saharan Africa: improving post-harvest management and storage technologies of smallholder farmers. UN World Food Programme, Kampala, Uganda.

Fischler, M., Berlin, R., Bokusheva, R., Finger, R., Marín, Y., País, F., Pavón, K. and Pérez, F. (2011). 5 year ex-post impact study: POSTCOSECHA programme Central America. *SDC*. Retrieved from https://www.shareweb.ch/site/Agriculture-and-Food-Security/focusareas/Documents/phm_ic_postcosecha_impact_study.pdf (accessed March 30, 2019).

Gustavsson, J., Cederberg, C., Sonesson, U., van Otterdijk, R. and Meybeck, A. (2011). Global food losses and food waste. Food and Agriculture Organization of the United Nations, Rome, Italy. Retrieved from http://www.fao.org/docrep/014/mb060e/mb060e.pdf (accessed March 30, 2019).

Food and Agriculture Organization (2008). Household metal silos: key allies in FAO's fight against hunger. Retrieved from http://www.fao.org/fileadmin/user_upload/ags/publications/silos_E_light.pdf (accessed on March 30, 2019).

Ipsos (2018) The YieldWise Initiative: 2017 supplementary survey report. Ipsos, Paris, France.

Jones, M., Alexander, C. and Lowenberg-DeBoer, J. (2011). An initial investigation of the potential for hermetic Purdue Improved Crop Storage (PICS) bags to improve incomes for maize producers in sub-Saharan Africa. Working Paper No. 11-3. Department of Agricultural Economics, Purdue University, USA. Retrieved from http://ageconsearch.umn.edu/bitstream/115554/2/11-3PICS.Maize-2.pdf (accessed on March 30, 2019).

Lowenberg-Deboer, J. (2014). Purdue Improved Crop Storage (PICS) in West and Central Africa. Unpublished overview paper, Purdue University.

Mendoza, N. B. (2016). https://www.devex.com/news/reducing-post-harvest-losses-what-is-the-next-breakthrough-88080 (accessed on March 04, 2019).

Rockefeller Foundation (2018). YieldWise food loss. Retrieved from https://www.rockefellerfoundation.org/our-work/initiatives/yieldwise/ (accessed on March 10, 2019).

Shah, S., Agarwal, R. and Sonka, S. (2017). A time and place: non-profit engagement in the creation of markets and industry emergence. Retrieved from http://ssrn.com/abstract=2959714 (accessed March 28, 2019).

Sonka, S. T. (2018). Tracking gains to small-hold farmers from reducing post-harvest losses. Retrieved from http://edsnidercenter.org/wp-content/uploads/2018/03/Tracking-Gains-03-15-18-Final.pdf (accessed on March 15, 2018).

Sonka, S. T. (2020). Personal communication with maize supply chain participants. Iringa, Tanzania.

Sonka, S. T. (Forthcoming, 2020). Measuring to Manage. *Proceedings of the Conference on Reduction of Food Loss and Waste*, November 11–12, 2019. Pontifical Academy of Sciences, Vatican City.

Sonka, S. T., Agarwal, R. and Shah, S. K. (2018). To reduce postharvest loss, start with a sustaining end in mind. *Pyxera Global Engagement Forum* (Fall/Winter), 44–7.

Swiss Agency for Development and Cooperation [SDC] (2012). Food security in Central America: small silos that make a big difference. *Latin Brief*, November 2012. Retrieved from https://www.shareweb.ch/site/Agriculture-and-Food-Security/focusareas/Documents/phm_sdc_latin_brief_2012_nov.pdf (accessed March 30, 2019).

Van Vark, C. (2012). https://www.theguardian.com/global-development-professionals-network/2012/dec/14/post-harvest-loss-food-security (accessed on March 04, 2019).

WRI. 2019. *Reducing Food Loss and Waste: Getting Across the Tipping Point* (forthcoming).

# Chapter 2

## Advances in understanding fungal contamination in cereals

*Kizito Nishimwe, Iowa State University, USA and University of Rwanda, Rwanda; Julie Aiza L. Mandap, Iowa State University, USA and University of The Philippines Los Baños, The Philippines; and Gary P. Munkvold, Iowa State University, USA*

1 Introduction
2 Fungi affecting cereal grains
3 Fungal contamination in maize
4 Fungal contamination in rice
5 Fungal contamination in wheat
6 Conclusion
7 References

### 1 Introduction

In 2050, the world population is estimated to increase by 2 billion people and reach more than 9 billion people. Feeding the world will require an increase in the current production and a significant reduction in food losses and wastes (FLW). However, increase in the production means more land dedicated to agriculture, and impacts on biodiversity, which affects the environment (Mc Carthy et al., 2018). On the other hand, one-third of the current production is estimated to be lost or wasted (FAO, 2020). Ultimately, solutions for feeding the world will most likely focus on the reduction of FLW.

Initially, the term cereal came from the ancient Greek and was named after the ancient Greek goddess, '*ceres*', the goddess of agriculture and fertility. Cereals are cultivated grasses that produce fruits (grains or seeds) after pollination (Wrigley, 2010). Cereals are food security crops because of their high nutritive values for energy (carbohydrate-based energy) and other nutrients (proteins) and their long-term storability (Jespersen and Munck, 2009). They constitute the main food staples in many parts of the world and are primary animal feed ingredients (Batey, 2017). Cereal production is lost and wasted due to insects, rodents, and fungal infestations (Mason, 2018). For the latter, fungi cause cereal losses in three ways: by reducing grain yields, by spoiling cereals and rendering them unfit for

http://dx.doi.org/10.19103/AS.2020.0072.02

human and animal consumption, or by producing secondary metabolites like mycotoxins, which are harmful to humans and animals (Miller, 1995).

Fungi colonize cereal grains because they are heterotrophs and cannot make their nutrients from inorganic substances. They take advantage of cereal nutrients to maintain their metabolism and reproduction requirements for survival and reproduction. Since this chapter will discuss the fungal contamination in cereals, it is worthwhile to understand the cereal ecosystem and its importance for fungal growth. Also, the understanding of fungal ecology is the first step in successful fungal contamination control in cereals. This chapter will discuss (i) fungal ecology and (ii) storage fungal contamination in three major types of cereals (maize, wheat, and rice) which constitute more than 90% of total world cereal production (Fig. 1).

### *1.1 Humankind and cereals*

Cereals provide energy (starch) and other nutrients (protein, vitamins, and oil) to humans and livestock (Wrigley, 2017). They constitute the pillar for human civilization because of their long-term storability and direct use with minimal processing (Rosentrater and Evers, 2017; Wrigley, 2010; Batey, 2017). For instance, wheat is from the Tigris and Euphrates River Valley, and its cultivation began approximately 10 000 years B.C.E (Finnie and William, 2017). For millennia, they evolved through the selection process of improved seeds for superior attributes to maximize yield; during the green revolution in 1950+-1960+, Dr. Norman Borlaug and others developed disease-resistant varieties of wheat with high yields and disseminated them across the world

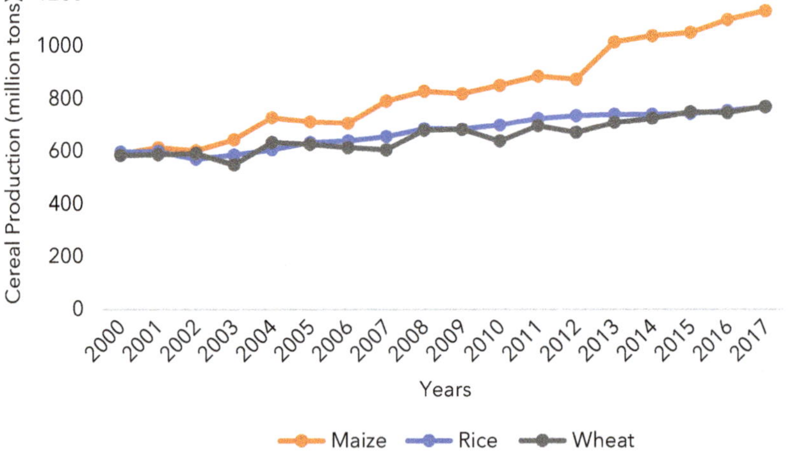


**Figure 1** Total cereal production in the world from 2000 to 2017 (FAOSTAT).

(Kung, 1993). Cereals are part of different cultures around the world employed to celebrate the harvest. For example, the Thanksgiving Day celebrated in the United States (USA) with the celebration of the grain harvest and was brought to the US by European migrants (Smith, 2003). In Eastern Europe, a similar ceremony, known as *'Dozynki'*, celebrated the grain harvest. Likewise, some spiritual commemorations in the Middle East use bread, a cereal-based product, for their celebration. For instance, Jewish Passover commemorates the Jewish departure from slavery in Egypt (Wrigley, 2010). In Rwanda, the end of harvest is celebrated in August, and known as *'Umuganura'* day (Grayson, 2017).

Today, cereal use differs across the globe, depending on their availability or the local habits. In many tropical countries, cereals (e.g. maize) are produced for human consumption and constitute the primary staple food for the local population. In industrialized countries, some cereals are mostly intended for animal feed or industrial uses. In the United States, maize is used to produce biofuel and livestock benefits from processing by-products that are used to make feeds. Rice contributes significantly to food nutrition security and remains an important staple across the globe especially in Asia and Africa as a dietary energy source. The United States primarily exports paddy rice and Asia produces and exports milled rice (Abera et al., 2018).

Given the importance of world cereal production, this chapter will focus on maize, wheat, and rice (Fig. 1).

Pollination is the transfer of pollen from anther (a male part) to stigma (a female part) of a plant. Pollination and subsequent fertilization of the ovum result in the development of seeds which contain valuable nutrients and the embryo for the next generation of plants (Wrigley, 2010).

From the taxonomy standpoint, cereals are monocotyledonous belonging to the Poaceae family with specific and easily recognizable characteristics (Wrigley, 2017). Plants have nodes for each stem. For each node, there is a single leaf, which is located in opposite ranks. The reproduction parts are collectively known as the inflorescence. Cereals have two distinct reproduction organs; a female organ, the pistil, and, a male organ, the stamen. The plants considered cereals are maize, rice, wheat, sorghum, millet, barley, oats, rye, and triticale. However, the three major types of cereals produced in the world (more than 90%) are maize, wheat, and rice (Wrigley, 2010; Perreta et al., 2011).

Pseudocereals can be confused with cereals because of many resemblances. Like cereals, their grains are rich in starch, contain proteins, and dried grains can be used alternatively to cereals. However, they are not monocotyledonous, but are dicotyledonous and belong to different families. For example, buckwheat belongs to the Polygonaceae family, quinoa and grain amaranth belong to Amaranthaceae family, respectively (Rosentrater and Evers, 2017).

## 2 Fungi affecting cereal grains

Among all plant pathogens, fungi are considered the most ubiquitous and diverse group of microorganisms, causing diseases in a wide array of hosts. Recent estimates reached a total number that ranged from 5.1 million fungal species (O'Brien et al., 2005) to 6 million species, including endophytes that are mostly unculturable (Schoch et al., 2014). However, because of the overwhelming species richness of fungi, less than 5% have been described (Mueller and Schmit, 2007).

Kingdom Fungi (Mycota) is a monophyletic group in the domain Eukarya and is now believed to be more closely related to animals than to plants (Pitt and Hocking, 2009a). Fungi share similar morphological features such as profuse production of spores and filamentous mycelia. Fungi broadly include mushrooms, mildews, yeasts, smuts, rusts, and molds - terms that are of no taxonomic significance, but conveniently used in describing the most predominant groups of fungi. Traditionally, fungi have been identified by cultural and morphology-based taxonomy. In recent decades, identification and disease diagnosis have been revolutionized through the development of molecular techniques and protocols for rapid and unambiguous detection of fungal species (Links et al., 2012; Schoch et al., 2014; Crous et al., 2015).

More than 150 species of filamentous fungi and yeasts have been isolated from cereal grains (Christensen and Kaufmann, 1969). Fungi are responsible for the highest fraction of loss of grain in storage (Christensen and Kaufmann, 1969; Wicklow, 1995), and are regarded as the most important cause of spoilage in stored grain and cereal food products (Christensen and Kaufmann, 1969; Pitt and Hocking, 2009a).

### *2.1 Ecophysiology of fungi*

Fungi associated with grain can be seed-borne, soilborne, airborne, or found as surface contaminants. Fungi produce numerous spores that are easily disseminated by wind, water, or vectors such as insects. Prevalence and dominance of fungi are greatly dependent on the interactions of abiotic and biotic factors, that is, the presence of other microflora and fauna. Fungi are present in grains that are typically coexistent with insects and other microflora such as bacteria (Magan et al., 2003). Fungi are heterogeneous but can be grouped based on different characteristics, among which are their adaptation to a particular environmental condition. Fungal species, in general, can be grouped based on their affinity and capability to survive at different levels of temperature and moisture content (Lacey et al., 1994; Magan and Aldred, 2005; Fleurat-Lessard, 2017). Fungi can be mesophilic, thermophilic, xerophilic, or hydrophilic species. These ecophysiological categories are also used to describe other eukaryotes or prokaryotes, such as bacteria, but with a different range of conditions per group. Each ecophysiological group based on temperature and moisture are further described below and illustrated in Fig. 2:

**Thermotolerants and thermophiles** - species capable of surviving high temperatures specifically. These two terms are arbitrarily differentiated by minimum or maximum temperatures that permit their growth. Thermotolerants have a temperature range of growth from below 20°C to a maximum of ≤50°C, whereas thermophiles have a minimum of ≥20°C and a maximum at or above 50°C (Cooney and Emerson, 1964). Examples of filamentous fungi under this group are *Aspergillus flavus*, and *A. nidulans*.

**Cryotolerants (syn. Psychrophiles)** - these are fungi capable of growth at low temperatures. This group includes *Penicillium verrucosum*, *Paecilomyces* spp., and other *Penicillia*.

**Mesophiles** - grow best at moderate temperatures (20°C–45°C) such as some species of *Alternaria*, *Fusarium*, *Mucor*, *Ulocladium*, *Epicoccum*, and *Arthrobotrys*. This group includes most of the pathogens of terrestrial plants.

**Hygrophiles** - fungi that are unable to develop at or below 0.90 water activity ($a_w$) (Richard-Molard et al., 1985) such as *Fusarium* spp., *Alternaria* spp., *Claviceps* spp., *Chrysogenum* spp., *Epicoccum* spp., *Rhizopus* spp., and *Mucor* spp. These species remain viable in dry grain only after 2–4 months of storage. Extended dry conditions result in ceased growth and non-viable spores.

**Xerophiles** - include fungal species that are capable of growth below 0.80–0.85 $a_w$. These are drought-tolerant species that comprise most storage

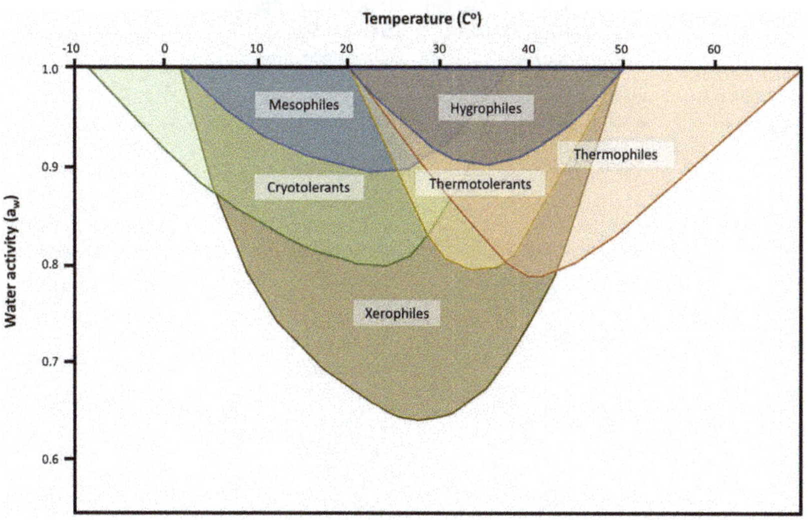


**Figure 2** Ecophysiological groups of fungal species in pre- or postharvest cereals depending on initial water activity and temperature (adapted from Lacey et al., 1980).

microflora colonizing grain after harvest when grain moisture content reaches 15–17%. Mainly are the genera *Aspergillus*, *Eurotium*, and *Penicillium* ( Richard-Molard et al., 1985). *A. glaucus* and *A. restrictus* grow at less than 0.75 $a_{w.}$

## 2.2 Field and storage fungi

Fungi infecting cereal grains can be broadly categorized based on two ecological groups, either field fungi or storage fungi (Christensen, 1957). The ecophysiological grouping previously described converges in these two broader categories (Table 1). This grouping has long been known and used, although in some cases, a definite distinction between the two groups is impossible since growth of fungi may occur in the field and during storage as well. This categorization is based mainly on the moisture requirements of the species and hence it is not a valid taxonomic identification. Available moisture for microbial growth can be measured as water activity ($a_w$) or expressed in percentage as the equilibrium relative humidity (ERH). Relative humidity is the percentage of water vapor available in air at a given temperature. When

**Table 1** Fungal species in three ecological groups found in common cereal grains (adapted from Lau and Sheridan, 1975; Pelhate, 1988; Cahagnier, 1998).

| Habitat | Maize | Wheat | Rice |
|---|---|---|---|
| Field mycoflora | *Acremonium strictum*<br>*Alternaria* spp.<br>*Cephalosporium* sp.<br>*Epicoccum purpurascens*<br>*Fusarium verticillioides*<br>*F. graminearum* | *Alternaria tenuissima*<br>*E. purpurascens*<br>*F. graminearum*<br>*Ulocladium atrum*<br>*Geniculifera* sp.<br>*Chyrsosporium* sp. | *Alternaria alternata*<br>*Bipolaris oryzae*<br>*Drechslera oryzae*<br>*F. oxysporum*<br>*Pyricularia grisea*<br>*Ustilaginoidea virens* |
| Intermediate mycoflora | *Aureobasidium pullulans*<br>*Chaetomium* spp.<br>*Cladosporium cladosporioides*<br>*Mucor hyemalis*<br>*Papulaspora* sp.<br>*Rhizopus stolonifer*<br>*Trichoderma* sp.<br>*Verticillium lecanii* | *A. pullulans*<br>*Absidia glauca*<br>*C. cladosporioides*<br>*Chaetomium globosum*<br>*M. hyemalis*<br>*M. racemosus*<br>*V. lecanii* | *A. pullulans*<br>*C. globosum*<br>*Cladosporium* spp.<br>*Curvularia lunata*<br>*Rhizopus* sp.<br>*Syncephalastrum* spp.<br>*Trichoderma viride*<br>*Verticillium* spp. |
| Storage mycoflora | *Aspergillus candidus*<br>*A. flavus*<br>*A. ochraceus*<br>*Eurotium chevalieri*<br>*Eurotium* spp.<br>*Penicillium chrysogenum*<br>*P. funiculosum*<br>*P. rubrum*<br>*P. verrucosum* | *A. candidus*<br>*A. parasiticus*<br>*A. versicolor*<br>*Eurotium amstelodami*<br>*E. chevalieri*<br>*P. cyclopium*<br>*P. stoloniferum*<br>*P. verrucosum* | *A. flavus*<br>*A. niger*<br>*A. ochraceus*<br>*A. versicolor*<br>*P. aurantiogriseum*<br>*P. islandicum*<br>*P. citreonigrum*<br>*P. commune* |

RH of the enclosed air, in a storage container, for example, reaches a state of equilibrium with the moisture content of the grain, equilibrium relative humidity (ERH) is attained (Bothast, 1978; Atanda et al., 2011).

**Field fungi.** Include the hygrophilic species that cause minor damage only in storage and can be dealt with proper drying prior to storage. Fungi attack on mature or developing seeds that have at least 20% moisture or at a level of ERH between 90% and 100%. Examples are *Fusarium*, *Alternaria*, *Cladosporium*, and *Helminthosporium*, with all the species commonly occurring in freshly harvested seeds. *Fusarium* can proceed infection in storage until the water available becomes insufficient for its survival. Species that attack near harvest are also grouped as field fungi and these include *Curvularia*, *Stemphylium*, *Epicoccum*, and *Nigrospora* (Bothast, 1978). Field fungi become more prevalent when weather is relatively wet during normal seed development, but the presence of a fungus does not automatically imply loss. For example, *Alternaria* is often present under the outer pericarp of wheat, but causes no symptoms unless wheat grain is subjected to continuous wetness or rain.

**Storage fungi.** Postharvest conditions are more defined and stable compared to field conditions, making mycoflora relatively similar for all stored grain. Abiotic and biotic factors can be assessed and are easier to be manipulated in storage. Storage fungi are those that can tolerate and germinate at lower $a_w$ levels compared to those in the field. Attack begins at storage when moisture levels of grain are between 13% and 20% or at ERH levels of 70–90%. Species belonging to genera *Aspergillus*, *Penicillium*, *Eurotium* are the most common storage fungi (Bothast, 1978). *Aspergillus* spp. may compete to grow even at 65% RH and frequently initiates spoilage which is then followed by the attack of other molds. Certain *Penicillium* species also thrive prior to harvest and thus are considered as field fungi. But *Penicillium* is more commonly considered a storage genus since it predominates more in storage. *A. flavus* and *A. niger* were also found to be present both in stored and freshly harvested grain (Jedidi et al., 2018), hence, the invasion of these species might have started from the field and were carried over through storage. Contamination by storage fungi at preharvest differs depending on the geographical site, prevailing weather conditions, and host. *A. glaucus* and *Penicillium* spp. were found to be present in surface-disinfected wheat and barley kernels (Flannigan, 1978); *P. verrucosum* in barley (Lillehoj and Goransson, 1980); and *P. aurantiogriseum* and *P. verrucosum* in cereal grains in England (Mills et al., 1992).

## 2.3 Genera of major storage fungi

***Aspergillus.*** Species of *Aspergillus* and its teleomorph *Eurotium* can thrive in low moisture conditions (≤14%) and elevated temperatures (≥30°C) making them the predominating spoilage fungi in the tropics. Colonies are macroscopically observed as a cluster of fruiting bodies with globose heads commonly yellow and green in color, with some species being blue, black, or grey in color (Pitt and Hocking, 2009b). Together with the Penicillia group, *Aspergillus* is known to exploit the entire resource available in each kernel, usually observed as heavy sporulation on the surface of a seed without the mycelium extending into the surrounding seeds (Wicklow, 1995). Microscopically, this group is characterized by conidiophore formation with large and swollen apex called a vesicle. A vesicle bears copious phialides simultaneously, unlike *Penicillium* and the other related genera (Pitt and Hocking, 2009b). Abundant chains of conidia that are distinctly spherical are formed from the phialides. Major species include *A. flavus*, *A. ochraceous*, *A. restrictus*, *A. candidus*, and the *A. glaucus* group (Sauer et al., 1992). *Aspergillus* spp., especially *A. flavus*, produce aflatoxins, the most important mycotoxins in grain storage.

***Penicillium***. It is similar to *Aspergillus* in terms of exploitation of the entire resource. Unlike *Aspergillus*, *Penicillium* spp. mostly requires higher moisture, that is, 16.5–20% in starchy cereal grains; 17–20% in soybeans; and 10–15% in sunflower, peanuts, and copra (Sauer et al., 1992). *Penicillium* spp. are more morphologically and ecologically diverse than *Aspergillus* spp. with the color of fruiting bodies ranging from shades of blue to green and grey making it commonly referred to as the blue or green mold (Wallace, 1973; Pitt and Hocking, 2009b). It causes mustiness, caking, and discoloration of germs, whole seeds, and kernels. This genus may produce ochratoxin A and other toxins. Genera related to *Penicillium* based on morphology are *Scopulariopsis, Paecilomyces,* and *Geosmithia,* which also causes spoilage in different commodities (Pitt and Hocking, 2009b).

Other molds that cause seed spoilage belong to the group Mucorales that requires at least 88% relative humidity to proliferate and are not typically initiators of grain spoilage, for example, *Absidia, Mucor,* and *Rhizopus* (Wallace, 1973).

## 2.4 Conditions favoring fungal activity in storage

Proliferation of molds is governed by a mix of biotic and abiotic factors, specified into three main considerations: (1) grain type and nutrient composition; (2) moisture availability and temperature conditions; (3) biotic factors such as insect infestation. Maximum mold growth in stored cereals occurs at temperatures between 30°C and 35°C and >15% moisture content (Muir and White, 2001). Generally, the lower limit that allows fungal growth is lower than the minimum limit that allows mycotoxin production (Cahagnier et al., 1993, 2005; Pardo et al., 2006; Magan et al., 2014). This, of course, varies specifically down to species level of fungi chiefly in response to changes in $a_w$, temperature, and gas composition (Table 2).

**Table 2** Moisture content and minimum water activity ($a_w$) necessary for fungal growth and mycotoxin production (adapted from Pelhate, 1988; Cahagnier, 1998)

| Fungi | Mycotoxin[1] | $a_w$ limit for growth | Related MC (%) | $a_w$ limit for toxin production | Related MC (%) |
|---|---|---|---|---|---|
| *Aspergillus flavus* | AFB1 | 0.78-0.84 | 17.6-19.0 | 0.84 | 19.2 |
| *A. parasiticus* | AFB1 | 0.84 | 19.0 | 0.87 | 20.2 |
| *A. ochraceus* | OTA | 0.77 | 16.5 | 0.85 | 19.6 |
| *Penicillium patulum* | Patulin | 0.81 | 18.0 | 0.95 | 27 |
| *P. expansum* | Patulin | 0.82-0.84 | 18.3-19 | 0.99 | 30.0 |
| *P. verrucosum* | OTA | 0.80-0.81 | 18 | 0.83-0.86 | 18.8-19.8 |

[1]AFB1 - Aflatoxin B1; OTA - Ochratoxin A.

**Water activity.** Lacey and Magan (1991) considered this concept as the most practical expression of the amount of available water for microbial growth. Water is a fundamental requirement to the growth of microorganisms and for fungi, water is particularly necessary for germination in any host. Water activity ($a_w$) is the ratio of partial pressure of water vapor of the test material and the saturation vapor pressure of pure water under the same temperature. When vapor and temperature are in a state of equilibrium, $a_w$ of a sample is equivalent to the relative humidity of air surrounding the sample in an enclosed environment. Hence, $a_w$ is numerically the same as the equilibrium relative humidity (ERH) expressed as a decimal. Generally, storage fungi require $a_w$ level of at least 0.65, equivalent to 65% ERH, equating to an equilibrium moisture content of around 13% in cereal grain and broadly at temperatures 10–40°C (Atanda et al., 2011). The availability of water in grain also influences the tolerance of fungi in varying $CO_2$ and $O_2$ levels.

**Temperature.** Fungal development and host degradation involve biochemical reactions that are related to temperature. In general, lower temperature results in slower development of fungi hence slower degradation of grain. However, there's always an exception such as *A. glaucus* and several species of *Penicillium*, *Cladosporium*, *Fusarium*, and *Mucor* that were shown to be capable of growth even from −5°C to −8°C (Burrell, 1974). Nevertheless, it remains safe to store grain at low temperatures given that the moisture content is at a safe level. The opposite happens at room or higher temperatures where most fungal enzymes are activated, and growth can be at optimum. This is the reason why the temperature range of 4–10°C is considered safe for storage. There is a wide range of temperatures over which fungi can grow with each species varying extensively. For example, *Penicillium* and *Cladosporium* are cold-tolerant but some species of *Penicillium* can also grow at temperatures as high as >30°C. Molds can continue to grow in a desired temperature unless conditions change that can thermally inactivate enzymes, or when oxygen and nutrient sources are depleted, and accumulation of carbon dioxide becomes limited for the molds present.

**Gas composition.** Obligate aerobes are organisms that require oxygen to grow and become limited with a carbon dioxide build-up of around 14%. With yeasts as an exception, most fungi that causes biodeterioration in grain are considered obligate aerobes, to be more specific, they are microaerophiles that survive in environments with lower levels of oxygen, ranging from 2 to 10% $O_2$. For example, *A. flavus* growth is significantly inhibited when oxygen is reduced from 5% down to 1% (Detroy et al., 1971). The most tolerant storage mold species in low oxygen conditions were *P. roqueforti* and *A. candidus* (Magan and Lacey, 1984). In some cases, storage fungi can tolerate high $CO_2$ levels with a very low oxygen available such as *P. roqueforti* which can proliferate in grain at >80% $CO_2$ and at least 4% $O_2$.

## 2.5 Mechanisms of fungal invasion

A prerequisite to any fungal invasion is spore germination under the prevailing conditions followed by successful penetration into the host's surface. Fungal propagules that are inherently present in air or on grain penetrate the coat before invasion of mycelium inside the grain tissues. The establishment of fungi within the host can lead to accumulation of mycotoxins with or without visible mold growth on the surface. It is usually 3–7 weeks before harvest when molds start to infect and spread in cereal grains (Gherbawy and Voigt, 2010). Fungi have specialized structures that may persist in the soil for several years such as the sclerotia of *Aspergillus* spp. The sclerotia are capable of germination and subsequently produce conidia. Other fungal propagules such as spores and mycelia may also survive in plant debris left in the field after harvest and thus serve as an inoculum source (Payne and Widstrom, 1992; Wicklow, 1994).

The seed coat (testa) is the most important barrier protecting the grain (e.g. embryo and endosperm) from fungal attack. Storage fungi can kill the embryo even before discoloration is evident. By the time mold is visible on the grain surface, mustiness or off-odors can be observed. Dense formation of mycelia over grain kernels can produce webbing and cake the grain. Fungal species that are known to invade the embryo include *A. glaucus, A restrictus, A. candidatus, A. ochraceus, A. versicolor,* and *P. viridicatum* (Christensen and Kaufmann, 1974; Tsuruta et al., 1981).

In a whole grain bin perspective, fungal growth and invasion may appear stratified due to uneven distribution of moisture throughout the bin favoring the most competitive species at a specific moisture level. For example, in maize in a metal bin, *A. flavus* is typically found below the layer of *P. palitans,* which usually grows alone in creating a mat in the upper 300 mm of maize in the bin (Semeniuk, 1954). Stratification becomes difficult to observe if the typically dense and cottony mycelium of *Absidia*, *Mucor*, and *Rhizopus* also colonize the upper stratified zones.

Damage in grain provides an entry point for fungi to commence invasion. Mechanical damage gives a conducive entry of fungi in insufficiently dried grain. Some molds benefit well from insect activity which may serve as disseminators or vectors of fungal propagules, attractants or repellants (Magan et al., 2003), aside from creating damage that serves as entry points. Different species of molds utilize several ways to start invading a particular plant part and the physiology of the host plays an important role in fungal colonization. For example, maize in the field is contaminated mainly in two ways: (1) airborne or insect-vectored conidia reaches the silks and grow into the ear, and; (2) mechanically damaged or insect/bird damaged kernels are colonized by the fungus. *A. flavus* is not an aggressive invader of maize. Unlike *Fusarium* spp., it takes advantage of the crop under stressed conditions and enters through natural openings or damaged tissues. *A. flavus* has been shown to disperse to adjacent kernels through cracks or mycelia entering the kernel through the rachis after a wound is instigated (Smart et al., 1990). Higher aflatoxin concentration was observed in kernels where the infection started. *A. flavus* can also be present internally in the husk aside from the ears and silks and these tissues become more colonized as the plant enters maturity (Siriacha et al., 1991).

## 2.6 Ecological succession of molds in storage

Succession begins as a result of grain moisture changes during storage. Grain that is artificially dried often is placed into storage at a moisture content that is too low to allow spore germination or hyphal growth. Once the grain moisture content rises, some growth of xerotolerant species commences (Sinha, 1992). The mesophilic group replaces the hygrophilic species after harvest and will continue to persist during storage if grain remains at more than 18% moisture content. At more than 20% moisture of grain under prolonged humid conditions in storage, Mucorales fungi (e.g. *Rhizopus* spp., *Mucor* spp.) may also proliferate.

On the other hand, xerotolerant species from the *Aspergillus* and *Penicillium* groups will take over the storage mycoflora when grain moisture after around 4–6 months is at 15.5–17% (Fig. 3). Changes in the initial conditions are influenced by the metabolic activities of the initial fungal assemblage producing water that consequently elevates moisture content to a point where the second wave of fungal species can start to invade (Christensen and Kaufmann, 1974; Bothast, 1978). The key also lies on the capability of fungal propagules, particularly spores, to remain dormant and become viable once the conditions became favorable.

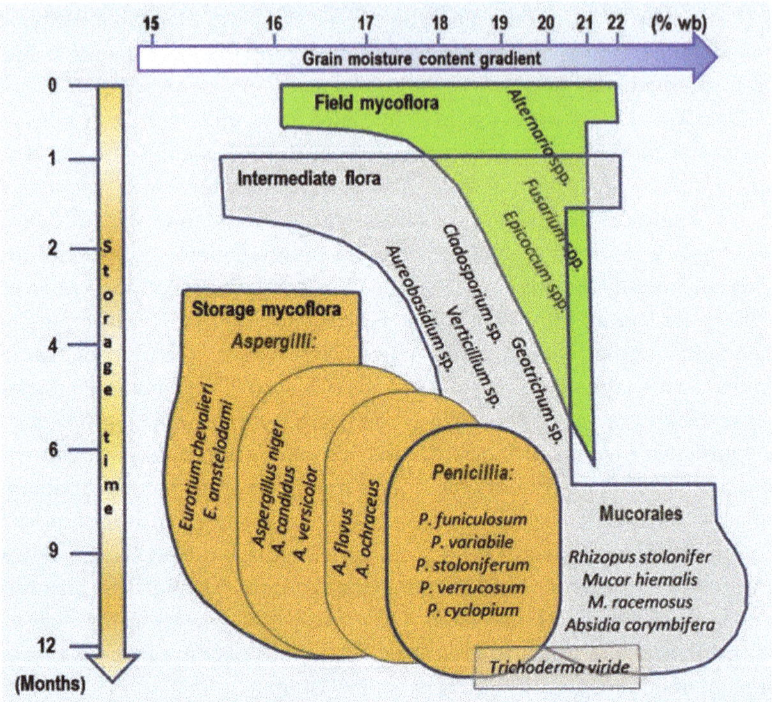


**Figure 3** Succession of storage fungi from initial storage to prolonged storage periods (adapted from Pelhate, 1988).

Studies on ecological succession in the past decade were revolutionized by polymerase chain reaction (PCR)-based assays and other molecular diagnostic techniques that allowed rapid species-specific identification. Metagenomics studies of microbial communities in different ecological niches are also now feasible through PCR techniques that utilize targeted primers to amplify fungal Deoxyribonucleic acid (DNA) including the internal transcribed spacer (ITS) sequenced for identification of fungal species or develop fungal phylogenies (Schoch et al., 2014; Links et al., 2012). Succession was visualized by multivariate analysis of the temporal changes of each of the fungal community (Fleurat-Lessard et al., 2010).

## 3 Fungal contamination in maize

In global cereal grain production, maize (*Zea mays* L.) predominates, followed by wheat and rice (Wrigley, 2010). Maize production has overtaken wheat and rice production over the past decade after being ranked third in global

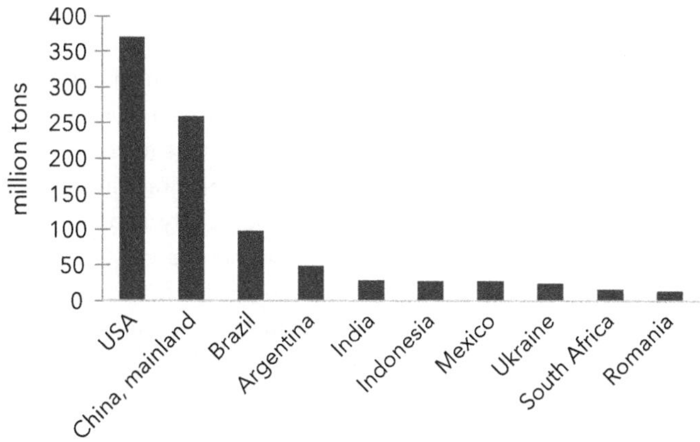


**Figure 4** Maize production of different countries in 2017 (FAOSTAT).

production in 2004 (Golob et al., 2004). The United States is the highest producer of maize (371M tons), then China and Brazil with 259M and 98M tons, respectively (Fig. 4) (FAOSTAT). Maize, indigenous to the western hemisphere, is a major staple food grain throughout the world, particularly in sub-Saharan Africa, Latin America, and Asia (Smale et al., 2011). Aside from being used as feedstock, maize is utilized as a raw material in many industrial and value-added products such as bioethanol, sorbitol, dextrose, glucose, and oil. Agronomic management in maize grain fields is similar to those producing a commercial seed crop but the latter is more intensive since the value of good seed is higher than grain (Beck, 2002).

Like any other cereal grain, maize seed has three major parts: pericarp, endosperm, and embryo. The part that primarily serves as protection from fungal invasion is the pericarp. Mature kernels of maize have a highly compressed and thin pericarp which is 5–6% of the kernel by weight (Wolf et al., 1952; Watson, 1987). The pericarp of maize influences the drying rate of grain because of its relatively low permeability to water (Syarief et al., 1987). It has been known that increased damage to the pericarp results in greater susceptibility to storage molds with higher severity observed if damage occurs near the germ (Tuite et al., 1985). The endosperm is the nutritive tissue composed of starch, minerals, proteins, and other compounds enveloping the embryo. Storage molds typically attack the embryo at the lower limit of moisture content that permits fungal invasion and also damages the endosperm at higher moisture contents (Sauer et al., 1992). Fungi rank second after insects as the cause of deterioration and loss in maize (Ominski et al., 1994), reported to cause around 50–80% damage in maize during storage period if conditions are in favor of their growth (Kossou and Aho (1993), as cited by Fandohan et al., 2003).

### 3.1 Maize storage

Maize seed and grain are stored separately, although in most developing countries, grain and seed are not clearly distinguished from each other and they are stored together (Pingali and Pandey, 2001). The main objective in grain storage is to preserve the quality and maintain the grain in good condition, meeting the negotiated nutritional and product safety expectations for both marketing and processing to minimize economic losses (Chakraverty et al., 2003; Zorya et al., 2011). The safe moisture levels for maize grain storage are below 13.5%, but in areas with limited sunshine hours for drying, harvested grain is often stored at intermediate moisture levels between 15% and 21% (Wilson and Desmarchelier, 1994).

Long-term storage of maize involves cooling off the grain as soon as possible to reach the initial target temperature of around 15°C. Aeration is continued to uniformly cool the grain and maintain a dry condition to avoid migration of internal moisture. Cool and dry storage conditions thereby discourage mold development and insect activity. Maize grain can be stored safely at a temperature of 18°C or less and not between 21°C and 27°C, which is the most conducive for the majority of storage molds.

In sub-Saharan Africa, the common practice is to sun-dry maize by laying cobs on the ground for 5–7 days and then transfer into traditional granaries for storage such as open-air cribs made of woven bamboo/twigs, mud silos, or a raised platform made up of tree poles and thatch grasses (Markham et al., 1994; Kankolongo et al., 2009). These systems are often inadequate in providing a good dry environment for maize especially in humid and semi-humid zones (Fandohan et al., 2003). In developed countries, maize is mostly stored in commercial infrastructures with better control of abiotic factors to control pests. Grain can also be stored in bags such as woven polypropylene (PP) that are widely available and used. PP bags are affordable but the use of fumigants along with the bags increases the cost. In recent years, the use of hermetic storage devices is becoming common in Asia and Africa (De Groote et al., 2013; Njoroge et al., 2014). In hermetic storage, gas exchange is limited creating an internal atmosphere in the device that becomes depleted with oxygen. Low oxygen level is detrimental to the growth and development of aerobic microorganisms and insects. Several studies have shown the benefits of hermetic storage through reduced aflatoxin accumulation, maize weevil and larger grain borer (LGB) (*Prostephanus truncatus*) infestations, and grain weight loss (Njoroge et al., 2014; Walker et al., 2018).

### 3.2 Molds and their associated damage in maize

***Aspergillus***. Species belonging to this group are not considered aggressive invaders, but opportunistic fungi that take advantage of a host that are under stressed conditions or have pre-existing damage. In maize, it can cause discoloration and rot of germs and whole kernels (Sauer et al., 1992). About 185 species of *Aspergillus* are taxonomically identified (Kirk et al., 2001) and those that are considered important and frequently reported in maize are further described below:

- ***A. restrictus*** - grows very slowly in maize and wheat at 14.0–14.5% moisture content and 10–25°C causing no detectable increase in grain temperature. It kills and causes germ discoloration characterized by a purplish-black cast and mustiness in both grain and flour. It is not known to produce toxins detrimental to humans and animals (Sauer et al., 1992).
- ***A. glaucus*** - favored by 14.5–15.0% grain moisture content. Causes mustiness and caking by the time molds are evident on the surface. It can still grow at low temperatures from 2 to 5°C and when growth is dense it can cause heating of the grain up to 35°C. It is considered as the most common fungi that infect grains and seeds followed by *A. flavus* (Sauer et al., 1992).
- ***A. candidus*** - germ is killed rapidly and turns into jet black favored at 15.5-16% moisture content. It may cause heating up to 55°C and considered together with *A. flavus* as the main fungi capable of increasing grain temperature up to 50–55°C (Sauer et al., 1992). Minimum $a_w$ for growth reported as 0.75 and is tolerant at low $O_2$ of about 0.45% (Magan and Lacey, 1984). Morphologically, it is the only *Aspergillus* species producing consistently white conidia (Pitt and Hocking, 2009b).
- ***A. flavus*** - forms bright yellow-green colonies on invaded grain, morphologically different from *A. parasiticus* by variably shaped and sized conidia. It rapidly grows at 25–37°C; minimum $a_w$ for growth at 25°C is 0.84 (Pitt and Hocking, 1977) and 0.78$a_w$ at 33°C (Ayerst, 1969). Considered the most important storage fungi in terms of aflatoxin contamination that has been a particular problem in maize and by-products (Garcia and Heredia, 2006; Amaike and Keller, 2011). Infection and consequent aflatoxin production may start in the field and can be detected in freshly harvested maize (Lillehoj et al., 1976). Preharvest invasion of *A. flavus* in maize grown in cooler areas in the midwestern maize belt in the United States occurs less often than in warmer regions (Lillehoj et al., 1980).
- ***A. versicolor*** - a mesophile with optimum growth at 27°C and 0.98$a_w$ (Smith and Hill, 1982) and growth is weak or absent at 37°C. It produces green conidia on kernels and grows rather slowly compared to other species (Pitt and Hocking, 2009b).
- **Other reported species** - *A. carbonarius*, *A. clavatus*, *A. niger*, *A. ochraceous* (Magnoli et al., 2006; Pitt and Hocking, 1997).

*A. flavus* produces aflatoxins which are carcinogenic to humans and animal (IARC, 2015). The U.S Food and Drug Administration (FDA) has set the maximum limit for 20 μg/kg in food intended for human consumption.

***Penicillium***. Species belonging to this genus are opportunistic saprophytes and able to grow in almost any environment. Most of the accounts on *Penicillium* incidence in food products are not identified down to species level, likely because the species are commonly similar in color and colony appearance (blue to gray). Masses of *Penicillium* spores cause the so-called blue-eye damage in maize, which resulted from the heavy sporulation on the surface of the germ just below the pericarp. However, if maize is stored for a long time at 14.0–14.5%, the blue-eye damage is caused by the heavy sporulation of *A. restrictus* and *A. glaucus*. An increase in grain temperature indicates spoilage is in progress, and the problem is that blue-eye develops without causing the temperature to rise (Sauer et al., 1992). Some of the frequently identified *Penicillium* species in maize are listed below:

- ***P. citrinum*** - colonies appear with clear to yellow or brown exudates in the center. It is a mesophile capable of growth between 5°C and 37°C (Pitt, 1973) optimum at 26–30°C (Domsch et al., 1980). Minimum $a_w$ at 25°C is 0.80–0.84 (Hocking and Pitt, 1979). Produces citrinin that is moderately toxic but has been reported to cause kidney degeneration in poultry. *P. citrinum* was the most frequently isolated species in Thai, Indonesian, and Philippine maize (Pitt et al., 1993, 1998; Wicklow et al., 1998) and was reported that it even persisted in milled maize in Argentina (Broggi et al., 2002).
- ***P. aurantiogriseum*** - can still grow at around −2°C, at a maximum of 30°C, and at least 0.81$a_w$ (Mislivec and Tuite, 1970a,b). At carbon dioxide levels of 10% above growth is stimulated (Magan and Lacey, 1984). It is a common species in drying cereals and cereal products (Pitt and Hocking, 1997) and has also been reported in maize flour in Kenya (Muriuki and Siboe, 1995).
- ***P. oxalicum*** - temperature range for growth is wide (8°C–40°C) with optimum temperature near 30°C so incidence in tropical areas is widespread. Overlaps with *A. flavus* niche except at very low $a_w$. Found to be present in both pre- and postharvest maize; chiefly in preharvest maize (Pitt and Hocking, 1997) and lower incidence in stored maize (Askun, 2006).
- ***P. viridicatum*** - considered a psychrophile that grows optimally around 23°C with minimum $a_w$ of 0.80–0.81 at 23–25°C. Commonly confused with *P. verrucosum* but distinguished from it by anthraquinone toxins that are only produced by *P. viridicatum*. May also be present in the field but mostly associated with stored maize (Mislivec and Tuite, 1970a,b).
- ***P. verrucosum*** - optimum temperature is 20°C with a minimum of 0.80$a_w$ for growth and germination (Pardo et al., 2006). The majority of isolates

are ochratoxin A producers, some are citrinin producers (Frisvad and Samson, 2004). Ochratoxin A is a chronic nephrotoxin damaging several kidney tissues (Bayman and Baker, 2006). More common in maize grown in cool temperate zones (Pitt and Hocking, 2009b).

- **Other species**: *P. glabrum, P. commune, P. digitatum, P. variabile* (Pitt and Hocking, 2009b).

***Eurotium***. All *Eurotium* species found associated in maize are xerophiles, which means that they can thrive at low moisture conditions (≤ 0.80 $a_w$) and may begin colonizing grain at 15–17% moisture content. This genus is one of the teleomorph (sexual) genera of *Aspergillus* (Pitt and Samson, 2007). Barron and Lichtwardt (1959) identified the following *Eurotium* species in dried and stored maize in Iowa: *E. amstelodami*, *E. chevalieri,* and *E. rubrum*. These are the same species recovered in maize samples in Thailand, Indonesia, and Philippines (Pitt et al., 1993; Pitt et al., 1998; Wicklow et al., 1998) with *E. amstelodami* as the least common in Southeast Asian commodities (Pitt and Hocking, 2009b).

***Fusarium***. Although more common as field fungi, some species were found to persist in stored maize (Pitt et al., 1993). These species include *F. verticillioides*, *F. semitectum*, and *F. proliferatum*. *F. verticillioides* was found in maize stored for up to 2 years at a temperature range of 10–40°C and relative humidity of 40–88% (Wicklow et al., 1998). Molds and their toxins that persisted from field and carried through storage may continuously persist even in maize products such as grits, chips, flour (Pitt and Hocking, 1997). *Fusarium* spp. fungal biomass in kernels was found to be highly correlated with fumonisin accumulation in maize kernels (Waalwijk et al., 2008).

## 3.3 Factors affecting mold growth in stored maize

In the previous section, the influence of abiotic factors such as temperature, $a_w$, and gas composition on mold colonization during storage has been discussed. Every mold species attacking maize in storage has specific requirements to initiate infection, proliferate, and subsequently cause spoilage if conditions are continuously in favor of their growth. These conditions are largely influenced by the storage structure - a good one should be able to keep the grains dry; temperature is maintained uniformly; and insects, birds, and rodents are completely excluded (Sanchis and Magan, 2004). In stored maize, the influence of insect activity on mold development and mycotoxin contamination has been reported but not as well-studied as in field conditions. Nevertheless, there's an established association between insect activity and mold growth. Being vectors of fungal propagules is a major role of insects, and this happens simultaneously with the establishment of entry points for molds through the feeding sites (Fennell et al., 1975; Fennell et al., 1978). This largely explains the long-standing positive correlation of mycotoxin contamination with insect populations (Lillehoj et al., 1980; Sobek and Munkvold, 1999; Bowers et al., 2013).

The movement of maize weevil (*Sitophilus zeamais*) in grain bins has been observed to match the pattern *A. flavus* sporulation (Dix, 1984), indicating that the insect is capable of passively carrying spores on its body from one point to another. Maize weevil was also noted to enhance mold proliferation likely during its oviposition and other metabolic activities (Sinha, 1984; Dunkel, 1988). Insect metabolic activities, in general, result in increased relative humidity, temperature, and moisture content of grains (Sinha, 1971; Hambleton, 1979; Lacey et al., 1980), which promotes germination and development of postharvest fungi in the intergranular spaces of the grain (Lacey et al., 1980). Subsequently, not only the insects but also the molds contribute to heating which may continue even after insects are killed (Sauer et al., 1992). Moreover, the presence of volatiles, body fragments of insects, and fungal metabolites simultaneously contributes to these changes (Sone, 2001). The areas that have increased temperature due to insect-mold activities are called hot spots.

In certain cases, mold growth is inhibited by the presence of an insect. Yeh (1979) showed that in the presence of confused flour beetle (*Tribolium confusum*), no *A. flavus* incidence in the bins containing maize was observed, but *A. flavus* incidence was 3.3% if the bin has foreign grain beetle; probably due to the quinones secreted by the confused flour beetle (Van Wyk et al., 1959). Some stored-grain insects are also capable of feeding on spores, such as the larger black flour beetle (*Cynaeus angustus*) on *Cladosporium*, *Fusarium moniliforme*, and *Scopulariopsis brevicaulis* (Sinha, 1971). Some can either feed solely on molds or on grain to complete their life cycle, for example, red flour beetle (*Tribolium castaneum)*, while some only feed on the grain and not the molds, for example, maize weevil (Ferreira-Castro et al., 2012).

## 4 Fungal contamination in rice

Rice is an important food staple around the world, mainly in Asia and Africa, and provides more than 27% of dietary energy supply and 20% of dietary protein intake (FAO, 2004). In 2017, worldwide rice production was estimated at 759 million tons (503.9 million tons, milled basis), and China is the leading rice-producing country. Other major rice-producing countries are India, Indonesia, Bangladesh, Myanmar, Thailand, and Vietnam (FAO, 2018). Originating from Asia, rice was domesticated and has played an essential role in human nutrition for more than 10000 years. Rice is a semi-aquatic annual grass plant and grows in subtropical climates characterized by warm and humid conditions. Rice production demands specific requirements for growth: during daytime a maximum of 28°C and at nighttime a minimum of 22°C (Lai et al., 2015).

The rice taxonomy is very diverse. Rice belongs to the genus *Oryza* and contains seven species namely *O. sativa* (rice), *O. barthii* (Barth's rice), *O. glaberrima* (African rice), *O. latifolia* (broadleaf rice), *O. longistaminata* (longstamen rice), *O. punctata* (red rice), and *O. rufipogon* (brownbeard rice) (USDA, 2006). Rice commonly refers to *O. sativa* and is the predominant rice species cultivated around the world. Based on the grain shape and texture, rice is divided into two subspecies, namely *indica* and *japonica*. The indica varieties are characterized by long grain, less amylopectin and less sticky, and are grown in submerged regions while japonica is short grain and sticky, and requires less water (Wei and Huang, 2019; Atungulu et al., 2019).

Rice is rich in nutrients, including vitamins and minerals. However, carbohydrates are the major component of the rice grain (70–80%) in the form of starch. Amylose and amylopectin are the significant components of starch; and amylopectin constitutes 65–85% of the starch granules. Starch composition plays a crucial role in different rice properties; it determines the cooking properties and influences the interaction with other components in the rice endosperm (proteins, lipids, water) (Bao, 2019). The protein content in rice is low and is estimated to be 7% (Kawakatsu and Takaiwa, 2019). Finally, rice is not considered a source of lipids, and the endogenous lipids are minor. The rice bran contains 15–25% of lipids, and is very rich in unsaturated (oleic and linoleic) acids (Tong and Bao, 2019).

## 4.1 Storage and harvest of rice

In the field, rice grains are enveloped by husk and are called paddy rice or rough rice with moisture content at 19-25%. The rice processing consists of three distinct operations: harvest, drying, and milling.

- **Harvest:** Operation consists of cutting, collecting threshing the rice crop, separating, and cleaning the rice grains. The process is straightforward and can be completed manually. However, the harvest losses are much higher with traditional methods (up to 15%), compared to postharvest losses with advanced equipment (2%).
- **Drying:** After the harvest, rice grains are dried to reduce the moisture content to 12–13%. The reduction of moisture content aims to prolong rice storability and safety, which ultimately prevents deterioration of rice quality due to mold and insect infestations. Different methods are used to dry rice, including sun drying, ambient air drying, and heated air drying. At this step, rice grains are enveloped into the husk and are called paddy rice.

- **Milling:** White rice is a final product for consumption obtained at the end of the milling process. First, the husk is removed from paddy rice, which results in brown rice, the husk as a by-product. The whitening process removes the bran layer from the brown rice and results in white rice and rice bran. Any remaining loose bran is removed by the polishing process to obtain refined rice. Finally, broken kernels are separated from whole rice kernels (Khir and Pan, 2019).

## 4.2 Fungal infection in rice kernels during storage

After harvest, rice grains are dried and have to be stored in appropriate conditions. Any inappropriate storage conditions can lead to fungal colonization of rice grains and therefore can compromise rice quality (Lai et al., 2015). On the other hand, proper storage (low moisture content and low temperature) prevents or reduces rice respiration significantly; otherwise rice will lose in freshness and dry matter. Like any living organism, rough rice retains the respiration process with the oxygen absorption and conversion of carbohydrates into $CO_2$, water, and the reaction generates heat. The generated energy and water heat the grain mass and promote fungal growth (Atungulu et al., 2019).

The fungal contamination in stored rice grains takes advantage of any kernel cracks and fractures to invade the internal tissues of kernels. The fungal colonization leads to the enzyme production, which digests starch grains and creates cracks and discoloration easily identifiable with a naked eye (Reddy et al., 2008). For example, 'Shoshin-kakke' or yellow rice in Japan is caused by *Penicillium* spp. contamination in rice (Kushiro, 2015).

Rice grains are susceptible to fungal contamination if conditions are favorable during cultivation and storage. Many fungi are potential mycotoxin producers and can subsequently contaminate rice with toxic effects on consumers. Fungal contamination of rice may vary from one study to another. For instance, *Aspergillus* spp. and *Penicillium* spp. were found to be predominant molds in Turkish rice (Aydin et al., 2011), *Aspergillus* spp. dominant in India (Reddy et al., 2009) and *Fusarium* genus in South Korea. Other fungal genera found are *Penicillium*, *Phoma*, *Myrothecium*, and *Cladosporium* (Ok et al., 2014; J. Lee et al., 2009; T. Lee et al., 2011).

## 4.3 Mycotoxin producing fungi in stored rice

Different fungi are known to infect rice in the field or during storage, and three genera are considered the most significant not only for the mycotoxin production but also for their ubiquitous nature. *Fusarium* species infect rice while in the field while *Aspergillus* and *Penicillium* species contamination occurs during storage (Park et al., 2005). Since the chapter focuses on fungal contamination during storage, the following paragraphs will discuss *Aspergillus* and *Penicillium*.

- ***Aspergillus*** – *Aspergillus* spp. mainly *Aspergillus flavus* produce aflatoxins. The aflatoxin contamination is recurrent in tropical countries because of high relative humidity coupled with high temperatures. The contamination starts in the field and worsens with poor postharvest conditions. During postharvest, aflatoxin contamination is primarily influenced by heat, water activity, and air composition. If not well dried (more than 12% moisture content), rice can be still colonized by *A. flavus*, which grows at 10–43°C and water activity 0.82–0.99, with an optimum at 30°C. *A. flavus* grows better on brown rice compared to white rice. Brown rice has the embryo and the bran layers, rich in oils, favoring *A. flavus*. Paddy seeds are found to be highly contaminated if compared with milled kernels suggesting that the milling process removes the majority of fungi (Gonçalves et al., 2019). There are four different types of aflatoxins: aflatoxin B1, aflatoxin B2, aflatoxin G1, and aflatoxin G2, and they are very toxic to humans and animals at low concentrations (μg/kg).The U.S Food and Drug Administration (FDA) has set the maximum limit as 20 μg/kg in food intended for human consumption. The International Agency for Research on Cancer (IARC) classified aflatoxins as carcinogenic to humans, and they are associated with an increase of developing liver cancer (IARC, 2015). Ochratoxin A and sterigmatocystin are additional toxins produced by *Aspergillus* species (*A. ochraceus* and *A. versicolor*) and are also reported to contaminate rice (Gonçalves et al., 2019).
- ***Penicillium* spp.** – *Penicillium* spp. also produce mycotoxins in stored rice. *Penicillium* colonization in rice induces color change because of their enzymatic actions on the rice. 'Yellow rice' or 'Shoshin-kakke' is well known and there are documented cases of several victims in Japan and other Asian countries after consumption of 'yellow' polished rice. Investigations showed that *P. citreonigrum* produces a mycotoxin, citrinin – a neurotoxic, which was the primary causal agent of 'yellow rice' disease (Uraguchi, 1969; Kushiro, 2015; Scott, 2004). Other *Penicillium* fungi produce mycotoxins and induce rice color change, as well. *P. rugulosum* produces rugulosin and change rice grains to yellow. Penicillic acid is produced by *P. aurantiogriseum* and turns the contaminated grain rice to yellow. *P. islandicum* and *P. commune* generate brown color in contaminated rice and produce cyclochlorotin and cyclopiazonic acid, respectively (Kushiro, 2015).

## 5 Fungal contamination in wheat

After maize, wheat is the second most-produced cereal worldwide. In 2017, worldwide wheat production achieved 771 million tons (Fig. 1). In 2016, wheat was grown on an area estimated at 216 million hectares in diverse environments (Schmidt et al., 2016), but more favorable at the latitude range of 30°-60°N and 27°-40°S (Tadesse et al., 2016). Wheat constitutes a significant source of calories and proteins for humans and livestock. In some countries (for instance Egypt, Algeria, Israel), the consumption exceeds 150 kg per capita per year. The export countries include Argentina, Australia, Canada, Russia, Ukraine and the United States (Cendoya et al., 2018). Because of its unique viscoelastic properties, wheat is generally used in the baking industry to make bread, cakes, biscuits, pasta, and so on (Wrigley, 2015).

From the taxonomic standpoint, cultivated wheat belongs to the genus *Triticum*. The einkorn groups (*Triticum monococcum)* are diploid ($2n = 2x = 14$), emmers groups (*T. turgidum)* are teraploids ($2n = 4x = 28$), and dinkels (T. aestivum) are hexaploids ($2n = 6x = 42$) (Tadesse et al., 2016). Hexaploid wheat is the most common and accounts for 95% of all wheat produced. Based on the color, kernel hardness, and planting season, hexaploid wheat forms five classes: hard red winter, hard red spring, soft red winter, hard white, and soft white. Tetraploid durum wheat is used for the semolina production used in pasta and macaroni products or couscous in Maghreb countries (Finnie and William, 2017).

The wheat kernel consists of three different parts. An outer layer, the bran, which constitutes 13-17% of kernel weight, protects wheat kernels. The germ, an embryonic plant, forms 2-3% of the kernel and is rich in lipids and essential nutrients. Lastly, the starchy part, the endosperm, is the major component of the kernel (80-85%) and contains high amounts of starch and protein. Between endosperm and bran, there is a specialized layer, aleurone, which contains enzymes and is removed during the milling process (Finnie and William, 2017).

### *5.1 Fungal infection in wheat kernels during storage*

During storage, wheat kernels respire and allow $O_2$ and $CO_2$ exchanges. Respiration process keeps the germ alive and inhibits deterioration of kernel quality, such as oxidative rancidity. However, it is essential to maintain minimal respiration because of heat produced and weight losses during the process, which result in deterioration of kernel quality and favor the infection of different pests (insect, fungi, and bacteria). Temperature, moisture content, and oxygen quantity are parameters that influence respiration during storage. During storage, the moisture content is maintained below 14% which inhibits respiration and preserves kernels (Finnie and William, 2017). If not preserved, fungi can grow on wheat kernels and cause damage to wheat kernels. Three different fungal genera, *Aspergillus* spp., *Fusarium* spp., and *Penicillium* spp., commonly cause infection during wheat storage (Birck et al., 2006) and the following section will discuss their impact.

***Fusarium* spp.** The genus *Fusarium* includes the most damaging fungi in wheat. The infection starts in the field and continues during storage. The primary disease caused by *Fusarium* is commonly known as '*Fusarium* Head Blight' (FHB) or scab. There are more than 16 species involved in FHB. However, *F. graminearum* (Dweba et al., 2017) and *F. culmorum* (Schmidt et al., 2016) are most prominent causes of FHB. Other *Fusarium* spp., like *F. asiaticum*, and *F. boothii* are involved in FHB (Backhouse, 2014). The disease causes not only yield reduction because *Fusarium* attack starch granules, protein, and cell walls, but also these *Fusarium* species produce mycotoxins very toxic to humans and livestock (Snijders, 1990).

Abiotic factors influence *Fusarium* spp. growth. For instance, it was observed that *F. proliferatum* has a maximum growth at the highest $a_w$ (0.995) and 30°C. The growth decreases with $a_w$ decrease (Cendoya et al., 2014). *Fusarium* spp. produce different types of mycotoxins, but the most notorious mycotoxins produced are type B trichothecenes, deoxynivalenol (DON), nivalenol (NIV) and their derivatives particularly 3-acetyl and 15-acetyl deoxynivalenol (3ADON and 15ADON) and 4-acetyl nivalenol (4ANIV), and zearalenone (ZON) (Dweba et al., 2017).

***Aspergillus* spp.** As discussed previously, *Aspergillus* spp. are storage molds, and contaminate cereal kernels if growth conditions allow and produce mycotoxins especially aflatoxins, harmful to humans and animals. The U.S Food and Drug Administration (FDA) has set the maximum limit as 20 μg/kg in food intended for human consumption. *Aspergillus* spp. can still grow in low moisture content (≤14%) and high temperature (≥30°C) (Pitt and Hocking, 2009b). Ochratoxin A (OTA), a nephrotoxicmycotoxin, is produced particularly by *A. ochraceus* and *P. verrucosum* and is known to contaminate wheat kernels. *A. ochraceus* predominates in warm and tropic climates (Aish et al., 2004).

***Penicillium* spp.** *Penicillium* spp. are storage fungi and regulary contaminate grains. *P. verrucosum* is the most reported *Penicillium* spp. which contaminates wheat grains and produces mycotoxins, ochratoxin A and citrin (Jacobsen, 2014). The optimum growth of *P. verrucosum* on wheat grain is 0.95 $a_w$ and 25°C with a minimum 0.8 $a_w$. *P. verrucosum* mainly occurs in temperate climate (Cairns-Fuller et al., 2005).

## 6 Conclusion

Cereals, like any other crop, are attacked by a wide array of pests and diseases causing losses around one-third of the production. In this chapter, we focused on the three major cereals produced across the world, that is, maize, rice, and wheat, and the predominant fungal species contaminating their grain. Losses due to fungal contamination can be attributed to two main events: first, fungi cause grain spoilage that renders the grain unfit for human and animal consumption. Second, several species of fungi may contaminate grain with toxic secondary metabolites called mycotoxins, which are harmful to humans and animals especially when consumed. The main fungal genera involved in spoiling grains and producing mycotoxins are *Aspergillus* spp. and *Penicillium* spp. Abiotic and biotic factors greatly influence the degree of spoilage and mycotoxin production due to their colonization during storage.

In maize, long-term storage would require cooling-off and drying the grain as soon as possible to prevent the initiation of fungal colonization. There are more than nine species of *Aspergillus* and *Penicillium* attacking maize, and some *Fusarium* species that were able to persist even at drier conditions. *A. flavus* is considered as the most important storage fungi colonizing maize due to its aflatoxin production favored by high moisture and high-temperature conditions. Insect activity has also been related to fungal colonization, in which higher contamination is observed in insect-infested maize compared to maize without insect damage. The pericarp of maize provides resistance to fungal invasion. However, if damaged, can easily serve as a point of entry for fungal attack. Moreover, higher colonization can be observed if the damage is near the germ since it is where most nutrients are available for absorption.

Rice plays an important role in the human diet, mainly in Asia and Africa. After harvest, the paddy rice is dried to reduce the moisture content to 12–13% and help reduce the fungal colonization. Paddy rice is further milled to obtain white as the final product. The fungal colonization in rice takes advantage of broken kernels, and leads to color discoloration easily identifiable. *Fusarium* spp. (which infect rice in the field), *Penicillum* spp. and *Aspergillus* spp. (occur during storage) are predominant fungi in rice contamination. If not well preserved, the fungal colonization in rice will consequently lead to mycotoxin production, harmful to consumers.

Wheat is the second most-produced cereal after maize. Depending on the region, wheat is used for the semolina production, pasta and macaroni or couscous. During the storage, the moisture content is maintained below 14% to preserve the wheat kernel from fungal colonization. Three different fungal genera, *Aspergillus* spp., *Fusarium* spp., and *Penicillium* spp., commonly cause infection during wheat storage and produce mycotoxins. *Fusarium* spp. are involved in

FHB disease, which cause not only yield reduction but also mycotoxin production that is very toxic to humans and animals. *Aspergillus* spp. produce aflatoxins (*A. flavus*) and ochratoxin A (*A. ochraceus* and *P. verrucosum*). *Penicillum* spp. frequently contaminates wheat kernels and produce ochratoxin A and citrin.

In general, fungal propagules often are already present in the grain but stay dormant until conditions become suitable for their development. Proper drying of grains before storage (<13.5% moisture content which inhibits fungal growth) is a forefront management practice to control fungal growth, thus minimizing mycotoxin contamination in grains as well. Maintaining anaerobic conditions (e.g. using hermetic storage) during storage also disadvantage fungal growth. Understanding the growth and development of fungi with respect to the interrelations between abiotic and biotic factors during storage are the core elements that should be considered to successfully manage storage fungi or molds.

## 7 References


Abera, B. B., Terefe, B., Baye, K. and Covic, N. 2018. Rice contribution to food and nutrition security and leveraging opportunities for sustainability, nutrition and health outcomes. In: Ferranti, P., Berry, Elliot, M. and Anderson, J. R. (Eds) *Encyclopedia of Food Security and Sustainability*. Elsevier, Amsterdam, Netherlands. 257–63. doi:10.1016/B978-0-08-100596-5.21540-0.

Aish, J. L., Rippon, E. H., Barlow, T. and Hattersley, S. J. 2004. Ochratoxin A. In: Magan, N. and Olsen, M. (Eds) *Mycotoxins in Food Detection and Control*. Woodhead Publishing, Cambridge, pp. 307–38. doi:10.1533/9781855739086.3.307.

Amaike, S. and Keller, N. P. 2011. *Aspergillus flavus*. *Annual Review of Phytopathology* 49(1), 107–33. doi:10.1146/annurev-phyto-072910-095221.

Askun, T. 2006. Investigation of fungal species diversity of maize kernels. *Journal of Biological Sciences* 6(2), 275–81. doi:10.3923/jbs.2006.275.281.

Atanda, S. A., Pessu, P. O., Agoda, S., Isong, I. U., Adekalu, O. A., Echendu, M. A. and Falade, T. C. 2011. Fungi and mycotoxins in stored foods. *African Journal of Microbiology Research* 5(25), 4373–82. doi:10.5897/AJMR11.487.

Atungulu, G. G., Kolb, R. E., Karcher, J. and Mohammadi Shad, Z. 2019. Postharvest technology: rice storage and cooling conservation. In: Bao, J. (Ed.), *Rice Chemistry and Technology*. Elsevier Inc. in cooperation with AACC International, pp. 517–55. doi:10.1016/B978-0-12-811508-4.00016-2.

Aydin, A., Aksu, H. and Gunsen, U. 2011. Mycotoxin levels and incidence of mould in Turkish rice. *Environmental Monitoring and Assessment* 178(1–4), 271–80. doi:10.1007/s10661-010-1688-9.

Ayerst, G. 1969. The effects of moisture and temperature on growth and spore germination in some fungi. *Journal of Stored Products Research* 5(2), 127–41. doi:10.1016/0022-474X(69)90055-1.

Backhouse, D. 2014. Global distribution of *Fusarium graminearum*, *F. asiaticum* and *F. boothii* from wheat in relation to climate. *European Journal of Plant Pathology* 139(1), 161–73. doi:10.1007/s10658-013-0374-5.

Bao, J. 2019. Rice starch. In: Bao, J. (Ed.), *Rice*. Elsevier Inc. in cooperation with AACC International, pp. 55–108. doi:10.1016/b978-0-12-811508-4.00003-4.

Barron, G. L. and Lichtwardt, R. W. 1959. Quantitative estimationsof the fungi associated with deterioration of stored corn in Iowa. *Iowa State College Journal of Science* 34, 147–55.

Batey, I. 2017. The diversity of uses for cereal grains. In: Wrigley, C. W. and Batey, I. L. (Eds) *Cereal Grains: Assessing and Managing Quality* (2nd edn.). Woodhead Publishing Limited, Sawston, Cambridge. pp. 45–56. doi:10.1016/B978-0-08-100719-8.00003-6.

Bayman, P. and Baker, J. L. 2006. Ochratoxins: a global perspective. *Mycopathologia* 162(3), 215–23. doi:10.1007/s11046-006-0055-4.

Beck, D. L. 2002. *Management of Hybrid Maize Seed Production*. CIMMYT, Mexico, DF.

Birck, N. M., Lorini, I. and Scussel, V. M. 2006. Fungus and mycotoxins in wheat grain at post harvest. *9th International Working Conference on Stored Product Protection*, pp. 198–205.

Bothast, R. J. 1978. Fungal deterioration and related phenomena in cereals, legumes and oilseeds. In: Hultin, H. O. and Milner, M. (Eds) *Post-Harvest Biology and Biotechnology*. Food Nutrition Press Inc., Westport, CT, 462p.

Bowers, E., Hellmich, R. and Munkvold, G. 2013. Vip3Aa and Cry1Ab proteins in maize reduce Fusarium ear rot and fumonisins by deterring kernel injury from multiple Lepidopteran pests. *World Mycotoxin Journal* 6(2), 127–35. doi:10.3920/WMJ2012.1510.

Broggi, L. E., Gonzalez, H. H. L., Resnik, S. L. and Pacin, A. M. 2002. Mycoflora distribution in dry-milled fractions of corn in Argentina. *Cereal Chemistry Journal* 79(5), 741–4. doi:10.1094/CCHEM.2002.79.5.741.

Burrell, N. J. 1974. Chilling. In: Christensen, C. M. (Ed.), *Storage of Cereal Grain and Their Products*. American Association of Cereal Chemists, Inc., St. Paul, MN, pp. 420–53.

Cahagnier, B., Lesage, L. and Richard-Molard, D. 1993. Mould growth and conidiation in cereal grains as affected by water activity and temperature. *Letters in Applied Microbiology* 17(1), 7–13. doi:10.1111/j.1472-765X.1993.tb01423.x.

Cahagnier, B. 1998. *Moisissures des Aliments peu Hydrates*. Lavoisier Tec & Doc, Paris, France, p. 228.

Cahagnier, B., Jacobsen, E. E. and Fleurat-Lessard, F. 2005. Des moisissures aux mycotoxines: signification ecophysiologique de marqueurs biochimiques de croissance fongique et prevision des durees de conservation sans risque deterioration. In: Fleurat-Lessard, F., Ndiaye, A. and Knight, J. D. (Eds) *Stored Malting Barley: Management of Quality Using an Expert System*. INRA Editions, Paris, pp. 121–36. *Colloque INRA* n_ 101, Reims, 28–29 May 2001.

Cairns-Fuller, V., Aldred, D. and Magan, N. 2005. Water , temperature and gas composition interactions affect growth and ochratoxin A production by isolates of *Penicillium verrucosum* on wheat grain. *Journal of Applied Microbiology* 99(5), 1215–21. doi:10.1111/j.1365-2672.2005.02695.x.

Cendoya, E., Farnochi, M. C., Chulze, S. N. and Ramirez, M. L. 2014. Two-dimensional environmental profiles of growth and fumonisin production by *Fusarium proliferatum* on a wheat-based substrate. *International Journal of Food Microbiology* 182–3, 9–17. doi:10.1016/j.ijfoodmicro.2014.04.028.

Cendoya, E., Chiotta, M. L., Zachetti, V., Chulze, S. N. and Ramirez, M. L. 2018. Fumonisins and fumonisin-producing fusarium occurrence in wheat and wheat by products: a review. *Journal of Cereal Science* 80, 158–66. doi:10.1016/j.jcs.2018.02.010.

Chakraverty, A., Mujumdar, A. S. and Ramaswamy, H. S. 2003. *Handbook of Postharvest Technology: Cereals, Fruits, Vegetables, Tea, and Spices* (1st edn., vol. 93). CRC Press, New York : Marcel Dekker. 912p.

Christensen, C. M. and Kaufmann, H. H. 1969. *Grain Storage: the Role of Fungi in Quality Loss*. University of Minnesota Press, Minneapolis, MN.

Christensen, C. M. and Kaufmann, H. H. 1974. Microflora. In: Christensen, C. M. (Ed.), *Storage of Cereal Grain and Their Products*. American Association of Cereal Chemists, Inc., St. Paul, MN. 219–40 .

Christensen, C. M. 1957. Deterioration of stored grain by fungi. *The Botanical Review* 23(2), 108–34. doi:10.1007/BF02930521.

Cooney, D. G. and Emerson, R. 1964. *Thermophilic Fungi: an Account of Their Biology, Activities, and Classification*. W. H. Freeman and Company, San Francisco.

Crous, P. W., Hawksworth, D. L. and Wingfield, M. J. 2015. Identifying and naming plant-pathogenic fungi: past, present, and future. *Annual Review of Phytopathology* 53, 247-67. doi:10.1146/annurev-phyto-080614-120245.

De Groote, H., Kimenju, S. C., Likhayo, P., Kanampiu, F., Tefera, T. and Hellin, J. 2013. Effectiveness of hermetic systems in controlling maize storage pests in Kenya. *Journal of Stored Products Research* 53, 27–36. doi:10.1016/j.jspr.2013.01.001.

Detroy, R. W., Lillehoj, E. B. and Ciegler, A. 1971. Aflatoxin and related compounds. In: Ciegler, A., Kadis, S. and Ajl, S. J. (Eds) *Microbial Toxins* (vol. 6). Academic Press, New York, NY.

Dix, D. E. 1984. *Interactive Bionomics of the Maize Weevil, Sitophilus zeamais Motschulsky, and Aspergillus flavus Link*. University of Georgia, Athens, GA, 129p.

Domsch, K. H., Gams, W. and Anderson, T. H. 1980. *Compendium of Soil Fungi* (2 vols.). Academic Press, London, UK.

Dunkel, F. V. 1988. The relationship of insects to the deterioration of stored grains by fungi. *International Journal of Food Microbiology* 7(3), 227–44. doi:10.1016/0168-1605(88)90042-6.

Dweba, C. C., Figlan, S., Shimelis, H. A., Motaung, T. E., Sydenham, S., Mwadzingeni, L. and Tsilo, T. J. 2017. *Fusarium* head blight of wheat: pathogenesis and control strategies. *Crop Protection* 91, 114–22. doi:10.1016/j.cropro.2016.10.002.

Fandohan, P., Hell, K., Marasas, W. F. O. and Wingfield, M. J. 2003. Infection of maize by *Fusarium* species and contamination with fumonisin in Africa. *African Journal of Biotechnology* 2(12), 570–9. doi:10.5897/AJB2003.000-1110.

FAO. 2004. Rice is life: international year of rice 2004. Available at: http://www.fao.org/rice2004/en/rice2.htm (accessed on September 29, 2019).

FAO. 2018. FAO Rice Market Monitor (RMM). FAO. Rice Market Monitoring XXI(1), 38.

FAO. 2020. Food loss and food waste. Available at: http://www.fao.org/food-loss-and-food-waste/en/ (accessed on 16 March 2020).

Fennell, D. I., Lillehoj, E. B. and Kwolek, W. F. 1975. *Aspergillus flavus* and other fungi associated with insect-damaged field corn. *Cereal Chemistry* 52, 314–21.

Fennell, D. I., Lillehoj, E. B., Kwolek, W. F., Guthrie, W. D., Sheeley, R., Sparks, A. N., Widstrom, N. W. and Adams, G. L. 1978. Insect larval activity on developing corn ears and subsequent aflatoxin contamination of seed. *Journal of Economic Entomology* 71(4), 624–8. doi:10.1093/jee/71.4.624.

Ferreira-Castro, F. L., Potenza, M. R., Rocha, L. O. and Correa, B. 2012. Interaction between toxigenic fungi and weevils in corn grain samples. *Food Control* 26(2), 594–600. doi:10.1016/j.foodcont.2012.02.016.

Finnie, S. and Atwell, W. A. 2017. Wheat. In: Finnie, S. and Atwell, W. A. *Wheat Flour*, Elsevier Inc. in cooperation with AACC International, St. Paul, Minnesota, U.S.A. 1–16. doi:10.1016/B978-1-891127-90-8.50001-2.

Flannigan, B. 1978. Primary contamination of barley and wheat grain by storage fungi. *Transactions of the British Mycological Society* 71(1), 37–42. doi:10.1016/S0007-1536(78)80005-9.

Fleurat-Lessard, F., Fourar-Belaifa, R. and Bouznad, Z. 2010. Multivariate analysis of the temporal changes of fungal communities in unsafe storage conditions of some common wheat varieties in relation to relative humidity level and rice weevil infestation. In: Carvalho, O. M., Fields, P. G., Adler, C. S., Arthur, F. H., Athanassiou, C. G., Campbell, J. F., Fleurat-Lessard, F., Flinn, P. W., Hodges, R. J., Isikber, A. A., et al. (Eds) *Proceedings of the 10th International Working Conference on Stored Product Protection, Estoril, Portugal*. Julius Kühn-Institut, Berlin and Germany. 27 June–2 July 2010, pp. 520–8.

Fleurat-Lessard, F. 2017. Integrated management of the risks of stored grain spoilage by seedborne fungi and contamination by storage mould mycotoxins - an update. *Journal of Stored Products Research* 71, 22–40. doi:10.1016/j.jspr.2016.10.002.

Frisvad, J. C. and Samson, R. A. 2004. Polyphasic taxonomy of *Penicillium* subgenus *Penicillium*. A guide to identification of food and air-borne terverticillate *Penicillia* and their mycotoxins. *Studies in Mycology* 49, 1–173.

Garcia, S. and Heredia, N. 2006. Mycotoxins in Mexico: epidemiology, management, and control strategies. *Mycopathologia* 162(3), 255–64. doi:10.1007/s11046-006-0058-1.

Gherbawy, Y. and Voigt, K. (Eds) 2010. *Molecular Identification of Fungi*. Springer-Verlag, Berlin Heidelberg. doi:10.1007/978-3-642-05042-8_11.

Golob, P., Kutukwa, N., Devereau, A., Bartosik, R. E. and Rodriguez, J. C. 2004. Chapter Two: Maize. In: Hodges, R. and Farrell, G. (Eds) *Crop Post-Harvest: Science and Technology* (vol. 2). Blackwell Publishing Ltd, Ames, IA.

Gonçalves, A., Gkrillas, A., Dorne, J. L., Dall'Asta, C., Palumbo, R., Lima, N., Battilani, P., Venâncio, A. and Giorni, P. 2019. Pre- and postharvest strategies to minimize mycotoxin contamination in the rice food chain. *Comprehensive Reviews in Food Science and Food Safety* 18(2), 441–54. doi:10.1111/1541-4337.12420.

Grayson, H. 2017. A place for individuals: positive growth in Rwanda. *Eastern African Literary and Cultural Studies* 3(2–4), 107–30. doi:10.1080/23277408.2017.1374052.

Hambleton, L. D. 1979. Spontaneous heating of grain. In: *Grain Storage*. Technical Publication Trust, Perth, Australia, pp. 94–107.

Hocking, A. D. and Pitt, J. I. 1979. Water relations of some *Penicillium* species at 25°C. *Transactions of the British Mycological Society* 73(1), 141–5. doi:10.1016/S0007-1536(79)80084-4.

IARC. 2015. IARC (International Agency for Research on Cancer) Mycotoxin control in low- and middle income countries. IARC Press, Lyon, France. doi:10.2514/6.2008-1946.

Jacobsen, B. J. 2014. Good agricultural and harvest practices to reduce mycotoxin contamination in wheat in temperate countries. In: Leslie, J. F. and Logrieco, A. F. (Eds) *Mycotoxin Reduction in Grain Chains*. John Wiley & Sons, Ltd. Ames, Iowa 50010, USA. doi:10.1002/9781118832790.

Jedidi, I., Soldevilla, C., Lahouar,A., Marín, P., González-Jaén, M. T. and Saida, S. 2018. Mycoflora isolation and molecular characterization of *Aspergillus* and *Fusarium* species in Tunisian cereals. *Saudi Journal of Biological Sciences* 25(5), 868–74. doi:10.1016/j.sjbs.2017.11.050.

Jespersen, B. M. and Munck, L. 2009. Cereals and cereal products. *Infrared Spectroscopy for Food Quality Analysis and Control.*, 275–319. doi:10.1016/B978-0-12-374136-3.00011-0.

Kankolongo, M. A., Hell, K. and Nawa, I. N. 2009. Assessment for fungal, mycotoxin and insect spoilage in maize stored for human consumption in Zambia. *Journal of the Science of Food and Agriculture* 89(8), 1366–75. doi:10.1002/jsfa.3596.

Kawakatsu, T. and Takaiwa, F. 2019. Rice proteins and essential amino acids. In: Bao, J. (Ed.), *Rice*. Elsevier Inc. in cooperation with AACC International, Woodhead Publish, pp. 109–30. doi:10.1016/B978-0-12-811508-4.00004-6.

Khir, R. and Pan, Z. 2019. Chapter 2: Rice. In: Pan, Z., Zhang, R. and Zicari, S. (Eds) *Integrated Processing Technologies for Food and Agricultural By-Products*. Elsevier Inc., pp. 21–58. doi:10.1016/B978-0-12-814138-0.00002-2.

Kirk, P. M., Cannon, P. F., David, J. C. and Stalpers, J. A. 2001. *Dictionary of the Fungi* (9th edn.). CAB International, Wallingford, UK.

Kossou, K. D. and Aho, N. 1993. *Stockage et conservation des grains alimentaires tropicaux*. Les Editions du Flamboyant, Cotonou Bénin, 50pp.

Kung, S. 1993. Introduction: from green revolution to gene revolution. In: Shain-dow Kung, *Transgenic Plants*. Academic Press, Inc. Cambridge, Massachusetts. xvii–xxxi doi:10.1016/b978-0-12-428782-2.50006-6.

Kushiro, M. 2015. Historical review of researches on yellow rice and mycotoxigenic fungi adherent to rice in Japan. *Mycotoxins* 65(1), 19–23. doi:10.2520/myco.65.19.

Lacey, J., Hill, S. T. and Edwards, M. A. 1980. Microorganisms in stored grains: their enumeration and significance. *Tropical Stored Products Information* 39, 19–33.

Lacey, J. and Magan, N. 1991. Fungi in cereal grains: their occurrence, water and temperature relationships. In: Chelkowski, J. (Ed.), *Cereal Grain. Mycotoxins, Fungi and Quality in Drying and Storage*. Elsevier, Amsterdam, the Netherlands, pp. 77–117.

Lacey, J., Hamer, A. and Magan, N. 1994. Respiration and losses in stored wheat under different environmental conditions. In: Highley, E., Wright, E. J., Banks, H. J. and Champ, B. R. (Eds) *Stored Product Protection. Proceedings of the 6th International Working Conference on Stored Product Protection*, Canberra, Australia, 6–11 April 1994 (vol. II). CABI Publishing, Wallingford (UK), pp. 1007–13.

Lai, X., Liu, R., Ruan, C., Zhang, H. and Liu, C. 2015. Occurrence of aflatoxins and ochratoxin A in rice samples from six provinces in China. *Food Control* 50, 401–4. doi:10.1016/j.foodcont.2014.09.029.

Lau, K. H. and Sheridan, J. E. 1975. Mycoflora of rice (*Oryza sativa* L.) seed in New Zealand. *New Zealand Journal of Agricultural Research* 18(3), 237–50. doi:10.1080/00288233.1975.10423639.

Lee, J., Chang, I. Y., Kim, H., Yun, S. H., Leslie, J. F. and Lee, Y. W. 2009. Genetic diversity and fitness of *Fusarium graminearum* populations from rice in Korea. *Applied and Environmental Microbiology* 75(10), 3289–95. doi:10.1128/AEM.02287-08.

Lee, T., Lee, S. H., Lee, S. H., Shin, J. Y., Yun, J. C., Lee, Y. W. and Ryu, J. G. 2011. Occurrence of *Fusarium* mycotoxins in rice and its milling by-products in Korea. *Journal of Food Protection* 74(7), 1169–74. doi:10.4315/0362-028X.JFP-10-564.

Lillehoj, E. B., Fennell, D. I. and Kwolek, W. F. 1976. *Aspergillus flavus* and aflatoxin in Iowa corn before harvest. *Science* 193(4252), 495–6. doi:10.1126/science.821144.

Lillehoj, E. B. and Goransson, B. 1980. Occurrence of ochratoxin- and citrinin- producing fungi on developing Danish barley grain. *Acta Pathologica et Microbiologica*

*Scandinavica. Section B, Microbiology* 88(3), 133–7. doi:10.1111/j.1699-0463.1980.tb02618.x.

Lillehoj, E. B., Kwolek, W. F., Horner, E. S., Widstrom, N. W., Josephson, L. M., Franz, A. O. and Catalano, E. A. 1980. Aflatoxin contamination of preharvest corn: role of *Aspergillus flavus* inoculum and insect damage. *Cereal Chemistry* 57, 255–7.

Links, M. G., Demeke, T., Gr€afenhan, T., Hill, J. E., Hemmingsen, S. M. and Dumonceaux, T. J. 2012. Simultaneous profiling of seed-associated bacteria and fungi reveals antagonistic interactions between microorganisms within a shared epiphytic microbiome on Triticum and Brassica seeds. *New Phytologist* 202, 542–53.

Magan, N. and Lacey, J. 1984. Effects of gas composition and water activity on growth of field and storage fungi and their interactions. *Transactions of the British Mycological Society* 82(2), 305–14. doi:10.1016/S0007-1536(84)80074-1.

Magan, N., Hope, R., Cairns, V. and Aldred, D. 2003. Post-harvest fungal ecology: impact of fungal growth and mycotoxin accumulation in stored grain. *European Journal of Plant Pathology* 109(7), 723–30. doi:10.1023/A:1026082425177.

Magan, N. and Aldred, D. 2005. Conditions of formation of ochratoxin A in drying, transport and in different commodities. *Food Additives and Contaminants* 22 (Suppl. 1), 10–6. doi:10.1080/02652030500412154.

Magan, N., Aldred, D. and Baxter, E. S. 2014. Good postharvest storage practices for wheat grain. In: Leslie, J. F. and Logrieco, A. F. (Eds) *Mycotoxin Reduction in Grain Chains*. Wiley-Blackwell, Ames, pp. 258–67.

Magnoli, C., Hallak, C., Astoreca, A., Ponsone, L., Chiacchiera, S. and Dalcero, A. M. 2006. Occurrence of ochratoxin A producing fungi in commercial corn kernels in Argentina. *Mycopathologia* 161(1), 53–8. doi:10.1007/s11046-005-0237-5.

Markham, R. H., Bosque-Perez, N. A., Borgemeister, C. and Meikle, W. G. 1994. Developing pest management strategies for *Sitophilus zeamais* and *Prostephanus truncates* in the tropics. *FAO Plant Protection Bulletin* 42(3), 97–116. Available at: http://agris.fao.org/agris-search/search.do?recordID=XF9657657.

Mason, L. J. 2018. *Effect and Control of Insects, Molds and Rodents Affecting Corn Quality*. In: Sergio Serna-Saldivar (Ed.), *Corn: Chemistry and Technology* (3rd edn.). Elsevier Inc., Sawston, Cambridge. p. 690. doi:10.1016/B978-0-12-811971-6.00008-5.

McCarthy, U., Uysal, I., Badia-Melis, R., Mercier, S., O'Donnell, C. and Ktenioudaki, A. 2018. Global food security – issues, challenges and technological solutions. *Trends in Food Science and Technology* 77(May), 11–20. doi:10.1016/j.tifs.2018.05.002.

Miller, J. D. 1995. Fungi and mycotoxins in grains: implication for stored roducts research. *Journal of Stored Products Research* 31(1), 1–16. Available at: https://ac.els-cdn.com/0022474X9400039V/1-s2.0-0022474X9400039V-main.pdf?_tid=b51cfb5d-6589-4fbd-af12-d4e48122268b&acdnat=1548347680_9e581aef0cc8da1a253b927646de1f62. doi:10.1016/0022-474X(94)00039-V.

Mills, J. T., Seifert, K. A., Frisvad, J. C., Abramson, D. and Louis-Seize, G. 1992. Relative importance of *Penicillium* species producing nephrotoxins in Canadian grain products. In: Jayas, D. S., White, N. D. G., Muir, W. E. and Sinha, R. N. (Eds) *Extended Abstracts: International Symposium on Stored-Grain Ecosystems*. Department of Agricultural Engineering, University of Manitoba, Winnipeg, Manitoba, Canada, pp. 29–31.

Mislivec, P. B. and Tuite, J. 1970a. Species of Penicillium occurring in freshly harvested and in stored dent corn kernels. *Mycologia* 62(1), 67–74.

Mislivec, P. B. and Tuite, J. 1970b. Temperature and relative humidity requirements of species of *Penicillium* isolated from yellow dent corn kernels. *Mycologia* 62(1), 75–88.

Mueller, G. M. and Schmit, J. P. 2007. Fungal biodiversity: what do we know? What can we predict? *Biodiversity and Conservation* 16(1), 1–5. doi:10.1007/s10531-006-9117-7.

Muir, W. E. and White, N. D. G. 2001. Microorganisms in stored grain. In: Muir, W. E. (Ed.), *Grain Preservation Biosystems*. Department of Biosystems Engineering, University of Manitoba, Winnipeg, pp. 28–42.

Muriuki, G. K. and Siboe, G. M. 1995. Maize flour contaminated with toxigenic fungi and mycotoxins in Kenya. *African Journal of Health Sciences* 2(1), 236–41.

Njoroge, A. W., Affognon, H. D., Mutungi, C. M., Manono, J., Lamuka, P. O. and Murdock, L. L. 2014. Triple bag hermetic storage delivers a lethal punch to *Prostephanus truncates* (Horn)(Coleoptera: Bostrichidae) in stored maize. *Journal of Stored Products Research* 58, 12–9. doi:10.1016/j.jspr.2014.02.005.

O'Brien, H. E., Parrent, J. L., Jackson, J. A., Moncalvo, J. M. and Vilgalys, R. 2005. Fungal community analysis by large-scale sequencing of environmental samples. *Applied and Environmental Microbiology* 71(9), 5544–50. doi:10.1128/AEM.71.9.5544-5550.2005.

Ok, H. E., Kim, D. M., Kim, D., Chung, S. H., Chung, M. S., Park, K. H. and Chun, H. S. 2014. Mycobiota and natural occurrence of aflatoxin, deoxynivalenol, nivalenol and zearalenone in rice freshly harvested in South Korea. *Food Control* 37(1), 284–91. doi:10.1016/j.foodcont.2013.09.020.

Ominski, K. H., Marquardt, R. R., Sinha, R. N. and Abramson, D. 1994. Ecological aspects of growth and mycotoxin production by storage fungi. In: Miller, J. D. and Trenholm, H. L. (Eds) *Mycotoxins in Grains. Compounds Other than Aflatoxin*. Eagen Press, USA, pp. 287–305.

Pardo, E., Malet, M., Marín, S., Sanchis, V. and Ramos, A. J. 2006. Effects of water activity and temperature on germination and growth profiles of ochratoxigenic *Penicillium verrucosum* isolates on barley meal extract agar. *International Journal of Food Microbiology* 106(1), 25–31. doi:10.1016/j.ijfoodmicro.2005.07.002.

Park, J. W., Choi, S. Y., Hwang, H. J. and Kim, Y. B. 2005. Fungal mycoflora and mycotoxins in Korean polished rice destined for humans. *International Journal of Food Microbiology* 103(3), 305–14. doi:10.1016/J.IJFOODMICRO.2005.02.001.

Payne, G. A. and Widstrom, N. W. 1992. Aflatoxin in maize. *Critical Reviews in Plant Sciences* 10(5), 423–40. doi:10.1080/07352689209382320.

Pelhate, J. 1988. Ecology of the microflora of grains and seeds. In: Multon, J. L. (Ed.), *Preservation and Storage of Grains Seeds and Their By-Products*. Lavoisier publishing Inc., New York, pp. 244–62.

Perreta, M., Ramos, J., Tivano, J. C. and Vegetti, A. 2011. Descriptive characters of growth form in Poaceae-an overview. *Flora - Morphology, Distribution, Functional Ecology of Plants* 206(4), 283–93. doi:10.1016/j.flora.2010.04.022.

Pingali, P. L. and Pandey, S. 2001. Meeting world maize needs: technological opportunities and priorities for the public sector. In: Pingali, P. L. (Ed.) *CIMMYT 1999-2000 World Maize Facts and Trends*. CIMMYT, Mexico City.

Pitt, J. I. 1973. An appraisal of identification methods for *Penicillium* species: novel taxonomic criteria based on temperature and water relations. *Mycologia* 65(5), 1135–57. doi:10.1080/00275514.1973.12019533.

Pitt, J. I. and Hocking, A. D. 1977. Influence of solute and hydrogen ion concentration on the water relations of some xerophilic fungi. *Journal of General Microbiology* 101(1), 35–40. doi:10.1099/00221287-101-1-35.

Pitt, J. I., Hocking, A. D., Bhudhasamai, K., Miscamble, B. F., Wheeler, K. A. and Tanboon-Ek, P. 1993. The normal mycoflora of commodities from Thailand. 1. Nuts and oilseeds. *International Journal of Food Microbiology* 20(4), 211–26. doi:10.1016/0168-1605(93)90166-e.

Pitt, J. I. and Hocking, A. D. 1997. *Fungi and Food Spoilage* (2nd edn.). Blackie Academic and Professional, London.

Pitt, J. I., Hocking, A. D., Miscamble, B. F., Dharmaputra, O. S., Kuswanto, K. R., Rahayu, E. S. and Sardjono. 1998. The mycoflora of food commodities from Indonesia. *Journal of Food Mycology* 1, 41–60.

Pitt, J. I. and Samson, R. A. 2007. Nomenclatural considerations in naming species of *Aspergillus* and its teleomorphs. *Studies in Mycology* 59, 67–70. doi:10.3114/sim.2007.59.08.

Pitt, J. I. and Hocking, A. D. 2009a. *Aspergillus* and related teleomorphs. In: Pitt, J. I. and Hocking, A. D. (Eds) *Fungi and Food Spoilage* (3rd edn.). Springer, New York, 520pp.

Pitt, J. I. and Hocking, A. D. 2009b. *Fungi and Food Spoilage* (3rd edn.). Springer, New York, p. 520.

Reddy, K. R. N., Reddy, C. S., Abbas, H. K., Abel, C. A. and Muralidharan, K. 2008. Mycotoxigenic fungi, mycotoxins, and management of rice grains. *Toxin Reviews* 27(3–4), 287–317. doi:10.1080/15569540802432308.

Reddy, K. R. N., Reddy, C. S. and Muralidharan, K. 2009. Detection of Aspergillus spp. and aflatoxin b1 in rice in India. *Food Microbiology* 26(1), 27–31. doi:10.1016/j.fm.2008.07.013.

Richard-Molard, D. Lesage, L. and Cahagnier, B. 1985. Effect of water activity on mold growth and mycotoxins. *Properties of Water in Foods*, Vol. 90, 273–92.

Rosentrater, K. A. and Evers, A. D. 2017. Introduction to cereals and pseudocereals and their production. *Kent's Technology of Cereals*. doi:10.1016/b978-0-08-100529-3.00001-3.

Sanchis, V. and Magan, N. 2004. Environmental conditions affecting mycotoxins. In: Magan, N. and Olsen, M. (Eds) *Mycotoxins in Food: Detection and Control*. Woodhead Publishing Ltd., Cambridge, pp. 74–189.

Sauer, D. B., Meronuck, R. A. and Christensen, C. M. 1992. Microflora. In: Sauer, D. B. (Ed.) *Storage of Cereal Grains and Their Products* (4th edn.). American Association of Cereal Chemists, Inc, St. Paul, MN, pp. 313–40.

Schmidt, M., Horstmann, S., De Colli, L., Danaher, M., Speer, K., Zannini, E. and Arendt, E. K. 2016. Impact of fungal contamination of wheat on grain quality criteria. *Journal of Cereal Science* 69, 95–103. doi:10.1016/j.jcs.2016.02.010.

Schoch, C. L., Robbertse, B., Robert, V., Vu, D., Cardinali, G., Irinyi, L., Meyer, W., Nilsson, R. H., Hughes, K., Miller, A. N., Kirk, P. M., Abarenkov, K., Aime, M. C., Ariyawansa, H. A., Bidartondo, M., Boekhout, T., Buyck, B., Cai, Q., Chen, J., Crespo, A., Crous, P. W., Damm, U., De Beer, Z. W., Dentinger, B. T., Divakar, P. K., Dueñas, M., Feau, N., Fliegerova, K., García, M. A., Ge, Z. W., Griffith, G. W., Groenewald, J. Z., Groenewald, M., Grube, M., Gryzenhout, M., Gueidan, C., Guo, L., Hambleton, S., Hamelin, R., Hansen, K., Hofstetter, V., Hong, S. B., Houbraken, J., Hyde, K. D., Inderbitzin, P., Johnston, P. R., Karunarathna, S. C., Kõljalg, U., Kovács, G. M., Kraichak, E., Krizsan, K., Kurtzman, C. P., Larsson, K. H., Leavitt, S., Letcher, P. M., Liimatainen, K., Liu, J.

K., Lodge, D. J., Luangsa-ard, J. J., Lumbsch, H. T., Maharachchikumbura, S. S., Manamgoda, D., Martín, M. P., Minnis, A. M., Moncalvo, J. M., Mulè, G., Nakasone, K. K., Niskanen, T., Olariaga, I., Papp, T., Petkovits, T., Pino-Bodas, R., Powell, M. J., Raja, H. A., Redecker, D., Sarmiento-Ramirez, J. M., Seifert, K. A., Shrestha, B., Stenroos, S., Stielow, B., Suh, S. O., Tanaka, K., Tedersoo, L., Telleria, M. T., Udayanga, D., Untereiner, W. A., Diéguez Uribeondo, J., Subbarao, K. V., Vágvölgyi, C., Visagie, C., Voigt, K., Walker, D. M., Weir, B. S., Weiß, M., Wijayawardene, N. N., Wingfield, M. J., Xu, J. P., Yang, Z. L., Zhang, N., Zhuang, W. Y. and Federhen, S. 2014. Finding needles in haystacks: linking scientific names, reference specimens and molecular data for Fungi. *Database: The Journal of Biological Databases and Curation*. Database, 1-21 2014. doi:10.1093/database/bau061.

Scott, P. M. 2004. *Other Mycotoxins. Mycotoxins in Food*. In: N. Magan and M. Olsen (Ed.), Mycotoxins in Food: Detection and Control, Woodhead Publishing Ltd, Sawston, Cambridg. doi:10.1533/9781855739086.3.406.

Semeniuk, G. 1954. Microflora. In: Anderson, J. A. and Alcock, A. W. (Eds) *Storage of Cereal Grains and Their Products*. American Association of Cereal Chemists, St. Paul, MN.

Sinha, R. N. 1971. Fungus as food for some stored product insects. *Journal of Economic Entomology* 64(1), 3-6. doi:10.1093/jee/64.1.3.

Sinha, R. N. 1984. Effects of weevil (Coleoptera: Curculionidae) infestation on abiotic and biotic quality of stored wheat. *Journal of Economic Entomology* 77(6), 1483-8. doi:10.1093/jee/77.6.1483.

Sinha, R. N. 1992. The fungal community in the stored grain ecosystem. In: Caroll, G. C. and Wicklow, D. T. (Eds) *The Fungal Community* (2nd edn.). Marcel Dekker, Inc., New York, 976pp.

Siriacha, P., Tanboon-Ek, P. and Buangsuwon, D. 1991. Aflatoxin in maize in Thailand. *Aciar Proceedings* 36, 187-93.

Smale, M., Derek, B. and Thom, J. 2011. Maize revolutions in Sub-Saharan Africa. Policy research working paper 5659. The World Bank. Development research group. Agriculture and rural development team. Available at: http://wwwwds.worldbank.org/servlet/WDSContentServer/WDSP/IB/2011/05/11/000158349_20110511144001/Rendered/PDF/WPS5659.pdf.

Smart, M. G., Wicklow, D. T. and Caldwell, R. W. 1990. Pathogenesis in *Aspergillus* ear rot of maize: light microscopy of fungal spread from wounds. *Phytopathology* 80, 1287-94.

Smith, S. L. and Hill, S. T. 1982. Influence of temperature and water activity on germination and growth of *Aspergillus restrictus* and *A. versicolor*. *Transactions of the British Mycological Society* 79(3), 558-60. doi:10.1016/S0007-1536(82)80058-2.

Smith, A. F. 2003. The first thanksgiving. *Gastronomica: The Journal of Food and Culture* 3(4), 79-85. Available at: https://www.jstor.org/stable/10.1525/gfc.2003.3.4.79?seq=1.

Snijders, C. H. A. 1990. *Fusarium* Head Blight and Mycotoxin Contamination of Wheat, a Review. *Netherlands Journal of Plant Pathology* 96(4), 187-98. doi:10.1007/BF01974256.

Sobek, E. A. and Munkvold, G. P. 1999. European corn borer (Lepidoptera: Pyralidae) larvae as vectors of *Fusarium moniliforme*, causing kernel rot and symptomless infection of maize kernels. *Journal of Economic Entomology* 92(3), 503-9. doi:10.1093/jee/92.3.503.

Sone, J. 2001. Mold growth in maize storage as affected by compound factors: different levels of maize weevils, broken corn and foreign materials, and moisture contents. *Journal of Asia-Pacific Entomology* 4(1), 17-21. doi:10.1016/S1226-8615(08)60096-5.

Syarief, A. M., Gustafson, R. J. and Morey, R. V. 1987. Moisture diffusion coefficients for yellow-dent corn components. *Transactions of the ASAE. American Society of Agricultural Engineers* 30, 522–8.

Tadesse, W., Amri, A., Ogbonnaya, F. C., Sanchez-garcia, M., Sohail, Q. and Baum, M. 2016. *2.1 Introduction*. doi:10.1016/B978-0-12-802000-5/00002-2.

Tong, C. and Bao, J. 2019. *Rice Lipids and Rice Bran Oil*. A.A.C.C International, Rice. Published by Elsevier Inc. in cooperation with AACC International. doi:10.1016/b978-0-12-811508-4.00005-8.

Tsuruta, O., Gohara, S. and Saito, M. 1981. Scanning electron microscopic observations of a fungal invasion of corn kernels. *Transactions of the Mycological Society of Japan* 22, 121–6.

Tuite, J., Koh-Knox, C., Stroshine, R., Cantone, F. A. and Bauman, L. F. 1985. The effect of various kinds of physical damage to corn kernels on the development of *Penicillium* species and *Aspergillus glaucus* in storage. *Phytopathology* 75(10), 1137–40. doi:10.1094/Phyto-75-1137.

Uraguchi, K. 1969. Mycotoxic origin of cardiac beriberi. *Journal of Stored Products Research* 5(3), 227–36. doi:10.1016/0022-474X(69)90037-X.

USDA 2006. Classification. *USDA Plants*. Available at: https://www.plants.usda.gov/java/ClassificationServlet?source=profile&symbol=ORYZA&display=31 (Accessed on October 8, 2019).

Van Wyk, J. H., Hodson, A. C. and Christensen, C. M. 1959. Microflora associated with the confused flour beetle, *Tribolium confusum*. *Annals of the Entomological Society of America* 52(4), 452–63. doi:10.1093/aesa/52.4.452.

Waalwijk, C., Koch, S. H., Ncube, E., Allwood, J., Flett, B., de Vries, I. and Kema, G. H. J. 2008. Quantitative detection of *Fusarium* spp. and its correlation with fumonisin content in maize from South African subsistence farmers. *World Mycotoxin Journal* 1(1), 1875–0710.

Walker, S., Jaime, R., Kagot, V. and Probst, C. 2018. Comparative effects of hermetic and traditional storage devices on maize grain: mycotoxin development, insect infestation and grain quality. *Journal of Stored Products Research* 77, 34–44. doi:10.1016/j.jspr.2018.02.002.

Wallace, H. A. H. 1973. Fungi associated with stored grain. In: Sinha, R. N. and Muir, W. E. (Eds) *Grain Storage: Part of a System*. Avi Publishing Co., Westport, CT.

Watson, S. A. 1987. Structure and composition. In: *Corn: Chemistry and Technology. S.A. Watson and P. E. Ramstad*, eds. American Association of Cereal Chemistry, St. Paul, MN, pp. 52–82.

Wei, X. and Huang, X. 2019. *Origin, Taxonomy, and Phylogenetics of Rice*. A.A.C.C international, Rice. Published by Elsevier Inc. in cooperation with AACC International. doi:10.1016/b978-0-12-811508-4.00001-0.

Wicklow, D. T. 1994. Preharvest origins of toxigenic fungi in stored grain. In: Highley, E., Wright, E. J., Banks, H. J. and Champ, B. R. (Eds) *Stored Product Protection*. CAB International, Wallingford, England, pp. 1075–81.

Wicklow, D. T. 1995. The mycology of stored grain: an ecological perspective. In: Jayas, D. S., White, N. D. G. and Muir, W. E. (Eds) *Mycology of Stored Grain Ecosystems*. Dekker, New York, pp. 197–249.

Wicklow, D. T., Weaver, D. K. and Throne, J. E. 1998. Fungal colonists of maize grain conditioned at constant temperatures and humidities. *Journal of Stored Products Research* 34(4), 355–61. doi:10.1016/S0022-474X(98)00021-6.

Wilson, S. G. and Desmarchelier, J. M. 1994. Aeration according to seed wet-bulb temperature. *Journal of Stored Products Research* 30(1), 45–60. doi:10.1016/0022-474X(94)90271-2.

Wolf, M. J., Buzan, C. L., MacMasters, M. M. and Rist, C. E. 1952. Structure of the mature corn kernel. II. Microscopic structure of the pericarp, seed coat, and hilar layer of dent corn. *Cereal Chemistry* 29, 334–34.

Wrigley, C. W. 2010. An introduction to the cereal grains: major providers for mankind's food needs. In: *Cereal Grains: Assessing and Managing Quality*. Woodhead Publishing Limited. Sawston, Cambridge. doi:10.1533/9781845699529.1.3.

Wrigley, C. W. 2015. *An Overview of the Family of Cereal Grains Prominent in World Agriculture. Encyclopedia of Food Grains* (2nd edn., Vol. 1–4). Elsevier Ltd. doi:10.1016/B978-0-12-394437-5.00006-1.

Wrigley, C. W. 2017. *Cereal-Grain Morphology and Composition. Cereal Grains: Assessing and Managing Quality* (2nd edn.). Elsevier Ltd. Sawston, Cambridge. doi:10.1016/B978-0-08-100719-8.00004-8.

Yeh, M. Y. 1979. *Dissemination of Toxigenic Fungi in Shelled Corn by Ahasverus advena and Tribolium confusum*. University of Minnesota, Minneapolis, MN, 71pp.

Zorya, S., Morgan, N. and Rios, L. D. 2011. *Missing Food: The Case of Postharvest Grain Losses in Sub-Saharan Africa*. World Bank, Washington, DC.

# Chapter 3

## Advances in detection and management of phosphine resistance in stored grain pests

*David Schlipalius and Paul Ebert, University of Queensland, Australia*


1 Introduction
2 The biochemical and genetic basis of resistance
3 Alternative fumigants
4 Conclusion and future trends
5 References


### 1 Introduction

Phosphine ($PH_3$) is a toxic-reducing gas and the primary disinfestation treatment for stored grain worldwide. It is accepted by most major markets as a residue-free fumigant. Other fumigants suffer from practical limitations including high cost, reduced efficacy against eggs or reduced efficacy at low temperatures less than 20°C, poor penetration of the grain bulk due to sorption, low volatility or high density (Table 1). Another factor is that some fumigants, such as methyl bromide, have unacceptable environmental consequences (United Nations Environment Programme, 1992). As a result, phosphine has become the fumigant of choice for pest managers. While phosphine is rather slow acting, the low cost of phosphine, as well as the fact that phosphine does not leave toxic residues on the grain, makes it possible to use high concentrations of the gas to decrease the time required for an effective fumigation. In the last 20 years strong resistance to phosphine has been detected in all major grain pest species (Collins et al., 2005; Pimentel et al., 2009; Jagadeesan et al., 2012; Nayak et al., 2013, 2017; Nguyen et al., 2015; Holloway et al., 2016; Collins et al., 2017). Despite the option of using increased concentrations of the gas to achieve adequate pest control, resistance is the biggest threat to the continued routine use of phosphine globally. This chapter will discuss advances in our understanding of phosphine resistance: how it arises, how it works and how it affects insect management practices.

http://dx.doi.org/10.19103/AS.2020.0072.03

**Table 1** Qualitative comparison of available fumigants

| Fumigant | Chemical symbol | Synonyms | Primary use | Seed viability | Speed of action | Ease of application[1] | Residues | Low sorption | Rapid diffusion | Environmentally friendly |
|---|---|---|---|---|---|---|---|---|---|---|
| Phosphine | $PH_3$ | Phostoxin | Routine disinfestation | ✓ | ✓ | ✓✓ | ✓ | ✓ | ✓ | ✓✓ |
| Sulfuryl fluoride | $S_2F_2$ | ProFume | Resistance breaker/ structural | ✓ | ✓ | ✓ | ✗ | ✓ | ✓ | ✗ |
| Methyl bromide | $CH_3Br$ | MeBr | Quarantine/export | ✓ | ✓ | ✓ | ✓ | ✓ | ✓ | ✗ |
| Carbonyl sulfide | OCS | COS, COSmic | Currently unregistered | ✓ | ✓ | ✗ | ✓ | ✓ | ✓ | ✓ |
| Ethanedinitrile | $C_2N_2$ | EDN, Sterigas | Structural treatments | ✗ | ✓ | ✓ | ✗ | ✓ | ✓ | ✓ |
| Ethyl formate | $C_3H_6O_2$ | Vapormate | Small scale | ✓ | ✓ | ✓ | ✓ | ✗ | ✗ | ✓✓ |
| Carbon dioxide | $CO_2$ | | certified organic | ✓ | ✗ | ✓ | ✓ | ✓ | ✓ | ✓✓ |
| Nitrogen | $N_2$ | | certified organic | ✓ | ✗ | ✓ | ✓ | ✓ | ✓ | ✓✓ |

[1]Includes commercial availability.

### *1.1 What is phosphine resistance?*

Phosphine gas is applied in storages to kill pest insects that infest and damage the product, whether that product is cereal grains or other foodstuff. The dose required to kill pest insects is dependent on how sensitive they are to the fumigant. For fumigants, doses are a product of time and concentration of gas at temperatures (usually above 15°C) where the insects are likely to be active. (Generally, the higher the temperature, the more effective the fumigant). The parameters that can be most easily controlled are time and concentration of fumigant. Insects that are resistant can survive doses (time × concentration) that will normally kill a susceptible population of insects at the temperatures they are applied.

When phosphine was first used, the vast majority of the individuals in an insect population were susceptible and resistance was relatively rare (Champ and Dyte, 1977). At that time, the concentration of gas and time of exposure required for an effective fumigation was much lower than what is required now. The ease of controlling susceptible insects led to poor practices such as fumigating grain in unsealed storages. Insects could easily escape before being exposed to a lethal dose of gas, which enabled some insects to survive and pass on any resistance they had to their offspring. In more recent years, phosphine usage has increased, which has led to widespread strong resistance in multiple stored product pests including red flour beetle (*Tribolium castaneum*) (Nayak et al., 2017), lesser grain borer (*Rhyzopertha dominica)* (Collins et al., 2017), rice weevil (*Sitophilus oryzae*) (Holloway et al., 2016), rusty grain beetle (*Cryptolestes ferrugineus*) (Nayak et al., 2013), Khapra beetle (*Trogoderma granarium*) (Sher et al., 2003) and Psocids (Nayak et al., 1998; Nayak and Collins, 2008).

### *1.2 What does resistance mean?*

A dose of phosphine is the product of two parameters: concentration (c) and time (t). Resistance to phosphine either increases the effective concentration of phosphine or slows its rate of action. Resistance can be overcome by either increasing the concentration of gas that is applied or by increasing the time of exposure to the gas. This simple strategy has been used to maintain the effectiveness of phosphine for more than four decades and has only been possible because phosphine gas does not leave toxic residues on the grain. However, the current level of resistance in some species has nearly reached the practical limits of the strategy.

Unlike most fumigants, phosphine exhibits a non-linear concentration-time relationship (Winks and Waterford, 1986; Collins et al., 2001). Thus, rapid control of pest insects necessitates the use of a very high concentration of gas, making complete control in short time frames increasingly expensive.

Conversely, a lower concentration of gas can be effective, but only if it is maintained for an extended period of time, which is increasingly difficult to accomplish with most grain storage structures. Commercial grain delivery schedules can also constrain the time available for fumigation. The longer time frames and higher doses required to control resistant insects have started to reach the practical limits of current pest management practices. Therefore, resistance management has come to require a greater understanding of the biological basis of phosphine toxicity and resistance.

## 1.3 Practical conditions for pest management

There are practical advantages that can be leveraged for resistance management in grain storages that can be fumigated (i.e. gas-tight). They are essentially closed systems that can maintain gas doses at high levels for long periods of time. The Australian standard for gas tightness is measured by a half-life negative pressure test (described in Australian Standard AS2628-2010). A storage is considered gas-tight and can maintain gas concentrations well if the half-life is over five minutes. This allows even strongly resistant insects to be controlled if the doses are maintained correctly (Collins et al., 2005). Thus, it is possible to employ resistance management strategies that could not be used in field crop situations, such as the 'high dose' strategy (Denholm and Rowland, 1992; Gryspeirt and Grégoire, 2012). The strategy consists of maintaining a high dose of fumigant for a long enough time to reduce the populations of all insects including resistant ones. For example, a dose rate to control for strongly resistant *C. ferrugineus*, the most strongly resistant pest so far reported (Table 2), is 1.0 $mgL^{-1}$ (740 ppm) maintained for at least 19 days at 25°C, or for 14 days at 35°C (Kaur and Nayak, 2015). Increasing the maintained dose decreases the time required for population extinction but this is often difficult to maintain.

This high-dose approach is not generally applied to field crops because it is relatively easy for insects to escape the region of pesticide application in the open environment. In such situations, the increase in dose would simply increase selective pressure, facilitating the development of high-level resistance. During fumigation in gas-tight structures, however, insects cannot readily escape and so control is effective. Furthermore, insects outside the structure will be unexposed to phosphine and therefore will not be under selective pressure, providing a refugium for susceptible insects. This may have delayed resistance development as a high dose/refuge strategy would favour dilution of resistant genotypes with untreated susceptible ones once a grain store was opened (Lilford et al., 2009; Shi et al., 2012, 2013). Interbreeding between the two genotypes would result in progeny that would be susceptible to phosphine and would slow the progression of

**Table 2** Comparison of resistance ratios in various stored product pests

| Common name | Species | $LC_{50}$ Susceptible (95% FL) ($mgL^{-1}$) | $LC_{99.9}$ Susceptible ($mgL^{-1}$) | $LD_{50}$ Resistant ($mgL^{-1}$) | $LD_{99.9}$ Resistant (mg/L) | Time tested[1] | Resistance ratio reported | References |
|---|---|---|---|---|---|---|---|---|
| Lesser grain borer | *Rhyzopertha dominica* | 0.00174 (0.00167–0.00180) | 0.00398 | 1.04866 (0.96763–1.13491) | 11.65 | 20 hours | ~600 | Collins et al., 2002 |
| | | | | 0.412 (0.398–0.426) | 2.09 | 48 hours | ~240 | Mau et al., 2012 |
| Rust red flour beetle | *Tribolium castaneum* | 0.009 (0.008–0.009) | 0.02 | 3.885 (3.33–4.452) | 36.99 | 20 hours | 431 | Jagadeesan et al., 2012 |
| Rice weevil | *Sitophilus oryzae* | 0.005 (0.005–0.006) | NR | 0.277 (0.245–0.308) | NR | 48 hours | 52 (27.23–97.45) | Nguyen et al., 2015 |
| Rusty grain beetle | *Cryptolestes ferrugineus* | 0.005 (0.004–0.005) | 0.009 (0.008–.01) | 6.57 (5.8–7.34) | 37.68 (29.99–51.2) | 48 hours | 1341 (741–2447) | Nayak et al., 2013 |
| Cigarette beetle | *Lasioderma serricorne* | 13.0 (7.11–20.2) ppm | NR | >10 000 ppm | NR | 24 hours | >770 | Hori and Kasaishi, 2005 |
| Khapra beetle | *Trogoderma granarium* | 4.2 | NR | 8 | NR | 24 hours | ~2 | Bell and Wilson, 1995, Sher et al., 2003 |
| Psocids (all life stages) | *Liposcelis bostrychophila* | NR | 0.03 | NR | 2.00 | 6 days | 67 | Nayak and Collins, 2008 |

[1]All tests were conducted at a standard temperature of 25°C.

resistance (Rafter et al., 2017). Another benefit of the high-dose fumigation strategy is that the very low rates of pest insect survival extend the length of time before the next fumigation is required. Even though the strategy favours survival of resistant insects, if the distribution of fumigant is not uniform or if the concentration of gas is not maintained properly, this potential problem is manageable if the number of fumigations is kept to a minimum and grain is not held for long periods of time (greater than a year). Guidelines for the Australian industry recommend that no more than three consecutive fumigations for the same parcel of grain without changing the fumigant should be used (Collins, 2009).

There are physical control options that should be part of any serious grain-handling operation for augmenting resistance management. Cooling of grain bulks by aeration can be used to reduce insect population growth and temperatures held below 15°C can significantly reduce population growth (Flinn et al., 1997; Arthur and Casada, 2010). The fact that the structures suitable for fumigation are closed systems also allows a reduction of reinfestation of the grain by insect movement from outside to inside the storage structure. Incursion can also be minimised by reducing the local population size through good facility sanitation, for example, minimising grain spills, eliminating waste dumps, cleaning machinery and treating empty storage facilities prior to introduction of new grain. By limiting insect incursion at the handling facility, where the resident insect may harbour resistance, the main source of insects injected into the system is the field, where the insects are carried in with the harvest and are unlikely to have been exposed to phosphine.

## 1.4 How do resistance mechanisms impact management?

Now that adjusting phosphine fumigation protocols to counter the increase in resistance levels is reaching its practical limit (Nayak et al., 2013), a more detailed understanding of resistance is required to develop pest control strategies of greater sophistication. While the resistance mechanisms are not exactly simple, the genetic basis is clearly understood (Schlipalius et al., 2012, 2018). From a pest management standpoint, resistance results from existing variants in DNA that are inherited and passed on to offspring. While new variants can result from new mutation events, this is a very rare occurrence. The mode of inheritance of phosphine resistance genes are similar to other insecticide resistance mutations in that they are *recessive* and the susceptible form of the genes are *dominant* (Schlipalius et al., 2002). This means that, of the genes inherited from either parent, if only one copy of the resistance gene and one copy of the susceptible type gene are present, then that insect is susceptible to the fumigant but is also a 'carrier' of resistance (Fig. 1). Two resistant forms of the gene must be present within the same insect (i.e. homozygosity) for the

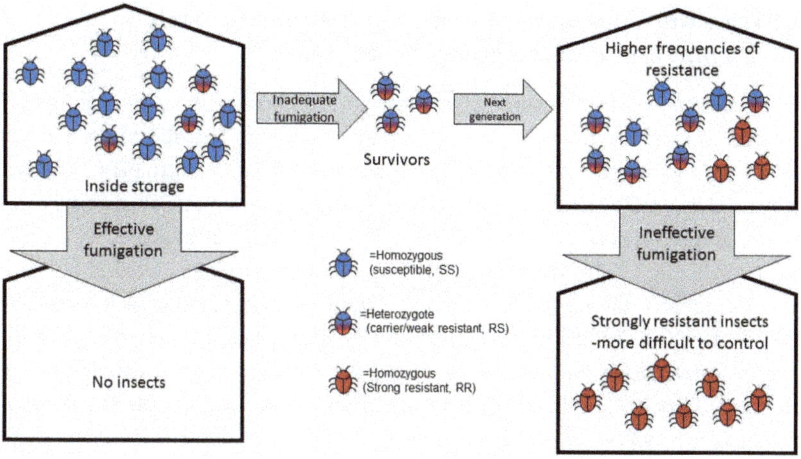


**Figure 1** The genetics of resistance inheritance in the context of fumigation. Inadequate fumigations that allow even a low number of insects to survive will cause an increase in resistance in a population. Homozygous individuals are more difficult to control possibly requiring the use of 'break fumigants' to reduce resistance frequency. High-dose effective fumigations that kill all insects will not lead to an increase in resistance gene frequency.

resistance trait to be observed (Schlipalius et al., 2002; D. I. Schlipalius et al., 2008). This directly impacts how resistance can be managed. We can use this information on inheritance to disrupt or reduce the number of resistance mutations being passed on to successive generations.

There are two major levels of resistance: for short (20–48 hour) exposures 'weak' resistance is 5–30 times the basal tolerance level of a fully susceptible reference strain (Daglish et al., 2002; D. I. Schlipalius et al., 2008; Jagadeesan et al., 2012), whereas 'strong' resistance is usually more than 100 times the tolerance of the susceptible strain. 'Weak' resistance has been observed since the 1970s and is caused by homozygosity of a resistance mutation in a single gene, *rph1* (*resistance to phosphine 1*) (Schlipalius et al., 2002; D. I. Schlipalius et al., 2008). Strong resistance requires homozygosity of a resistance variant of a second gene, *rph2*, together with homozygosity of *rph1* in the same individual. The *rph2* gene can confer only weak resistance by itself (5–12 times the basal tolerance) (D. I. Schlipalius et al., 2008) but, importantly, synergistically enhances resistance of *rph1* to produce the strong resistance trait.

It has been found that the same genes (*rph1* and *rph2*) are responsible for strong phosphine resistance in all pest species that have been studied globally (Schlipalius et al., 2012, 2019; Nguyen et al., 2016). Therefore, if a means of overcoming resistance caused by either *rph1* or *rph2* can be found, the result will be weakly resistant insects that can be easily managed.

## 1.5 How does our understanding of the genetic basis change our understanding of resistance?

The Australian grain industry has carried out the most extensive ongoing post-harvest resistance-monitoring programme in the world. The effort to monitor the evolution and spread of phosphine resistance has been ongoing for decades (Nayak et al., 2017; Collins et al., 2017). Other more limited studies have also been conducted elsewhere in the world (Pimentel et al., 2009; Chen et al., 2015; Koçak et al., 2015; Ramya et al., 2018). This monitoring has involved collecting live insects from the field and determining the level of resistance to phosphine of the collected insects or their progeny. Typically, live insects are exposed to discriminating doses of phosphine for a specific period of time, usually between 20 hours (Champ and Dyte, 1977) and 48 hours, but sometimes more (Collins et al., 2001; Nayak et al., 2013), and at specific concentrations. The discriminating doses used depend on the species and are usually set at the approximate $LC_{99.9}$ for the resistance level to be discriminated from (Table 2), that is, the susceptible or 'weak' resistance type for that species. This is usually performed in gas-tight chambers using specific doses of phosphine injected through a septum. The concentration of the source gas is usually measured with a gas spectrophotometer or phosphine sensor. After fumigation, the insects are held for a 2-week recovery period and the proportion of insects that survive is calculated (called a bioassay). The bioassay requires no previous knowledge of how resistance works for determining whether resistant insects are present in the sample being tested.

The bioassay does however have some limitations. As a minimum, dose-response monitoring requires approximately 100 insects to be able to detect resistant homozygotes in a population (Daglish and Collins, 1999). It is often difficult to collect so many insects from storage facilities, so breeding in the laboratory may be required to increase insect numbers. This extends the time required for the assay and while it overcomes the insensitivity of the assay, the number of individuals from the field that are being surveyed does not change. While trapping of insects can be used to increase the number of insects collected, the insects are often killed during trapping (Stevens et al., 2011) and so are unsuitable for bioassays. Thus, bioassays limit studies of field resistance to insects directly collected from storage facilities or their progeny.

A rapid alternative to the bioassay has also been developed. This assay, referred to as a 'knockdown' test, determines the time required for a dose of phosphine to render the insects unconscious, which causes them to drop off a filter paper strip. The time to knockdown roughly correlates with resistance as determined in the bioassay (Nayak et al., 2013; Agrafioti et al., 2019; Athanassiou et al., 2019). The strength of the bioassay and the knockdown test is that they do not rely on knowing the resistance mechanism at all as they are purely empirical.

## 1.6 Can knowing genes provide a better way to detect resistance?

We can now use our genetic knowledge of resistance to develop assays to detect the resistance mutations directly in the DNA of pest insects, thus providing three distinct advantages. Because DNA testing can identify resistance variants directly in individual insects, it can detect resistance gene variants that are present at very low frequencies. This makes the test effective even if only a few insects have been collected. The DNA assay can even allow detection of resistance gene variants that are undetectable using previous resistance detection methods. For example, since resistance is recessive, heterozygotes or 'carriers' of the resistance gene variants are essentially invisible to the traditional testing methods, but this is straightforward with a DNA test. The DNA test also allows the detection of resistance in dead insects (D. Schlipalius et al., 2008; Chen et al., 2015; Schlipalius et al., 2019). As a result, the DNA assay can test for resistance gene frequencies in trapped insects that are typically killed as they are collected. Being able to test the trap catches facilitates the determination of resistance frequencies in insects that are resident at a storage site, but not currently residing within a storage structure.

The ability of the DNA assay to detect resistance mutations at low frequencies and in the heterozygous state (Schlipalius et al., 2019) provides a very distinct advantage to pest managers. The DNA test allows resistance to be detected at sites years earlier than they can be detected by bioassay. This gives us an ability to see potential resistance problems emerging well before they cause fumigation failures. This advance warning can allow pest managers to implement resistance management practices proactively. Importantly, it provides a tool that allows resistance management strategies to be accurately monitored so that the effectiveness of current management practices or reactive interventions can be assessed.

## 1.7 Genetic assays for phosphine resistance

Two genes, *rph1* and *rph2*, contribute to strong phosphine resistance. On the face of it, it would seem necessary to develop a DNA assay that could detect resistance variants of both genes, but this is not the case. Resistance variants of the *rph2* gene alone need to be detected because resistance variants of this gene correlate much more with strong resistance than do variants of *rph1* (Chen et al., 2015). This is because resistance variants of the *rph1* gene originated first and are already very widespread (Mau et al., 2012; Holloway et al., 2016; Nayak et al., 2017; Collins et al., 2017). Thus, resistance variants of *rph1* are likely to appear in most populations already, and *rph2* variants rarely appear without *rph1* variants. Resistance variants of *rph1* that occur in insects that lack *rph2* are only weakly resistant to phosphine, which is an easily managed problem.

Most current DNA testing focusses on single specific resistance variants in the *rph2* gene that occur at a grain storage facility or in a region that is being investigated (Kaur et al., 2013a,b; Chen et al., 2015; Kaur et al., 2015; Koçak et al., 2015). The focus on a specific variant greatly simplifies the assay, allowing it to be readily carried out in simple molecular laboratories as can be found in almost any country. In such a narrowly focussed configuration, the assay is relatively straightforward and cost-effective. Focussing on just a single variant, however, has the potential for omitting other locally arising variants resulting in partial information. Surprisingly, this is not necessarily a major limitation. In the specific resistance variant assays, one variant appears to be more common worldwide than any other - a variant labelled P45S in *T. castaneum* but also called P49S in *R. dominica* (Schlipalius et al., 2012). This single resistance variant dominates the storage facilities across India. By assaying for this single variant, the magnitude of the resistance problem in *R. dominica* and *T. castaneum* was readily apparent (Kaur et al., 2015). This particular variant is also detected in Australian, American and Turkish populations (Chen et al., 2015; Koçak et al., 2015; Schlipalius et al., 2019). In storages where the great majority of resistance can be explained by this one variant, focussing on it requires only a simple assay system.

In some situations, it may be useful to detect every known, or even unknown, resistance variants of the *rph2* gene. In this situation, DNA is extracted from individual insects and multiple regions of the *rph2* gene are amplified by the polymerase chain reaction. These DNA fragments are then sequenced using the latest next-generation sequencing technology, which allows DNA from thousands of insects to be analysed at the same time. Prior to DNA sequencing, however, each DNA fragment must be tagged so that it can be traced back to the insect from which it originated (Schlipalius et al., 2019). The sequence data from each individual insect can then be compared to a reference sequence of the *rph2* gene and each variant in the *rph2* gene of each insect can then be detected bioinformatically. This assay provides a much more detailed picture of the development and spread of resistance, especially in areas where multiple variants coexist.

There are very few variants of the *rph2* gene that cause phosphine resistance, with only nine detected to date (Schlipalius et al., 2012; Nguyen et al., 2016). In contrast, the *rph1* gene has a much more complicated set of variants that confer resistance. This complexity makes development of genetic assays more difficult. Also, given that weak resistance is already very widespread in areas where phosphine is used extensively (Mau et al., 2012; Holloway et al., 2016; Nayak and Daglish, 2017; Collins et al., 2017), detection of the *rph1* variants that are responsible for the weak resistance is not really practically useful as such variants are likely to exist in most populations already. Thus, the technical issues associated with detecting resistance variants of *rph1* is not really much of a problem.

### 1.7.1 Complementary resistance assays: bioassays

While the molecular assay is much more sensitive than mortality testing, direct mortality testing of insects remains the gold standard reference when validating the molecular assay. Molecular genetic assays are performed by extracting DNA from individual insects and performing a specific polymerase chain reaction (PCR) amplification of the *rph2* gene specifically. Then the DNA can be either sequenced to detect the mutations (Schlipalius et al., 2019) or tested via a specific assay such as cleaved amplified polymorphic sequence (CAPS) marker assay (R. Kaur et al., 2013a,b; Chen et al., 2015; Koçak et al., 2015). Two types of data are provided by the molecular assay: known resistance mutations and potential additional resistance mutations (Schlipalius et al., 2019). However, each new mutation identified by the molecular assay must be confirmed as being responsible for resistance via mortality testing. By treating the molecular assay and mortality testing as complementary assay systems, we are able to improve the molecular assay through the identification and confirmation of new resistance mutations as a side-effect of simply using the assay.

Molecular testing of resistance genes directly from individual insects has the potential to transform resistance management in several ways. First, it allows area-wide comprehensive resistance mapping to identify problem sites or management practices, which will facilitate focussed intervention (D. Schlipalius et al., 2008; Schlipalius et al., 2019). Secondly, it can provide a 'before and after' site assessment to monitor resistance management efforts such as 'eradication' or population reduction. Lastly, it can provide a readily accessible 'high-tech' tool for extension officers and post-harvest consultants.

### 1.7.2 Strategic management of resistance

Molecular testing, or resistance testing in general, is better for long-term effective strategic management than as a tactical quick response. Resistance testing to facilitate cost-cutting by dose minimisation is more of a problem than a solution as doses should be maintained at recommended levels for the designated period to achieve effective pest control (Collins et al., 2001). Any reduction in dosage rates will allow resistant insects to survive and breed and will accelerate future resistance outbreak issues (see Fig. 1) (Shi et al., 2013).

## 1.8 What has genetic testing told us?

There have been several surveys of the spread and frequency of *rph2* resistance gene variants in Australia, the United States, India and Turkey

(Kaur et al., 2013a,b; Chen et al., 2015; Kaur et al., 2015; Koçak et al., 2015; Schlipalius et al., 2019). The patterns of resistance-variant occurrence and spread are vastly different between countries. In India, for example, there is mostly one major genetic variant (P45S/P49S) that explains most of the resistance observed in *T. castaneum* and *R. dominica* collected from bag-stack storage godowns and warehouses. The frequencies of strong resistance are generally quite high (>70%) (Kaur et al., 2015; Rafter et al., 2019). The fumigation practice in India is to fumigate individual bag stacks under plastic tarps using phosphine pellets.

In Australia, the patterns of *rph2* resistance gene variants are quite different. In a survey of *R. dominica* collected from farms and central storages across eastern Australia (Schlipalius et al., 2019) it was observed that frequencies of resistance in farm storages were generally quite low and the highest frequencies of resistance gene variants were observed in central storages. It is notable that the greatest resistance-gene variant frequencies were observed in 'bunker' type storages, where fumigation occurs under plastic sheeting that is sealed over large temporary accumulations of grain. This is one demonstration of how genetic testing can detect the impact of storage structure types on the probability of developing strong resistance.

The Australian survey also took into account multiple variants of the *rph2* gene and showed that resistance had arisen independently at least three times in different localities (Schlipalius et al., 2019). The different resistances had then spread locally across wide areas (hundreds to thousands of kilometres), presumably by a combination of factors such as insect movement to other farms and local transport to central storages. In the Australian context, resistance frequencies were quite low compared to those observed in surveys in other countries, especially on farms some distance from central storages. The Australian industry has had a best practice policy of high-dose rates and fumigating in storages as sealed as possible in order to manage phosphine resistance, which likely contributed to the effective resistance management.

## 1.9 How do insect populations (resistance gene variants) move?

The latest research indicates that many species of pest insects that infest grain are adept at flying and can move many kilometres in search of food and mates (Edde et al., 2006; Mahroof et al., 2010; Ridley et al., 2016; Daglish et al., 2017). Trap catch studies have shown that *T. castaneum*, *R. dominica* and *C. ferrugineus* can be caught many kilometres away from grain storages (Ridley et al., 2011, 2016; Holloway et al., 2018). However, population genetics studies have also shown that movement by humans during grain transport or on equipment likely plays a large part in insect gene flow and probable movement of resistance genes (Ridley et al., 2016; Toon et al., 2016, 2018; McCulloch et al., 2019).

Further studies have shown that some insects will leave storages containing ample food and start flying (Gurdasani et al., 2018, 2019; Rafter et al., 2019). These insects will have been exposed to fumigant, so potentially will carry resistance gene variants. The implication of this is that any exposed grain that is infested with insects on a farm or near a storage facility must be assumed to be a reservoir of resistant insects and a source of reinfestation. Hygiene of grain storage sites is very important as elimination of spilled or spoiled grain will limit the size of remnant populations of potentially resistant insects.

Insect flight is also seasonal and highly dependent on prevailing weather conditions and temperature (Ridley et al., 2011, 2016; Daglish et al., 2017; Holloway et al., 2018). Therefore, clean-up of insect populations may have larger impacts in the cooler months when the insects are not flying and are less likely to reinfest storages. Regardless of seasonal variation, insect movement across the local landscape has implications for the spread of resistance in the local area. Your problems are also your neighbours' problems (and *vice versa*), whether they know it or not.

## *1.10 Fitness costs associated with phosphine resistance*

Many insecticide resistance-management strategies that have been developed for field crops rely on the observation that resistance variants that help insects survive exposure to an insecticide have a measurable fitness cost in the absence of insecticide exposure. A fitness cost is any physiological trait that reduces the number of offspring, dispersal or survival of resistant insects. This means that only resistant insects have a reproductive advantage over susceptible insects when the pesticide is present. In the absence of pesticide, the susceptible insects reproduce more effectively than resistant insects. As a result, in the absence of pesticide selection, the frequency of the resistance gene variant will decrease in the population over time. Several factors will have to be considered to determine whether/how this management strategy can be applied to the protection of stored grain. Does phosphine resistance incur such fitness costs?

### *1.10.1 Flight*

Resistance variants of the *rph2* gene do appear to have some associated fitness cost when two copies of the resistance variant are present (i.e. when they are homozygous). The impact of being homozygous for the resistance variant of *rph2* mostly affects flight and movement (Malekpour et al., 2016; Gurdasani et al., 2019). These effects may have something to do with the role of the *rph2* gene in energy production. In India, trapping studies in and around storage sheds (i.e. godowns) confirm that insects homozygous for *rph2* resistance variants may initiate flight less frequently and possibly fly less distance than carriers or susceptible insects (Malekpour et al., 2016, 2018; Rafter et al., 2019). Laboratory studies conducted on flight initiation are consistent with the conclusions that strongly resistant individuals may travel less, although how far they are capable of travelling is still unclear. Implications for this in insect management practice are that resistant insects may be less likely to fly away and so remnant populations maintained around storage sites (e.g. in waste grain dumps) may be more likely to retain resistant insects that can potentially reinfest disinfested grain.

### 1.10.2 Fecundity

Fecundity, the number of offspring that an individual produces, will affect how resistance genes will be passed on to succeeding generations. Presently there are varying reports of fecundity changes in pest insects. Some strongly resistant populations of *Oryzaephilus surinamensis, R. dominica, T. castaneum* (Sousa et al., 2009) and *Sitophilus zeamais* (Pimentel et al., 2009) have been seen to have variable fecundity; however, it is unclear whether this is due to background population genetics or the resistance genes themselves. Other studies conducted on laboratory strains showed no significant fecundity differences due to resistance in *R. dominica* or *T. castaneum* (Daglish et al., 2014, 2015).

However, fecundity temporarily changes after exposure to phosphine. Resistant insects treated with sublethal doses of phosphine (that would normally kill susceptible insects) show a significant reduction in offspring production (Ridley et al., 2012a,b) for about three weeks. After this period, however, the offspring production appears to recover to pre-exposure rates. This indicates that if fumigations are insufficient to kill all resistant insects, selection for strong resistance can occur fairly quickly.

## 2 The biochemical and genetic basis of resistance

Faced with increasing problems due to resistance and fumigation failures, we need to understand what the underlying biological mechanisms are that cause resistance. We may be able to use this information to overcome resistance directly. To understand phosphine resistance we have to start with what we know about the underlying genetics. We know that there are two major resistance factors called *rph1* and *rph2*. They are recessive, which means that heterozygotes ('carriers') do not exhibit the resistance trait. Moreover, these two genes can cause weak resistance when they are homozygous by themselves but can work together synergistically to cause a strong resistance trait when both genes are homozygous for resistance variants.

### 2.1 Gene identities: *rph1*

The *rph1* gene encodes an enzyme of fat metabolism, Cytochrome-b5-related protein (*Cyt-b5-r*), a fatty acid desaturase that essentially puts double bonds into fat molecules to make them more fluid (Schlipalius et al., 2018). The same double bonds that make membranes more fluid also make insects vulnerable to phosphine. Thus, gene variants that completely eliminate the activity of the enzyme cause resistance, but the decrease in double bonds in the lipids of the membrane does not appear to harm the insect. In fact, most of the resistant variants that have been found have large deletions in this gene which would make it non-functional (Schlipalius et al., 2018). It has been found that this gene is not essential for life or particularly detrimental to a healthy lifecycle and does not appear to have any fitness costs (Kaur et al., 2013a,b; Daglish et al., 2015). The vulnerability to phosphine that the normally functioning *rph1/cyt-b5-r* confers is likely because fats with double bonds (i.e. desaturated) are more sensitive to oxidative stress and reactive oxygen species (ROS). ROS is a major mediator of phosphine toxicity (Bolter and Chefurka, 1990), and oxidised fats are always seen after phosphine exposure (Chaudhry and Price, 1992; Valmas et al., 2008; Liu et al., 2015; Schlipalius et al., 2018).

## 2.2 Gene identities: rph2

The enzyme encoded by the *rph2* gene is dihydrolipoamide dehydrogenase (DLD). The DLD enzyme is a key participant in multiple enzyme complexes involved in energy metabolism (Brautigam et al., 2006; Schlipalius et al., 2012), so it is absolutely essential to life. The critical role of the DLD enzyme means that changes to the enzyme are constrained to those that do not greatly perturb normal functionality in energy metabolism (Tretter and Adam-Vizi, 2005; McLain et al., 2011b). The requirement for the enzyme to function normally limits the number of changes that can occur within the *rph2* gene to cause resistance. To date, only nine phosphine resistance mutations have been catalogued and they all occur around the active site of the DLD protein (Schlipalius et al., 2012; Nguyen et al., 2016). The small number of changes to the gene that can cause resistance to phosphine while preserving normal enzyme function, explains why it has been much slower to evolve to phosphine resistance than the *rph1* gene (Schlipalius et al., 2018).

## 2.3 Models of phosphine toxicity

The identity and function of the known resistance genes suggest a synergistic mechanism of phosphine toxicity. One of the major features of phosphine toxicity is ROS production and subsequent fat oxidation (i.e. lipid peroxidation) (Bolter and Chefurka, 1990; Chaudhry and Price, 1992; Valmas et al., 2008). Lipid peroxides are quite toxic in their own right and can interfere with energy metabolism (McLain et al., 2011a,b). It seems likely that the *rph1* gene increases the sensitivity of fat molecules of the membrane to oxidative damage unless gene variants impair the function of the encoded enzyme. The impaired function of resistance variants of *rph1* decreases the susceptibility of cellular membranes to oxidative damage, which delays lethality due to phosphine exposure. Longer fumigations extend exposure to phosphine, providing time for the toxic oxidatively damaged molecules to accumulate and interfere with additional cellular processes. This indirect mechanism would explain why phosphine is more toxic at lower concentrations than other toxic gases. It would also suggest why time is very important in achieving mortality because it is difficult to achieve a desired mortality rate (>99.9%) in short time frames (1–2 days), especially for resistant insects (Daglish and Collins, 1999).

Knowing the identities and functional roles of these phosphine resistance genes opens up opportunities to exploit previously unknown biological vulnerabilities that mutations in these genes may confer. For instance, this functional information may assist in developing alternative or synergistic chemicals and treatments. To take one example, arsenic is significantly more toxic to resistant insects and nematodes than susceptible ones (Chaudhry and Price, 1991; Schlipalius et al., 2012). Although arsenic is inherently unsuitable for use on food grains the chemistry of this interaction with resistance genes may be studied to identify phosphine synergists or alternative fumigants that can preferentially target resistant insects (Schlipalius et al., 2012).

## 3 Alternative fumigants

An alternative strategy to treat commodities that may have high frequencies of phosphine-resistant insects is to use a different fumigant. Any alternative chemical must be cost-effective and have to pass the regulatory hurdles that will make it acceptable to domestic markets and export destinations. While it is unlikely that a chemical with all of the beneficial properties of phosphine will be found, there are ways that alternative fumigants can be beneficial none-the-less. One option is to use such a chemical infrequently as a 'resistance breaker' (Nayak et al., 2016) that can be used to reduce the number of phosphine-resistant insects at a storage site. A second option is to co-fumigate with a chemical that complements the two primary weaknesses of phosphine, that is, slow action and reduced efficacy against resistant insects. A third option is to identify a chemical that synergistically enhances the toxicity of phosphine. In all cases a chemical that is more effective against phosphine-resistant insects than against phosphine-susceptible insects will be ideal as that will provide a means of driving down the frequency of phosphine-resistant insects in the population.

The number of potential alternative fumigants available for disinfesting storages is limited because of the physical, chemical, toxicological and environmental properties that are required (see Table 1) (Hagstrum and Subramanyam, 2006). The fumigant must be a gas at the fumigation temperature with a density similar to that of air, though moderate differences in density can be accommodated with forced recirculation of the gas. The chemical cannot be corrosive. It also must not adsorb onto the grain if it is to penetrate the grain bulk effectively and not leave chemical residues on the commodity (Hagstrum and Subramanyam, 2006). The chemical must be effective against all life stages of all insect pests as well as against all resistant strains of each species. The fumigant cannot leave residues that harm humans or livestock or that alter the properties of the grain for bakers, brewers or other end-users. The chemical must not persist in the environment or damage non-target species, such as aquatic organisms. The chemical should also not harm the atmosphere such as causing damage to the protective ozone layer or acting as a greenhouse gas (United Nations Environment Programme, 1992).

## 3.1 Resistance 'breakers'

### 3.1.1 Methyl bromide (MeBr)

Methyl bromide (MeBr) remains as one of the most used fumigants worldwide. However, due to its persistence in the environment and its ozone-depleting properties, its usage is now limited by the Montreal Protocol (United Nations Environment Programme, 1992). Exemptions to this restriction include its use as a quarantine treatment to ensure the insect-free status of grain that is traded internationally. It is no longer approved as a fumigant for routine use on-farm. Despite being more expensive than phosphine, it acts very quickly and is effective against all species and life stages of insect pests of stored grain. Not only is there no cross-resistance between methyl bromide and phosphine (Rajendran, 1992; Kim et al., 2019), but resistance to methyl bromide has rarely been reported (Champ and Dyte, 1977).

### 3.1.2 Sulfuryl fluoride (SF)

Sulfuryl fluoride (SF) is a promising limited-use alternative to phosphine. There is no cross-resistance between phosphine and SF (Jagadeesan and Nayak, 2017). While SF is relatively fast-acting against adults, its efficacy against eggs is less than other developmental stages. As a result, fumigations are usually extended to ensure that eggs have had time to hatch, exposing the larvae to a lethal dose of phosphine. Due to fluoride residues that remain on grains following fumigation, the usage of SF is currently limited to one fumigation per parcel of grain. SF is also reported to be a greenhouse gas that is 4000 to 5000 times more potent than $CO_2$ and persists in the atmosphere with a half-life of 30–40 years (Papadimitriou et al., 2008; Mühle et al., 2009). As this fumigant increases in importance from a break fumigant in a few countries to a chemical of widespread use, public concern about its greenhouse gas properties may increase. SF is comparatively expensive and does not enhance phosphine toxicity. It does show promise as a co-fumigant with phosphine, however, due to the complementary nature of the two compounds: phosphine is inexpensive and effective against eggs, whereas SF is effective against phosphine-resistant insects and is faster acting than phosphine (Jagadeesan et al., 2018).

### 3.1.3 Ethyl formate

Ethyl formate has been mostly developed for high-value commodities such as dried and fresh fruit. While it can be used against cereal grain pests, the physical properties of ethyl formate such as poor grain penetration, high grain sorption, high concentrations required and flammability make it difficult to use on large grain bulks (Haritos et al., 2006; Dojchinov et al., 2010). As a result, its use is focussed on farm storages of up to 200 tonnes. Ethyl formate is used at high concentrations and is extremely flammable. A mixture with $CO_2$, commercially known as Vapormate (Damcevski et al., 2010), reduces flammability and improves efficacy. In this form, it can provide an on-farm option as a break fumigant.

### 3.1.4 Carbonyl sulfide (COS) and ethanedinitrile (EDN)

Carbonyl sulfide (COS) and ethanedinitrile (EDN) were patented as grain fumigants (now lapsed) by CSIRO (Zettler et al., 1997; Bell, 2000; Weller and Morton, 2001), but were never taken through the registration process required for international adoption. These two chemicals may become important if other chemicals become ineffective or are restricted in their use by regulations. EDN devitalises grain (Ryan et al., 2006), which means that it is inappropriate for the fumigation of barley that is used for malting or for seed grain that must be able to grow when planted. Devitalisation can sometimes be an asset; for example, when grain is imported for feed and for quarantine purposes, the grain and any contaminating weed seed must be killed prior to the grain being allowed into the importing country.

### 3.1.5 Carbon dioxide ($CO_2$) and nitrogen ($N_2$)

Carbon dioxide ($CO_2$) and nitrogen ($N_2$) have been used in certified organic storage and are normal constituents of the earth's atmosphere. They are not highly toxic gases like the other fumigants and kill mostly due to asphyxiation (lack of oxygen). As such they need to be maintained at very high concentrations to displace oxygen and thus are considered 'modified atmospheres' rather than toxic fumigants. For example, $N_2$ needs to be maintained at greater than 99% concentration for more than 7–10 days to suppress insect growth (Aulicky et al., 2017). This requirement for maintaining very high concentrations for long periods of time makes them relatively expensive and suitable only for the highest quality storages as it increases the requirements for gas-tightness. This makes them less suitable as 'resistance breakers', as the requirements would mostly exclude the types of storages that are most likely to have problems, such as temporary storage types, for example, bunkers (Schlipalius et al., 2019).

## 4 Conclusion and future trends

Resistance to phosphine is a major problem due to the heavy reliance on this fumigant for pest control in stored grain. We now have DNA assays that allow us to monitor and evaluate the effectiveness of resistance management practices. While the development of these tools may seem to have occurred too late to be of much use in some instances, regions that have very dry or very cold weather, or that have employed excellent pest management practices, will have less pest pressure and therefore less phosphine usage and fewer resistance problems. Furthermore, it should be remembered that fresh grain that enters the storage system with each harvest will contain pest insects that have never been exposed to phosphine and should therefore be fully susceptible. This source of insects may serve the function of refugia as used in resistance management of pests of field crops (Gryspeirt and Grégoire, 2012). Naive insects from the field are expected to be limited in number, so they will only be able to significantly reduce the frequency of resistance at a storage facility if the resident insect populations are low. Best practice to limit insect numbers include drying and cooling of grain, maintaining excellent facility hygiene, and eliminating carry-over of grain from year to year. The same best practice management will be equally valid for any pest control system that is developed to replace phosphine once phosphine reaches the end of its useful life.

When phosphine is no longer useful, alternative chemicals will be necessary for pest control. Existing alternatives, such as methyl bromide, SF and ethyl formate, have limitations that restrict them from being used as general fumigants. Instead, they are currently deployed as break fumigants to limit the number of sequential phosphine fumigations or to control pests such as strongly resistant *C. ferrugineus* when phosphine fumigation fails. There are also examples of gases that act synergistically to improve the efficacy of phosphine. These are still experimental and as yet unpublished, but they provide hope that phosphine will be a critical contributor to pest management in stored grain for many, many years to come.

## 5 References

Agrafioti, P., Athanassiou, C. G. and Nayak, M. K. 2019. Detection of phosphine resistance in major stored-product insects in Greece and evaluation of a field resistance test kit. *Journal of Stored Products Research* 82, 40–7. doi:10.1016/J.JSPR.2019.02.004.

Arthur, F. H. and Casada, M. E. 2010. Directional flow of summer aeration to manage insect pests in stored wheat. *Applied Engineering in Agriculture* 26(1), 115–22. doi:10.13031/2013.29462.

Athanassiou, C. G., Kavallieratos, N. G., Brabec, D. L., Oppert, B., Guedes, R. N. C. and Campbell, J. F. 2019. From immobilization to recovery: towards the development of a rapid diagnostic indicator for phosphine resistance. *Journal of Stored Products Research* 80, 28–33. doi:10.1016/J.JSPR.2018.10.004.

Bell, C. H. 2000. Fumigation in the 21st century. *Crop Protection* 19(8–10), 563–9. doi:10.1016/S0261-2194(00)00073-9.

Bolter, C. J. and Chefurka, W. 1990. Extramitochondrial release of hydrogen peroxide from insect and mouse liver mitochondria using the respiratory inhibitors phosphine, myxothiazol, and antimycin and spectral analysis of inhibited cytochromes. *Archives of Biochemistry and Biophysics* 278(1), 65–72. doi:10.1016/0003-9861(90)90232-n.

Brautigam, C. A., Wynn, R. M., Chuang, J. L., Machius, M., Tomchick, D. R. and Chuang, D. T. 2006. Structural insight into interactions between dihydrolipoamide dehydrogenase (E3) and E3 binding protein of human pyruvate dehydrogenase complex. *Structure* 14(3), 611–21. doi:10.1016/J.STR.2006.01.001.

Champ, B. R. and Dyte, C. E. 1977. Fao global survey of pesticide susceptibility of stored grain pests. *FAO Plant Protection Bulletin* 25(2), 49–67.

Chaudhry, M. Q. and Price, N. R. 1991. Comparative toxicity of arsine and stibine to susceptible and phosphine-resistant strains of Rhyzopertha-Dominica (F) (Coleoptera, Bostrychidae). *Comparative Biochemistry and Physiology Part C: Comparative Pharmacology* 99(1–2), 41–5. doi:10.1016/0742-8413(91)90072-2.

Chaudhry, M. Q. and Price, N. R. 1992. Comparison of the oxidant damage induced by phosphine and the uptake and tracheal exchange of P-32 radiolabeled phosphine in the susceptible and resistant strains of Rhyzopertha-Dominica (F) (Coleoptera, Bostrychidae). *Pesticide Biochemistry and Physiology* 42(2), 167–79. doi:10.1016/0048-3575(92)90063-6.

Chen, Z., et al. 2015. Diagnostic molecular markers for phosphine resistance in US populations of *Tribolium castaneum* and *Rhyzopertha dominica*. *PLoS ONE* 10(3). doi:10.1371/journal.pone.0121343.

Collins, P. J. 2009. *Strategy to Manage Resistance to Phosphine in the Australian Grain Industry, Cooperative Research Centre for National Plant Biosecurity*. Available at: https://www.graintrade.org.au/sites/default/files/file/NWPGP/NWPGP2017/Chemical. ResistenceManagementStrategy-June2017.pdf.

Collins, P. J., et al. 2001. Combating resistance to phosphine in Australia. In: Donahaye, E. J., Navarro, S. and Leesch, J. G. (Eds), *Proceedings of the International Conference Controlled Atmosphere and Fumigation in Stored Products*. Executive Printing Services, Fresno, CA, Clovis, CA, USA, pp. 593–607. Available at: https://www.researchgate.net/publication/228515160 (Accessed on 24 June 2019).

Collins, P. J., Daglish, G. J., Pavic, H. and Kopittke, R. A. 2005. Response of mixed-age cultures of phosphine-resistant and susceptible strains of lesser grain borer, Rhyzopertha dominica, to phosphine at a range of concentrations and exposure periods. *Journal of Stored Products Research* 41(4), 373–85. doi:10.1016/j.jspr.2004.05.002.

Collins, P. J., Falk, M. G., Nayak, M. K., Emery, R. N. and Holloway, J. C. 2017. Monitoring resistance to phosphine in the lesser grain borer, Rhyzopertha dominica, in Australia: a national analysis of trends, storage types and geography in relation to resistance detections. *Journal of Stored Products Research* 70, 25–36. doi:10.1016/J.JSPR.2016.10.006.

Daglish, G. J. and Collins, P. J. 1999. Improving the relevance of assays for phosphine resistance. In: Zuxun, J., et al. (Eds), *Proceedings of the 7th International Working Conference on Stored-Product Protection*. Sichuan Publishing House of Science and Technology, Beijing, China, Chengdu and China, pp. 584–93.

Daglish, G. J., Collins, P. J., Pavic, H. and Kopittke, R. A. 2002. Effects of time and concentration on mortality of phosphine-resistant Sitophilus oryzae (L) fumigated with phosphine. *Pest Management Science* 58(10), 1015–21. doi:10.1002/ps.532.

Daglish, G. J., Nayak, M. K. and Pavic, H. 2014. Phosphine resistance in Sitophilus oryzae (L.) from eastern Australia: inheritance, fitness and prevalence. *Journal of Stored Products Research* 59, 237–44. doi:10.1016/j.jspr.2014.03.007.

Daglish, G. J., Nayak, M. K., Pavic, H. and Smith, L. W. 2015. Prevalence and potential fitness cost of weak phosphine resistance in Tribolium castaneum (Herbst) in eastern Australia. *Journal of Stored Products Research* 61, 54–8. doi:10.1016/j.jspr.2014.11.005.

Daglish, G. J., Ridley, A. W., Reid, R. and Walter, G. H. 2017. Testing the consistency of spatio-temporal patterns of flight activity in the stored grain beetles Tribolium castaneum (Herbst) and Rhyzopertha dominica (f.). *Journal of Stored Products Research* 72, 68–74. doi:10.1016/j.jspr.2017.03.005.

Damcevski, K. A., Dojchinov, G., Woodman, J. D. and Haritos, V. S. 2010. Efficacy of vaporised ethyl formate/carbon dioxide formulation against stored-grain insects: effect of fumigant concentration, exposure time and two grain temperatures. *Pest Management Science* 66(4), 432–8. doi:10.1002/Ps.1895.

Denholm, I. and Rowland, M. W. 1992. Tactics for managing pesticide resistance in arthropods: theory and practice. *Annual Review of Entomology* 37(1), 91–112. doi:10.1146/annurev.en.37.010192.000515.

Dojchinov, G., Damcevski, K. A., Woodman, J. D. and Haritos, V. S. 2010. Field evaluation of vaporised ethyl formate and carbon dioxide for fumigation of stored wheat. *Pest Management Science* 66(4), 417–24. doi:10.1002/Ps.1893.

Edde, P. A., Phillips, T. W., Nansen, C. and Payton, M. E. 2006. Flight activity of the lesser grain borer, Rhyzopertha dominica F. (Coleoptera : Bostrichidae). *Environmental Entomology* 35(3), 616–24. doi:10.1603/0046-225X-35.3.616.

Flinn, P. W., Hagstrum, D. W. and Muir, W. E. 1997. Effects of time of aeration, bin size, and latitude on insect populations in stored wheat: a Simulation Study. *Journal of Economic Entomology* 90(2), 646–51. doi:10.1093/jee/90.2.646.

Gryspeirt, A. and Grégoire, J. C. 2012. Effectiveness of the high dose/refuge strategy for managing pest resistance to Bacillus thuringiensis (Bt) plants expressing one or two toxins. *Toxins* 4(10), 810–35. doi:10.3390/toxins4100810.

Gurdasani, K., Rafter, M. A., Daglish, G. J. and Walter, G. H. 2018. Characterising the variables associated with Tribolium castaneum adults that initiate flight in laboratory tests - generating predictions for the field. *Journal of Stored Products Research* 79, 123–31. doi:10.1016/J.JSPR.2018.10.005.

Gurdasani, K., Rafter, M. A., Daglish, G. J. and Walter, G. H. 2019. The dispersal flight of Tribolium castaneum – a field test of laboratory generated predictions. *Journal of Stored Products Research* 83, 25–33. doi:10.1016/J.JSPR.2019.05.009.

Hagstrum, D. W. and Subramanyam, B. 2006. Fumigants and modified atmospheres. In: *Fundamentals of Stored-Product. Entomology*. Elsevier, pp. 193–9. doi:10.1016/b978-1-891127-50-2.50024-7.

Haritos, V. S., Damcevski, K. A. and Dojchinov, G. 2006. Improved efficacy of ethyl formate against stored grain insects by combination with carbon dioxide in a "dynamic" application. *Pest Management Science* 62(4), 325–33. doi:10.1002/ps.1167.

Holloway, J. C., Falk, M. G., Emery, R. N., Collins, P. J. and Nayak, M. K. 2016. Resistance to phosphine in Sitophilus oryzae in Australia: a national analysis of trends and frequencies over time and geographical spread. *Journal of Stored Products Research* 69, 129–37. doi:10.1016/j.jspr.2016.07.004.

Holloway, J. C., Mayer, D. G. and Daglish, G. J. 2018. Flight activity of Cryptolestes ferrugineus in southern New South Wales, Australia. *Journal of Pest Science* 91(4), 1353–62. doi:10.1007/s10340-018-0981-1.

Jagadeesan, R. and Nayak, M. K. 2017. Phosphine resistance does not confer cross-resistance to sulfuryl fluoride in four major stored grain insect pests. *Pest Management Science* 73(7), 1391–401. doi:10.1002/ps.4468.

Jagadeesan, R., Collins, P. J., Daglish, G. J., Ebert, P. R. and Schlipalius, D. I. 2012. Phosphine resistance in the rust red flour beetle, Tribolium castaneum (Coleoptera: Tenebrionidae): inheritance, gene interactions and fitness costs. *PLoS ONE* 7(2), e31582. doi:10.1371/journal.pone.0031582.

Jagadeesan, R., Singarayan, V. T., Chandra, K., Ebert, P. R. and Nayak, M. K. 2018. Potential of co-fumigation with phosphine (PH3) and sulfuryl fluoride (SO2F2) for the management of strongly phosphine-resistant insect pests of stored grain. *Journal of Economic Entomology* 111(6), 2956–65. doi:10.1093/jee/toy269.

Kaur, R. and Nayak, M. K. 2015. Developing effective fumigation protocols to manage strongly phosphine-resistant *Cryptolestes ferrugineus* (Stephens) (Coleoptera: Laemophloeidae). *Pest Management Science* 71(9), 1297–302. doi:10.1002/ps.3926.

Kaur, R., Daniels, E. V., Nayak, M. K., Ebert, P. R. and Schlipalius, D. I. 2013a. Determining changes in the distribution and abundance of a Rhyzopertha dominica phosphine resistance allele in farm grain storages using a DNA marker. *Pest Management Science* 69(6), 685–8. doi:10.1002/ps.3514.

Kaur, R., Ebert, P. R., Walter, G. H., Swain, A. J. and Schlipalius, D. I. 2013b. Do phosphine resistance genes influence movement and dispersal under starvation? *Journal of Economic Entomology* 106(5), 2259–66. doi:10.1603/EC13234.

Kaur, R., Subbarayalu, M., Jagadeesan, R., Daglish, G. J., Nayak, M. K., Naik, H. R., Ramasamy, S., Subramanian, C., Ebert, P. R. and Schlipalius, D. I. 2015. Phosphine resistance in India is characterised by a dihydrolipoamide dehydrogenase variant that is otherwise unobserved in eukaryotes. *Heredity* 115(3), 188–94. doi:10.1038/hdy.2015.24.

Kim, B. S., Song, J. E., Park, J. S., Park, Y., Shin, E. M. and Yang, J. 2019. Insecticidal effects of fumigants (EF, MB, and 3) towards phosphine-susceptible and -resistant *Sitophilus oryzae* (Coleoptera: Curculionidae). *Insects* 10(10). doi:10.3390/insects10100327.

Koçak, E., Schlipalius, D., Kaur, R., Tuck, A., Ebert, P., Collins, P. and Yılmaz, A. 2015. Determining phosphine resistance in rust red flour beetle, *Tribolium castaneum* (Herbst.) (Coleoptera : Tenebrionidae) populations from Turkey. *Turkish Journal of Entomology* 39(2). doi:10.16970/ted.17464.

Lilford, K., et al. 2009. Fumigation of stored-grain insects – a two locus model of phosphine resistance. *18th World Imacs Congress and Modsim09 International Congress on Modelling and Simulation*, pp. 540–6.

Liu, T., Li, L., Zhang, F. and Wang, Y. 2015. Transcriptional inhibition of the Catalase gene in phosphine-induced oxidative stress in *Drosophila melanogaster*. *Pesticide Biochemistry and Physiology* 124, 1–7. doi:10.1016/j.pestbp.2015.05.005.

Mahroof, R. M., Edde, P. A., Robertson, B., Puckette, J. A. and Phillips, T. W. 2010. Dispersal of *Rhyzopertha dominica* (Coleoptera: Bostrichidae) in different habitats. *Environmental Entomology* 39(3), 930–8. doi:10.1603/EN09243.

Malekpour, R., Rafter, M. A., Daglish, G. J. and Walter, G. H. 2016. Influence of phosphine resistance genes on flight propensity and resource location in *Tribolium castaneum* (Herbst) (Coleoptera: Tenebrionidae): the landscape for selection. *Biological Journal of the Linnean Society* 119(2), 348–58. doi:10.1111/bij.12817.

Malekpour, R., Rafter, M. A., Daglish, G. J. and Walter, G. H. 2018. The movement abilities and resource location behaviour of *Tribolium castaneum*: phosphine resistance and its genetic influences. *Journal of Pest Science* 91(2), 739–49. doi:10.1007/s10340-017-0935-z.

Mau, Y. S., Collins, P. J., Daglish, G. J., Nayak, M. K., Pavic, H. and Ebert, P. R. 2012. The rph1 gene is a common contributor to the evolution of phosphine resistance in independent field isolates of *Rhyzopertha dominica*. *PLoS ONE* 7(2), e31541. doi:10.1371/journal.pone.0031541.

McCulloch, G. A., Mohankumar, S., Subramanian, S., Rajan, T. S., Rahul, C., Surendran, R., Gaurav, R., Chandrasekaran, S., Daglish, G. J. and Walter, G. H. 2019. Contrasting patterns of phylogeographic structuring in two key beetle pests of stored grain in India and Australia. *Journal of Pest Science* 92(3), 1249–59. doi:10.1007/s10340-019-01121-9.

McLain, A. L., Humphries, K. M., Stofan, D., Kinter, M. T. and Szweda, L. I. 2011a. Inactivation of alpha-Ketoglutarate dehydrogenase by 4-Hydroxy-2-nonenal; modification and metabolic fate of the cofactor lipoic acid. *Free Radical Biology and Medicine* 51, S30. doi:10.1016/j.freeradbiomed.2011.10.073.

McLain, A. L., Szweda, P. A. and Szweda, L. I. 2011b. alpha-Ketoglutarate dehydrogenase: a mitochondrial redox sensor. *Free Radical Research* 45(1), 29–36. doi:10.3109/10715762.2010.534163.

Mühle, J., Huang, J., Weiss, R. F., Prinn, R. G., Miller, B. R., Salameh, P. K., Harth, C. M., Fraser, P. J., Porter, L. W., Greally, B. R., O'Doherty, S. and Simmonds, P. G. 2009. Sulfuryl fluoride in the global atmosphere. *Journal of Geophysical Research* 114(D5), D05306. doi:10.1029/2008JD011162.

Nayak, M. K. and Collins, P. J. 2008. Influence of concentration, temperature and humidity on the toxicity of phosphine to the strongly phosphine-resistant psocid *Liposcelis bostrychophila* Badonnel (Psocoptera : Liposcelididae). *Pest Management Science* 64(9), 971–6. doi:10.1002/Ps.1586.

Nayak, M. K. and Daglish, G. J. 2017. Base-line susceptibility of field populations of *Rhyzopertha dominica* (F.) to spinosad in Australia. *Journal of Stored Products Research* 70, 1–6. doi:10.1016/j.jspr.2016.10.005.

Nayak, M. K., Collins, P. J. and Reid, S. R. 1998. Efficacy of grain protectants and phosphine against *Liposcelis bostrychophila*, *L-entomophila*, and *L-paeta* (Psocoptera : Liposcelidae). *Journal of Economic Entomology* 91(5), 1208–12. doi:10.1093/jee/91.5.1208.

Nayak, M. K., Holloway, J. C., Emery, R. N., Pavic, H., Bartlet, J. and Collins, P. J. 2013. Strong resistance to phosphine in the rusty grain beetle, *Cryptolestes ferrugineus* (Stephens) (Coleoptera: Laemophloeidae): its characterisation, a rapid assay for diagnosis and its distribution in Australia. *Pest Management Science* 69(1), 48–53. doi:10.1002/ps.3360.

Nayak, M. K., Jagadeesan, R., Kaur, R., Daglish, G. J., Reid, R., Pavic, H., Smith, L. W. and Collins, P. J. 2016. Use of sulfuryl fluoride in the management of strongly phosphine-resistant insect pest populations in bulk grain storages in Australia. *Indian Journal of Entomology* 78(special), 100. doi:10.5958/0974-8172.2016.00030.4.

Nayak, M. K., Falk, M. G., Emery, R. N., Collins, P. J. and Holloway, J. C. 2017. An analysis of trends, frequencies and factors influencing the development of resistance to phosphine in the red flour beetle *Tribolium castaneum* (Herbst) in Australia. *Journal of Stored Products Research* 72, 35–48. doi:10.1016/J.JSPR.2017.03.004.

Nguyen, T. T., Collins, P. J. and Ebert, P. R. 2015. Inheritance and characterization of strong resistance to phosphine in *Sitophilus oryzae* (L.). *PLoS ONE* 10(4), e0124335. doi:10.1371/journal.pone.0124335.

Nguyen, T. T., Collins, P. J., Duong, T. M., Schlipalius, D. I. and Ebert, P. R. 2016. Genetic conservation of phosphine resistance in the rice weevil *Sitophilus oryzae* (L.). *The Journal of Heredity* 107(3), 228–37. doi:10.1093/jhered/esw001.

Papadimitriou, V. C., Portmann, R. W., Fahey, D. W., Mühle, J., Weiss, R. F. and Burkholder, J. B. 2008. Experimental and theoretical study of the atmospheric chemistry and global warming potential of $SO_2F_2$. *The Journal of Physical Chemistry: A* 112(49), 12657–66. doi:10.1021/jp806368u.

Pimentel, M. A. G., Faroni, L. R. D., Guedes, R. N. C., Sousa, A. H. and Tótola, M. R. 2009. Phosphine resistance in Brazilian populations of *Sitophilus zeamais* Motschulsky (Coleoptera: Curculionidae). *Journal of Stored Products Research* 45(1), 71–4. doi:10.1016/j.jspr.2008.09.001.

Radek, A., Vlastimil, K., Jan, P. and Vaclav, S. 2017. Field efficacy of brief exposure of adults of six storage pests to nitrogen-controlled atmospheres. *Plant Protection Science* 53(3), 169–76. doi:10.17221/136/2016-PPS.

Rafter, M. A., McCulloch, G. A., Daglish, G. J. and Walter, G. H. 2017. Progression of phosphine resistance in susceptible *Tribolium castaneum* (Herbst) populations under different immigration regimes and selection pressures. *Evolutionary Applications* 10(9), 907–18. doi:10.1111/eva.12493.

Rafter, M. A., Muralitharan, V., Chandrasekaran, S., Mohankumar, S., Daglish, G. J., Loganathan, M. and Walter, G. H. 2019. Behaviour in the presence of resource excess–flight of *Tribolium castaneum* around heavily-infested grain storage facilities. *Journal of Pest Science* 92(3), 1227–38. doi:10.1007/s10340-019-01085-w.

Rajendran, S. 1992. Selection for resistance to phosphine or methyl-bromide in *Tribolium-castaneum* (Coleoptera, Tenebrionidae). *Bulletin of Entomological Research* 82(1), 119–24. doi:10.1017/S000748530005152X.

Ramya, R. S., Srivastava, C. and Subramanian, S. 2018. Monitoring of phosphine resistance in Indian populations of *Tribolium castaneum* (Herbst) from stored wheat. *Indian Journal of Entomology* 80(1), 19–23. doi:10.5958/0974-8172.2018.00005.6.

Ridley, A. W., Hereward, J. P., Daglish, G. J., Raghu, S., Collins, P. J. and Walter, G. H. 2011. The spatiotemporal dynamics of *Tribolium castaneum* (Herbst): adult flight and gene flow. *Molecular Ecology* 20(8), 1635–46. doi:10.1111/j.1365-294X.2011.05049.x.

Ridley, A. W., Magabe, S., Schlipalius, D. I., Rafter, M. A. and Collins, P. J. 2012a. Sublethal exposure to phosphine decreases offspring production in strongly phosphine resistant female red flour beetles, *Tribolium castaneum* (Herbst). *PLoS ONE* 7(12), e53356. doi:10.1371/journal.pone.0053356.

Ridley, A. W., Schlipalius, D. I. and Daglish, G. J. 2012b. Reproduction of phosphine resistant *Rhyzopertha dominica* (F.) following sublethal exposure to phosphine. *Journal of Stored Products Research* 48, 106–10. doi:10.1016/j.jspr.2011.10.005.

Ridley, A. W., Hereward, J. P., Daglish, G. J., Raghu, S., McCulloch, G. A. and Walter, G. H. 2016. Flight of *Rhyzopertha dominica* (Coleoptera: Bostrichidae)-a spatio-temporal analysis with pheromone trapping and population genetics. *Journal of Economic Entomology* 109(6), 2561-71. doi:10.1093/jee/tow226.

Ryan, R., et al. 2006. Sterigas™ and Cosmic™: update on proposed new fumigants. In: Lorini, I., et al. (Eds), *Proceedings of the 9th International Working Conference on Stored Product Protection*, Sao Paulo, pp. 624–9. Available at: http://bru.gmprc.ksu.edu/proj/iwcspp/pdf2/9/6165.pdf (Accessed on 24 June 2019).

Schlipalius, D., et al. 2008. DNA testing for phosphine resistance- the future of resistance monitoring and management. In: *International Conference on Controlled Atmospheres and Fumigation in Stored Products*, Chengdu, China, pp. 595–8.

Schlipalius, D. I., Cheng, Q., Reilly, P. E., Collins, P. J. and Ebert, P. R. 2002. Genetic linkage analysis of the lesser grain borer *Rhyzopertha dominica* identifies two loci that confer high-level resistance to the fumigant phosphine. *Genetics* 161(2), 773–82.

Schlipalius, D. I., Chen, W., Collins, P. J., Nguyen, T., Reilly, P. E. and Ebert, P. R. 2008. Gene interactions constrain the course of evolution of phosphine resistance in the lesser grain borer, *Rhyzopertha dominica*. *Heredity* 100(5), 506–16. doi:10.1038/Hdy.2008.4.

Schlipalius, D. I., Valmas, N., Tuck, A. G., Jagadeesan, R., Ma, L., Kaur, R., Goldinger, A., Anderson, C., Kuang, J., Zuryn, S., Mau, Y. S., Cheng, Q., Collins, P. J., Nayak, M. K., Schirra, H. J., Hilliard, M. A. and Ebert, P. R. 2012. A core metabolic enzyme mediates resistance to phosphine gas. *Science* 338(6108), 807–10. doi:10.1126/science.1224951.

Schlipalius, D. I., Tuck, A. G., Jagadeesan, R., Nguyen, T., Kaur, R., Subramanian, S., Barrero, R., Nayak, M. and Ebert, P. R. 2018. Variant linkage analysis using *de novo* transcriptome sequencing identifies a conserved phosphine resistance gene in insects. *Genetics* 209(1), 281-90. doi:10.1534/genetics.118.300688.

Schlipalius, D. I., Tuck, A. G., Pavic, H., Daglish, G. J., Nayak, M. K. and Ebert, P. R. 2019. A high-throughput system used to determine frequency and distribution of phosphine resistance across large geographical regions. *Pest Management Science* 75(4), 1091–8. doi:10.1002/ps.5221.

Sher, F., Shakoori, A. and Ali, S. 2003. Toxicity of phosphine against 4 instar larvae of *Trogoderma granarium* collected from different godowns of Punjab. In: *Proceedings of the Pakistan Congress of Zoology Volume 23*. University Arid Agr, Rawalpindi, Pakistan, pp. 67-25.

Shi, M. R., Collins, P. J., Ridsdill-Smith, J. and Renton, M. 2012. Individual-based modelling of the efficacy of fumigation tactics to control lesser grain borer (*Rhyzopertha dominica*) in stored grain. *Journal of Stored Products Research* 51, 23-32. doi:10.1016/j.jspr.2012.06.003.

Shi, M. R., Collins, P. J., Ridsdill-Smith, T. J., Emery, R. N. and Renton, M. 2013. Dosage consistency is the key factor in avoiding evolution of resistance to phosphine and population increase in stored-grain pests. *Pest Management Science* 69(9), 1049-60. doi:10.1002/ps.3457.

Sousa, A. H., Faroni, L. R. D., Pimentel, M. A. G. and Guedes, R. N. C. 2009. Developmental and population growth rates of phosphine-resistant and -susceptible populations of stored-product insect pests. *Journal of Stored Products Research* 45(4), 241-6. doi:10.1016/j.jspr.2009.04.003.

Stevens, M. M., Warren, G. N., Mo, J. and Schlipalius, D. I. 2011. Maintaining DNA quality in stored-grain beetles caught in Lindgren funnel traps. *Journal of Stored Products Research* 47(2), 69-75. doi:10.1016/j.jspr.2010.10.002.

Toon, A., Daglish, G. J., Ridley, A. W., Emery, R. N., Holloway, J. C. and Walter, G. H. 2016. Random mating between two widely divergent mitochondrial lineages of *Cryptolestes ferrugineus* (Coleoptera: Laemophloeidae): a test of species limits in a phosphine-resistant stored product pest. *Journal of Economic Entomology* 109(5), 2221-8. doi:10.1093/jee/tow178.

Toon, A., Daglish, G. J., Ridley, A. W., Emery, R. N., Holloway, J. C. and Walter, G. H. 2018. Significant population structure in Australian *Cryptolestes ferrugineus* and interpreting the potential spread of phosphine resistance. *Journal of Stored Products Research* 77, 219-24. doi:10.1016/j.jspr.2018.05.004.

Tretter, L. and Adam-Vizi, V. 2005. Alpha-ketoglutarate dehydrogenase: a target and generator of oxidative stress. *Philosophical Transactions of the Royal Society of London: Series B, Biological Sciences* 360(1464), 2335-45. doi:10.1098/rstb.2005.1764.

United Nations Environment Programme. 1992. *Montreal Protocol on Substances that Deplete the Ozone Layer*. United Nations Environment Programme, Nairobi, Kenya.

Valmas, N., Zuryn, S. and Ebert, P. R. 2008. Mitochondrial uncouplers act synergistically with the fumigant phosphine to disrupt mitochondrial membrane potential and cause cell death. *Toxicology* 252(1-3), 33-9. doi:10.1016/j.tox.2008.07.060.

Weller, G. L. and Morton, R. 2001. Fumigation with carbonyl sulfide: a model for the interaction of concentration, time and temperature. *Journal of Stored Products Research* 37(4), 383-98. Available at: http://www.ncbi.nlm.nih.gov/pubmed/11463400 (Accessed on 12 December 2018). doi:10.1016/s0022-474x(00)00041-2.

Winks, R. G. and Waterford, C. J. 1986. The relationship between concentration and time in the toxicity of phosphine to adults of a resistant strain of *Tribolium castaneum* (Herbst). *Journal of Stored Products Research* 22(2), 85-92. doi:10.1016/0022-474X(86)90024-X.

Zettler, L. J., Leesch, J. G., Gill, R. F. and Mackey, B. E. 1997. Toxicity of carbonyl sulfide to stored product insects. *Journal of Economic Entomology* 90(3), 832-6. doi:10.1093/jee/90.3.832.

# Chapter 4

## Advances in understanding rodent pests affecting cereal grains

*Peter R. Brown, CSIRO Health and Biosecurity, Australia; Grant R. Singleton, International Rice Research Institute, The Philippines; Steven R. Belmain, Natural Resources Institute (NRI) – University of Greenwich, UK; Nyo Me Htwe, Plant Protection Division – Myanma Agriculture Service, Myanmar; Loth Mulungu, Sokoine University of Agriculture, Tanzania; Mashaka Mdangi, Ministry of Agriculture, Tanzania; and Regino Cavia, University of Buenos Aires, Argentina*

1 Introduction
2 Impact of rodents on stored grain
3 Rodent pest species and their biology
4 Types of rodent damage
5 Strategies to reduce rodent damage
6 Case studies
7 Conclusion
8 Future trends in research
9 Where to look for further information
10 References

## 1 Introduction

Rodents have been eating our crops, ravaging our grain stores and taking food from our tables for thousands of years. The damage that rodents cause to cereal crops in the field is reasonably well known (pre-harvest losses are 5–15% and up to 100% in some situations; Singleton et al., 2010; John, 2014; Brown et al., 2017), but there is very little understanding about the level of damage that rodents cause to cereals post-harvest; estimates vary widely. This is in contrast to the well-studied post-harvest impacts caused by insect pests (see other chapters in this volume). In this chapter, we review the current knowledge of post-harvest storage impacts by rodents, describe the main pest species, the types of damage they cause, and consider some strategies to manage rodents. We mainly focus on post-harvest impacts by rodents in developing countries, because this is where their impacts are greatest.

http://dx.doi.org/10.19103/AS.2020.0072.04

Two earlier reviews of post-harvest losses caused by rodents in cereal systems in sub-tropical and tropical systems (Hopf et al., 1976) and rice in Asia (Singleton, 2003) highlighted that losses are likely to be as high as pre-harvest losses, but there was a dearth of quantitative studies on post-harvest losses. One global estimate was that vertebrates (mainly rodents) caused 5% (approximately 500 million tons) of post-harvest food losses (Cao et al., 2002). Post-harvest losses in some regions may be greater than pre-harvest losses, and reports of 20% losses caused by rodents to grain after harvest are not unusual (Aplin et al., 2003). The FAO estimated that rodent-related post-harvest losses vary profoundly (3–50%) depending on the specific local conditions (Brooks and Fiedler, 1999). Whatever the figure, farmers have gone to the effort of growing their crop, harvesting it, threshing it and then transporting it to grain stores, so the impact can be much greater because the extra time, effort and expense invested have been wasted.

Lund (2015) summarised post-harvest losses for a range of commodities in many tropical and subtropical countries based on earlier work published by Hopf et al. (1976). The main types of storage were stacks, sacks, cribs, warehouses, granaries and temporary structures. The main grain commodities were 'cereals', but rice, maize and barley were listed. The rates of damage or loss ranged from 0.5% to 25% and averaged around 6% (Hopf et al., 1976; Cao et al., 2002; Lund, 2015). Shankar and Abrol (2012) provide a much lower estimate of rodent losses of 2.5% during post-harvest operations (out of a total yield loss of 9.33%); of note is that rodent losses were nearly equal to losses caused by insects (at 2.55%). Given the similar levels of damage with insects, it is surprising that very little attention has been paid to understanding rodent pests in grain storage systems compared to the well-established work on insect pests (in terms of quantifying damage and research and development for improved management). This is further compounded by the difficulty of accurately identifying rodent damage and that farmers are often not aware of the magnitude of chronic losses to their grain stores caused by rodents because they have always lived with such losses.

Rodents cause a range of types of damage to stored grain. Commensal rats and mice are the primary culprits (Table 1). Rodent damage to stored foodstuffs far exceeds the actual value of what they consume. There are direct losses to grain (direct consumption), but there is also spoilage (contamination of grain from hair, urine and faeces) and partial damage which opens opportunities for other pests and diseases to cause further damage (such as insects and moulds). Moreover, rodent gnawing in seed/grain storage structures increases the moisture content of seeds and reduces germination viability of seeds. Spoilage from rodent hair, urine and faeces also increases the risk of disease transfer to humans or livestock (Lovera et al., 2017). Rodents

**Table 1** Rodent damage/losses to grains in storage facilities

| Country | Species | Damage or loss | Grain and storage structure type | Source |
|---|---|---|---|---|
| **Asia** | | | | |
| Bangladesh | *B. bengalensis* | 13% | Lowland | Aplin et al. (2003) |
| | *B. bengalensis* | 5–10% | Irrigated | http://india.leisa.info/index.php?url=getblob.php&o_id =204244&a_id=211&a_seq=0 |
| India | *R. rattus* | 8.62–9.5% | Nadia | Santra and Manna (2008) |
| | *R. rattus*<br>*M. musculus* | 5.75–6.5% | Burdwan | Santra and Manna (2008) |
| | Not stated | 25–30% | Lowland | Kushwaha (1986) |
| | *R. rattus*<br>*M. musculus*<br>*B. bengalensis* | 2.5% | Storage: residential premises and farm-level storage | Rao (2003) |
| | *R. rattus*<br>*M. musculus*<br>*B. bengalensis* | Not stated | Rural, urban, warehouses and godowns | Shankar and Abrol (2012) |
| | Not stated | 25–30% | | Hart (2001) |
| Indonesia | *R. rattus diardii*<br>*M. musculus*<br>(+ *Suncus murinus*, an insectivore) | Not stated | | Sidik et al. (1986) |
| | Not stated | 5–10% | Lowland rice (Sulawesi) | Baco et al. (2010) |
| Korea | *R. norvegicus*<br>*R. rattus*<br>*M. molissimus* | 20% widespread in rural areas | | Pest Control Section, Plant Protection Division, Agricultural Production Bureau, Ministry of Agriculture and Forestry, Seoul, Korea (Singleton 2003) |
| Laos | *Rattus* spp.<br>*Mus* spp. | Up to 10% | Widespread | Direction de'l Agriculture, Vientiane, Laos (Singleton 2003) |
| | *R. rattus* | 7.4–10% | | Brown et al. (2013) |
| Malaysia | *R. exulans* | Not stated | Common | Crop Protection, Department of Agriculture, Kuala Lumpur, Malaysia (Singleton 2003) |
| | *R. rattus diardii*<br>*R. exulans*<br>*R. norvegicus*<br>*M. musculus* | Not stated | | Muda (1986) |

(*Continued*)

**Table 1** (*Continued*)

| Country | Species | Damage or loss | Grain and storage structure type | Source |
|---|---|---|---|---|
| Myanmar | *B. bengalensis*<br>*R. rattus* | 7-14% | | Belmain at al. (2015), Htwe et al. (2017) |
| | *R. rattus*<br>*R. exulans*<br>*M. musculus*<br>*B. bengalensis*<br>*B. indica* | 1-14% | | Htwe et al. (2017) |
| Philippines | *R. norvegicus*<br>*M. musculus* | 5% | | Bureau of Plant Industry, Region VI, Iloilo City, Philippines (Singleton 2003) |
| Sri Lanka | Still in progress | 8% | | Htwe (unpublished data) |
| Thailand | *R. norvegicus*<br>*R. rattus*<br>*R. exulans*<br>*Bandicota* sp.<br>*Mus* spp. | 5% | Widespread | Plant Pest Control Research Centre, Plant Industry Division, Department of Agriculture, Bangkhen, Thailand (Singleton 2003) |
| **Africa** | | | | |
| Egypt | Not stated | 50% | Maize, wheat, rice, cottonseed in houses and stores | Hopf et al. (1976) |
| Ethiopia | Not stated | 5-15% | Grains in huts on stilts and underground bags | Hopf et al. (1976) |
| | Rodents and other pests (not insects) | 12-20% | Maize, sorghum, teff, barley and wheat in bags in houses | Hengsdijk and de Boer (2017) |
| Ghana | Not stated | 2-3% | Grains, maize, rice | Hopf et al. (1976) |
| Malawi | Not stated | 0.5-1.5% | Maize, groundnuts, sorghum, millet in woven cane bins, grass baskets | Hopf et al. (1976) |
| | Not stated | 15% | Cob maize in cribs | Hopf et al. (1976) |
| Sierra Leone | Not stated | 1-10% | Rice in cane baskets | Hopf et al. (1976) |
| | Not stated | 10-100% | Rice in sacks | Hopf et al. (1976) |
| | Not stated | 2-3% | Rice and maize in roof and cribs | Hopf et al. (1976) |
| Tanzania | *R. rattus, M. natalensis* | 20-60% | Rural household stores | Mdangi et al., (2013, 2016); Mulungu et al. (2015) |

| Country | Species | Damage or loss | Grain and storage structure type | Source |
|---|---|---|---|---|
| Zaire | Not stated | 3% | Rice and maize in bags in roof | Hopf et al. (1976) |
| Zambia | Not stated | 10% | Cob maize, sorghum, millet in farm cribs | Hopf et al. (1976) |
| **Latin America** | | | | |
| Argentina | Not stated | Not stated | Large silo bags (200-ton capacity) (corn, soybean, forage silage and wheat grains) (survey of stakeholder perceptions: rodents ranked second after armadillos; 14% of respondents claimed that rodents were the main harmful species) | Zufiaurre et al. (2019) |
| Brazil | Not stated | 4–8% | Rice, maize and beans in stacks, sacks and cribs | Hopf et al. (1976) |
| Mexico | Not stated | 5–10% | Rice and maize in cribs, sacks in roofs | Hopf et al. (1976) |
| Latin America (in general) | *Rattus* spp. *Mus* spp. *Heteromys* spp. *Peromyscus* spp. plus others (see Table 2 for details) | Not stated | Grain stores, farmer's houses | Elias (1984), Rodriguez Muñoz (1993), Brooks and Fiedler (1999) |

*Source*: rearranged and updated after Buckle and Smith (1994), Hopf et al. (1976) and Singleton (2003).

can transfer around 60 diseases to humans including typhoid, paratyphoid, trichinosis, scabies, plague and haemorrhagic fevers like Lassa fever and leptospirosis (Cao et al., 2002; Meerburg et al., 2009). This is particularly dramatic in the context of a developing country; if the main income earner becomes sick, the family can fall into a debt trap (Singleton et al., 2010). This is further compounded because poorly resourced health systems cannot guarantee an appropriate action.

Rodents are one of the few orders of mammals that have continually growing incisor teeth (the other being lagomorphs such as rabbits and hares). To wear down their continuously growing incisor teeth, rodents can gnaw (chew) anything that is not harder than their teeth, which includes copper or iron. These sharp incisor teeth enable rodents to not only eat tough seeds but also cause damage to storage structures made of wood, plastic, plaster and some metals.

The greatest impact of rodent damage occurs in developing countries, particularly in tropical and sub-tropical regions, where facilities are not well developed and where this damage directly affects the livelihoods and food security of rural households and communities (Singleton et al., 2010). In many of these countries, the infrastructure for grain storage is rudimentary and can be as simple as a woven basket in a thatched house or a simple wooden or corrugated iron grain store, which is no match for rodents. The main staple cereal grain stores are all affected by rodents and include all the major grain types, including maize, rice, wheat, barley, sorghum, millet, oats, rye and triticale (other raw or processed food commodities are also affected, including grain pulses, root crops, fruit, vegetables and nuts). Grain is stored either in houses or in adjacent small grain storage structures made with locally available materials. The extent of stored grain and seed losses during storage depends on the farmers' practices of harvesting; storage structure; sanitation around the storage area; the duration of storage; and the distribution, abundance and species composition of rodents in the area (Belmain et al., 2006; Htwe et al., 2017). Losses by rodents in developed countries can also be high, but, again, there are very few quantitative studies. In Australia, occasional outbreaks of house mouse populations can lead to problems of contamination of wheat with carcasses and droppings. In some years the wheat that is uploaded to ships for export has to pass through fine screens, which increases considerably the time taken to complete a shipment. The cost associated with this is considerable. New developments in storage technologies are becoming more common and include large silo bags (60-m long plastic tubular bag laid on the ground with 200-ton capacity) and have been adopted in more than 50 countries (Abadía et al., 2013). These pose new risks by rodents and other vertebrates (Zufiaurre et al., 2019).

There are only a small number of studies where rodent damage has been directly measured, and these have been mainly undertaken in the last 10 years or so. These studies compare losses from a known quantity of grain samples from paired containers allowing (open) or denying (closed) access by rodents. Thus, in a comparative way, losses over time can be determined. These are expanded in the following section. Nonetheless, a huge gap remains in our understanding of post-harvest impacts caused by rodents.

## 2 Impact of rodents on stored grain

There have been different approaches to look at the impact of rodents on grain stores. Surveys of farmers have explored post-harvest impacts of rodents, and more recently there have been comparative studies on quantitative losses from grain stores. Some measurements are incidental observations made during assessments for insect pests. As Singleton (2003) stated in a review of impacts of rodents on rice production in Asia, post-harvest losses are probably of a similar magnitude to pre-harvest losses. However, the data were patchy and there have been few studies of the impacts of rodents on post-harvest storage of rice at the time of writing. According to Singleton (2003), it was clear that rodents played a significant role in influencing food security and poverty alleviation programmes for the rural poor in Asia. Furthermore, in many Asian countries, farmers simply accept post-harvest losses, partly due to the lack of simple and effective methods of control (Singleton, 2003). A similar scenario occurs with poultry farmers in Argentina (Cavia et al., 2019). There are also unquantified impacts on household nutrition and health through potential transmission of gastroenteric diseases and zoonoses to humans and livestock (Belmain et al., 2015). It is likely this situation is similar in other regions. Unfortunately, there are few comprehensive studies, and so there are many gaps in our knowledge.

Singleton and Petch (1994) conducted a review of rodent pest impacts through Southeast (SE) Asia with a primary focus on pre-harvest impacts, but they considered post-harvest impacts too. They reported on responses by government staff from many SE Asian countries. Many responses confirmed that rodents were a problem in grain stores, but they were not sure which species were involved. In Laos, there was a major problem in the uplands (hilly regions of Laos). For Malaysia, there was no information available. For the Philippines, *Rattus* spp. were responsible for post-harvest damage. There were no estimates of losses provided.

Ahmad et al. (1995) studied rat populations (*R. rattus*) in wholesale grain markets in Pakistan. The estimated population size ranged from 5 to 61 rats per grain shop. They determined in the laboratory that rats ate 12.7 g of rice each night and estimated the annual grain loss per shop was 740 kg (consumption, contamination, spillage and wastage). After scaling up to all markets across the country, annual losses would be approximately 4000 mt/year or 0.3% of the estimated 1.225 million mt that passed through the markets each year.

Parshad (1999) reviewed the rodent impacts to agriculture in India and summarised rodent losses post-harvest. There were 18 rodent species identified, but only 4 species seriously affected grain stores (in houses and godowns/warehouse; *B. bengalensis*, *Rattus norvegicus*, *R. rattus* complex, *M. musculus*). Parshad (1999) summarised many quantitative assessments of rodent infestations in a range of houses and market premises with around 8 house rats per premise and 11 rats per godown, with much higher estimates in some cases (thousands of rodents). Levels of damage were less in situations where grain was stored in metallic bins (0.1%), compared to samples of spilled grain collected from flour mills, grain stores and houses (4.7%). High levels of damage were attributed to inadequate maintenance of buildings, combined with a lack of hygiene and poor handling of food materials, leading to spillage and neglect of rodent proofing. Many grain stores were made from traditional materials such as clay, wood, bamboo, straw, jute bags and bricks, which were considered vulnerable to rodent attack.

Cao et al. (2002) calculated that, globally, vertebrate pests accounted for 5% of post-harvest food losses, mostly attributed to rodents. They stated that the levels of loss were lower in developed countries because of well-designed food storage and processing systems. There are very few studies from developed countries. However, one study, by Wildey (2002), reported the results from a farm survey in the United Kingdom that rodents infested about 70% of grain stores.

Von Maltitz et al. (2003) surveyed farmers in Limpopo province of South Africa to ask about post-harvest impacts, including rodents. Rodents were often ranked as the most important post-harvest constraint for a range of stored crops in Limpopo. Rodents were ranked number 2 after insects affecting stored sorghum and maize.

In a chapter in the book *Integrated pest management in stored grains*, Shankar and Abrol (2012) largely focussed on insect pests, but they summarised some relevant material for the post-harvest impacts of rodents in India. They reported that rodents caused losses of 2.5% (out of a total of 9.3%). The main species attributed to damaging stored grains in India were *R. rattus*, *M. musculus*, *B. bengalensis* and *B. indica*. In contrast, Hart (2001) reported that overall losses of grain to rodents in India were approximately 25% in pre-harvest and 25–30% in post-harvest situations, bringing the loss to at least US$5 billion annually in stored food and seed grains in India.

Brown et al. (2013) set out known amounts of rice grain in open and closed metal bowls in farmers' grain stores in Lao villages and monitored weight loss over time. Weight loss attributed to rodents was 10.3% in the dry season and 7.4% in the wet season, enough grain loss that could have fed a Lao household for 1.5 months. There was a weak but positive relationship between rodent droppings and rodent losses. Unfortunately, Brown et al. (2013) were unable to determine if rodent damage was associated with any particular environmental condition in or around the grain stores.

Mdangi et al. (2013) studied rodent damage to stored maize in smallholder farms in Tanzania. There were significant differences in damage, loss and contamination between different storage structures. They ranged from 0% in closed sacks to 7.9% in closed cribs, 17.7% in open sacks and up to 40.4% in open cribs. There was a strong positive relationship between the number of rodent droppings per month and damage to stored maize. *R. rattus* was the main rodent species.

Belmain et al. (2015) set out known amounts of rice grain in open and closed baskets in farmers' grain stores in rural Bangladesh and Myanmar. Comparisons were made between areas where rodent control was carried out and areas where it was not. Where no rodent control was conducted, losses were 2.5% in Bangladesh and 17% in Myanmar. Where rodent control was conducted, losses were 0.5% in Bangladesh and 5% in Myanmar. Rural households in Bangladesh were losing more than 70 kg of rice per year, which was enough to feed a family for 3–4 months. Daily rat trapping reduced losses.

In Myanmar, Htwe et al. (2017) estimated post-harvest losses by rodents of 4–14%. The loss of 481 kg of rice by rodents from grain stores was considered enough to feed a family for 4 months. The gnawing habit of rodents caused an additional loss; the average loss of rice seed stored in different bags was 4.5 ± 2.1%, and germination loss was 43.1% by rodents gnawing different storage bags in Myanmar (Maw, 2017). In Myanmar, rodents' faeces were counted from grain collected near the surface of the stored grain (stored as loose grain in the grain store) and the estimated contamination rate was 0.01–0.02% (Htwe et al., 2017). See the case study for Myanmar below.

Swanepoel et al. (2017) carried out a systematic review of rodent pest impacts to quantify and identify trends in the impact of rodent pest research on small-holder agriculture in the Afro-Malagasy region. This work identified 19 out of 162 publications with data on post-harvest losses, from which the authors calculated median storage losses of 7.9% by rodents from across the studies.

Edoh Ognakossan et al. (2018) examined rodent impacts on stored maize in Kenya and found the cumulative weight loss ranged from 2.2% to 6.9% in shelled cobs and from 5.2% to 18.3% in dehusked cobs after storage for 3 months. *R. rattus* was the only species trapped. The study confirmed that there were significant increases in mould, *Fusarium* spp., and aflatoxin contamination in rodent-damaged grain.

For Latin America, it is not clear what species cause losses or the extent of the damage because there are very few studies of the post-harvest impacts by rodents to grain storages (Table 1). One study published by the FAO (Kravets, 1991) stated that most grain stores (silos) were well constructed and grains pass through quickly. The presence of rodents was observed in general around the silos but mainly fed on spilled grains in loading areas. The rodent species were not identified, and no estimates of damage or loss were provided. A survey of stakeholder perceptions in Argentina identified that rodents were the second-largest concern for farmers in terms of damage to large silo bags (after armadillos; Zufiaurre et al. 2019). Nearly 60% of respondents reported that rodents caused damage to large silo bags and that 14% of respondents claimed that rodents were the main harmful species (Zufiaurre et al., 2019). The rodent species were not identified. There are also several other studies in Argentina reporting impacts on pig, dairy and poultry farms (*R. norvegicus*, *R. rattus* and *M. musculus*; Miño et al., 2007; Fraschina et al., 2014; Lovera et al., 2019). Therefore, by association, it is presumed these are also the main rodent pests in post-harvest grain stores since their distribution within farms is associated to livestock food sources (Lovera et al., 2015; Montes de Oca et al., 2017; Lovera et al., 2019). The likely post-harvest pest species for Latin America are further considered in Section 3.

## 3 Rodent pest species and their biology

The key rodent pests of grain stores are the '*R. rattus*' complex (see below), *R. exulans*, *M. musculus*, *B. bengalensis* and *B. indica* (Tables 1 and 2), but this depends on the country (Table 1). The '*R. rattus*' complex is responsible for major post-harvest losses in many countries (Aplin et al., 2003). The '*R. rattus*' complex and *M. musculus* have a world-wide distribution, and it is interesting to note that the brown rat (*R. norvegicus*) is not really considered an important post-harvest grain storage pest, although it is implicated in some publications (e.g. in India, Parshad, 1999). '*Rattus rattus*' complex, *R. exulans* and *M. musculus* are excellent climbers and are well suited to invading grain stores. The multimammate rat (*Mastomys natalensis*) is implicated in post-harvest losses in some parts of sub-Saharan Africa, but captures are less than the '*R. rattus*' complex (Mdangi et al., 2013).

The biological and ecological features for three of the main rodent pest species are shown in Table 3.

There is much scientific debate about the origins and number of species belonging to the highly invasive '*R. rattus*' complex (which goes by numerous common names: black rat, ship rat and house rat but are all part of the '*R. rattus*' complex). The story about this species is a story about human history, evolution and expansion itself. While it is generally considered that the Indian peninsula is the original home of *R. rattus*, there have been multiple origins of commensalism, and thus there is variation in the rat's genome, hence the nomen '*R. rattus*' complex (Aplin et al., 2003, 2011; Baig et al., 2019). There are four major lineages, each showing a different expansion and evolution, which mirror patterns of human dispersal and trade over the last hundreds to thousands of years (Aplin et al., 2011). There are similar stories about the other main commensals *M. musculus*, *R. norvegicus* and *R. exulans*.

Identifying the main rodent pest species affecting post-harvest grain in Latin America is difficult. There are several FAO publications summarising the likely species (e.g. Elias, 1984; Rodriguez Muñoz, 1993), but there is confusion about some of the taxonomy and few studies have measured damage or

**Table 2** Key rodent pest species that cause damage to stored grains

| Region | Rodent species | Common name | Remarks |
|---|---|---|---|
| Worldwide | *Rattus 'rattus'* complex | Black rat, ship rat, house rat | Widespread and common |
| | *R. norvegicus* | Brown rat, sewer rat | Widespread and common |
| | *Mus musculus* | House mouse | Widespread and common |
| Asia | *Bandicota bengalensis* | Lesser bandicoot rat | Commensal |
| | *Rattus exulans* | Little house rat, pacific rat | Commensal |
| Africa | *Mastomys natalensis* | Multimammate rat | Commensal |
| | *Acomys cahirinus* | Spiny mouse | Commensal |
| | *Arvicanthis niloticus* | African grass rat | Peri-domestic |
| | *Gerbilliscus* spp. | Gerbils | Peri-domestic |
| Latin America (updated from Brooks and Fiedler 1999) | *Calomys laucha* | Vesper mouse | Wild, sylvan and peri-domestic |
| | *C. musculinus* | Vesper mouse | Wild, sylvan and peri-domestic |
| | *C. callosus* | Vesper mouse | Wild, sylvan and peri-domestic |
| | *Akodon azarae* | Grass mouse | Wild, sylvan and peri-domestic |
| | *Sigmodon hispidus* | Cotton rat | Peri-domestic |
| | *Oligoryzomys longicaudatus* | Rice rat | Peri-domestic |
| Latin America (updated from Elias 1984; Rodriguez Muñoz 1993) | *Heteromys anomalus* | Trinidad spiny pocket mouse | Wild and sylvan |
| | *Peromyscus* spp. | Deer mice | Peri-domestic |
| | *Ototylomys* spp. | Big-eared climbing rat | Wild and sylvan |
| | *Eligmodontia* spp. | Gerbil mice, laucha | Grasslands |
| | *Neotoma* spp. | Packrat, woodrat | Wild, but some found in houses |
| | *Proechimys* spp. | Spiny rat | Wild and sylvan |

*Source*: adapted and updated from Elias (1984), Rodriguez Muñoz (1993), and Brooks and Fiedler (1999). Note the different species for Latin America as listed by Brooks and Fiedler (1999), Elias (1984) and Rodriguez Muñoz (1993) (see text for details).

losses. Elias (1984) summarised rodent species in Latin America. Of the 124 genera (593 species), 41 were implicated as pests and 8 were implicated as pests in stored products. The four genera recognised in Latin America as pests of stored products are *Heteromys*, *Peromyscus*, *Rattus* and *Mus* (Table 2). Other genera were suspected including *Ototylomys*, *Eligmodontia*, *Neotoma* and *Proechimys* because they have appeared in places such as houses and warehouses (Elias, 1984; Table 2). Rodriguez Muñoz (1993) also listed these

**Table 3** Summary characteristics of the three main widespread and common rodent pest species

| Characteristic | *Mus musculus* | *Rattus rattus* complex | *Rattus norvegicus* |
|---|---|---|---|
| Common name | House mouse | Black rat, roof rat | Brown rat, sewer rat |
| Adult size<br>HB = head + body,<br>T = tail, W = weight | HB = 60-95 mm<br>T = 75-95 mm<br>W = 10-20 g | HB = 160-205 mm<br>T = 185-245 mm<br>W = 95-340 g | HB = 180-255 mm<br>T = 150-215 mm<br>W = 200-400 g |
| Description | Small size. Tail length about same as H + B. Wide range of body colours | Body slightly smaller than *R. norvegicus*, large ears. Tail length > H + B | Body slightly larger than *R. rattus*, large head with small ears. Tail length < H + B |
| Litter size | 1-10 (average 5-6) | 1-10 (average 6-7) | 1-12 (average 6-7) |
| Diet | Omnivorous | Omnivorous | Omnivorous |
| Nesting habitat | Subterranean, buildings | Buildings (especially roofs) | Subterranean |
| Gestation period (days) | 19-21 | 20-21 | 20-21 |
| Age at sexual maturity (weeks) | 5-8 | 8-10 | 8-10 |
| Feeding habitat | Fields and buildings | Trees, fields and buildings | Buildings |
| Neophobic (fear of new objects) | No. Mice will readily explore new items found in their territory and try new food. | Yes. Rat will not explore new items or new foods readily. | Yes. Same as for *R. rattus*. |
| Communication of food preferences to other animals | Rarely | Not known | Yes |
| Effectiveness of 1st generation anticoagulants | Low | Low | High |
| Colour vision | Colour blind | Colour blind | Colour blind |
| Sense of smell | Acute | Acute | Acute |

species. In contrast, Brooks and Fiedler (1999) listed a different set of rodent species that are likely to cause post-harvest losses (*Calomys laucha*, *C. musculinus*, *C. callosus*, *Akodon azarae*, *Sigmodon hispidus* and *Oligoryzomys longicaudatus*) (Table 2). Elias and Fall (1988) mentioned *Akodon* and *Calomys*, and Gómez Villafañe et al. (2005) mentioned *C. laucha* as probable pests on stored grain. It is clear, therefore, that further research work is needed to understand the species and levels of damage or losses caused by rodents to grain stores in Latin America.

The commensal species (species that are highly adapted to modified human environments and live along-side humans) are well adapted to highly modified environments of grain stores. In many countries, the pest rodents identified in the field are different to those species identified in grain stores. For example, in Indonesia, the main field pest is the ricefield rat, *Rattus argentiventer*, but it is likely that the main grain storage pests belong to the '*R. rattus*' complex. In Tanzania, the main field pest is *M. natalensis*, but, again, a species belonging to the '*R. rattus*' complex is the main pest in grain stores (Mdangi et al., 2013).

Whilst the species or rodent might be known, information about its biology (breeding, habitat use, movements, etc.) is not well understood in post-harvest situations. There are a few studies examining the habitat use and movements of rodents in and around grain stores. One of these is by Aplin et al. (2003), who studied movements of a species of the '*R. rattus*' complex in and around rice fields and village environments in upland regions of Laos, where grain stores are built within village environments. The '*R. rattus*' complex species is the dominant pest in both village and field habitats. Many individual rats sheltered in piles of freshly harvested rice straw and Job's tears stalks (*Coix lacryma-jobi* or adley millet) immediately post-harvest. Numbers of rats increased in villages shortly after the harvest was complete. In essence, these rats were following food resources from rice fields into the village areas where the grain was being stored (Aplin et al., 2006). The implication is that the management of grain stores needs to be considered in the light of surrounding habitats (crops, houses, intensive livestock, etc.).

Ahaduzzaman and Sarker (2010) reported from farmer surveys that rodent numbers during a bamboo flowering (and masting) event peaked in Chittagong Hill Tract (Bangladesh) around the time of rice harvesting, with high numbers continuing into the post-harvest period.

In Myanmar, some rodent species, particularly *B. bengalensis*, are well known for their hoarding behaviour. The amount of stored grain in the burrows under rice piles was 8.7 ± 5.7 kg measured at 4 weeks after piling. This was equivalent to 3% of the total rice yield (Htwe et al., 2017).

A good knowledge of the changes in abundance, movement patterns, breeding dynamics and nesting site locations would enable improved management strategies to be developed. This would allow managers to better target the location and timing of control strategies. Until more is known, managers will continue to apply control practices indiscriminately. Rodent management strategies are outlined in Section 5.

## 4 Types of rodent damage

Upon close inspection, rodent damage to grains stores should be obvious. There should be ample evidence of rodent faeces/droppings and, in heavy infestations, smear marks along walls and runways. There might also be evidence of gnawing along walls, floors or ceilings, especially on wood. Damaged grain can be discerned by relatively obvious teeth marks from rodent gnawing.

Under normal conditions, rodents are wasteful eaters, biting out small pieces of grain and often discarding the remainder. Thus, the potential loss to crops and to stored produce by rodents is much higher than they have actually consumed (Cavia et al., 2019). Some rodent species, especially from the genus *Rattus*, tend to only sample small amounts of food from numerous food types. In general they are neophobic, meaning they tend to avoid new objects or food stuff. In reality, this may not necessarily be true in grain stores, where there is often a super-abundance of one or more food types, so these species are likely to readily consume these grains.

Rodents typically hold a cereal grain in their front paws, holding it at each end and then rapidly de-husking the grain with their teeth. They then turn it and eat it like a banana. As they nibble, they rotate the grain. Not all of the grain is eaten; some parts are discarded and appear to be coarsely ground (kibbled). Rodents that are hungry tend to eat more of the grain than those that are not hungry; they tend to take more time handling the grain but still discard grains after only a few bites.

Rodents tend to concentrate on the germ of seeds where there is a rich source of unsaturated fatty acids and proteins with the best amino acids (Edoh Ognakossan et al., 2018). By removing the germ rodents reduce the nutritional value of the grain and its potential for germination (Edoh Ognakossan et al., 2018).

Contamination and spoilage from hair, urine and faeces is a real concern. A single rat can leave behind around 25 000 droppings each year and produce over 3 litres of urine each year. There is also structural damage not only to grain stores or facilities (including physical damage to wood or metal structures) but also to bags (spilling contents) and to electrical wiring, water pipes and other infrastructure. The costs of rodent damage are thus far greater than the economic loss from the damaged grain itself.

Apart from the obvious direct consumption of grain and from spoilage, impacts of rodents on stored grain for smallholder farmers in developing countries can lead to less food availability in the market, increased price of food, people worrying whether they will have enough food to eat as time progresses post-harvest, lower quality seed in the market for subsequent crops and changes affecting lives or short-term risks to lives (Ahaduzzaman and Sarker, 2010).

## 5 Strategies to reduce rodent damage

### *5.1 Principles for rodent management in grain stores*

There are some basic principles about reducing damage by rodents to grain storages that should be followed. The best approach is to use an integrated pest control strategy using preventative and control measures that consider environmental, ecological and socio-economic factors to effectively control pests. These can be modified according to the situation and can be applied in a broad sense in developing and developed countries alike.

1 *Deny access* - prevent rodents from accessing the food grains in the first place. In reality, it is difficult to make grain stores rodent proof, particularly if they are old or in poor condition. Choose dry, tight storage facilities (Cao et al., 2002). The tight storage units require solid doors, ventilation and construction. In the smallholder context in Africa, Mdangi et al. (2013) suggest using closed storage (bags or cribs) and improved storage structures to reduce rodent damage. Sacks were more expensive and required wire mesh protection, but they led to fewer losses and a higher cost-benefit ratio.
2 *Prevent borrowing and nesting sites* - ensure there are no opportunities for rodents to live within the confines of storage structures. Raise grain stores off the ground, open up cavity walls and so on.
3 *Use a combination of control strategies* - which include rodenticides and physical methods. A range of rodenticides and physical control strategies are outlined in the following sections. Avoid using cats and dogs. There are no experimental studies demonstrating the success of these predators in grain stores. Also, cats and dogs are at risk of secondary poisoning from rodenticides if they are used as a control strategy, and both are carriers of diseases that affect humans. Continuous trapping using inexpensive kill-traps was shown to reduce rodent damage to grain stores in villages in Myanmar and Bangladesh (Belmain et al., 2015). However, this approach is less acceptable in Myanmar because killing animals is not readily accepted by Buddhists; Buddhism is the most commonly practised religion in Myanmar.
4 *Prevent access to food resources* - general hygiene around grain stores (removal of rodent harbour such as plants and piles of rubbish) and good cleanliness are the key here. If there are no spillages, then there is no free food for the rodents.
5 *Monitor effectiveness of programmes* - how do you know you are being successful? Conduct routine inspections and use a simple method for detecting rodent activity or abundance (trapping, foot tracks in powder, etc.) before and after management is implemented.

## 5.2 Rodenticides

A wide range of rodenticides are available. All must be placed in bait stations or boxes to prevent accidental spillage or access by children, domestic pets or livestock. They can be categorised into four main groups as follows:

1 *Sub-acute* (single or multiple doses, slower-acting).
2 *Acute* (single-dose, fast-acting; hours to 1–2 days).
3 *First-generation anticoagulants* (multiple dose, slow-acting, can take up to 10 days for the animal to die).
4 *Second-generation anticoagulants* (single-dose, highly potent, slow-acting: death of rodents occurs at around 4–7 days).

Although rodenticides are often used as the primary method for controlling rodents in grain stores, a combination of control strategies are required. One of the significant drawbacks of using rodenticides is that the rodenticide baits need to be sufficiently attractive in order for a rodent to take and eat them. This is very difficult if high-quality alternative food is present, in the form of the stored grain.

Von Maltitz et al. (2003) reported from farmer surveys focused on rodent problems that 40% of respondents used a preventative method against rodent damage. The most common method was the occasional single-dose use of a chronic rodenticide in and around the house and food store. Rodent species identified were *R. rattus*, *R. norvegicus* and *Mastomys* spp., with *R. rattus* being the dominant species trapped at the time of the survey.

Meerburg et al. (2004) indicated the following four factors that determine the uptake of a rodenticide bait:

1 Whether the rodents are neophobic or neophilic.
2 The population structure of the target rodent population.
3 Bait palatability.
4 Habitat structure, including access to alternative food.

There are a range of repellents and fumigants also registered for rodent management but are not discussed here in detail because they are not very effective in managing rodent populations.

Sub-acute rodenticides are beneficial if a lethal dose has been ingested by the target rodents before the onset of anorexia, but there is a disadvantage if a sub-lethal dose is taken (Buckle and Eason, 2015).

The anticoagulants were originally designed to circumvent the neophobic feeding behaviour of rodents (particularly the species of the '*R. rattus*' complex and *R. norvegicus*), where they sample only small amounts of new food types. The first-generation anticoagulants required multiple feeds, but the second-generation anticoagulants are more potent and need only a single feed. As the name suggests, the anticoagulants inhibit the clotting of blood, and death normally occurs through organ failure and internal bleeding and can occur at 4–10 days, depending on the active ingredient and dose. In general, anticoagulant rodenticides have a high efficacy, are easy to use, are inexpensive and, if applied correctly, have a minimal environmental risk when adequate mitigation measures are properly applied (Jacob and Buckle, 2018). There is a risk of secondary poisoning with the anticoagulants - they can accumulate through the food chain. Animals that accidentally ingest anticoagulants can be treated with vitamin K, which is readily available.

The anticoagulants are now an integral component of rodent management around the world and in a wide range of situations from food processing (in bait stations), grain storage, intensive livestock systems, households, to commercial properties, but genetic resistance has developed in many countries (Rost et al., 2009). Testing for genetic resistance in rodents is routine in Europe (Pelz et al., 2005; Prescott, 2003) but not in many other countries globally. Furthermore, the effects of rodenticide on wildlife via primary and secondary poisoning have not been adequately tested (Lohr and Davis, 2018).

Cellulose baits (powdered corn cob) have been suggested as an alternative to traditional rodenticides (Schmolz, 2010), but there is no credible evidence that cellulose baits work. Schmolz (2010) reported that all test animals died in no-choice tests, and in choice trials 11 of 12 rats survived. If alternative food is available (such as in grain stores), these baits do not work.

Rodenticide baits need to be placed in bait stations, but these need to be where rodents are likely to encounter them. In developed countries because of the requirement for compliance with regulations, bait stations in warehouses and commercial premises are serviced under contract by pest control companies and visited every 4, 6 or 8 weeks to replenish baits. Bait stations are necessary to reduce spillage and minimise non-target poisoning. Endepols et al. (2003) described a scheme to compare the effectiveness of allocating rodenticide baiting points to specific structural elements or only where obvious signs of rat activity occurred. They looked at 25 farms and monitored activity before and after treatment. Rat control was often conducted irregularly and with poor preparation and documentation (especially on small- and medium-sized farms). Complete rat eradication was achieved on many farms. A ≥75% level of implementation of the control plan always resulted in complete control success.

## 5.3 Physical and other management practices

A key component of any pest management is to manage the environment to reduce pest incidence, essentially to deny them access to nesting sites and to expose them to increased predation pressure. This can be achieved through habitat management and physical management. It is often impractical in grain stores to have rodent-proof buildings and eliminate food sources, but reducing habitat complexity might disadvantage rat populations, through increased exposure to predation or reduced opportunities for nesting. Rats actively avoid predators and prefer to stay close to cover when moving between nesting and feeding sites (Lambert et al., 2008; Jones et al., 2017). In grain stores food is freely available for rats throughout the year. It is likely that the home range of rodents in grain stores will be small (Gómez Villafañe et al., 2008), so they do not need to move far to find food. Habitat management is important, but Brown et al. (2013) could not determine the key attributes that led to rodent losses in grain stores in Laos.

Targeted habitat management has the potential to reduce the size of rat populations and should be used as part of an integrated management approach for rats (Lambert et al., 2008). It is recommended that at least 20–30 m (corresponding to the average home range size) should be maintained free from vegetation and other harbourage around storage structures where food resources are located (Montes de Oca et al., 2017). Wire netting can also be installed to exclude rodents (hole size < 6 mm) from electrical boxes and access through holes and gaps in walls and so on (see Mdangi et al., 2013).

There are different types of traps commercially available. These can be grouped into single-catch snap traps or cage traps and multiple capture live-traps. Each of these requires time and effort to set up and check. They also need to be baited with an attractive food type (e.g. peanut butter and bacon are good lures), and they need to be positioned so that rodents encounter them (i.e. along edges of walls or on roof beams). Unfortunately, because of the neophobic behaviour of rats, it is necessary to move traps around and perhaps leave them baited but unset for several days to reduce the neophobic response of rodents to new objects in their environment. Glue boards are not recommended.

Cats are important predators of rodents and have exerted strong selective pressure on the behaviour and physiology of rats. Cats are sometimes suggested to help control rodents. Mice and rats caught by these cats may be intermediate hosts for parasites such as *Toxoplasma gondii* (Meerburg et al., 2004). Excrement from infected cats can then pose a hazard to the health of farm animals and humans. Apart from the health risk presented by cats, there is no sound evidence that cats regulate rodent populations. In one study, Mahlaba et al. (2017) found that rodent activity decreased in farm homesteads in Africa where cats were present, but there was no decline in rodent abundance.

## 6 Case studies

### 6.1 Case study 1: community village-level rodent management in Myanmar

The study was conducted in Maubin Township (Ayeyarwaddy region) and Daik U Township (Bago region) in 2015. A community village-level rodent management system including trapping, sanitation around storage houses and installing rodent-proofing (installing nylon open mesh at the access route of rodents) was conducted in three villages. Sanitation included regular sweeping, the removal of materials under and around the storage structure (breeding sites for rodents) and cutting tree branches that can provide access routes for rodents to grain stores. Plastic kill-traps inside the stores and bamboo snare traps outside the stores were set for two consecutive nights every 2 weeks. Evidence of rodent contamination and weight loss in two standard 8 kg baskets (rodents had access to one basket but not the other) was quantified. This management strategy commenced at the beginning of the time of storage of rice until 2 weeks after the stores were depleted of grain. This system reduced rodent losses from 4% (28 kg) to 1% (7 kg) (Fig. 1) (Htwe et al., 2017). The cost for managing rodents was 16 250 kyats (US$14/house); farmers could benefit by 94 770 kyats (US$81/house).

In 2016, a farmer-led village-level rodent management system was conducted in Maubin Township (n = 40 families). A focus group discussion was conducted at the end of the storage season to document the farmers' adaptive practices. All farmers' stores installed rodent-proof nylon (open mesh); 50% of farmers followed the sanitation recommendations, but only 10% of farmers implemented trapping. The recommended management practices were simple to follow. However, only 10% of farmers followed the entire package of recommended practices and 90% identified a range of challenges; which included the following:

1 Farmers have their individual storage house, and they do not tend to work together as a community for rodent management.

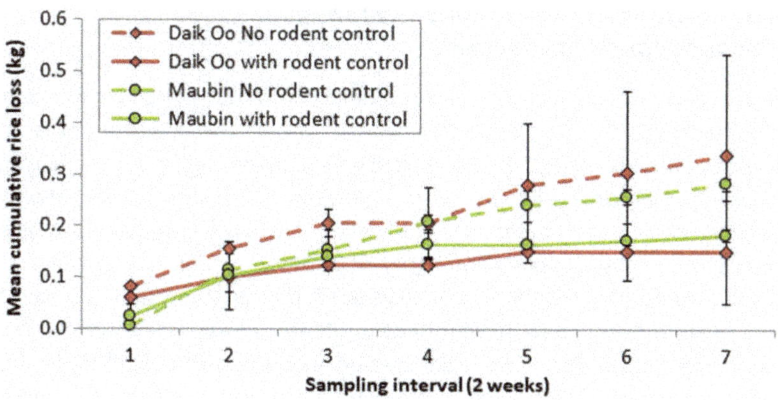


**Figure 1** Cumulative increase in rodent damage to rice in grain stores in Myanmar (dashed lines) in comparison to sites where rodent control was undertaken (solid lines).

2 Their store house was also used as a storage area for farm machinery, and cleanliness in and around stores remained poor.
3 Most farmers were reluctant to use the snap traps as it was culturally unacceptable because of religious beliefs. As a footnote, there are landless people in rural Myanmar who make their living as rat-hunters; they sell rats for human consumption. On some occasions they are employed by villagers to capture rodents in and around their stores and fields.

There are always challenges to develop pest management systems that can be adaptable and feasible for farmers to apply. Social dimension plays an important role in developing the adoption pathways for broad-scale rodent management approaches.

Future directions for developing village ecologically based rodent management (EBRM) in Myanmar for grain store include:

- Build capacity and strengthen local champions.
- Be aware of the social dimension in different regions that may constrain farmers from following all of the management recommendations.
- Build knowledge on the biology (population and breeding ecology and spatial behaviour) of the target species.
- Develop a farmer participatory approach for different crops and different agroecosystems in Myanmar.

There was an average of 45% reduction in rodent damage to rice when rodent control was undertaken in rice stores in Myanmar (Fig. 1).

## 6.2 Case study 2: rural community rodent management in Tanzania

The study was conducted in rural communities in Tanzania in order to examine rodent damage loss and contamination in stored maize in smallholder farms in East Africa (Mdangi et al., 2013). Different novel techniques for assessing rodent damage, namely open and closed structures (cribs and sacks), were employed in a treatment-control trial design replicated across different households and hamlets within the Berega community of central Tanzania. Six farmers were selected randomly, each from five different hamlets. Three farmers used the sacks to store their maize, while the other three used cribs. Before the maize was placed in bags and cribs, it was treated with an insecticide (Actellic Super Dust, manufactured by Syngenta) according to the instructions on the package to prevent insect damage.

Sampling of maize was conducted every month for 7 months. In each month, sampling from each replicate was done using a 0.25-kg container of maize (n = 4), making a total of 1 kg of maize seeds. These samples were taken from the middle and the periphery from each storage structure. Damaged and undamaged seeds were counted and weighed, and the percentage of grain damage and the amount of weight loss were calculated. Similarly, samples taken each month were also used to assess the level of contamination and number of rodent droppings.

To determine the rodent species and population abundance in the households, five farmers from five different hamlets were randomly selected. Equal numbers of live-traps (five locally made traps and five Sherman live-traps) were baited with peanut butter and were placed against walls and in the corners of the house. Trapped animals were identified to species level following the guidelines given in Kingdon (1997).

The results from this study showed significant correlations between the monthly rates of rodent-damaged maize seeds, maize weight loss and the number of rodent droppings (Mdangi et al., 2016). Similarly, significant differences were observed for damage, loss and contamination between different storage structures (open and closed cribs and sacks). The mean monthly rate of damage was 40.4%, 7.9%, 17.7% and 0% in open cribs, closed cribs, open sacks and closed sacks, respectively.

These findings suggest that reducing rodent infestation through the use of improved storage structures could lead to major savings in the amount and quality of stored cereals available to households, thus increasing food security.

Future research directions for rodents and post-harvest losses in Tanzania include conducting more research on the factors influencing these losses, because there is a scarcity of documented information available. There are many technologies which smallholder farmers are using to protect their produce, but most are yet to be evaluated and documented.

## 7 Conclusion

Based on a handful of studies in Asia and eastern Africa over the past decade, we now know more about the type and level of post-harvest damage and losses caused by rodents to cereal staples, but further work is required to study the ecology of the different pest species under different situations. The species causing damage and the level of losses or damage caused by rodents to post-harvest grain stores in Latin America also needs to be a focus of research.

More research is required to address questions such as: Where do the rodents nest? How far do they range in their daily and seasonal movements?

When do they breed? These are just some of the research questions to be asked.

Recent research in the developing countries has demonstrated that the losses caused by rodents can be reduced (see previous case studies). Improving rodent management strategies reduces damage and losses. If losses were reduced by half from around 10% to 5%, that would be a significant increase in the amount of grain available for human consumption. This would be especially important for the food security of smallholder farming households. It is virtually impossible to be free from rodents, but their impact can be greatly reduced.

## 8 Future trends in research

There is now a greater understanding of the impacts of rodents on cereals in grain stores, but there are still many gaps in our knowledge. More research needs to be conducted on the ecology of rodents in and around grain stores. Several questions remain:

- Which are the key pest species?
- Are the species different from those that occur in the neighbouring fields, or do they simply follow where the cereal is accumulated?
- Where do they nest and breed?
- How far do they move?
- How do they access grain stores?
- What control strategies work best?
- Can management methods take advantage of the "landscape of fear" that influences the movement of rodents around the agricultural landscape (see Krijger et al., 2017)?

There is ongoing work to identify new types of rodenticides, but many of these are still some years away. Making the most of currently available strategies remains the best approach. A combination of methods is recommended (see Section 5). There is research work looking at developing fertility control agents, but these will likely work best in combination with other management methods.

A rapidly expanding field of detection and reporting systems is becoming available. Some devices include passive infra-red motion detectors, which are placed in a box similar in design to a bait station, to detect the presence and/or activity of rodents. The device then sends alerts to mobile or smart phones via an app. This can then be used to trigger some management intervention or be used to confirm that the activity has declined after intervention. Monitoring the effectiveness of management is always important, and such systems will be more readily available in the future.

*Sample of rice grain from a rice store in Laos. Note the faeces mixed in with the rice grain. Photo: Alex McWilliam (CSIRO)*

*Rice grain stored in a grain store, Laos. The rodent faeces are easily seen. Photo: Alex McWilliam (CSIRO)*

*Typical grain store within a rural village, Laos. Note the metal sleeves to try and prevent access by rats. The grain is stored loosely in the woven bamboo structure. Photo: Alex McWilliam (CSIRO)*

*Typical grain store within a rural village, Laos. Note the wide 'rat guards' at the top of the posts to try and prevent access by rats. The grain is stored loosely in the woven bamboo structure. Photo: Alex McWilliam (CSIRO)*

*Rodent damage to maize kernels. Note the gnawing of the kernels to remove the germ. Photo: Steven Belmain (NRI)*

*Rodent damage to maize cobs, Tanzania. Photo: Steven Belmain (NRI)*

Maize drying crib, Tanzania. Photo: Steven Belmain (NRI)

Grain storage in Africa. Photo: Steven Belmain (NRI)

Untidy grain storage, Bangladesh (prior to intervention). Photo: Steven Belmain (NRI)

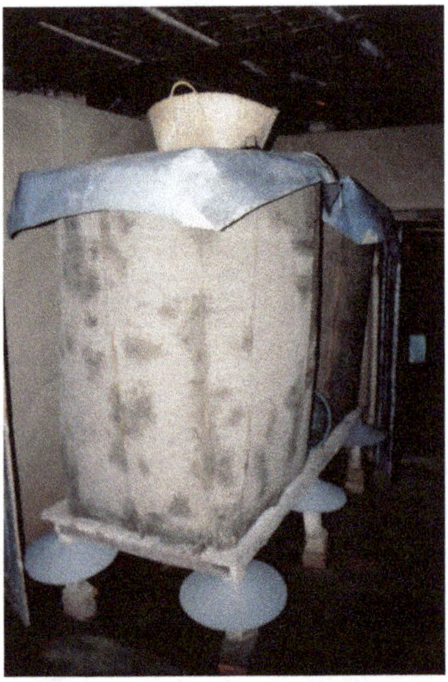

Tidy grain storage, Bangladesh (after intervention). Note the metal flashing on the legs and the metal cover over the top of the grain store. Photo: Steven Belmain (NRI)

## 9 Where to look for further information

There are only a few resources available specifically for managing rodents in grain stores.

- The second edition of Buckle and Smith's (2015) *Rodent Pests and Their Control* contains a chapter by Lund (2015) which summarised impacts by rodents.
- The *Encyclopedia of Pest Management* (Cao et al., 2002) often summarises what is known about post-harvest impacts, but this tends to be only in summary form.
- The EcoRodMan website (https://ecorodman.nri.org/) provides some current information about the ongoing studies looking at rodent damage to stored grains and provides some results of recent trials of storage systems.

## 10 References

Abadía, B., Bartosik, R., Cardoso, L., de la Torre, D. and Giorda, J. J. 2013. Manejo eficiente del grano en la poscosecha. Efficient management of the grain in the post-harvest. In: Abadía, B. and Bartosik, R. (Eds), *Manual de buenas prácticas en poscosecha granos. Hacia el agregado de valor en origen la producción primaria*. Ediciones INTA, Buenos Aires, Argentina, pp. 13-94.

Ahaduzzaman, S. K. M. and Sarker, S. K. 2010. The Chittagong story: a regional damage assessment during a rodent population outbreak. In: Singleton, G. R., Belmain, S. R., Brown, P. R. and Hardy, B. (Eds), *Rodent Outbreaks: Ecology and Impacts*. International Rice Research Institute, Los Baños, Philippines.

Ahmad, E., Hussain, I. and Brooks, J. E. 1995. Losses of stored foods due to rats at grain markets in Pakistan. *International Biodeterioration and Biodegradation* 36(1-2), 125-33. doi:10.1016/0964-8305(95)00087-9.

Aplin, K. P., Brown, P. R., Jacob, J., Krebs, C. J. and Singleton, G. R. 2003. Field methods for rodent studies in Asia and the Indo-Pacific. ACIAR Monograph 100. Australian Centre for International Agricultural Research, Canberra, Australia, 223p.

Aplin, K. P., Brown, P. R., Singleton, G. R., Douangboupha, B. and Khamphoukeo, K. 2006. Rodents in the rice environments of Laos. In: Schiller, J. M., Chanphengxay, M. B., Linquist, B. and Appa Rao, S. (Eds), *Rice in Laos*. International Rice Research Institute, Los Banos, Philippines, pp. 291-308.

Aplin, K. P., Suzuki, H., Chinen, A. A., Chesser, R. T., ten Have, J., Donnellan, S. C., Austin, J., Frost, A., Gonzalez, J. P., Herbreteau, V., Catzeflis, F., Soubrier, J., Fang, Y. P., Robins, J., Matisoo-Smith, E., Bastos, A. D. S., Maryanto, I., Sinaga, M. H., Denys, C., Van Den Bussche, R. A., Conroy, C., Rowe, K. and Cooper, A. 2011. Multiple geographic origins of commensalism and complex dispersal history of black rats. *PLoS ONE* 6(11), e26357. doi:10.1371/journal.pone.0026357.

Baco, D., Razak, N. and Juddawi, H. 2010. Rodent outbreaks in South Sulawesi, Indonesia: the importance understanding cultural norms. In: Singleton, G. R., Belmain, S. R., Brown, P. R. and Hardy, B. (Eds), *Rodent Outbreaks: Ecology and Impacts*. International Rice Research Institute, Los Baños, Philippines, pp. 129-37.

Baig, M., Khan, S., Eager, H., Atkulwar, A. and Searle, J. B. 2019. Phylogeography of the black rat *Rattus rattus* in India and the implications for its dispersal history in Eurasia. *Biological Invasions* 21(2), 417–33. doi:10.1007/s10530-018-1830-0.

Belmain, S. R., Meyer, A. N., Kamal, N. Q., Shafali, R. B., Singleton, G. R., Aplin, K. and Mian, Y. 2006. Ecologically based rodent management for rice-based cropping systems in Bangladesh. In: Sweetmore, A., Kimmins, F. and Silverside, P. (Eds), *Perspectives on Pests II: Achievements of Research under the UK Department for International Development Crop Protection Programme, 2000–05*. Natural Resources International, Aylesford, UK, pp. 179–81.

Belmain, S. R., Htwe, N. M., Kamal, N. Q. and Singleton, G. R. 2015. Estimating rodent losses to stored rice as a means to assess efficacy of rodent management. *Wildlife Research* 42(2), 132–42. doi:10.1071/WR14189.

Brooks, J. E. and Fiedler, L. A. 1999. Vertebrate pests: damage on stored foods. In: Mejia, D. and Lewis, B. (Eds), *Compendium of Post-Harvest Operations*. Food and Agriculture Organization (AGSI-FAO), p. 26. Available from: http://www.fao.org/3/a-av014e.pdf.

Brown, P. R., McWilliam, A. and Khamphoukeo, K. 2013. Post-harvest damage to stored grain by rodents in village environments in Laos. *International Biodeterioration and Biodegradation* 82, 104–9. doi:10.1016/j.ibiod.2012.12.018.

Brown, P. R., Douangboupha, B., Htwe, N. M., Jacob, J., Mulungu, L. S., My Phung, N. T., Singleton, G. R., Stuart, A. M. and Sudarmaji. 2017. Control of rodents pests in rice cultivation. In: Sasaki, T. (Ed.), *Achieving Sustainable Cultivation of Rice Volume 2: Cultivation, Pest and Disease Management*. Burleigh Dodds Science Publishing, Cambridge, UK, pp. 343–76.

Buckle, A. P. and Eason, C. T. 2015. Control methods: chemical. In: Buckle, A. P. and Smith, R. H. (Eds), *Rodent Pests and Their Control* (2nd edn.). CAB International, Oxford, UK, pp. 123–54.

Buckle, A. P. and Smith, R. H. (Eds). 1994. *Rodent Pests and Their Control*. CAB International, Wallingford, UK.

Buckle, A. P. and Smith, R. H. (Eds). 2015. *Rodent Pests and Their Control* (2nd edn.). CAB International, Oxford, UK.

Cao, D., Pimentel, D. and Hart, K. 2002. Postharvest food losses (vertebrates). In: Pimintel, D. (Ed.), *Encyclopedia of Pest Management*. Marcel Dekker, Boca Raton, FL, pp. 648–9.

Cavia, R., Guidobono, J. S., Fraschina, J. and Busch, M. 2019. Effects of physical barriers and eradication on recolonization of rodents in poultry farms. *International Journal of Pest Management* 65(4), 370–80. doi:10.1080/09670874.2018.1530485.

Edoh Ognakossan, K., Mutungi, C. M., Otieno, T. O., Affognon, H. D., Sila, D. N. and Owino, W. O. 2018. Quantitative and quality losses caused by rodents in on-farm stored maize: a case study in the low land tropical zone of Kenya. *Food Security* 10(6), 1525–37. doi:10.1007/s12571-018-0861-9.

Elias, D. J. 1984. Roedores como plagas de productos almacenados; control y manejo (in Spanish: English translation: Rodents as pests of stored products; control and management). FAO Regional Office for Latin America and the Caribbean. Post-Harvest Technology 3. FAO, Santiago, Chile. Available at: http://www.fao.org/3/x5052s/x5052S02.htm#2.%20Roedores%20en%20am%C3%A9rica%20latina.

Elias, D. J. and Fall, M. W. 1988. The rodent problem in Latin America. In: Prakash, I. (Ed.), *Rodent Pest Management*. CRC Press, Boca Raton, FL, pp. 13–28.

Endepols, S., Klemann, N., Pelz, H. J. and Ziebell, K. L. 2003. A scheme for the placement of rodenticide baits for rat eradication on confinement livestock farms. *Preventive Veterinary Medicine* 58(3–4), 115–23. doi:10.1016/S0167-5877(03)00024-2.

Fraschina, J., León, V. A. and Busch, M. 2014. Role of landscape scale in the distribution of rodents in an agroecosystem of Argentina. *Journal of Agricultural Science* 6(12), 22–35. doi:10.5539/jas.v6n12p22.

Gómez Villafañe, I. E., Miño, M. H., Cavia, R., Hodara, K., Courtalón, P., Suárez, O. V. and Busch, M. 2005. *Roedores Guía de la Provincia de Buenos Aires* (In Spanish; English Translation: *Field Guide of Rodents of the Buenos Aires Province*). LOLA, Buenos Aires, Argentina, 97p.

Gómez Villafañe, I. E., Muschetto, E. and Busch, M. 2008. Movements of Norway rats (*Rattus norvegicus*) in two poultry farms, Exaltación de la Cruz, Buenos Aires, Argentina. *Mastozoología Neotropical* 15, 203–8.

Hart, K. 2001. Post harvest losses. In: Pimentel, D. (Ed.), *Encyclopedia of Pest Management*. Marcel Dekker, New York, NY, pp. 723–30.

Hengsdijk, H. and de Boer, W. J. 2017. Post-harvest management and post-harvest losses of cereals in Ethiopia. *Food Security* 9(5), 945–58. doi:10.1007/s12571-017-0714-y.

Hopf, H. S., Morley, G. E. J. and Humphries, J. R. O. 1976. *Rodent Damage to Growing Crops and to Farm and Village Storage in Tropical and Subtropical Regions*. Centre for Overseas Pest Research and Tropical Products Institute (COPR/TPI), London, UK.

Htwe, N. M., Singleton, G. R. and Maw, P. P. 2017. Post-harvest impacts of rodents in Myanmar; how much rice do they eat and damage? *Pest Management Science* 73(2), 318–24. doi:10.1002/ps.4292.

Jacob, J. and Buckle, A. P. 2018. Use of anticoagulant rodenticides in different applications around the world. In: van den Brink, N. W., Elliott, J. E., Shore, R. F. and Rattner, B. A. (Eds), *Anticoagulant Rodenticides and Wildlife. Emerging Topics in Ecotoxicology* (vol. 5). Springer, Cham, Switzerland, pp. 11–43. https://doi.org/10.1007/978-3-319-64377-9_2

John, A. 2014. Rodent outbreaks and rice pre-harvest losses in Southeast Asia. *Food Security* 6(2), 249–60. doi:10.1007/s12571-014-0338-4.

Jones, C. R., Lorica, R. P., Villegas, J. M., Ramal, A. F., Horgan, F. G., Singleton, G. R. and Stuart, A. M. 2017. The stadium effect: rodent damage patterns in rice fields explored using giving-up densities. *Integrative Zoology* 12(6), 438–45. doi:10.1111/1749-4877.12251.

Kingdon, J. 1997. *The Kingdon Field Guide to African Mammals*. Academic Press Publication, London, UK, 464p.

Kravets, F. O. 1991. Biología y control de roedores plaga en Argentina. In: *Biología y Control de Roedores en America Latina: Informe de Países* (in Spanish; English translation: *Rodent Biology and Control in Latin America: Countries Report*). FAO, Rome, Italy, p. 39.

Krijger, I. M., Belmain, S. R., Singleton, G. R., Groot Koerkamp, P. W. G. and Meerburg, B. G. 2017. The need to implement the landscape of fear within rodent pest management strategies. *Pest Management Science* 73(12), 2397–402. doi:10.1002/ps.4626.

Kushwaha, K. S. 1986. Rat control in rice. In: Maclean, J. and Banta, S. J. (Eds), *Progress in Rainfed Lowland Rice*. International Rice Research Institute, Los Baños, Philippines, pp. 371–9.

Lambert, M. S., Quy, R. J., Smith, R. H. and Cowan, D. P. 2008. The effect of habitat management on home-range size and survival of rural Norway rat populations. *Journal of Applied Ecology* 45(6), 1753–61. doi:10.1111/j.1365-2664.2008.01543.x.

Lohr, M. T. and Davis, R. A. 2018. Anticoagulant rodenticide use, non-target impacts and regulation: a case study from Australia. *Science of the Total Environment* 634, 1372–84. doi:10.1016/j.scitotenv.2018.04.069.

Lovera, R., Fernández, M.S., Cavia, R., 2015. Wild small mammals in intensive milk cattle and swine production systems. *Agriculture, Ecosystems and Environment* 202, 251–9. doi:10.1016/j.agee.2015.01.003.

Lovera, R., Fernandez, M. S., Jacob, J., Lucero, N., Morici, G. E., Brihuega, B., Farace, M. I., Caracostantogolo, J. L. and Cavia, R. 2017. Intrinsic and extrinsic factors related to pathogen infection in wild small mammals in intensive milk cattle and swine production systems. *PLoS Neglected Tropical Diseases* 11(6), e0005722. doi:10.1371/journal.pntd.0005722.

Lovera, R., Fernández, M. S. and Cavia, R. 2019. Small rodent species on pig and dairy farms: habitat selection and distribution. *Pest Management Science* 75(5), 1234–41. doi:10.1002/ps.5299.

Lund, M. 2015. Commensal rodents. In: Buckle, A. P. and Smith, R. H. (Eds), *Rodent Pests and Their Control* (2nd edn.). CABI Publishing, Wallingford, Oxfordshire, UK, pp. 19–32.

Mahlaba, T. A. M., Monadjem, A., McCleery, R. and Belmain, S. R. 2017. Domestic cats and dogs create a landscape of fear for pest rodents around rural homesteads. *PLoS ONE* 12(2), e0171593. doi:10.1371/journal.pone.0171593.

Maw, P. P. 2017. Impact of rodents on different rice seed storage structures. Masters Thesis. Yezin Agriculture University, Myanmar, 88p.

Mdangi, M., Mulungu, L. S., Massawe, A. W., Eiseb, S. J., Tutjavi, V., Kirsten, F., Mahlaba, T., Malebane, P., Maltitz, E. V., Monadjem, A., Dlamini, N., Makundi, R. H. and Belmain, S. R. 2013. Assessment of rodent damage to stored maize (*Zea mays* L.) on smallholder farms in Tanzania. *International Journal of Pest Management* 59(1), 55–62. doi:10.1080/09670874.2012.744495.

Mdangi, M., Sibuga, P. K., Massawe, A. W., Magadula, J. J., Ngowo, V., Mrosso, F. P., Mkangwa, Z. C., Kilonzo, B. S. and Mulungu, L. S. 2016. Farmers' knowledge, attitude and perceptions on the use of locally available plant material to prevent rodent damage to maize crop in Eastern Tanzania. *African Journal of Agricultural Research* 11(48), 4902–10. doi:10.5897/AJAR2016.11715.

Meerburg, B. G., Bonde, M., Brom, F. W. A., Endepols, S., Jensen, A. N., Leirs, H., Lodal, J., Singleton, G. R., Pelz, H. -J., Rodenburg, T. B. and Kijlstra, A. 2004. Towards sustainable management of rodents in organic animal husbandry. *NJAS – Wageningen Journal of Life Sciences* 52(2), 195–205. doi:10.1016/S1573-5214(04)80014-9.

Meerburg, B. G., Singleton, G. R. and Kijlstra, A. 2009. Rodent-borne diseases and their risks for public health. *Critical Reviews in Microbiology* 35(3), 221–70. doi:10.1080/10408410902989837.

Miño, M. H., Cavia, R., Villafañe, I. E. G., Bilenca, D. N. and Busch, M. 2007. Seasonal abundance and distribution among habitats of small rodents on poultry farms. A contribution for their control. *International Journal of Pest Management* 53(4), 311–6. doi:10.1080/09670870601105949.

Montes de Oca, D. P. M., Lovera, R. and Cavia, R. 2017. Where do Norway rats live? Movement patterns and habitat selection in livestock farms in Argentina. *Wildlife Research* 44(4), 324–33. doi:10.1071/WR16219.

Muda, A. 1986. Pest problems and the use of pesticides in grain storage in Malaysia. ACIAR Proceedings No. 14. Australian Centre for International Agricultural Research, Canberra, Australia, pp. 11–6.

Mulungu, L. S., Mrosso, F. P., Katakweba, A. A. S., Mdangi, M. E., Tesha, P. P. H., Ngowo, V., Mchomvu, M. and Kilonzo, B. S. 2015. Farmer's knowledge, attitude and practice on rodent management in lowland irrigated rice in Central-eastern Tanzania. *International Research Journal of Plant Science* 6, 7–14. doi:10.14303/irjps.2014.016.

Parshad, V. R. 1999. Rodent control in India. *Integrated Pest Management Reviews* 4(2), 97–126. doi:10.1023/A:1009622109901.

Pelz, H. J., Rost, S., Hünerberg, M., Fregin, A., Heiberg, A. C., Baert, K., MacNicoll, A. D., Prescott, C. V., Walker, A. S., Oldenburg, J. and Müller, C. R. 2005. The genetic basis of resistance to anticoagulants in rodents. *Genetics* 170(4), 1839–47. doi:10.1534/genetics.104.040360.

Prescott, C. V. 2003. A reappraisal of blood clotting response tests for anticoagulant resistance and a proposal for a standardised BCR test methodology. Technical Monograph 2003 of RRAC, Rodenticide Resistance Action Committee, Croplife International, Brussels, Belgium. Available at: http://www.croplife.org/website/pages/RRAC.aspx.

Rao, A. M. K. M. 2003. Rodent problems in India and strategies for their management. In: Singleton, G. R., Hinds, L. A., Krebs, C. J. and Spratt, D. M. (Eds), *Rats, Mice and People: Rodent Biology and Management*. ACIAR Monograph 96. Australian Centre for International Agricultural Research, Canberra, Australia, pp. 203–12.

Rodriguez Muñoz, J. A. 1993. *Roedores Plaga: Un Problema Permanente en America Latina y el Caribe* (in Spanish; English translation: *Plague Rodents: a Permanent Problem in Latin America and the Caribbean*). Primera Consulta Latinoamericana Sobre Biologia y Control de Roedores Plaga. Oficina Regional de la FAO Para America Latina y le Caribe. FAO, Santiago, Chile, 130p. Available at: http://repositorio.uchile.cl/handle/2250/145649.

Rost, S., Pelz, H. J., Menzel, S., MacNicoll, A. D., León, V., Song, K. J., Jäkel, T., Oldenburg, J. and Müller, C. R. 2009. Novel mutations in the VKORC1 gene of wild rats and mice – a response to 50 years of selection pressure by warfarin?. *BMC Genetics* 10, 4. doi:10.1186/1471-2156-10-4.

Santra, L. B. and Manna, C. K. 2008. Studies of some aspects of rodent ecology in the four districts of the Gangetic plain of West Bengal, India. *University Journal of Zoology, Rajshahi University* 27, 85–90. doi:10.3329/ujzru.v27i0.1961.

Schmolz, E. 2010. Efficacy of anticoagulant-free alternative bait products against house mice (*Mus musculus*) and brown rats (*Rattus norvegicus*). *Integrative Zoology* 5(1), 44–52. doi:10.1111/j.1749-4877.2010.00191.x.

Shankar, U. and Abrol, D. P. 2012. Integrated pest management in stored grains. In: Abrol, D. P. and Shankar, U. (Eds), *Integrated Pest Management*. CAB International, Wallingford, UK, pp. 386–407.

Sidik, M., Halid, H. and Pranata, R. I. 1986. Pest problems and the use of pesticides in grain storages in Indonesia. ACIAR Proceedings 14. Australian Centre for International Agricultural Research, Canberra, Australia, pp. 37–43.

Singleton, G. R. 2003. Impacts of rodents on rice production in Asia. IRRI Discussion Paper Series No. 43. International Rice Research Institute, Los Baños, Philippines, 30p.

Singleton, G. R. and Petch, D. A. 1994. A review of the biology and management of rodent pests in Southeast Asia. ACIAR Technical Reports No. 30. Australian Center for International Agricultural Research, Canberra, Australia, 65p.

Singleton, G. R., Belmain, S. R., Brown, P. R., Aplin, K. P. and Htwe, N. M. 2010. Impacts of rodent outbreaks on food security in Asia. *Wildlife Research* 37(5), 355–9. doi:10.1071/WR10084.

Swanepoel, L. H., Swanepoel, C. M., Brown, P. R., Eiseb, S. J., Goodman, S. M., Keith, M., Kirsten, F., Leirs, H., Mahlaba, T. A. M., Makundi, R. H., Melabane, P., von Maltitz, E. F., Massawe, A. W., Monadjem, A., Mulungu, L. S., Singleton, G. R., Taylor, P. J., Soarimalala, V. and Belmain, S. R. 2017. A systematic review of rodent pest research in Afro-Malagasy small-holder farming systems: are we asking the right questions? *PLoS ONE* 12, e0174554. doi:10.13140/RG.2.2.24795.77609.

Von Maltitz, E. F., Kirsten, F., Malebane, P. S., Belmain, S. R., Sandmann, E. R. I. C., Lundell-Magnuson, E., Mosala, M., Hlangwani, K. F., Mavasa, M. R., Mugogovhali, T. V., Nyamande, T. P., Ramugondo, R., Randela, R., Stathers, T. E. and Kleih, U. K. 2003. Developing a rodent management strategy for South Africa's Limpopo province. In: Singleton, G. R., Hinds, L. A., Krebs, C. J. and Spratt, D. M. (Eds), *Rats, Mice and People: Rodent Biology and Management*. ACIAR Monograph 96. Australian Center for International Agricultural Research, Canberra, Australia, pp. 418–21.

Wildey, K. 2002. UKs grain storage picture - production, storage and pests. In: Zdarkova, E., Wakefield, M., Lukas, J. and Hubert, J. (Eds), *Proceedings of the 2nd Meeting of the Working Group 4, Biocontrol of Arthropod Pests in Stored Products*, Prague, 30–31 May 2002. Research Institute of Crop Production, pp. 23–6.

Zufiaurre, E., Abba, A. M. and Bilenca, D. N. 2019. Assessment of stakeholder perceptions of the damage to silo bags by vertebrate species in Argentina. *Human Dimensions of Wildlife* 24(1), 80–6. doi:10.1080/10871209.2019.1538434.

# Part 2

## Storage technologies

# Chapter 5

## Advances in bulk storage of cereals and grains

*Michelle A. Friedmann and Dirk E. Maier, Iowa State University, USA*

1 Introduction
2 Storage structure types: upright silo structures
3 Horizontal building structures
4 On-the-ground piles
5 Maintaining and preserving quality and quantity: preventative strategies
6 Maintaining and preserving quality and quantity: intervention strategies
7 Best management practices: sanitation, loading, aeration and monitoring (SLAM)
8 Best management practices: dry, inspect, clean and examine (DICE)
9 Conclusion
10 Abbreviations
11 Where to look for further information
12 References

## 1 Introduction

Postharvest losses can result from poor storage conditions of grain and inadequate practices that cause insect infestation and growth of microorganisms. As the world's population has continued to rise, the production of cereals and grains has increased to meet the growing demand for food, feed, biofuel and industrial products. Over the last decade, advancement in technologies such as improved harvesting practices and genetically modified crops has helped generate the largest amounts of cereals and grains ever produced.

Cereals and grains are produced on a seasonal basis and are typically harvested only once a year. In order to continuously feed the world's population, grain must be stored from one harvest until the next and for periods varying from one month up to more than a year. The storage of grain plays a vital role in the economy of every country. The global grain trade mitigates typical fluctuations in the supply of grain but when severe shortages are anticipated,

http://dx.doi.org/10.19103/AS.2020.0072.05

governments generally step in to assure a stable demand throughout the year. In a number of countries namely China and India, governments still operate strategic grain reserves at the provincial and central government levels. Storing grain from one season to the next and releasing it back into the market throughout the year can smooth out fluctuations in market prices, which benefits both consumers and producers (Proctor et at., 1994). Proper grain storage allows consumption throughout the year, and surplus grain to be exported, which is of economic, political and social importance (Richard-Molard, 2003).

Commonly used storage methods include silos, bins and tanks, horizontal structures such as warehouses, and on-the-ground piles. The size of these structures (both on-farm and off-farm) has been increasing around the world in locations where large quantities of grain are produced, handled, dried, stored, transported, processed, exported and/or imported. Grain storage facilities are not only useful for distribution of grain, but they also provide other services such as initial processing (drying, cleaning, aeration, fumigation, blending), merchandising (quality assessment, weighing), and storage (Boumans, 1985).

Physical, chemical and biological factors are responsible for causing significant qualitative and quantitative losses in grain during storage (Pandharinath and Chandra, 2014). Environmental factors, type of storage structure used, length and purpose of storage, method of storing grains and so on are all things that impact the quality of the grain during storage (Pandharinath and Chandra, 2014). Grain stored in bulk is particularly difficult to monitor and preserve because of the large quantities of grain, but it is essential to preserve quality and minimize losses.

Numerous prevention/intervention strategies have been established to maintain and preserve the quality of grain during bulk storage. These strategies include ambient and/or chilled aeration, sealing and fumigation, modified atmosphere treatment, ozonation, cold plasma treatment and so on. Monitoring the quality (and quantity) of these larger bulk volumes during the storage period and initiating an intervention strategy based on an early warning signal is also challenging because the warning signal needs to be assessed within the context that a proposed intervention/treatment strategy has to be applied to (and thus will affect) the entire stored grain mass.

Best management practices (SLAM/DICE) are proven but need to be more widely known and adopted. These practices may also need to be adjusted and further refined for larger volumes of bulk storage. Sanitation, loading, aeration and monitoring (SLAM) was specifically developed to teach best management practices of bulk stored grain, and assumes grain is sufficiently dry (due to field, sun or mechanical drying) to be placed into storage close to its safe storage moisture content (14% or less for a storage period of up to a year). Dry, inspect, clean and examine (DICE) was specifically developed to teach best management practices of bag-stored grain whether at the smallholder farmer level or aggregated and stacked in a warehouse.

In this chapter, silos, horizontal building structures, and on-the-ground piles will be discussed as methods of bulk grain storage, along with best management practices. Methods to monitor and maintain the quality of grain to minimize losses during storage are presented in a separate chapter.

## 2 Storage structure types: upright silo structures

There are many benefits to storing grain in bulk such as lower overhead costs and labor requirements, rapid handling, efficient and effective fumigation, lower land area requirement and a higher possibility for mechanization (Pandharinath and Chandra, 2014). There are many choices when it comes to designing bulk grain storage, but the essential requirements to store grain safely remain the same. Bulk grain storage must keep grain free from water, insects and birds but must also allow the grain to be easily cooled, dried and disinfected (Proctor et al., 1994).

The first step in choosing the appropriate type of storage for a particular application is to identify the storage need. Storage options that can be considered are round or rectangular, tall or short, steel or concrete, flat floored or hoppered (Proctor et al., 1994). Tall or deep silos have a height/diameter ratio of $H > 5D$ and shallow silos or warehouses have a ratio of $H < 2D$ (Boumans, 1985). In some grain storage facilities, handling is the major function (such as export/import terminals). Handling facilities with short storage periods, high throughput must be designed for many consecutive filling and emptying operations. These high-throughput facilities generally utilize vertical silo construction to take advantage of gravity flow to increase the speed of operation. In other types of storage facilities, long-term storage is the major function and can require horizontal storage warehouses with limited height compared to silos (Boumans, 1985). While designing the appropriate storage facility, one must consider providing adequate volume and proper protection, meeting drying and cooling needs, and establishing satisfactory methods of filling, storage and discharge (Boumans, 1985).

Common types of storage for bulk cereals and grains that will be discussed in this chapter include upright silo structures, horizontal building structures (warehouse) and on-the-ground piles. This section covers upright silo structures.

Upright silo structures, sometimes referred to as tower silos or bins, are the most common structure used by farmers and aggregators to store grain in bulk (Boumans, 1985). Not all tower silos are the same as they are designed differently for particular applications. Round silos are designed for capacities ranging from 1000 metric tonnes (MT) and upwards. Higher storage capacities can be constructed by increasing the height of the silo or increasing the silo diameter depending on the design criteria and considerations. If the diameter of the silo is increased in hopper-bottomed silos, valuable storage space can be lost because the hopper in the bottom of the silo will need to be enlarged as well (Boumans, 1985).

Silos are commonly constructed from concrete or corrugated bolted or smooth wall welded steel although, in some circumstances, plastic silos are used as well. The choice between these materials depends upon a number of design considerations, but ultimately comes down to capital and operational costs (Proctor et al., 1994).

Essential features of silos include nonporous (to keep out moisture) and strong (to resist bursting), low friction (prevent air pockets) walls. Desirable features of silos include durability, storm resistance, fire-proof construction, convenience, aesthetically pleasing, frost resistance, simple construction, low-cost and low maintenance (Shedd and Foster, 2017).

Structural considerations should be made when designing proper bulk storage to allow for silos to resist all loads to which it could be exposed, such as dead loads (weight of structure and stored grain) and live loads (forces from flowing grain) (Boumans, 1985). Silos are typically round and less commonly rectangular-shaped. Uneven pressure of grain within a rectangular structure can cause cracking – especially at the corners (Proctor et al., 1994). Square silos require reinforcement and have to be made much heavier than a round silo of the same size to handle these extra stresses. This makes the round grain silo economically superior to the rectangular silo (Boumans, 1985).

Many considerations must be made when designing an upright silo structure. Construction materials of the silo, structure style, aeration, roofs, headspace, sealability and foundation are all factors that must be of concern when designing the appropriate structure.

## 2.1 Construction materials

### 2.1.1 Concrete

Concrete silos are constructed from cement and other materials such as steel rod reinforcements and various additives so as to improve the characteristics of concrete. Concrete silos are most typically used for applications that require a long operation life and minimum maintenance (Boumans, 1985). They are typical for grain-processing facilities such as wheat flour mills and for river terminals and export elevators where large quantities of grain are handled and transferred. Concrete has many functions including holding the silo reinforcements in place, providing rigidity to the structure, protecting the grain from the outside environment (impermeable) and reducing the vertical wall pressures within the silo (Boumans, 1985). Concrete silos can range in height from 30 to 50 m and even more (Boumans, 1985).

The benefits of using a concrete silo over other building materials are that concrete:

1. can breathe and adapt to surrounding environmental conditions reducing the risk of condensation, concrete is a poor conductor of heat,
2. is easy to construct with, can be poured in any shape,
3. is strong, has a large load-bearing capacity for both vertical and radial loads,
4. is water and fire proof,
5. is hardly influenced by chemical agents, and
6. requires little maintenance (Boumans, 1985).

Although there are many advantages to using concrete for silo construction, the limitations are worthy to take note of:

1. heavy, has to be supported by strong soil and a strong foundation,
2. installing conveyors and other accessories is more difficult and time-consuming than with steel construction,
3. inside finish of the walls is rough,
4. increased wall thickness for concrete compared to steel results in a loss of volume, and
5. overall building time and cost are significantly higher compared to steel. (Boumans, 1985).

### 2.1.2 Corrugated bolted and smooth-wall welded steel

Steel is also commonly used for the construction of bulk grain silos because of its high tensile strength and low cost. Steel silos are economically viable for bulk storage of grain (> 25 MT), but are often too costly for small-scale storage (Proctor et al., 1994). These silos are constructed in the shape of a cylinder from either smooth or corrugated galvanized steel (Proctor et al., 1994).

Steel silos are placed on foundations (typically concrete slabs) to support and hold them in place, and facilitate grain handling. In a steel silo, the relatively thin steel sheets handle the pressure of the live load created by flowing grain. Center-corrugated steel sheets are pressed and cut from rolls of smooth-walled steel. They are more commonly used for wall construction rather than flat steel sheets because of their higher tensile and bending strengths. The horizontally corrugated sheets of the silo take tensile loads of the product mass while external steel stiffeners take vertical friction loads on the corrugated wall sheets. Corrugated sheets are heavier gauge (thicker) in the bottom rings and thinner near the top to minimize total dead weight and cost of the structure. In large diameter silos, tensile strength is gained by laminating multiple heavier gauge corrugated sheets for the respective rings from bottom to top.

In terms of construction, steel silos are erected by bolting together the top ring of corrugated sheets and attaching the slanted roof while sitting on the foundation. The silo is then lifted with hydraulic jacks just far enough off the ground to bolt the next ring to the bottom of the top ring. This process continues until all sheets for all rings are bolted together and the silo is fully erected. The vertical wall stiffeners, which should always be attached to the outside wall, are attached to the rings section by section. Smooth-wall silos are erected similarly but instead of bolts, steel sheets are welded together.

Relative to concrete, steel silos are comparatively light and can have a higher load-bearing capacity compared to concrete silos with the same outside surface area (Boumans, 1985). Steel silos are widely used from smaller scale on-farm storage of 10 MT or less to large-scale commercial silos holding 50 000 MT. The relatively low cost of steel silos is attractive to customers. Steel silos are easier to install with the smallest ones prefabricated in workshops. They can be installed in relatively short time, use readily available materials and the intense competition among manufacturers provides the customer with options and competitive pricing (Boumans, 1985).

The benefits of a steel silo over other building materials are:

1. smooth inside surface of the silo, lower coefficient of friction,
2. easy installation,
3. complicated designs are possible,
4. easily enlarged by adding rings to the bottom,
5. easy to add aeration for temperature control, and
6. easy to add level controls, man-holes, roof vents, exhaust air extractors and so on.

Although there are many advantages to using steel for silo construction, the limitations are worthy to take note of:

1. increased maintenance, painting,
2. corrosion from chemicals and rusting due to moisture,
3. care must be taken to evenly unload grain to avoid structural damage,
4. large diameter causes core flow discharge,
5. condensation on inside steel wall and roof surface can drip on grain and cause spoilage, and
6. difficult to seal and fumigate due to large number of bolts (8000–10 000 in a 1000 MT silo) and open eaves.

In general, steel silos do not last as long compared to concrete silos and require careful engineering design. A common issue with steel silos is failures due to large, uneven grain loads causing wall collapse, particularly during emptying. Proper engineering is necessary to assure strength of steel walls to prevent complex stresses at the vertical seams of bolted wall sheets (Proctor et al., 1994). When tall silos (> 25 m) are required, the cost of steel can become economically unfavorable compared to concrete because of the heavy-duty steel reinforcements necessary to prevent buckling of the walls (Proctor et al., 1994). Tall concrete silos only require enough steel to combat the hoop tension forces because the vertical compression load is carried by the concrete itself. In coastal areas, concrete is generally preferred for construction because of the higher risk of corrosion from salt-laden ocean air.

### *2.1.3 Plastic*

Synthetic materials such as glass-fiber reinforced polyester can also be used to create prefabricated bulk storage silos (Fig. 1). Many attempts have been made to construct small-scale storage silos out of plastic using synthetic materials such as rubber (O'Dowd, 1971) and high-density polyethylene (HDPE) (CFTRI, 1975). However, there were many issues with them including high cost per amount of storage capacity and damage by pests (Proctor et al., 1994).

Plastic silos can be designed to be clear, which allows the customer to see the fill level at any time. Assembly of plastic silos is quicker and easier when compared to concrete or steel silos. The smooth interior wall and hopper surfaces reduce bridging and eliminate grain-clogging caused by seams. Also, plastic silos do not rust or corrode and are easy to clean-out.

**Figure 1** Plastic grain silows promoted by the World Food Program to prevent post-harvest loss. Source: World Food Program (2020).

Recently, portable plastic silos have become available commercially in sub-Saharan Africa because of efforts by the World Food Program to expand the adoption of hermetic storage technology among smallholder farmers to help them reduce postharvest loss (WFP, 2019). Plastic silos holding 300–1000 kg are manufactured by the same companies that manufacture plastic water tanks which are ubiquitous in sub-Saharan Africa. The Kenya Bureau of Standards has introduced a standard for portable rigid plastic hermetic grain silos (KS 2874:2019) 'necessitated by the need to use hermetic storage technologies for storage of dry food commodities, their derived products and seeds in order to reduce postharvest losses and increase the shelf life without the need to use pesticides' (KEBS, 2019).

## 2.2 Structural style

When designing a silo, it is important to consider what structural style will best fit the needs. There are two types of silo bottoms: flat-bottomed and hopper-bottomed (Fig. 2). Both of these designs serve specific functions.

Hopper-bottomed silos are constructed with a sloping floor and are usually suspended completely above ground to avoid the risk of groundwater infiltration and to more easily add conveyors to the design (Proctor et al., 1994). The slope of the hopper bottom profile is essential to the silo-unloading process and must take into account the flow properties and mechanical behavior of the products being stored with the inside silo wall friction (Boumans, 1985). When high throughputs are required (such as import/export ports), it is desirable to have a hopper-bottom bin with a sloping floor for self-emptying purposes. To minimize the cost of hopper bottoms while maintaining the same volume, designing a taller hopper with a smaller diameter is generally more economical.

Flat-bottomed grain storage has increased volume capacity compared to hopper-bottomed bins. However, these bins are more difficult to empty because the grain is not discharged easily by gravity although many methods

**Figure 2** Flat-bottomed storage (left) and hopper-bottomed storage (right). Source (left): GSI Group (n.d.); Source (right): EWorldTrade (n.d.).

have been designed to facilitate grain unloading from flat-bottomed silos. For instance, portable conveying equipment, sweep conveyors, pneumatic systems and front-end loaders can be used. Sweep conveyors are commonly buried under the grain when the silo is full and automatically rotate around the center of the silo to draw grain to the unloading point once the gravity discharge is complete. The cost of installing permanent grain sweep conveyors is relatively high and may not be economically feasible if the silo is used for long-term storage and is only unloaded once or twice a year (Proctor et al., 1994).

## 2.3 Aeration

Another important aspect to preserve the quality of grain during storage is aeration. Aeration is the process of forcing air through grain for cooling and drying to achieve a sufficiently low-moisture content, retain the maximum grain quality, and minimize the detrimental impacts of insects and microorganisms during storage (Fig. 3) (Proctor et al., 1994). In sealed storage, aeration can also be used as effectively as in unsealed storage if the silo has an air-exhaust ventilator that can be sealed whenever fumigation is carried out (Proctor et al., 1994).

Aeration with the purpose of cooling grain is beneficial in many ways. Cooled and dried grain reduces the growth of insects and the rate of microbial and mold development. It can also preserve the germination viability of the grain and can even extend the effectiveness of grain protectants when used. It has been found that if the grain temperature can be reduced below 10°C wet bulb, insect growth can be stopped altogether (Proctor et al., 1994).

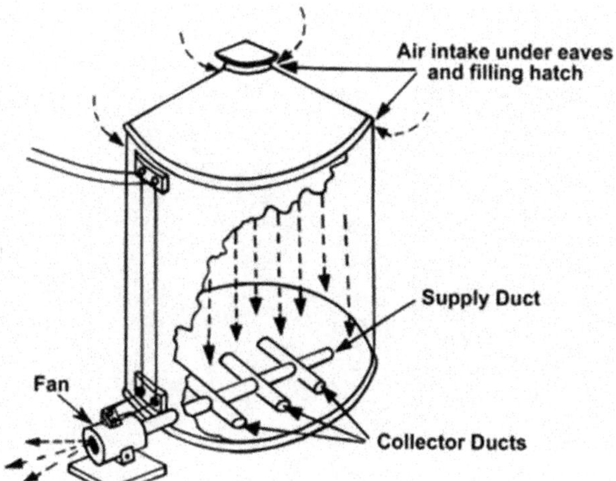


**Figure 3** Silo aeration diagram. Source: Worley & Hammond (2015).

Aeration systems must be designed properly to achieve effective and efficient cooling. The energy required to force air through the grain mass is dependent upon key factors including airflow rate, grain depth and grain physical properties such as surface and shape factors, kernel size distribution, moisture content, and quantity and nature of contamination, foreign material, broken grain and so on. (Proctor et al., 1994).

The requirements of ambient aeration in regards to design of the storage structure include some form of perforated ducting on the floor in which air can be blown into the grain. The ducts must be sized adequately for the required airflow rate to achieve the desired level of cooling and must be located to ensure equal distribution of air throughout the grain mass for uniform cooling. As a rule of thumb, ducts should be sized to limit air velocity to be no higher than 10 m/s within the duct, and the airflow into the grain should be limited to around 0.2 m/s (Proctor et al., 1994). The structure must also include venting above the grain to allow air to be expelled from the system.

In pressure aeration, fans are used to push ambient air from the bottom of the silo upwards (Fig. 4). In comparison, suction aeration reverses the fan to pull air from the top of the silo downwards. For most applications, pushing air up through the grain mass is preferable for several reasons. It typically takes several days to fill a large silo during the harvest season. As soon as the perforated floor or ducts are covered with grain, aeration fans should be turned on to take advantage of getting a cooling front started through the grain mass. By pushing air up, grain layers added are cooled while lower layers remain cool. By the time the silo is full, the cooling front will have moved through a substantial depth of the grain mass instead of just starting with the last layer filled. Additionally, most grain spoilage and infestation problems occur in the upper layers of the grain mass. Whenever such self-heating problems are discovered, the aeration fan

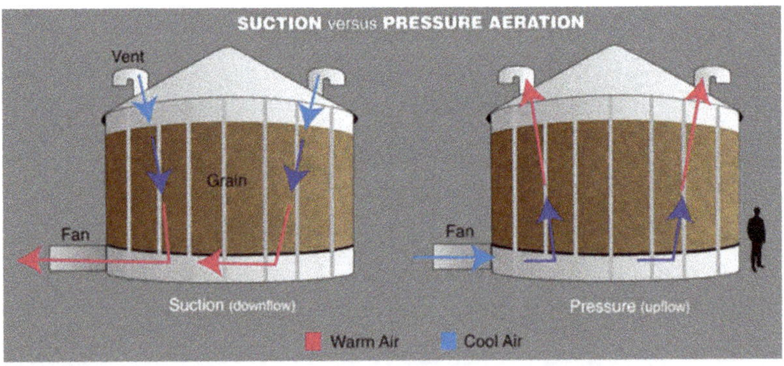


**Figure 4** Suction versus Pressure Aeration. Source: USDA Agricultural Research Service (2010).

can be turned on to push the associated heat and moisture accumulation out of the grain silo within a few hours rather than having to move them down through the entire grain mass. Recently, the USDA Agricultural Research Services (ARS) investigated the use of suction versus pressure aeration within a silo as means of reducing grain temperature to control insects (Fig. 4. The study consisted of two eight-month trials with six silos storing 1250 bushels of wheat. The results of the study showed that during the summer, suction aeration was more efficient at cooling the upper portion of the grain mass and that resulted in fewer insects within the silo. The researchers concluded that suction aeration is advantageous for rapid cooling of the top of the silo, where insects from the outside environment are more likely to initially infest when wheat is first placed in storage (Shipman, 2010).

## 2.4 Roof and headspace ventilation

The silo roof is critical in strengthening the silo, protecting the grain from the surrounding environment and needed for sealing the silo for fumigation purposes (Boumans, 1985). The silo roof should be designed to adequately withstand the weight of snow and the strongest winds that are expected for the area. It is also important that the roof be adequately vented to allow aeration air to exhaust and to relieve a shockwave in the event of a dust explosion. The material used for the roof should be fire-proof, strong and watertight. Condensation can occur because of temperature changes, so the roof should provide adequate venting and forced air extraction to prevent water from dripping onto the grain surface. Steel roofs on silos typically have a 30-degree slope to accommodate grain piling angles of repose that are less. Ventilation of the roof is needed (even in sealed silos) to withstand the pressure differentials that occur due to temperature variations between the outside and the silo headspace. Steel silos are particularly susceptible to these pressure variations because of the thermal conductivity of the steel. For sealed silos, the pressure differential can be minimized by painting the steel surfaces white to reduce the absorbance of the sun's radiation during the daytime (Boumans, 1985).

Silo roofs consist of multiple pie-shape steel panels that are cut and bent from steel coils with a smaller gauge thickness than the corrugated wall panels. The panels are assembled on the ground and bolted to the top silo ring, which is then raised to attach it to the next ring. To construct the roof for a concrete silo, first false work must be constructed to support the concrete until it sets and then it can be removed. A square twisted steel bar can be embedded in the concrete at the base of the roof and extending continuously around the silo to reinforce the structure and resist the outward pressure due to the weight of the roof. In addition to this a black wire mesh reinforcing should be used in the roof to increase stability (Shedd and Foster, 2017).

## 2.5 Sealed versus sealable

Silos can be constructed to be sealable with the purpose of creating a modified atmosphere. There are many benefits to sealing a silo including fumigation, prevention of contamination and dust migration, and the inclusion of inert gases for preservation of the grain or for extinguishing a fire within the grain mass (Boumans, 1985). When new silos are constructed, they should be designed to be sealable and ideally airtight for fumigation. The benefits of sealability are huge and the cost of creating a sealable silo is relatively small, and is lowest during the construction stage. There are no disadvantages to sealable stores and ventilation can easily be added to them when it is required (Proctor et al., 1994). Sealing of existing silos that were not designed to be airtight can be done but is labor intensive and more expensive, and thus can prove to be uneconomical (Proctor et al., 1994).

In a silo that is not sealed and atmosphere controlled, oxygen from the surrounding environment is always available to molds and insects for respiration. This respiration produces heat and moisture resulting in decreased quality of the grain. The only way that this process can be slowed down is by using aeration to remove heat and moisture from the silo, or stirring and blending to decrease the overall temperature and moisture content to a safer level (Richard-Molard, 2003).

The silos that are the easiest to seal are steel silos that are fully welded which forms an effective seal. In the past it was believed that it was difficult to maintain airtightness in concrete silos without extreme measures, however, concrete silos can be sealed easily with properly detailed joints and maintenance to prevent cracking of the silo walls. It is easier to maintain gas-tightness in independent silos compared to interconnect silos (such as in a battery) because of reduced stresses on the silo (Proctor et al., 1994). When designing a sealed silo for either type of material, it is essential to properly design the opening in the structure (such as grain inlets and discharge valves, doors, etc.) to be fitted with proper gaskets to ensure they can be sealed.

It is worth noting that concrete silos are not ideal when carbon dioxide is to be used for fumigation. This is because carbon dioxide reacts aggressively with the calcium hydroxide in the cement matrix of the concrete. The process is called carbonation and the reaction occurs naturally. It forms calcium carbonate and water. During fumigation with carbon dioxide, high concentrations are used and large amounts of gas are 'absorbed' by the concrete giving rise to loss of concentration. Also there is a risk of high negative pressures developing in the silo if it is well sealed. Carbonation is not detrimental to the concrete, but can increase corrosion of the reinforcing steel within the concrete by chemically neutralizing the alkalinity of the calcium hydroxide (Proctor et al., 1994).

The risk of condensation is something that must be considered in sealed storage, particularly when using steel structures. It has been hypothesized that there should be less risk of condensation in sealed storage versus unsealed storage when the stored grain is at or below a safe moisture content for storing. The fact is that condensation is a function of the difference between the dew point temperature of the headspace air and the temperature of the silo roof. As long as the dew point temperature is lower than the roof temperature, no condensation will occur, whether a silo is sealed or not.

Sealed storage should be tested for airtightness when first commissioned. Testing should also be conducted routinely and each time before fumigation is to take place (Proctor et al., 1994).

## 3 Horizontal building structures

Another common type of storage structure for bulk grain is horizontal buildings such as warehouses with flat or V-bottom floors. Warehouses are constructed for the purpose of storage and protection of goods. All horizontal warehouse structures consist of the same basic design elements including a floor, walls, roof and entrances. However, warehouses can vary considerably with the addition of different elements such as ventilators, windows, lighting, equipment and so on. The features selected for a particular warehouse should be selected carefully based on factors such as location, intended use, frequency of loading and unloading, and so on. Bulk grain is often stored in horizontal warehouses as opposed to silos for multiple reasons (Boumans, 1985) including:

1. Large amounts of grain must be stored for a long period of time with access to ventilation,
2. The soil does not have a high enough supporting capacity to allow the construction of vertical silos,
3. Local regulations do not allow high structures,
4. The building can be used for storage of other materials when grain is not being stored, and
5. Product that is cohesive and has a low flowability (e.g. soybean meal) must be stored.

There are also numerous disadvantages of using horizontal buildings to offset the positive reasons including:

1. The flowability of the product is limited by gravity and the outflow capacity is lower,
2. The loading and unloading of the warehouse is very labor-intensive generally requiring front-end bucket loaders,

3. More ground surface area is used for the same volume of storage in a vertical silo,
4. A larger roof is needed and increases the risk of water leaks, requiring more maintenance,
5. It is difficult to utilize automated equipment such as unloading conveyors,
6. Contamination between different types of grain in the same facility can occur,
7. Fumigation is more difficult, and
8. Large volumes of dust can float around inside the warehouse.

Even with these disadvantages, horizontal warehouses are frequently used to store recently harvested grain in bulk especially soybeans. Importance should always be placed on ensuring the harvest heat can be removed with aeration, and the quality of the grain can be maintained and will not be adversely impacted by physical factors such as moisture and heat. In Europe and Australia horizontal buildings are often equipped for in-store drying of harvested grain, which requires level-filling and air ducts to ensure uniform distribution of high airflow rates. In China, most of the strategic grain reserves of maize, soybeans, rice and wheat are stored for three to five years at a time in horizontal buildings. In Brazil and Argentina, horizontal buildings often have V-bottom floors that substantially increase the storage capacity.

Warehouses should be designed to protect grain against rodents and birds, as well as be suitable for fumigation when necessary. Two entrances are needed at a minimum to rotate stocks according to the 'first in, first out' principle. Other things to consider when designing a horizontal warehouse structure are the ease of maintenance and cleaning, as well as the quality of working conditions. Adequate lighting is essential for the safety of workers inside the warehouse, but is generally only justified when employees are regularly working during hours of darkness (Proctor et al., 1994).

## 3.1 Construction materials

### 3.1.1 Post-frame

Horizontal warehouse structures are typically constructed with either a steel or wood frame. The building frame provides structural support for wind and snow loads and strength to support other equipment such as grain conveyors, catwalks and lighting.

There is much debate related to the use of wood versus steel framing in construction, and careful consideration must be made to identify the best material for the project to fulfill the operational needs. Steel framing (Fig. 5) is stronger and is not flammable. Steel will not shrink, warp, split or mold compared to wood. This will result in lower maintenance costs over time. Wood framing (Fig. 6) is popular in North America because it is relatively inexpensive, readily available and easy to work with. However, wood must be of good quality and treated to protect from termites in order to last.

Steel and wood are generally cost-competitive although the construction cost of steel framing is normally higher due to the differences in labor costs. Wood is chosen for many applications because of the lower short-term costs and the quicker lead time to obtain the materials. Steel structures are popular in South America, Europe and China.

**Figure 5** Horizontal warehouse storage with steel framing, concrete walls for separation and fabric roof. Source: Legacy Building Solutions (2013).

**Figure 6** Horizontal warehouse storage with wood framing, steel roof and concrete sidewalls. Source: Shed Force (2020).

### 3.1.2 Concrete floor and sidewalls

The foundation for horizontal warehouse structures is typically constructed from reinforced concrete due to the strength and rigidity of concrete and the need to withstand large forces from the weight of the grain. In horizontal storage structures, reinforced concrete sidewalls around the perimeter of the structure allow stacking of grain to a greater depth compared to wooden walls. Concrete sidewalls extend 1.2 m underground to resist the lateral forces of the stored grain (https://bucklinbuildings.com/benefits-of-fabric-cover-grain-storage-buildings/) ('Benefits of Fabric Cover Grain Storage Buildings', n.d.). Wooden walls constructed with boards mounted against the wooden frame posts protect the steel walls of the building. Movable walls can also be constructed within the structure to accommodate multifunctional storage including sections or bunks for different grains or feed ingredients. These sidewalls have to withstand large forces due to the weight of grain they are to support, and thus need to be properly designed for loading and unloading. The flooring of horizontal warehouse structures can be concrete, asphalt or gravel depending upon what is most economical for the structure and the movement of unloading equipment.

### 3.1.3 Roofs

The roof of horizontal warehouse structures can be constructed from sheet metal or fabric panels. Fabric structures for grain storage create a water-tight, wind-tight roof that provides support to prevent sagging from a heavy snow load or windy conditions. Fabric is generally more economical, but less common and more temporary than sheet metal for roofing. It also allows more natural light into the facility (Fig. 5).

## 3.2 Structural style

Generally, warehouses are constructed with a flat bottom. However, this design presents many challenges when it comes to unloading stored grain. Flat-bottom facilities do not take advantage of gravity and are much more labor-intensive and time-consuming to unload. Equipment such as portable grain augers, pneumatic grain conveyors or human operated heavy machinery can be utilized to quicken the process, but are nevertheless less efficient than V-bottom buildings that utilize gravity to speed up the unloading process. In V-bottom buildings, the bottom of the warehouse is designed with the two long sides sloping toward the center of the structure. A conveyor tunnel is constructed beneath the flooring that allows grain to be discharged by gravity (Fig. 7).

V-bottom buildings are more costly and time-consuming to construct than flat-bottom buildings. They also require locations with low groundwater tables, and are therefore less common. Advantages include quicker load-out and more frequent load-unload cycles. For example, in Argentina V-bottom warehouses are constructed at ports where frequent loading and unloading of soybeans

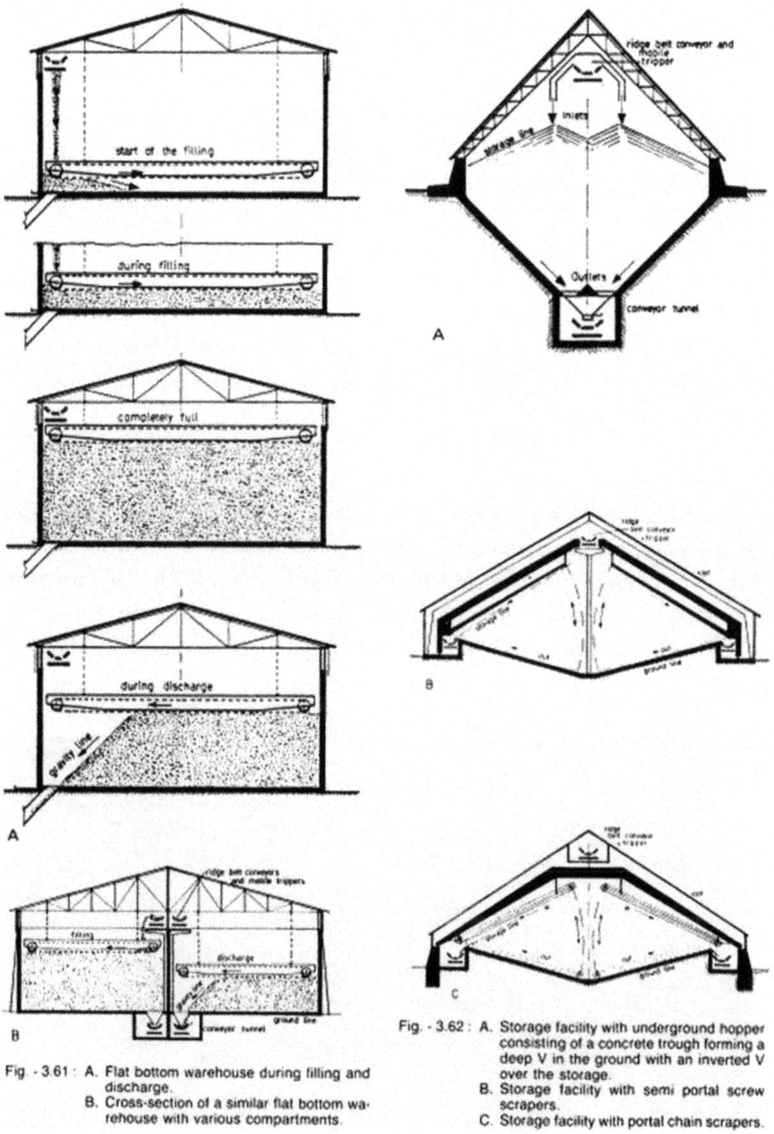


**Figure 7** Flat bottom warehouse versus underground V-bottom warehouse. Source: Boumans (1985).

occurs. If the primary purpose of the facility is for long-term storage with minimal unloading, a flat-bottomed foundation is most likely the most economical.

When designing the floor of a horizontal warehouse storage structure, ventilation is something that should be considered. In-floor aeration forces air through the grain mass and minimizes grain spoilage while reducing the horsepower of the aeration fan and saving electricity.

### 3.3 Aeration

Cooling grain with aeration is extremely important for proper storage. Cool temperatures minimize mold growth, limit moisture condensation and accumulation, and prevent insect development. Perforated ducts (full or half-round) placed on the floor work well for flat storage buildings (Fig. 8) and piles. If the grain mass is leveled, as is common in Europe but not in North America, duct spacing should be about equal to the pile depth. For a long triangular-shaped pile, one duct centered under the peak and running the length of the building may suffice as long as the building is not too wide. Unusual pile shapes make aeration design tricky. An experienced aeration system designer should be consulted in these cases.

Perforated steel aeration ducts (round or half-round) are the preferred choice, although plastic aeration tubes can also be used. Ordinary plastic drainage tile should be avoided because it does not have enough perforated surface area for good air movement. An aeration airflow rate of about 1/10–1/5 cubic feet per minute per bushel (CFM/bu) is recommended for dry grain (Maier and Wilcke, n.d.) As with silos, aeration fans can be operated automatically based on temperature and relative humidity of the air and monitoring of conditions in the grain mass.

**Figure 8** Plastic aeration tubes in a horizontal storage building. Source: Advanced Drainage Systems (2020).

### 3.4 Roof and headspace ventilation

Ventilation openings are essential within a warehouse to allow for the exchange of exhaust air with the outside environment. For horizontal storage buildings, positive pressure air blowing into the aeration ducts, and out of the top of the grain mass, is preferred. Ventilation openings should be located along the entire length of the roof ridge for air to escape. Ideally, high airflow fans should be installed at one end of the building above the grain surface to draw air from the other end of the building sweeping the underside of the roof and the grain surface to avoid moisture condensation especially at night when the dew point temperature of the headspace air drops below the temperature of the steel roof. As a rule of thumb, there should be 1 square foot of vent area for every 1000 cubic feet of air per minute from the aeration fans (Maier and Wilcke, n.d.). Roof vent openings should be covered with mesh screening to prevent birds and insects from entering the storage facility.

### 3.5 Sealed versus sealable

Horizontal warehouse buildings are more difficult to seal compared to silos. However, it is not impossible to seal them for fumigation purposes if special care is taken during the design and construction of these structures (Fig. 9). Attention to detail must be paid to ensuring all bolts are fitted with suitable sealing washers and all joints are carefully sealed with sealants. When

**Figure 9** Sealed horizontal storage building in Australia equipped for suction aeration (air drawn by fans installed on the sidewall above the level grain mass) and recirculation fumigation. Source: Dirk E. Maier (2000).

fumigating these structures, it is oftentimes more effective to cover the grain from within the structure. A plastic tarp can be placed over the grain mass and sealed adequately for fumigation. This is a typical practice in China where horizontal storage buildings used as part of the national grain reserve are built airtight, aeration and exhaust air ventilation systems are sealable, and the grain mass surface is leveled and covered with foam board for insulation and plastic tarp for fumigation.

## 4 On-the-ground piles

On-the-ground piles of grain are a temporary method for storing large volumes of grain when other storage options have been exhausted (Fig. 10). They are commonly seen at grain facilities in Australia, India, China, Eastern Europe, and North and South America during harvest times, especially when the harvest is unusually large. In Australia and South America they are referred to as 'Australian bunkers'; in North America, 'ground piles'. Grain can successfully be stored in bunkers and piles for medium- to longer-term 'temporary' storage if factors such as site preparation, layout, aeration and grain moisture content are taken into account. On-the-ground piles can be used until other permanent storage opens up for grain to be transferred. In reality, in the United States on-the-ground piles have become a permanent asset at most facilities and are used every season and for longer storage periods than is typically advisable. Covered piles are considered temporary storage while uncovered piles are only to be used for emergency storage.

Bunkers and piles are low cost in terms of construction materials and time to set up. The structure is minimal compared to silo and warehouse storage

**Figure 10** Corn stored in an uncovered pile on the ground which at commercial facilities in the United States is only allowed on an emergency basis. Source: Maulsby (2016).

and typically consists of side-wall support and a cover. Storage of grain in bunkers and piles is most successful when care is taken to ensure the quality of the grain is maintained. Attention to storage management and monitoring of the grain is required to minimize losses during storage. For example, corn that is sufficiently dried to 15% moisture or less can generally be stored in a covered pile and with aeration during the cooler fall and winter weather in the Midwestern United States. If the grain is carried into the spring, it needs to be dried to 14% moisture content to avoid high damage levels due to warm and humid air being drawn into the piled corn (Maier and Wilcke, n.d.).

Uncovered piles are the cheapest form of temporary grain storage but the risk of damage to the grain is high due to the exposure to water, insects, birds, animals, molds and so on. Grain should only be stored in uncovered piles at the end of harvest when the crop is the driest and the weather is the coolest and should be considered as emergency storage and moved to permanent storage as soon as possible. By waiting until the end of harvest to store grain in an uncovered pile, the duration that grain will be exposed to rain is shortened. Low grain temperatures are preferred because colder temperatures minimize mold growth, limit moisture movement and condensation, and control insect development.

It is not practical to check grain temperature and moisture content in bunkers and piles. Instead, carbon dioxide monitors can be used to check the exhaust air from fans that draw air through pile to keep the cover from blowing off. This should be done at several locations every two or three weeks to monitor the quality of the grain. Bunkers and piles are difficult to manage, but are unavoidable for stored grain managers, who often must deal with them in many geographies.

## 4.1 Construction materials

### 4.1.1 Pad

Before constructing an on-the-ground pile, the storage size must be calculated to design for the appropriate volume. The location selected for storage should be well drained (1–2% slope) and a storage pad can be constructed to additionally protect the grain. A pad can be constructed from a mixture of lime, fly ash, or cement in the soil that is compacted to reduce the water permeability. A plastic sheet on the surface of the pad can prevent ground moisture from wetting the grain but is not necessary. Tarp is placed on the bottom of Australian bunkers so stored wheat can be completely enclosed and ideally sealed for non-aerated storage of dry wheat that has to be fumigated intermittently. When on-the-ground piles are used year-after-year, asphalting the pad is a common practice that will reduce quality deterioration in the bottom layers.

### 4.1.2 Sidewall support

On-the-ground piles often consist of low retainer walls called 'sidewalls' to support the weight of the grain and to contain it within the bunker and pile. The sidewalls are usually arranged in a circular, elongated or rectangular manner. Sidewalls can be constructed out of a variety of materials. Self-supporting or free-standing walls are configured triangularly (see AE-84 and AE-92 for more details on their design) (Maier and Wilcke, n.d.). If these walls are made of screen-covered wood or perforated steel, they can also provide air into the pile for aeration (Fig. 11). The amount of storage will be maximized when the maximum slope of the pile is achieved. When building the pile, the drop distance from the spout of the conveyor to the pile should be kept at a minimum to achieve the maximum slope and prevent grain from rolling down the side of the pile.

### 4.1.3 Pile cover

Plastic tarp covering on a pile will protect the grain from rain and snow but is expensive, frustrating and labor-intensive undertaking. Tarp covers are highly recommended and should be used to protect the quality of the grain during the storage period (Fig. 12). An uncovered 'emergency' pile exposed to 2.5 cm of rainfall has the potential to rewet the top 30 cm of grain by an additional 9 percentage points of moisture (Mowitz, 2018). That layer will begin to mold and crust over within three to four weeks and result in considerable physical and odor damage to the rest of the pile. The cover will carry rainwater and melting snow away from the piled grain and prevents wetting.

**Figure 11** On-the-ground pile of grain with sidewall supports being loaded with a mobile conveyor and covered with a heavy duty plastic tarp. Source: https://www.youtube.com/watch?v=LeiKenNLHbE.

**Figure 12** Grain pile before being covered (left) and grain pile after being covered (right). Source: ALCO Custom Covers (2020).

Without aeration, condensation under the plastic tarp may cause severe problems and reduce the quality of the grain. When continuous aeration is used, condensation is unlikely to occur on the underside of the cover because the moving air keeps the grain near ambient temperature. It is essential that the airflow be near the plastic tarp to carry the moisture away and reduce condensation. Aeration also creates negative pressure in the grain mass intended to keep the tarp in place.

## 4.2 Structural style

### 4.2.1 Australian bunker

Australian bunkers are typically rectangular and have solid sidewalls as they are rarely aerated (Fig. 13). If aeration is to be applied, aeration ducts should be placed inside the structure so that the bunker can be sealed to achieve aeration cooling and to assure gas-tight fumigation. The same type of tarp covering the piled grain is placed inside the bunker to cover the pad and the sidewalls so a gas-tight seal for fumigation can be achieved. Successful fumigation of grain is critically important in Australia because their export regulations call for zero live insects in grain.

### 4.2.2 US grain pile

In the United States on-the-ground piles are circular, elongated or rectangular. In the Great Plains region where corn and sorghum are typically dry at harvest time, piles may not have sidewalls nor are they covered or equipped with aeration. In the Midwestern region where corn is typically artificially dried, piles have sidewalls that may or may not be perforated (screened), are covered, and are equipped with aeration. Aeration fans run continuously to keep the tarp

**Figure 13** Australian bunker filled with wheat covered with heavy duty tarp tightly secured to prevent wind from blowing it off. Same tarp covers the floor and sidewalls to allow for tight-sealed fumigation. Source: Dirk E. Maier (2000).

in place and as a result draw multiple cooling and warming cycles through the stored grain. If sidewalls are solid, then small-diameter aeration tubing is placed on the pile and under the tarp to allow for air to be drawn into the pile. If sidewalls are perforated (screened), then air is drawn into the pile and out of the perforated on-ground aeration ducts.

### *4.2.3 Argentine silobag*

A silobag is a plastic tube made of a triple layer of thick laminated extruded plastic, typically polyethylene. The bag is filled with a special bagging machine and is long and tubular shaped when it is fully loaded with grain (Fig. 14). The two outer layers are white with UV-protective coating, to reflect the sun's energy in the daytime and maintain the temperature. The third inner layer in contact with the grain is black, to keep temperature low. Once the bag is full, it is sealed to create conditions that preserves stored grain and prevents microorganisms and insects. Unfortunately, silobags are often damaged by rodents or field stubble and thus are not sufficiently airtight to kill insects by allowing carbon dioxide to build up and oxygen to decrease. When infested, silobags can be fumigated.

Silobags are particularly popular in Argentina where they are used to store more than 50% of all harvested cereals and grains. Over 60% of silobags used

**Figure 14** Silobags. Source: Ipesa Silo (n.d.).

worldwide are produced in Argentina, from where they are exported to 54 countries ('What are Silo bags? - Arpaco', 2018).

## *4.3 Aerated versus non-aerated*

Aeration is not necessary when dry grain is stored during the fall and winter months, but is necessary in covered on-the-ground piles that are intended for long-term storage. The problem of getting adequate distribution of aeration air to grain in a large pile is obvious. Aeration systems maintain the temperature of the grain and hold down the tarp on the pile. However, since grain is a good insulator, the temperature of the center of the pile will not change much if the pile core is not aerated. Aeration systems create negative pressure in a covered pile to generate the suction required for keeping the tarp in place. Low-velocity fans are used to aerate and provide about 0.1 $m^3$ of air per minute per tonne of grain that is under 15% moisture (Mowitz, 2018). Typical aeration fans are controlled automatically by wind anemometers that can change the suction based on wind velocity. The aeration necessary to hold down the tarp is generally sufficient for maintaining the grain at the average ambient temperature. If the tarp is held too tightly against the grain surface, insufficient air will enter the grain pile, and no aeration will be accomplished.

The general recommendation on duct spacing is that ducts should be spaced no farther apart than the grain is deep. To prevent the center of the grain pile from heating up, aeration ducts can be positioned in parallel along the axis of a rectangular pile. This allows the core of the pile to cool and makes it easier to remove the corn later. Ducts that are located at the front and back ends of the pile should extend about 21 m beyond the grain pile. For large piles that have a length longer than 61 m, T-shaped ducts are suitable for aerating the core. Ducts can be run in from the sides and intersect at the center with a long duct running parallel to the long axis (Mowitz, 2018).

The use of perforated bulkheads to contain piles of grain offers an aeration option that has proven successful. If a center tower is employed to fill the storage, aeration ducts and exhaust fans can be part of the center structure. The aeration air is drawn in through the perforated retainer wall and exhausted out of the center tower. This may provide the best opportunity to uniformly aerate a tarp covered pile. Air can be drawn along the underside of the tarp and ideally reduce the extent of water condensation caused by rapid swings in nighttime and daytime temperatures (see ASAE Paper 88-6055 for more details as referenced in Maier and Wilcke, n.d.).

## 5 Maintaining and preserving quality and quantity: preventative strategies

During storage, grain quality (and quantity) can be significantly negatively impacted if the proper precautions are not taken. Grain quality is influenced by many factors including environmental conditions during the growing season, cultivation practices, weather conditions at harvest and harvesting technique (Proctor et al., 1994). In general, quality cannot be improved during storage. In fact, quality is easily lost during the storage process and can only be maintained if the correct steps are taken. In most situations, the value of the stored grain (in dollars per MT) is usually higher than the cost of the structure in which it is stored. Investing money in improving the quality of the grain can quickly be recovered if the losses are considerably reduced due to preventative measures (Proctor et al., 1994). Correct handling, regular inspections of grain quality and knowledge of pest control strategies should be practiced by every person that stores grain to minimize losses (Pandharinath Said and Chandra Pradhan, 2014).

The definition of good quality grain varies based on who is concerned with the storage and utilization of the grain. For example, grain storage facilities prefer grain that is undamaged, dry and free of insects whereas millers prefer grain that results in a high percentage of good quality flour, and consumers are more concerned with appearance, flavor and cooking qualities of the grain (Proctor et al., 1994).

Compared to other foods such as meat and produce, grains are relatively easy to store and if done properly will keep for many months with minimal loss of quality and nutritional value. Storing grain becomes difficult when storage temperatures are high, insects are present and the moisture content is above safe levels (Proctor et al., 1994). When these circumstances are present, the correct storage conditions are critical to preserve the quality of the grain.

According to Grain Storage Pest Control Guide, effective grain hygiene and aeration cooling can overcome 85% of pest problems (GRDC, 2013).

Preventative strategies for maintaining grain quality include sealing the structure for exclusion of pests and rodents, and aerating the grain to cool it. Intervention strategies for grain that is infested with insects include gas fumigation, modified atmosphere treatment (hermetic storage) and ozonation (cold plasma treatment). Some of these strategies will be discussed in more detail in this section while others are discussed in more detail in other chapters of this book. This section focuses on the range of preventative strategies.

### 5.1 Sealing of structures and exclusion of pests

Grain is at risk of being exposed to pests and rodents during storage if the storage structure is not sealed properly. There are numerous benefits to sealing grain storage structures such as keeping out water, fumigation, resistance to infestation from insects and rodents, and prevention of rust (in steel silos) (Fowler, 2015). Pests are difficult to control and an integrated approach to manage infestations is the key to success. Sealing of grain silos is one way in which infestation is prevented. Thus, grain storage structures should be designed with sealing in mind.

Certain aspects of grain storage such as aeration mechanisms (drying fans and vents), conveyors and entrances are difficult to seal and are not designed to keep out insects (Fowler, 2015). Closing the openings and sealing cracks can prevent insects from entering the grain and can make the structure as airtight as possible in case fumigation is necessary (Anon, 2004). Sealing vents of adjacent or connected grain silos can also help prevent insects traveling from one silo to the next.

Rodents are prone to entering grain storage and can damage the quality and decrease the quantity of stored products. There are several structural features that can exclude rodents from storage structures. Attention should be placed on rodent-proofing doors, holes and opening, vents and windows, and foundations and floors. Openings greater than a ¼ inch can be sealed using cement, masonry or metal collars (Anon, 2004). Galvanized hardware cloth can be used to screen vents and windows. The gap between the bottoms of doors should be no more than a ¼ inch and the door should be sized appropriately to fit tightly. Sometimes rodents can burrow beneath the foundation of storage structures, so it is recommended that footings extend at least 1 m below the ground's surface. To minimize the number of rodents living near stored grain, replacing weeds and grass with gravel near the perimeter can prevent rodents from burrowing nearby (Anon, 2004).

Sealing grain storage structures is essential to limiting the infestation from pests and rodents after harvest. In Australia, sealing of storage structures has been implemented more so than anywhere else. Hopper silos up to 20 MT are pre-assembled as sealed silos and are then delivered to and erected at on-farm locations. Larger silos and horizontal buildings have to be sealed while being constructed which is labor- and time-consuming.

## 5.2 Aeration with ambient air

Aeration of storage structures is a commonly used method for preserving stored grain by preventing insect development, mold growth and maintaining grain quality (Fig. 15) (Morales-Quiros et al., 2016). Aeration consists of the forced movement of air (ambient or chilled) through a grain mass for the improvement of grain quality and storability (Pandharinath and Chandra, 2014). Typically, aeration is done mechanically by the use of fans and plays an important and effective role in preserving grains (Navarro and Noyes, 2002).

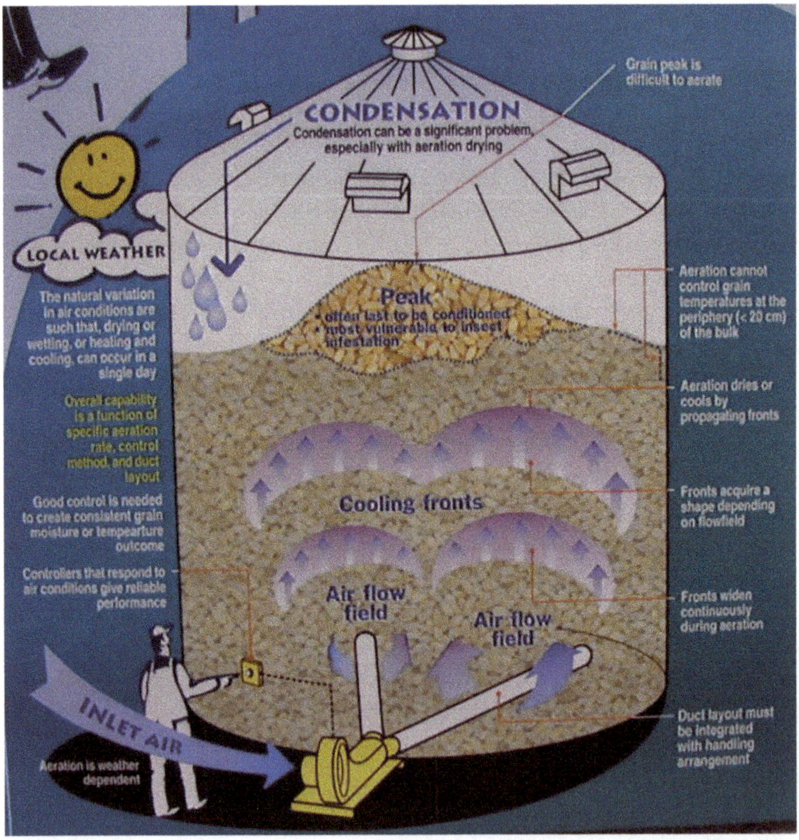


**Figure 15** Aeration and storage management guidance promoting improved stored grain management practices among Australian farmers and bulk grain handlers. Source: CSIRO (2000).

Aeration lowers and homogenizes the temperature of the grain, preventing moisture migration and condensation (Worley and Hammond, 2015). It is important to note that aeration is not the same as drying. During aeration, the moisture content of the stored product changes very little due to the low volume of air that is used. Drying is generally a rapid process where large amounts of heated air are forced through the grain to decrease the moisture content considerably during a short period of time. While grain drying is also used for preservation, it is different than aeration and should not be confused (Worley and Hammond, 2015).

Aeration functions to cool grain and lower its temperature to an acceptable level for proper storage. Maintaining grain at a low temperature protects stored grain from insect infestations for months if it is sufficiently dried prior to storage (TSGC, n.d.). In order to prevent condensation, adequate air must reach all areas of the stored grain to cool it.

Airflow rate is the primary factor that impacts the effectiveness of aeration (Worley and Hammond, 2015). Lower airflow rates are used for upright grain silos compared to horizontal warehouse structures because air distribution is generally more uniform in upright silos. Recommended airflow rates range from 0.05 $m^3$/min/MT to 0.1 $m^3$/min/MT for upright storage and 0.1 $m^3$/min/MT to 0.2 $m^3$/min/MT (Worley and Hammond, 2015).

Heat produced by grain during storage does not flow easily to the surface of the grain mass and must be removed by evaporative cooling to prevent the temperature from increasing (Boumans, 1985). Research shows that grain can be 30°C and higher when it is harvested. This is an ideal breeding temperature for pests within the grain (GRDC, 2013). During upward aeration, warm air rises to the top of the silo and some condensation may occur when the dew point temperature is higher than the temperature of the roof. During downward aeration, the upper layers of the grain are cooled first, which reduces the potential of moisture condensation (Worley and Hammond, 2015).

Fully perforated floors are frequently used with in-bin drying systems and are ideal to distribute air uniformly for aeration. It is not feasible to install fully perforated floors in large diameter silos (Worley and Hammond, 2015). Instead, in-floor trenches or above-floor ducts are placed in silos. A disadvantage of above-floor ducts is they interfere with unloading conveyors (Worley and Hammond, 2015). Therefore, above-floor ducts are typically used in horizontal storage structures. They are typically round or half-round shape and constructed from wood, steel, concrete, plastic or a combination of these materials. The ducts must be designed to be strong enough to support the grain load regardless of shape or material used (Worley and Hammond, 2015).

According to the GRDC, studies have shown that rust-red flour beetles stop breeding at 20°C, lesser grain borer at 18°C and below 15°C all storage pests stop breeding (GRDC, 2013). They recommend that grain be stored at temperatures less than 23°C during summer and less than 15°C during winter. They state that best practices include operating aeration fans continuously for the first 2–3 days of storage to create uniform moisture conditions, then to run the fans for 9–12 hours a day (during the coolest hours) for the next 3–5 days to push a second cooling front through the grain. It is recommended to not operate the fans when the relative humidity of the outside air is higher than 85% because there is a risk that this will add moisture to the grain and cause additional damage. It is important to have even cooling within the grain to prevent the formation of hot spots that favor insect growth and development (Hodgson and Holscher, 2009).

### 5.3 Aeration with chilled air

If the ambient air conditions are not cool enough to effectively and efficiently reduce the grain temperature, air can be chilled to improve its functionality. Research has shown that grain temperature is one of the biggest factors in maintaining the quality of stored grain (TSGC, n.d.). Chilled aeration allows grain to be cooled to safe storage levels regardless of the ambient conditions of the surrounding environment. This practice is particularly applicable in locations where the ambient conditions are not suitable for ambient aeration, including subtropical regions or summer months in the United States (Morales-Quiros et al., 2016). Grain chillers are connected to existing aeration ducts on storage silos and operate using three-phase power (Morales-Quiros et al., 2016).

In the United States, particularly in the Great Plains region, once grain is cooled in the summer, it will not need to be re-chilled. This is because grain is a good insulator and will maintain the reduced temperature until ambient aeration can be used in the fall months. Research has found that chilled aeration significantly lowers insect populations, shows competitive usage costs compared to conventional aeration in conjunction with fumigation, and can cool grain more quickly than conventional ambient aeration (Maier, 1992; Mason et al., 1997).

Aeration with chilled air can be an effective way to cool grain to safe storage temperatures in warm climates compared to ambient aeration.

## 6 Maintaining and preserving quality and quantity: intervention strategies

### 6.1 Gas fumigation

If grain is found to be infested with insects (such as weevils or lesser grain borer), there are three options: feed as is, sell at a discounted price, or kill all life stages of insects (through fumigation or modified atmosphere) (Hodgson and Holscher, 2009). Fumigation with a toxic gas is presently the most effective and commonly used method in integrated pest control strategies for stored grain (GRDC, 2013). Aluminum phosphide is the most common fumigant used on grain. This fumigant breaks down into phosphine gas which is a highly toxic, reactive and potentially explosive gas. Therefore, it is essential that careful planning and monitoring take place when fumigating (Langcuster, 2008). Regulations in most countries require users to have a fumigation plan that outlines the steps taken before, during and after fumigation to improve safety of employees (Langcuster, 2008). In many countries fumigants must be applied by a licensed professional because of their restricted use and risks they pose to people and the environment (Hodgson and Holscher, 2009).

Fumigation is most effective when sanitation, grain leveling, removal of fines and thorough silo sealing have been done in advance (Anon, 2004). The ability to seal a storage structure for fumigation is the first and most important step in achieving an effective fumigation (GRDC, 2013). The storage structure must be pressure-tested and sealed to prevent the gas from leaking and to provide a high enough dosage over a sufficient period of time to kill all life stages of infesting insects. In fact, the label requirement on phosphine products now requires the prevention of leaks (Anon, 2004). Unfortunately, it does not yet require monitoring the fumigation to track gas concentrations, which is a best practice.

Research shows that fumigation must be conducted in a pressure sealed structure to achieve a high enough concentration of fumigant over a long enough period of time to kill all life stages of the insects (GRDC, 2013). The main cause of unsuccessful fumigation is due to improper sealing of the storage. When fumigating horizontal warehouse buildings, it is advisable that a tarp be placed over the grain mass to prevent leaks from the structure which is often difficult to seal adequately (Anon, 2004). In the United States, there are many instances of fumigation failures from using steel silos that were not adequately sealed. Silos with many joints and holes as well as aeration systems such as drying fans, conveyors and sidewall doors are difficult to seal properly and are not designed to keep out insects (Fowler, 2015).

The GRDC recommends that a minimum of 300 parts per million (ppm) concentration of the fumigant gas be applied for seven days, or 200 ppm applied for a longer 10-day period to achieve effective fumigation. Research shows that in silos with small leaks, phosphine levels can be as low as 3 ppm near the leak and result in reduced concentration levels elsewhere within the silo (GRDC, 2013). To monitor and apply the fumigant, electronic monitoring systems are being developed and successfully practiced in different parts of the world and can be beneficial for the detection of leaks (Pandharinath and Chandra, 2014).

The GRDC also recommends grain to be ventilated with aeration fans for a minimum of one day after fumigation or five days if there are no aeration fans. After the ventilation period, a minimum withholding period of two days is required before the grain can be used for animal or human consumption.

Fumigants have no residual activity and stored grain is therefore susceptible to reinfestation within 72 hours after treatment (Hodgson and Holscher, 2009). It is essential to have a properly sealed grain storage structure to prevent insects from reinfesting fumigated grain. However, as a general rule, the silo should only be completely sealed while carrying out fumigation if the grain moisture content is above 8–12%. Otherwise, damage to grain could occur due to condensation. Regular monitoring of insect infestation and moisture condensation in the headspace is essential (GRDC, 2013).

Taking fumigation shortcuts may kill enough adult insects in grain so it passes delivery standards, but the repercussions of such practices are detrimental to the grain industry. If fumigation is not conducted properly (i.e., leaks, short application time, low dosage), the application may kill the live insects but not the larvae and eggs. This will result in reinfestation as soon as the larvae and eggs develop. Moreover, when poor fumigation is conducted, some insects with resistance may survive the treatment and pass this gene on to the next generation which jeopardizes effectiveness of the phosphine fumigation in the future (GRDC, 2013).

Fumigation can be difficult and dangerous so it is recommended that steps be taken to prevent the need for fumigation. Proper sanitation (cleaning and sanitizing of silos and equipment, removal of fines) prior to storing grain can minimize the risk of contamination and the need for fumigation (Hodgson and Holscher, 2009).

## *6.2 Modified atmosphere treatment*

While fumigation is a commonly used and effective method for controlling insect infestation within stored grain, excessive use of contact pesticides can lead to residues in the grain. This can have detrimental effects on human health and the environment as well as on the development of insect resistance (Cao et al., 2019; Cheng et al., 2012). An alternative to fumigation and contact insecticides for controlling insects and preventing mold growth is modified atmosphere treatment. The principle and effectiveness of MA storage has been studied extensively for more than 40 years (Navarro et al., 2012; Navarro, 2012). Modified atmosphere treatment provides an environmentally friendly, sustainable and cost-effective method for protecting stored grain that eliminates the need for chemical treatments, fumigants and climate control (Cao et al., 2019; Cheng et al., 2012; Villers et al., 2008).

The objective of MA treatment is to attain composition of atmospheric gases rich in $CO_2$ and low in $O_2$, or a combination of these two gases at normal or altered atmospheric pressure within the treatment enclosure, for the exposure time necessary to control the storage pests and preserve the quality of the commodity. MA is used as a general term for all cases in which the composition of atmospheric gases or their partial pressures in the treatment enclosure have been modified to create conditions favorable in achieving control of storage insects and preserve the quality of the commodity.

Navarro (2012) organizes MA treatments into three categories: hermetic storage, vacuum treatment and high-pressure carbon dioxide treatment.

1. **Hermetic storage** (also known as sealed storage or airtight storage) is a type of MA storage that is based on the principles that in a sealed structure, the respiration of living organisms (insects) will generate a bio-modified atmosphere that is carbon dioxide-enriched and oxygen-depleted (Villers et al., 2008).
2. **Vacuum treatment** is a type of MA storage that is based on the principles that in a low-pressure environment, there is a close correlation between partial pressure of the remaining oxygen and the rate of insect death (Navarro, 2012). Until recently, this treatment could only be carried out in specially constructed rigid and expensive vacuum chambers. A practical solution has been proposed that uses flexible liners. To achieve the low pressures in the flexible liners, sufficiently low pressures (25–50 mmHg absolute pressure) can be obtained (using a commercial vacuum pump) and maintained for indefinite periods (Navarro, 2012).
3. **High pressure carbon dioxide treatment** is a type of MA storage that is based on the principle that additional carbon dioxide in a high pressure environment (10–37 bars) can shorten the exposure time needed to achieve insect mortality (measured in hours and instead of days). This treatment can only be applied in specially designed metal chambers that can withstand the high pressure needed. These metal chambers are high in initial capital investment and are generally only practical for high-value products such as spices and nuts and are not typically used for treatment of bulk grain storage (Navarro, 2012).

The most common type of MA storage for grains is hermetic storage. Ambient atmospheric conditions are approximately 79% nitrogen, 20–21% oxygen, and 0.04% carbon dioxide (Cao et al., 2019). Through the natural metabolic process of insect respiration, oxygen levels can be depleted below 2% in properly sealed storage where mortality of all insect stages are achieved rapidly although lethal levels have also been reported in the range of 4–5% (Villers et al., 2008). High moisture commodities, such as freshly harvested corn, can successfully be stored hermetically prior to drying or processing because moisture levels stay constant which prevents the growth of molds which causes spoilage. Hermetic storage is more effective when the temperature of the grain is at room temperature (22°C) or above because of the increased metabolism of insects at sufficiently high temperatures. The temperature of the grain mass during hermetic storage significantly impacts the time required to achieve impact mortality. In large-scale hermetic storage, it may take a significant amount of time (weeks or months) for oxygen levels to decrease solely by respiration to levels that are low enough to kill the insects (below 4–5%), especially when infestation is low. In this situation, supplementing biomodified hermetic storage with additional control methods, such as adding carbon dioxide, nitrogen injection or fumigation can help reduce the time needed to achieve mortality (Maier, 2018).

All types of MA storage are only effective when the storage structure is properly sealed. The challenge of creating a gastight structure, especially when the structure is large, is a challenging and costly undertaking. It is extremely difficult to seal permanent structures such as concrete and metal silos to levels that are effective to achieve hermetic conditions. Sealed storage should not be confused with hermetic storage. Hermetic storage is designed for zero air exchange between the inside and outside of the structure, whereas sealed storage allows for some amount of leakage as long as fumigant concentrations do not fall below effective levels. It is also often necessary to ventilate concrete and metal silos (i.e., pressure venting) because of potential structural damage due to internal and external pressure fluctuations (Maier, 2018). Therefore, most hermetic storage systems are not used on permanent structures and alternatively take advantage of the low permeability of flexible plastic enclosures. These hermetic storage containers come in a variety of configurations and can store a range of dry commodities in the range of 60 kg–20 000 MT (Villers et al., 2010).

Airtight or hermetic storage has been successfully used to preserve grain for more than 30 years around the world (Calderon and Navarro, 1980). In the early 1990s, hermetic storage was first used on a large scale with the development of wheat in 'bunkers'. These 'bunkers' were able to handle capacities ranging from 10 000 MT to 30 000 MT (Villers et al., 2008). In Cyprus and Argentina, hermetic storage has been used successfully on a large scale in the form of underground pits covered with flexible roofs (Maier, 2018). In Australia, sealed commercial silos have been implemented successfully as well as sealed warehouses for bulk grain in China, hermetic bunker storage in Israel, and grain bag stacks enclosed within gastight liners (cocoons) in several African countries (Maier, 2018).

In China, Daolin et al. (2008) were able to successfully seal a warehouse and inject carbon dioxide into the building at concentrations ranging from 35% to 70%. They found that a storage time of 14 days was effective against each stage of the tested insects (*S. zeamais, R. dominica* and *T. castanuem*) (Daolin et al., 2008). Supplemental carbon dioxide had to be injected into the facility because the amount generated by the respiration of insects was not enough to be lethal due to the large volume of atmospheric air in the facility.

**Figure 16** GrainPro Ultra Hermetic Bunker™. Source: Chayil (2019).

For bulk storage, bunkers are the most common method for storing grain hermetically. The GrainPro Ultra Hermetic Bunker™ (Fig. 16) can handle capacities up to 30 000 MT with a 5-year life span. The cost of this type of storage is relatively minimal compared to traditional storage methods and has minimal operating expenses. The GrainPro Ultra Hermetic Bunker™ can be constructed simply by spreading out low-density polyethylene (LDPE) sheets on the bottom of a U-shaped compound consisting of an earthen ramp. An alternative (and superior) method is to cover the ground with asphalt or concrete and to replace the earthen ramp with a concrete or metal support wall. Once the grain is mechanically deposited into the pile, it is covered by large poly-vinyl chloride (PVC) sheets which come partly assembled to the site. The sheets are welded together on top of the pile with a manual heat welder. Ends of PVC sheet and LDPE sheet are rolled together and buried in a surrounding trench to obtain hermetic sealing (Chayil, 2019).

Hermetic storage is an effective way to preserve grain and protect it from damage by insects and molds. This type of storage is commonly used for smaller-scale storage (such as bags and plastic silos), but has successfully been implemented for bulk grain in bunker type storage.

## *6.3 Ozonation and cold plasma treatment*

From a physics standpoint, plasma refers to the fourth state of matter indicating the change in matter (from solid, liquid, gas and then plasma) by a gradual increase in energy (Fridman, 2008). Plasma consists of complete or partially ionized gas composed of photons, ions, free electrons, excited molecules and molecules with a neutral charge in their fundamental form (Pankaj et al., 2014). There are two different categories of plasma: (a) 'thermal' plasma, which refers to electrons and gas species that exist in thermodynamic equilibrium at a high temperature (20 000 K), and (b) 'non-thermal' or 'cold' plasma, in which electrons and gas species are not in equilibrium at a low temperature (<60°C). There are different techniques used to generate plasma ranging from corona discharges, microwave plasma, radio frequency plasma, dielectric barrier discharge (DBD) plasma and jet plasma, among others. DBD and jet plasma are commonly used because of their simple, versatile and adaptive designs and working. Different reactive species are generated in plasma and their composition depends on the gas initially used. For example, the exposed ambient air generates oxygen reactive species such as ozone, and nitrogen reactive species (Pankaj and Keener, 2017). Cold plasma is a novel technology used in the agriculture and food industry because of its numerous advantages. Cold plasma does not leave residues, requires a low energy input, and cold plasma has little if any impact on food matrices (Misra et al., 2016).

Cold plasma is used during postharvest storage to decontaminate grain surfaces for freshness, and to prevent mycotoxigenic fungi growth and insect infestation. Ohta (2016) reviewed available publications on the decontamination of grains using cold plasma. Previously, ozone was used to treat 8.9 MT of maize with 50 ppm ozone for 3 days and resulted in 92–100% mortality of adult red flour beetle, adult maize weevil and larval Indianmeal moth (Kells et al., 2001). However, another study showed that despite the ozone efficacy to disinfect grains, the use of ozone to treat grains contaminated with insects was not cost-effective (McDonough et al., 2011a,b). Another study reported the efficacy of atmospheric-pressure plasma jet to control Indianmeal moth, *Plodia interpunctella* (Abd El-Aziz et al., 2014).

In the grain industry, the primary food safety concern is mycotoxin contamination. Mycotoxins are secondary metabolites produced by fungi under favorable growth conditions, and they are harmful to humans and animals. Cold plasma technology has been successfully applied at lab-scale to decontaminate grains. A more detail review of different studies on mycotoxin and fungal decontamination in grain using cold plasma can be found in Misra et al. (2019). Reactive species react with mycotoxins and change their chemical structure, which results in less toxic compounds. Shi et al. (2017) treated AFB1 with high voltage atmospheric cold plasma and characterized AFB1 degradation products (DP). They showed that reactive species attack and degrade AFB1 at the C8–C9 double bond position and resulted in less DP.

Ozonation and cold plasma treatment have the potential to decontaminate grain surfaces but cold plasma technology does not yet exist commercially. However, ozone generators that are capable of treating grain at full scale have been built. Additional research and development work is needed to make cold plasma treatment physically and economically feasible for bulk grain storage.

## 7 Best management practices: sanitation, loading, aeration and monitoring (SLAM)

The SLAM strategy is a systematic approach of best management practices for successful storage of cereals and grains. The acronym stands for Sanitation, Loading, Aeration and Monitoring (Mason and Woloshuk, 2010). The main objective of SLAM is to maintain maximum postharvest quality by protecting stored grains and oilseeds from weather, rodents, insects, self-heating, molds, mycotoxins and pesticide residues. In addition, minimizing the deterioration process prevents spoilage, quality discounts, storage costs, and thus maximizes the return on every MT harvested, dried and stored.

### *7.1 Sanitation*

Handling equipment should be cleaned before and after harvest and should include combines, trucks, wagons, conveyors and so on. Sanitation in and around the storage silo is very important to reduce pest migration into the silo. Vegetation around the storage structure should be removed, and a weed-free facility should be maintained as weeds may act as a source of fungal infection. Spills should be avoided as they can attract and shelter pests in storage areas. Storage structures should be cleaned soon after emptying them. For long-term storage, empty storage structures should be treated with an approved insecticide inside and around the outside base before refilling. Stored maize can be protected from storage fungi with propionic acid and other organic acids, although their use is usually not recommended. Before silo filling, the unloading conveyor, conveyor tube opening, and side door opening should be sealed for sanitation and for keeping the grain from spilling out the silo (Leslie and Logrieco, 2014).

### *7.2 Loading*

Proper handling and loading of storage structures is important for safe storage. This process includes screening out fines and foreign material, coring and leveling the grain mass. Cleaning maize grain is critical to remove grain dust and broken kernels that insects and fungi thrive on and to improve aeration. Minimizing grain transfer operations reduces handling, broken kernels and fines. Coring will not remove the trash and fines, but repeated unloading helps to distribute this material throughout the silo rather than allowing it to accumulate in the center of the silo. Once the silo is full, the grain surface should be leveled, as peak loading of maize reduces airflow and increase airflow resistance. It may be necessary to operate a fan up to 50% longer to cool an overfilled silo than to cool a similar silo with leveled grain that is filled only to the eaves. Grain peaking also makes it more difficult and more dangerous to monitor stored grain. Utilizing a spreader to fill a silo or coring the grain mass also enables more uniform airflow (Leslie and Logrieco, 2014). Dangers of overfilling a silo include poor headspace ventilation, spoiled grain on the silo wall and difficulty monitoring the grain surface (GEAPS, n.d.).

### 7.3 Aeration

Aeration is key to maintaining low, uniform grain temperature and avoiding the localized hot spots that can result from the growth of storage fungi. Grain should be cooled uniformly to as low a temperature as reasonable for the geographic region and cooling maintained as long as possible during storage. Cooled grain will not remove fungal contaminants but will slow or halt fungal growth and mycotoxin biosynthesis while also reducing the insect population. Adequate exhaust vents must be installed to avoid condensation on the inside walls under the roof of the storage silo. Automatic fan controllers should be used to optimize aeration and run roof exhausters to avoid headspace condensation. When not operating, the fans should be sealed to prevent premature warming of the grain mass (Leslie and Logrieco, 2014).

### 7.4 Monitoring

Regular monitoring of stored grain especially for $CO_2$ is essential; to avoid fungal growth and mycotoxin biosynthesis, usually in hot spots during storage. Monitoring every 3–4 weeks during cold months and every 1–2 weeks during warm months usually suffices. Regular inspection and repair of leaky roofs and other holes is necessary to avoid water infiltration (Leslie and Logrieco, 2014).

## 8 Best management practices: dry, inspect, clean and examine (DICE)

The DICE strategy is another system approach of best management practices for successful storage, especially of cereals and grains stored in bags and stacked in warehouses. The acronym stands for Dry, Inspect, Clean and Examine. The DICE process provides a strategy to prevent damage to grain and increase its quality. The process ensures minimal wastage of grain so that postharvest loss could be decreased and profits increased. Each letter of the acronym will be discussed in more detail below:

### 8.1 Dry grain

Drying grain to a safe moisture content is important before putting it in storage. Safe moisture level for most grains is 12–14% or less. Certain grains, such as oilseed (e.g. soybeans, rapeseed and sunflower seeds) require a moisture content of 10% or less. Moisture content can be determined with digital meters.

### 8.2 Inspect

Once grain is dried to a safe level, the storage method must be selected. It is important to inspect the storage structure to ensure that it is in proper functioning condition. This includes sealing of cracks, covering potential inlets (such as doors and windows) with screens to prevent rodents and insects from entering, water damage and so on.

### 8.3 Clean

To protect your grain, the storage structure must be clean. This includes removing dirt, loose grain and weeds around the facility. Having a clean facility will reduce the risk of contamination and maintain grain quality.

### 8.4 Examine

Grain should be examined on a regular basis to eliminate small problems before they escalate further. Things to look for include mold, excess moisture, increased temperature, insect infestations, damage to storage structure and so on. If insects are found, it may be necessary to fumigate the grain or create a modified atmosphere environment. A regular and thorough examination of the stored grain can help reduce aflatoxin and mold problems and preserve the quality of the grain until end use.

If you follow the DICE process, grain damage will be prevented and quality will be maintained to reduce overall postharvest loss during storage.

## 9 Conclusion

Grain in large quantities is most efficiently handled, stored and transported in bulk rather than in bags. Equipment to handle, structures to store and vessels to transport are effectively deployed throughout the world to facilitate grain movement from farmers' fields to local and regional grain aggregation facilities, and from there into the domestic and export channels of grain-producing countries to the import channels of grain-deficit countries to meet food, feed, biofuel and industrial utilization needs.

The most significant current trend is increasing the size of new storage structures being built in grain-producing countries as a result of larger production in North and South America and the Black Sea Region of Eastern Europe, and in grain-importing countries as a result of greater demand for feed and animal source foods in Central America, Southeast and East Asia, Middle East, and North and sub-Saharan Africa. Greater amounts of grain stored for longer periods of time especially in subtropical and tropical climates requires well-built storage facilities that protect cereals and grains from quality deterioration and spoilage due to fungi, mycotoxins and insects, and well-trained operations managers with the knowledge and skills to apply state-of-the-art technologies and best management practices to protect and preserve the world's grain-based food and feed supply.

## 10 Abbreviations

The titles of ASABE journals are abbreviated as: *Trans. ASABE, Appl. Eng. Agric., J. Agric. Saf. Health,* and *Biol. Eng. Trans.*

For US states and Canadian provinces, use the two-letter postal codes.

Omit minor words (the, of, and, etc.) from the title unless surrounded by entire words.

For familiar organizations, use the initials: ASABE, ASAE, ASRAE, ASTM, CDC, CSA, CSSA, FAO, IEEE, ISO, JAWRA, JEQ, SAE, SSSA, SSSAJ, USDA, USDA-ARS, USDA-NRCS, USDA-NASS, USEPA, USGS and so on. Below are abbreviations commonly used in journal titles. For others, consult the ISO abbreviation list.

Advances, Advanced Adv.
Agriculture, Agricultural Agric.
Agronomy Agron.
American Am.
Annals Ann.
Annual Annu.
Applied, Applications Appl.
Association Assoc.
Biology, Biological Biol.
Biosystems Biosyst.
Chemistry, Chemical Chem.
Communications Commun.
Computers, Computing Comput.
Conference Conf.

Congress Congr.
Conservation Conserv.
Cooperative Coop.
Department Dep.
Division Div.
Drainage Drain.
Ecology, Ecological Ecol.
Electrical Electr.
Electronic Electron.
Engineers, Engineering Eng.
Environment, Environmental Environ.
Experiment, Experimental Exp.
Extension Ext.
Federation, Federal Fed.
Fundamentals Fundam.
Geographical, Geography Geogr.
Geological, Geology Geol.
Government Gov.
Horticulture Hortic.
Hydrology, Hydrological Hydrol.
Industry, Industrial Ind.
Institute Inst.
International Int.
Irrigation Irrig.
Journal J.
Literature Lit.
Management Manag.
Mathematics Math.
Mechanics, Mechanical Mech.
Medicine, Medical Med.
Modelling, Modeling Model.
National Natl.
Occupational Occup.
Physical, Physics Phys.
Proceedings Proc.
Processing, Processes Process.
Product, Production Prod.
Professional Prof.
Psychology Psychol.
Publication, Publishing Publ.
Quality Qual.
Quarterly Q.

Research Res.
Resource Resour.
Review Rev.
Safety Saf.
Science Sci.
Society Soc.
Software Softw.
Statistics Stat.
Supplement Suppl.
Symposium Symp.
System Syst.
Technical Tech.
Technology Technol.
Transactions Trans.
US, United States U.S.

## 11 Where to look for further information

In terms of future trends in research and development on advanced stored grain-storage technologies, preventative strategies for maintaining grain quality such as sealable structures and controlled aeration as well as intervention strategies such as gas fumigation, modified atmosphere, ozone and cold plasma treatments top the list. Improved storage technologies require utilization of best management practices such as SLAM (Sanitation, Loading, Aeration and Monitoring) and DICE (Dry Grain, Inspect, Clean and Examine) to achieve success in preserving and protecting stored grain quality and quantity.

The following websites offer extensive resource materials on drying, handling and storage of grains and cereals for further reference.

- https://www.extension.iastate.edu/grain/.
- https://www-mwps.sws.iastate.edu/catalog/grain-handling-storage.
- https://www.ag.ndsu.edu/graindrying.
- https://www.uaex.edu/farm-ranch/crops-commercial-horticulture/Grain_drying_and_storage/.
- https://inta.gob.ar/personas/bartosik.ricardo.

## 12 References

Abd El-Aziz, M. F., Mahmoud, E. A. and Elaragi, G. M. 2014. Non thermal plasma for control of the Indian meal moth, Plodia interpunctella (Lepidoptera: Pyralidae). *Journal of Stored Products Research* 59, 215–21. doi:10.1016/j.jspr.2014.03.002.

Advanced Drainage Systems. 2020. *Grain Aeration Pipe & Screen*. Available at: https://www.ads-pipe.com/products/pipes/grain-aeration-pipe-screen (Accessed on May 18, 2020).

ALCO Custom Covers. 2020. Grain bunker covers. Available at: https://alcocovers.com/stockpile-covers/grain-bunker-covers/ (Accessed on May 18, 2020).

Anon. *Pest Management for Grain Storage and Fumigation*. 2004. Available at: http://www.pesticides.montana.edu/documents/manuals/FumSeed.pdf.

Boumans, G. 1985. *Grain Handling and Storage*. Elsevier Science, Amsterdam, The Netherlands. Available at: https://www.sciencedirect.com/bookseries/developments-in-agricultural-engineering/vol/4/suppl/C.

Calderon, M. and Navarro, S. 1980. Synergistic effect of $CO_2$ and $O_2$ mixtures on two stored grain insect pests. In: *Developments in Agricultural Engineering* (Vol. 1, pp. 79–84).. doi:10.1016/B978-0-444-41939-2.50013-9.

Cao, Y., Xu, K., Zhu, X., Bai, Y., Yang, W. and Li, C. 2019. Role of modified atmosphere in pest control and mechanism of its effect on insects. *Frontiers in Physiology* 10(MAR), 206. doi:10.3389/fphys.2019.00206.

CFTRI. 1975. *Storage of Food Grains*. Mysore, India.

Chayil. 2019. Ultra hermetic Bunker™. Available at: http://www.chayilfoodsecurity.com/products/ultra-hermetic-bunker/ (Accessed on November 22, 2019).

Cheng, W., Lei, J., Ahn, J. E., Liu, T. X. and Zhu-Salzman, K. 2012. Effects of decreased $O_2$ and elevated $CO_2$ on survival, development, and gene expression in cowpea bruchids. *Journal of Insect Physiology* 58(6), 792–800. doi:10.1016/j.jinsphys.2012.02.005.

CSIRO. 2000. *Aerations Potential*. Education Poster, Stored Grains Research Laboratory.

Daolin, G., Shengbin, L., Ying, G., Chaoming, D., Zhao, Y., Jianwu, D., Jie, T., Honglin, M., Qile, H., Guiyong, L., Zhongping, M. and Liguang, L. 2008. Evaluation of large modern warehouse designed and constructed for application of carbon dioxide. *Proceedings of the of the 9th International Working Conference on Stored Product Protection* 95, pp. 463–9. Available at: https://pdfs.semanticscholar.org/7091/2a057cbf0df9a6dda0370c5e8f76dccd352a.pdf.

EWorldTrade. n.d. Steel Silos Turkish Silos Turkey Grain Storage Lipp Spiral Vertical 100Ton 500Ton 2000Ton Flat Bottom Hopper Bottom Silos Turkey. Available at: https://www.eworldtrade.com/pd/alderadisticaret/steel-silos-turkish-silos/46487/ (Accessed on May 18, 2020).

Fowler, W. 2015. Benefits of sealing a grain bin. *Blog Grain Bin Sealing*. Available at: https://sanitred.com/benefits-sealing-grain-bin-silo/.

Fridman, A. 2008. *Plasma Chemistry*. doi:10.1017/CBO9780511546075.

GEAPS. n.d. Chapter 10: S.L.A.M based quality grain management practices. In: *GEAPS 520 Grain Quality Management*. GEAPS.

GRDC. 2013. Grain storage fact sheet: western region grain storage pest control guide. Available at: www.grdc.com.a.

Group, G. n.d. Grain silo - metal / flat-bottom. Available at: https://www.agriexpo.online/prod/gsi-group/product-169172-12528.html (Accessed on May 18, 2020).

Hodgson, E. and Holscher, K. 2009. Think about stored grain pests Before harvest. Available at: https://crops.extension.iastate.edu/cropnews/2009/09/think-about-stored-grain-pests-harvest.

Ipesa Silo. n.d. Photo album. Available at: https://ipesasilo.com/photo-album/ (Accessed on May 18, 2020).

KEBS. 2019. Portable rigid plastic hermetic grain silo – specification. KS2874:2019. Kenya Bureau of Standards. Available at: https://members.wto.org/crnattachments/2019/TBT/KEN/19_1184_00_e.pdf (Accessed on February 2020).

Kells, S. A., Mason, L. J., Maier, D. E. and Woloshuk, C. P. 2001. Efficacy and fumigation characteristics of ozone in stored maize. *Journal of Stored Products Research* 37(4), 371–82. doi:10.1016/S0022-474X(00)00040-0.

Langcuster, J. 2008, June. Grain bin fumigation often necessary, risky. *Farm Progress*. Available at: https://www.farmprogress.com/grain-bin-fumigation-often-necessary-risky.

Legacy Building Solutions 2013. *Flat Grain Storage Solutions*. Available at: https://www.legacybuildingsolutions.com/flat-grain-storage-solutions/ (Accessed on May 18, 2020).

Leslie, J. F. and Logrieco, A. F. 2014. *Mycotoxin Reduction in Grain Chains*. Wiley Blackwell.

Maier, D. E. 1992. *The Chilled Aeration and Storage of Cereal Grains*. Michigan State University.

Maier, D. E. 2000. Personal pictures taken in Australia.

Maier, D. E. 2018, March. Grain Ops: silo bags as a storage option. *World Grain*. Available at: https://www.world-grain.com/articles/10299-grain-ops-silo-bags-as-a-storage-option.

Maier, D. E. and Wilcke, W. F. n.d. Temporary grain storage considerations. Available at: https://www.extension.purdue.edu/extmedia/GQ/GQTF38/GQTF-38.html (Accessed on November 21, 2019).

Mason, L. J. and Woloshuk, C. P. 2010. S L A M. post-harvest IPM. Available at: http://www.ces.purdue.edu/new.

Mason, L. J., Rulon, R. A. and Maier, D. E. 1997. Chilled and ambient aeration plus fumigation of stored popcorn. Part 2: pest management. *Journal of Stored Products Research* 33(1), 51–8. doi:10.1016/S0022-474X(96)00024-0.

Maulsby, D. 2016. Corn pile rises by grain bins in western Iowa stock photo. Available at: https://www.istockphoto.com/photo/corn-pile-rises-by-grain-bins-in-western-iowa-gm619052088-107880881 (Accessed on May 18, 2020).

McDonough, M. X., Campabadal, C. A., Mason, L. J., Maier, D. E., Denvir, A. and Woloshuk, C. 2011a. Ozone application in a modified screw conveyor to treat grain for insect pests, fungal contaminants, and mycotoxins. *Journal of Stored Products Research* 47(3), 249–54. doi:10.1016/j.jspr.2011.04.001.

McDonough, M. X., Campabadal, C. A., Mason, L. J., Maier, D. E., Denvir, A. and Woloshuk, C. 2011b. Ozone application in a modified screw conveyor to treat grain for insect pests, fungal contaminants, and mycotoxins. *Journal of Stored Products Research* 47(3), 249–54. doi:10.1016/j.jspr.2011.04.001.

Misra, N. N., Schluter, O. and Cullen, P. J. 2016. Cold plasma in food and agriculture. In: Misra, N. N., Schluter, Oliver K., and Cullen, P. J. (Eds) *Cold Plasma in Food and Agriculture*, Academic Press, 1–16. doi:10.1016/c2014-0-00009-3.

Misra, N. N., Yadav, B., Roopesh, M. S. and Jo, C. 2019. Cold plasma for effective fungal and mycotoxin control in foods: mechanisms, inactivation effects, and applications. *Comprehensive Reviews in Food Science and Food Safety* 18(1), 106–20. doi:10.1111/1541-4337.12398.

Morales-Quiros, A., Campabadal, C. A., Maier, D. E., Lazzari, S., Lazzari, F., Cook, S. and Phillips, T. W. 2016. Chilling aeration to control pests and maintain grain quality during in-bin storage of wheat in Kansas. *2016 American Society of Agricultural and Biological Engineers Annual International Meeting, ASABE 2016*. doi:10.13031/aim.20162448464.

Mowitz, D. 2018. Temporary grain storage solutions. *Successful Farming*. Available at: https://www.agriculture.com/machinery/grain-handling-and-equipment/tempary-grain-stage-solutions_211-ar49513.

Navarro, S. 2012, September. The use of modified and controlled atmospheres for the disinfestation of stored products. *Journal of Pest Science* 85(3), 301–22. doi:10.1007/s10340-012-0424-3.

Navarro, S. and Noyes, R. T. 2002. *The Mechanics and Physics of Modern Grain Aeration Management*. CRC Press.

Navarro, S., Timlick, B., Demianyk, C. J. and White, N. D. G. 2012. Controlled or modified atmospheres. In: Hagstrum, D. W., Phillips, T. W. and Cuperus, G. (Eds), *Stored Product Protection*. pp. 191–202. Available at: http://www.ksre.ksu.edu/bookstore/pub.

O'Dowd, E. T. 1971. Hermetic storage of cowpeas (Vigna unguiculata Walp.) in small granaries, silos and pits in northern Nigeria. *Samaru Miscellaneous Paper* 31, 1–38.

Ohta, T. 2016. Plasma in agriculture. In: *Cold Plasma in Food and Agriculture: Fundamentals and Applications* (pp. 205–21). doi:10.1016/B978-0-12-801365-6.00008-1.

Pandharinath, P. S. and Chandra, R. P. 2014. Food grain storage practices-A review. *Journal of Grain Processing and Storage*. 1-5. Available at: www.jakraya.com/journal/jgps.

Pankaj, S. K. and Keener, K. M. 2017, August 1. Cold plasma: background, applications and current trends. *Current Opinion in Food Science* 16, 49–52. doi:10.1016/j.cofs.2017.07.008.

Pankaj, S. K., Bueno-Ferrer, C., Misra, N. N., Milosavljević, V., O'Donnell, C. P., Bourke, P., Keener, K. M. and Cullen, P. J. 2014, January. Applications of cold plasma technology in food packaging. *Trends in Food Science and Technology* 35(1), 5–17. doi:10.1016/j.tifs.2013.10.009.

Proctor, D. L., FAO. and GASGA. 1994. *Grain Storage Techniques toEvolution and Trends in Developing Countries*. GASGA.

Richard-Molard, D. 2003. Cereals | Bulk Storage of Grain. In: *Encyclopedia of Food Sciences and Nutrition* (pp. 1014–8). doi:10.1016/b0-12-227055-x/00187-5.

Shed Force. 2020. *Storage Sheds*. Available at: https://www.shedforcewa.com.au/product/storage-sheds/ (Accessed on May 18, 2020).

Shedd, C. K. and Foster, W. A. 2017. Silo construction. 16, 189. Available at: https://www.canr.msu.edu/agriculture/delayed-planting-.

Shi, H., Cooper, B., Stroshine, R. L., Ileleji, K. E. and Keener, K. M. 2017. Structures of degradation products and degradation pathways of aflatoxin b1 by high-voltage atmospheric cold plasma (HVACP) treatment. *Journal of Agricultural and Food Chemistry* 65(30), 6222–30. doi:10.1021/acs.jafc.7b01604.

Shipman, K. 2010. ARS: pulling air down through bins gives better insect control. Available at: https://farmweeknow.com/story-ars-pulling-air-down-through-bins-gives-better-insect-control-4-41561 (Accessed on November 21, 2019).

TSGC. n.d. Grain bin temperature & moisture monitoring system. Available at: https://www.tsgcinc.com/ (Accessed on November 22, 2019).

USDA Agricultural Research Service. 2010. Controlling insects in stored grain. Available at: https://phys.org/news/2010-08-insects-grain.html (Accessed on May 18, 2020).

Villers, P., Navarro, S. and DeBruin, T. 2008. Development of hermetic storage technology in sealed flexible storage structures. *Citation for Controlled Atmosphere …*, (January). Available at: http://www.grainpro.com/pdf/PU2015PV0708-CHENGDU-PAPER-CAF-Presentation-Development-of-Hermetic-Storage-Technology.pdf.

Villers, P., Navarro, S. and Bruin, T. 2010. New Applications of Hermetic Storage for Grain Storage and Transport. Julius-Kühn-Archiv. 10.5073/jka.2010.425.086.

WFP. 2019. Post-harvest loss reduction. Available at: https://innovation.wfp.org/project/post-harvest-loss-prevention (Accessed on February 10, 2020).

What are Silo bags? – Arpaco. 2018. Available at: https://www.arpaco.com.pk/what-are-silo-bags/ (Accessed on November 21, 2019).

World Food Program. 2020. Post-harvest loss reduction program. Available at: https://innovation.wfp.org/project/post-harvest-loss-prevention (Accessed on May 18, 2020).

Worley, J. W. and Hammond, C. 2015. Aerating grain in storage. Available at: https://secure.caes.uga.edu/extension/publications/files/pdf/B 712_4.PDF.

# Chapter 6

# Developments in the use of hermetic bags for grain storage

*Dieudonne Baributsa, Purdue University, USA; and Ma Cristine Concepcion Ignacio, Iowa State University, USA and University of the Philippines Los Baños, The Philippines*

1 Introduction
2 Development of hermetic storage bags
3 Hermetic bags for grain storage
4 Dissemination of hermetic bags to reduce postharvest losses among smallholder farmers
5 Challenges in the use of hermetic storage bag technology
6 Case study: Purdue Improved Crop Storage (PICS) bag
7 Summary and future trends
8 Where to look for further information
9 References

## 1 Introduction

Postharvest management of agricultural products is a significant challenge in developing countries. Postharvest management, including handling and storage, play a vital role in keeping agricultural commodities safe from deterioration. Important staple crops such as maize, wheat, rice, and beans are susceptible to storage losses mostly due to insect pests. There are several grain storage techniques, including traditional/local methods, pesticides, and hermetic systems. The storage systems' most important feature is to preserve the integrity of the grain for a given period with minimal loss in quality and quantity.

Hermetic technologies as an alternative to traditional and chemical control methods have gained significant interest among farmers, the private sector, governments, and development agencies. Hermetic technologies owe their effectiveness to the airtight conditions created during storage. Biological processes such as respiration and metabolic activities are driven mostly by the presence of insects and other biological activities that lead to the depletion of oxygen and release carbon dioxide inside hermetic containers

http://dx.doi.org/10.19103/AS.2020.0072.06

(Murdock et al., 2012). Hence, the hypoxic environments these technologies create become unfavorable to the development and reproduction of insect pests and thus minimize or stop grain damage. Commonly used hermetic technologies include silos (metal and plastic), drums, cocoons, plastic bags, and other containers. These hermetic containers come in different forms and sizes.

Among hermetic technologies, hermetic bags are the most widely disseminated among smallholder farmers in sub-Saharan Africa (SSA) and Asia. The use of hermetic bags to store grain has significantly increased in the past 10 years, spearheaded by the development of the Purdue Improved Crop Storage (PICS) bag. The adoption of hermetic bags is driven by several factors, including: (i) the severity of storage losses at the farm level; (ii) the efficacy of the technologies, and (iii) other benefits such as being chemical-free, cost-effective, easy to use; and locally available. Hermetic bags significantly reduce food safety risks posed by the conventional method of treating stored grains with insecticides.

There are several commercially available types of hermetic bags: single-layer (e.g. SuperGrainbags™ bags manufactured by Grainpro Inc.), double-layer (e.g. AgroZ bags manufactured by A to Z), and triple-layer bags (e.g. PICS™ bags manufactured by several licensed plastic companies in SSA, Asia and Latin America) (Villers et al., 2006; Jonfia-Essien et al., 2010; Murdock and Baributsa, 2014; Coffi et al., 2016). These hermetic bags come in different forms and sizes, and most of them look similar to transport or storage containers (polypropylene [PP] woven bags and plastic polyethylene [PE] liners) that are regularly used by farmers. Hermetic bags are appropriate alternatives to traditional and conventional storage methods as they maintain the quality of stored products and allow smallholder farmers to be food secure and have the flexibility to sell their grain when prices are high.

## 2 Development of hermetic storage bags

Hermetic storage is not a new concept and has been recorded from prehistoric times when grain was stored underground in pits (Navarro et al., 1994). Hermetic bags are an adaptation of the traditional and other forms of hermetic storage systems such as underground pits, clay pots, jerrycans, silos, or drums. Underground pits were considered as a traditional method of storage that provides an airtight, chemical-free, and safe means to store dry crops (Navarro et al., 1994). The development of hermetic bags started in the 1980s with the PICS in Africa and the Cocoon in Israel (Kitch and Ntoukam, 1991; Navarro et al., 1993; Murdock and Baributsa, 2014). GrainPro's SuperGrainbags™ emerged out of this effort to develop solutions

Published by Burleigh Dodds Science Publishing Limited, 2020.

for smallholder farmers. Hermetic bags were developed to help smallholder farmers in developing countries reduce postharvest losses due to insects. Unlike some of the other hermetic storage methods such as jerrycans, or drums, hermetic bags can be easily scaled up cost-effectively.

The PICS technology was developed by Purdue University to address storage pest infestations on cowpeas (*Vigna inguiculata*, L. Walp.) in West and Central Africa (Baributsa et al., 2010, 2014a). Cowpea farmers in West Africa lose a substantial portion of their harvest (from 10% to as high as 100% after only a few months) during storage due to insect pests (*Callosobruchus maculatus*, Fabricius) (Moussa et al., 2011, 2014). PICS bags were developed to provide farmers with a chemical-free and cost-effective storage method. Existing storage methods such as granaries, ash, sand, botanicals, metal drums, jerrycans, and insecticides had shortcomings. Some of these methods were ineffective, costly, not scalable, and posed health risks. Providing a safe and cost-effective method to store grain would improve food security and also allow farmers to tap into better grain prices during the lean season. PICS bags come in three different sizes – 25 kg, 50 kg, and 100 kg bags. Each bag is made of two high-density polyethylene (HDPE) liners that are fitted inside a PP woven bag. PICS bags are manufactured by 15 plastic companies around the world. Eighty percent of PICS manufacturers are located in SSA, while the rest are in Asia and Latin America; as the latter are newly emerging markets for the PICS technology.

The development of hermetic bags by GrainPro started with the deployment of large-scale plastic storage structures (cocoons) for long-term grain storage. These cocoons come in different sizes of up to 300 tons. Over time, GrainPro developed intermediate and small storage systems, which are based on the hermetic principle. The small-scale hermetic storage system of GrainPro is the SuperGrainbags™, which are storage containers suitable for subsistence farmers (Jonfia-Essien et al., 2010). GrainPro hermetic systems became commercially available in the early 1990s, and today they are in use in more than 32 countries around the word (Villers et al., 2008). The SuperGrainbags™ were initially used for storing rice seeds as part of the 5-year project of the International Rice Research Institute (IRRI) in the Philippines (Villers et al., 2006). Each SuperGrainbags™ is a thin single multi-layered plastic liner, which is transparent and has low oxygen permeability. The capacities of SuperGrainbags™ can range from 60 kg to 1000 kg; 60–90 kg capacity being the most common (Villers et al., 2006, 2008). GrainPro bags are manufactured in Subic Bay in the Philippines and shipped around the world.

In the last 5 years, several other brands of hermetic bags (Table 1; Fig. 1) have emerged, innovating on or imitating existing ones. These include ZeroFly® hermetic storage bag, AgroZ bag, and Elite bag. AgroZ and Elite are

**Table 1** Characteristics of hermetic bags being sold in sub-Saharan Africa in 2019

| Type of hermetic bags | Woven bag (PP) | Liners | Examples |
|---|---|---|---|
| Triple bags | One PP bag | 2 HDPE inner liners | Purdue Improved Crop Storage (PICS) |
| | One PP bag | 2 HDPE inner liners | Elite Bag Light Duty Double Liner |
| Double bags | One PP bag | 1 liner with multiple layers | AgroZ |
| | One PP bag | 1 liner with multiple layers impregnated with insecticides | AgroZ Plus |
| | One PP bag impregnated with insecticides | 1 liner with multiple layers | ZeroFly Hermetic storage bag[a] |
| Single bag | No | 1 liner with multiple layers | GrainPro Bag Zipper |

[a]Polypropylene/woven bag is impregnated with Deltamethrin to prevent the infestation from outside the bag.

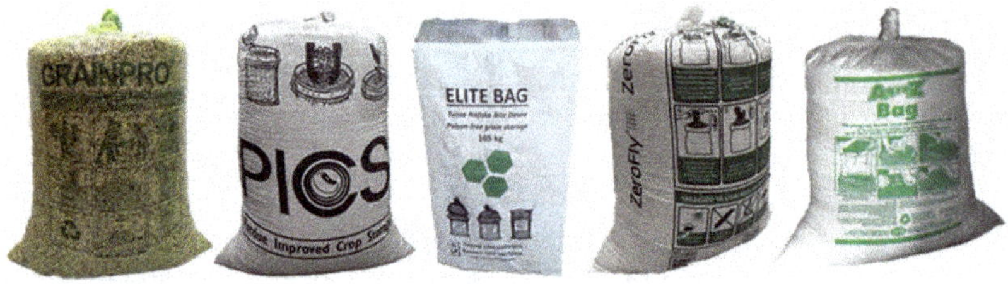


**Figure 1** Hermetic storage bags commercially available in Kenya in 2019.

manufactured in East Africa, while ZeroFly bags are imported from outside Africa. The development of these hermetic bags was mostly driven by the realization of the growing demand for safer and affordable storage technologies among smallholder farmers. AgroZ bags are double-layer bags (one-liner and one woven bag) developed by A to Z Textile Mills Ltd (Tanzania). The liner is co-extruded combining HDPE, Metallocene Linear Low-Density Polyethylene (MLLDPE), and low-permeability barrier layers limiting the passage of oxygen. Another hermetic bag developed by A to Z Textile Mills Ltd is the AgroZ bag Plus, which is a premium product, insecticide-treated hermetic storage bag, specifically designed to control larger grain borer (*Prostephanus truncates*, Horn). The ZeroFly hermetic storage bag is manufactured by Vestergaard and is composed of an outer PP bag impregnated with deltamethrin and an inner multilayered recyclable PE bag with a gas barrier. Elite Innovations (K) Ltd, based in Kenya, also developed a double-liner hermetic bag made from HDPE called Elite Bag Light Duty Double Liner.

Published by Burleigh Dodds Science Publishing Limited, 2020.

## 3 Hermetic bags for grain storage

### *3.1 How does it work?*

Most hermetic bags consist of two basic components: (i) an inner liner (single or double), and (ii) an outer bag. The inner liner could be an individual or multilayered film designed to provide a gas barrier, which is the most critical component of the hermetic bags. The outer bag is a PP bag designed to provide strength and protection during handling. The hermetic bags work under the principle that the bio-generated modified atmosphere created by the metabolic activities of organisms is present in the commodities. Insects and microorganisms consume oxygen and release carbon dioxide, hence generating a low-oxygen environment (Fig. 2). The created hypoxic conditions inhibit the growth and development of insects and mold, thus, preventing grain deterioration. The plastic inner liners of the hermetic bag are designed to limit or sensibly reduce the permeability of gases to ensure that the low-oxygen environment inside the bag is maintained to levels that hinder the development of all life stages of insects. When the inner liner is pierced, the technology loses its efficacy, and the bag should be replaced in the case of single, double or triple hermetic bags. For brands of hermetic bags that are triple bags, the second liner provides extra safety in case one of the two liners is damaged. Insects rarely damage the second liner of the triple-layer bags;

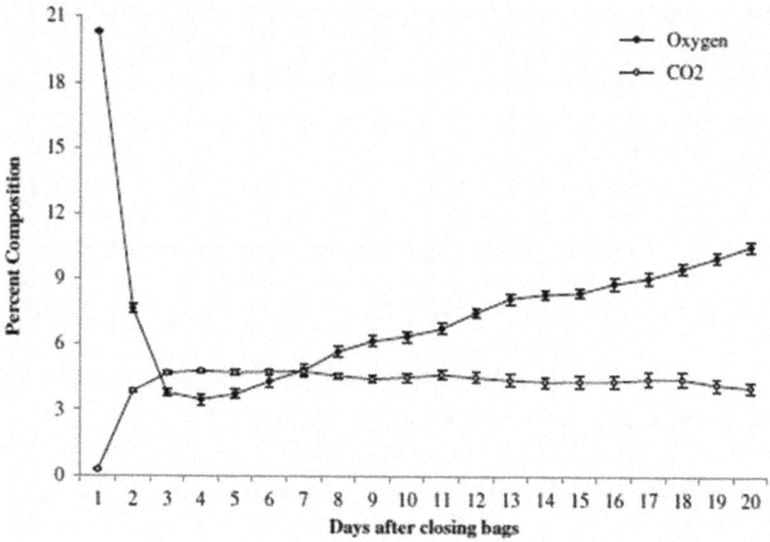


**Figure 2** Changes in the concentration of oxygen and carbon dioxide inside a PICS bag filled with mildly infested cowpea grain. Gas levels were monitored for 20 days using a Mocon Headspace analyzer (Murdock and Baoua, 2014).

hence, the technology can still provide protection during grain storage. Minor damages such as holes can be repaired with adhesive tape. Otherwise, the bag needs to be repurposed for other uses or recycled (Baributsa et al., 2014b).

All hermetic bags currently on the market rely on natural processes to achieve hypoxia. Research efforts to develop technologies that could deplete oxygen inside hermetic bags once the containers are closed have not been developed. This concept of adding oxygen scavengers to hermetic bags will become crucial as the use of these technologies expands to new crops or commodities that require a sharp reduction in oxygen levels not only to kill insects but also maintain food quality such as vitamins. A study conducted at Purdue University showed that adding oxygen scavengers into grain stored in PICS bags helped reduce the degradation of pro-vitamin A in biofortified orange maize (Nkhata et al., 2019). Also, the use of a controlled atmosphere to achieve a low-oxygen environment needs to be explored, especially for its application in developed countries where tolerance for insect presence in grains in minimal/low.

## 3.2 Efficacy of hermetic bags for grain storage

The efficacy of hermetic bags to protect grains and other stored products has been documented through research. Though there is a difference in the composition of hermetic bags (single, double or triple bags) and how liners are designed (single- or multilayered liners), little or no significant differences have been observed in the efficacy of these technologies. Published research on five hermetic bags currently being commercialized in Kenya and other countries in SSA is presented in Table 2. Studies have shown that hermetic bags are as effective as and sometimes better than insecticides in reducing storage losses due to insects (Baoua et al., 2012a; Baributsa and Njoroge, 2020; De Groote et al., 2013; Mutambuki et al., 2019).

Hermetic bags preserve grain and seed quality, but maize must be properly dried before storage in airtight conditions (Williams et al., 2014; Walker et al., 2018). Hermetic bags can achieve 100% insect mortality and reduce losses to less than 1% after several months of cereal (maize, *Zea mays* L., and paddy rice, *Oryza sativa* L.) storage (Baoua et al., 2014a; De Groote et al., 2013; Guenha et al., 2014; Njoroge et al., 2014) and legume crops such as common beans (*Phaseolus vulgaris*), mung beans (*Vigna radiata* (L.) Wilczek), pigeon peas (*Cajanus cajanus* (L.) Millsp), and Bambara nuts (*Vigna subterranea* (L.) Verdc.) (Mutungi et al., 2014, 2015; Baoua et al. 2012b, 2014b; Vales et al., 2014). For seed storage, research findings have shown that hermetic bags have a tolerable reduction in germination (Afzal et al., 2017; Guenha et al., 2014). Beyond insects, hermetic bags have shown to be effective at mitigating grain contaminants such as mycotoxins. Laboratory and field experiments have demonstrated that

**Table 2** Summary of research on hermetic bags being sold in sub-Saharan Africa in 2019

| | Focus of study | Type of hermetic bag | Type of crop | Source (location of study) |
|---|---|---|---|---|
| 1 | Grain damage due to insect infestation, insect mortality and seed viability | PICS bag | Cowpea | Sanon et al. (2011) (B. Faso); Baoua et al. (2012a,b) (Niger); Murdock et al. (2012) (Niger); Baoua et al. (2013) (Niger); Ibro et al. (2014) (Nigeria, Niger and B. Faso) |
| | | | Dry common beans | Mutungi et al. (2015) (Kenya) |
| | | | Maize | Baoua et al. (2014a) (Benin, Burkina Faso and Ghana); Njoroge et al. (2014) (Kenya); Ng'ang'a et al. (2016a) (Kenya); Afzal et al. (2017) (Pakistan); Williams et al. (2017) (USA); Mutambuki et al., (2019) (Kenya) |
| | | | Mungbean | Mutungi et al. (2014) (Kenya) |
| | | | Pigeon peas | Vales et al. (2014) (India) |
| | | | Rice (paddy) | Baoua et al. (2016) (B.Faso, Ghana and Niger) |
| | | | Wheat | Martin et al. (2015) (USA) |
| | | SuperGrainBag™ | Maize | De Groote et al. (2013) (Kenya); Chigoverah and Mvumi (2016) (Zimbabwe); Ndegwa et al. (2016) (Kenya); Likhayo et al. (2018) (Kenya) |
| | | | Rice (paddy) | Guenha et al. (2014) (Mozambique) |
| | | Zerofly bag | Maize | (Ghana)/Paudyal et al. (2017) (Ghana); Abass et al. (2018) (Tanzania) |
| 2 | Performance comparison of different hermetic bags | PICS and SuperGrainBag™ | Cowpea | Baoua et al. (2013) (Niger) |
| | | | Chickpea | Alemayehu et al. (2020) (Ethiopia) |
| | | | Maize | Chigoverah and Mvumi (2018) (Zimbabwe); Walker et al. (2018) (Kenya) |
| | | | Rice (paddy) | Awal et al. (2017) (Bangladesh) |
| | | PICS, SuperGrainBag™, Agrozbag and Agrozbag Plus | Maize | Coffi et al. (2016) (Tanzania) |
| 3 | Fungal contamination/ mycotoxin/aflatoxin | PICS bag | Maize | Williams et al. (2014) (USA); Ng'ang'a et al. (2016b) (Kenya); Tubbs et al. (2016) (USA); Maina et al. (2016) (Kenya); Lane et al., (2018) (USA) |
| | | SuperGrainBag™ | Maize | Kumari et al. (2017) (India) |
| 4 | Economic benefits | PICS bag | Common beans | Jones et al. (2011b) (Tanzania) |
| | | | Cowpea | Moussa et al. (2011, 2014) (West and Central Africa) |
| | | | Maize | Jones et al. (2011a,) (Malawi, Mozambique, Ghana and Tanzania) |

hermetic bags can be used to arrest mold growth and aflatoxin development (Williams et al., 2014; Tubbs et al., 2016; Lane et al., 2018). New applications include the use of hermetic bags to store and transport coffee and cacao to prevent insect attacks, mold, and the loss of aroma and flavor (Jonfia-essien et al., 2010; Donovan et al., 2019).

## 4 Dissemination of hermetic bags to reduce postharvest losses among smallholder farmers

In recent years, postharvest storage losses among smallholder farmers in SSA Africa have gained the attention of government, donors, and the development community. Several organizations and projects have made efforts to scale up the adoption of hermetic storage in SSA. With a grant from the Bill and Melinda Gates Foundation (BMGF), Purdue University (Indiana, USA) initiated large-scale dissemination of hermetic bags in SSA starting in 2007 (Baributsa et al., 2010). Then, the PICS technology was being disseminated in ten countries in West and Central Africa to reduce on-farm storage losses of cowpea. In 2011, the PICS bags were tested for the storage of grains beyond cowpea, including maize, common beans, and cassava chips (Baoua et al., 2014b; Hell, 2014; Njoroge et al., 2014). PICS bags proved to be effective in storing several dried products. The bags were still branded as PICS, but the 'C' was changed from 'Cowpea' to 'Crops'. PICS bags have been commercialized in SSA, Asia, Latin America, and the Caribbean (Baributsa, 2019). More details of this effort are discussed in the case study on PICS.

Several years after the initial launch of the PICS program, many public, private, and development agencies saw an opportunity to promote hermetic storage bags in SSA. GrainPro Inc., a private company that was promoting SuperGrainbags™ for rice storage in Southeast Asia, saw an opportunity to expand its market to Africa. In fact, in 2010, there was a discussion between Purdue University and GrainPro Inc. for a potential partnership to scale up the adoption of hermetic bags in SSA that resulted in a Memorandum of Understanding for collaboration in research between both organizations. As the interest of the private sector continued to grow, plastic companies innovating on or imitating existing hermetic bags came into the market. Countries like Kenya saw a substantial increase in the number of hermetic bags on the market due to a potential demand among farmers (Baributsa and Njoroge, 2020). The growth in the commercialization of hermetic bags to smallholder farmers was also propelled in part by investments from government and donors in disseminating hermetic technologies.

In 2013, PICS bags were introduced to farmers in Kenya through 850 demonstrations in villages, markets, and farmers groups (Baributsa and Njoroge, 2020). Further dissemination of PICS bags from 2013 to 2016 were implemented by several projects including the Kenya Agricultural Value Chains Enterprises Project (KAVES) (Foy and Wafula, 2016). Building on its initial successes, KAVES in collaboration with the Government of Kenya organized roadshows and media advertisements to increase awareness and competition among several hermetic bags, including PICS, AgroZ, SuperGrainbags™, Zerofly, and Elite. In addition, in 2014, AgResults was launched by a consortium of development partners to commercialize hermetic technologies (plastic bags, plastic tanks, and metal silos) to smallholder farmers in Kenya through private-sector competition (AgResults, 2019). Nine companies were selected to be part of the competition – African Farms and Markets Ltd/GrainPro, A to Z Textile Mills Ltd, Bell Industries Ltd/PICS, Corporate Business Forms Ltd/Vestergaard, Elite Innovations (K) Ltd, Ekima Engineering Workers, Kenya Promotions and Marketing Company Ltd, Kentainers Ltd, and Post Harvest Africa. The AgResults intervention included several hermetic bags : PICS, AgroZ, SuperGrainbags™, Zerofly, and Elite bags (AgResults, 2018a). Bags type varied by the manufacturer (Table 1) and the capacity ranged from 50 kg, 90 kg, 100 kg, and 250 kg. Between 2015 and 2018, about 1.4 million storage devices were sold, and the activities reached 220 000 smallholder farmers (AgResults, 2018b). At the end of the project, AgResults awarded over US$6.25 million to three companies that reached the threshold, including A to Z Textile Mills Limited, Bell Industries, and Elite Innovations (AgResults, 2019). It was estimated that AgResults likely increased the adoption of hermetic storage by 23 and 6 percentage point in Eastern Kenya and Rift Valley, respectively. The prize competition initiative to create a private sector-led market for postharvest storage technologies proved to be beneficial to smallholder farmers in Kenya.

Several other postharvest efforts were implemented in other countries, including the *Yieldwise Initiative*. This effort was funded by the Rockefeller Foundation to reduce postharvest losses in Nigeria (tomato), Kenya (mango), and Tanzania (maize). The Alliance for Green Revolution in Africa (AGRA) received funding to implement activities to reduce postharvest losses of maize in Tanzania (AGRA, 2015). Postharvest loss of maize in Tanzania is estimated at 20 to 30% but can reach as high as 40% (Rugumamu, 2004). *Yieldwise* planned to reduce postharvest losses along the maize value chain by clustering 100 000 farmers in groups and training them, linking them to input and output markets, and facilitating their access to finance. The program expected at least 60% of the 100 000 farmers to use loss reduction technologies. These technologies included hermetic storage devices such as cocoons, metal silos, and PICS bags. AGRA as the lead implementer worked with several partners including the government of Tanzania; the World Food Program (WFP); and several local nongovernmental organizations, input suppliers and dealers, commercial banks (e.g. Equity), farmers and farmers' organizations, maize traders and processors, and manufacturers of postharvest technologies (AGRA, 2017). Training of farmers was conducted by WFP in collaboration with local NGOs such as Rural Development Initiative (RUDI) in the Southern Highlands. Financial institutions such as Equity Bank were brought in to provide loans to farmers for increasing access to technologies and to the private sector for supplying inputs and storage technologies to farmers. Several meetings were held between financial institutions, NGOs, partners, and suppliers of inputs and PH technologies to build awareness about the financial products available at the Equity Bank. Data were not available for the impact of this program.

In summary, all these programs proved that there are opportunities to address postharvest losses among smallholder farmers in developing countries. Some of the perceived benefits of using hermetic storage include improved food security and income; and the reduction of insecticide usage. With the increase in the awareness, marketing, and distribution of hermetic bags by the private sector, there is an uptake leading to the reduction of grain storage losses and improved food security. Figure 3 shows that a typical family using four hermetic bags will have an 8.78 extra number of meals equivalent to US$7.20 food saved (ACDI/VOCA, 2017).

Hermetic storage devices could also provide smallholder farmers with the flexibility to store grain for several months while waiting for prices to increase (Swathi and Rajanikanth, 2017). In West Africa, farmers storing cowpea in 100 kg PICS bags made additional cash of US$27 (Moussa et al., 2014). In Kenya, the net gain for storing in hermetic bags ranges from US$18 for maize to US$24 for cowpea (Baributsa and Njoroge, 2020). A recent study conducted in Kenya showed that the major driver for adoption of hermetic bags appears to be health benefits (elimination of chemical use) and labor saving (AgResults, 2019). Additional evidence demonstrates that another health benefit of using hermetic storage technology is the mitigation of aflatoxin growth (ACDI/VOCA, 2017; Tubbs et al., 2016). Several studies have shown that the value of

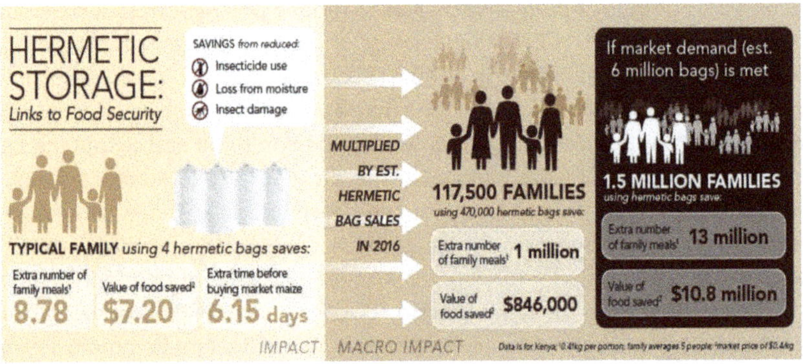


**Figure 3** Impacts of hermetic storage bag technology on food security (Source: ACDI/VOCA, 2017).

reduced storage loss or use of insecticides would easily justify the investment in purchasing hermetic bags (Ameri et al., 2018; Jones et al., 2011a, b; Moussa et al., 2011).

Several factors influence agricultural technology adoption among farmers, including technological, economic, institutional, and human aspects. With the large dissemination of hermetic bags, there are prospects to help smallholder farmers in developing countries improve their livelihood, including increased income, food security, and their health through chemical-free storage of grain and seed. Various authors have identified the following as determinants for the adoption of hermetic storage by smallholder farmers:

- Low cost and affordable
- Simple and easy to use
- Available for farmers
- Can be used to store small or large quantities of grain, seed, and other stored products
- Culturally acceptable
- Locally or regionally manufactured
- Durable or can be used for two or more storage seasons, if handled properly
- Do not affect the seed germination
- Do not affect the biochemical composition of stored products
- Mitigate the development of mold and mycotoxins

## 5 Challenges in the use of hermetic storage bag technology

Hermetic bags are simple and effective technologies adapted for smallholder farmers who are producing enough grain for home consumption and for sale but have challenges storing it. Though the demand for hermetic storage bags continues to grow with awareness building and training, there is a need to address some of the existing and emerging challenges to further increase adoption among smallholder farmers and other users. Treating smallholder farmers as customers is crucial for improving their experiences in using hermetic bags and for the development of a sustainable supply chain driven by private sector companies.

### 5.1 *Availability*

Awareness building and supply chain activities led by extension partners and the private sector have helped to increase the demand and availability of hermetic bags. Despite efforts to disseminate hermetic bags, developing a sustainable bag-distribution network that reaches farmers remains a challenge (Baributsa et al., 2010; Govereh et al., 2019; Nouhoheflin et al., 2017). In most countries, farmers often complain about the unavailability of hermetic bags in rural areas (Moussa et al., 2010). The unavailability of hermetic bags can be defined as bags not being available at all for farmers to purchase or being available late in the season after they have already used/purchased other storage technologies. There are areas that have benefited from demonstration activities but do not have vendors who are selling hermetic bags. Others have vendors that are less invested in the business; leading to shortages and late delivery of the hermetic bags to farmers. Challenges in getting hermetic bags at the last mile are linked to the hesitation of retailers and wholesalers to sell a new product, difficulties with enforcing contracts (when reliable vendors are not available), limitations on logistics (i.e. transportation), and, lastly, limited access to capital (Baributsa et al., 2010; Coulibaly et al., 2012). Improving awareness among the supply chain actors including manufacturers, distributors, and retailers could spur their interest in the hermetic bag business. Facilitating linkages to create trust and relations among the supply chain actors is vital to maintaining the flow of the bags to improve availability. The need to expand the production capacity of manufacturers or increase the volume of orders by distributors and vendors may require access to capital from financial institutions. To improve the availability of hermetic bags at the farm level (last mile), there is a need to look into non-traditional distribution systems such as the youth retailer model to sell PICS bags in rural markets or tap into cellphone technology to connect farmers with vendors; both models are currently being tested in Ethiopia (Sertse, 2016).

## 5.2 Affordability and access

The price of a hermetic bag ranges from US$2 to US$4, depending on the brand, capacity, region, country, and location within a country. While the cost of hermetic bags has not been a major complaint (Moussa et al., 2010), most smallholder farmers have tight budgets. Though most smallholder farmers often lack capital, those who have enough grain to store should be able to acquire hermetic bags by selling small quantities of grain and storing the remainder safely. Those with very little disposable income are more likely to invest in a product after they have had first-hand experience with its viability. Increasing awareness of hermetic bags has shown to increase the willingness to pay among farmers who are non-users (Channa et al., 2019). Thus, creating awareness through training and media activities to ensure everyone (farmers, grain merchants, vendors, and all potential customers in the grain value chain) knows about the benefits of the technology against its cost is still in need. In addition, there are smallholder farmers who often sell a large proportion of their grain at harvest, when prices are low, despite the desire to store. This is often the case for farmers who need cash to satisfy family's needs. For those farmers who sell immediately after harvest, and buy grain later in the season, there is a need to develop financial products to help them reap the benefits of using hermetic bags. Making loans available at harvest would provide farmers the flexibility to store and sell their crops at higher prices later. Using financing facilities such as the warehouse receipts, farmers can access the much-needed cash at harvest from lenders (e.g. microfinance institutions) and use their harvest (grain) as collateral for loans. Because lenders will have more confidence that the quality of grain can be maintained during storage, they would be willing to give loans to smallholder farmers. Access to such lines of credit can give farmers access to hermetic bags and would increase adoption. To further improve affordability, some private sector companies (e.g. in Ethiopia) have requested government exemption of Value Added Tax (VAT) on hermetic bags. Removing VAT (estimated at 18%) may result in discounts that could be transferred to smallholder farmers and increase adoption.

### 5.3 Usability and durability

Hermetic storage bags are easy to use and come in different sizes to facilitate handling. Most bags currently being sold on the market have 90–100 kg capacity, though smaller sizes are also available. Farmers prefer larger sizes because the cost of storage is cheaper, and they treat these hermetic bags as granaries (not for transport). Hermetic bags are perfectly scale neutral and can be used to store a few kilograms or a few tons of grain with a lower cost per kilogram than many other storage options currently available to smallholder farmers in many developing countries. Hermetic bags can be tied off just above the grain or filled to capacity as necessary. Other rigid-sided hermetic storage containers such as metal or plastic drums or plastic jugs can be used, but they must be filled completely to be effective; otherwise, they would still have oxygen in them, making the technology ineffective. Trainings through demonstrations or media (posters, radio, videos) have been used to teach farmers how to use hermetic bags properly. Unlike rigid hermetic containers, hermetic bags are susceptible to physical damage due to puncture during handling, abrasions, and perforations from insects and rodents (Manandhar et al., 2018). Proper handling must be employed during transportation and storage to prevent bursting and any physical damage that will compromise their hermeticity, thus, reducing their effectiveness. Trainings equip farmers with the knowledge needed to minimize damages due to handling and rodent attacks (Baributsa et al., 2014b; Baoua et al., 2018; Baributsa and Njoroge, 2020). Simple management practices such as storing clean grain without debris and storing bags on pallets in clean areas and away from direct sunlight will extend the shelf life and usage of the bags. If bags are damaged, farmers can patch small holes or tears with tape to maintain airtightness. The durability of hermetic bags may vary by manufacturer as some brands have single or double liners inside the woven bags. If hermetic bags with a single liner are damaged, their reusability will be limited. Research has shown that PICS bags can store grains for multiple seasons - at least for 3 years (Baributsa et al., 2014b; Baributsa and Njoroge, 2020; Moussa et al., 2014). Just like any other plastic material, mechanical properties of a bag liner are affected by the material composition and thickness. These properties can, in turn, affect the performance and reusability of hermetic bags. This might be one of the reasons hermetic bags have exhibited varying susceptibility to insects such as the large grain borer (Coffi et al., 2016; De Groote et al., 2013; Baoua et al., 2013).

## 5.4 Need for drying technology and moisture assessment

To maintain quality during storage in hermetic bags, grains must be dried to the recommended moisture content (e.g. 13–14% moisture content for maize). Moist grain/seed stored in hermetic storage bags will result in quality deterioration (e.g. germination loss), or grain will turn into silage (Williams et al., 2014; Tubbs et al., 2016). Mold growth rapidly increases if grains are not properly dried and stored, producing aflatoxin, which causes liver cancer and other health problems (Bankole et al., 2006; Likhayo et al., 2018). Drying and moisture assessment technologies available on the market are either expensive or not adapted to the conditions of smallholder farmers. Finding affordable and cost-effective drying methods and moisture assessment tools should be explored to improve the quality of grain stored in hermetic bags. Low-cost moisture assessment and drying devices have been developed (Ajao et al., 2018; Thompson et al., 2017; Tubbs et al., 2017; Walker and Davies, 2017) but need further refining or dissemination to be adopted by smallholder farmers.

## 5.5 Need for standards and environmental impact

The scale-up of hermetic bags has proved that smallholder farmers are potential markets for postharvest technologies. Consequently, different brands and qualities of hermetic bags are becoming available to farmers in developing countries; thanks to investments by the private sector. More products in the market will provide farmers with more alternatives and lead to competition. However, competition may eventually lead to the 'race to the bottom', where profit instead of quality products is the major driver of private companies (Baributsa and Njoroge, 2020). Developing a common international approach for testing and rating hermeticity is needed to protect farmers from buying low-quality bags. These proposed standards could use key properties of the plastic inner liner such as oxygen and water vapor transmission rates, and mechanical properties (impact failure, tear, and tensile strength). Overall, there is a need for advocacy to develop hermetic storage technology standards to address low-quality products that could compromise the hermetic bag markets. For instance, the Kenya Bureau of Standards and the East African Grain Council are working with several stakeholders in Kenya to develop hermetic storage bag standards (KEBS, 2019). In addition, government bans on plastic may prevent smallholder farmers from reaping the benefits of using hermetic storage bags. If ordinances banning or regulating the use of plastics are passed, they could reduce access to hermetic storage bags by smallholder farmers. There is a need to engage policy-makers to develop plans for reducing the potential impacts of hermetic bags on the environment. Several hermetic bag manufacturers are involved in recycling and could buy-back and recycle unusable hermetic bags. This would reduce the environmental impact of using plastic to produce hermetic bags.

## 6 Case study: Purdue Improved Crop Storage (PICS) bag

The PICS technology is a simple, low-cost (US$2.00–$3.00 100 kg bag) storage method for preserving grain without using insecticides. PICS technology is a storage solution for smallholder farmers made of a three-layer bag that includes two liners and an outer woven layer. By creating a hermetic (airtight) seal inside the bag, PICS bags eliminate insect pests, stop mold growth, and maintain grain and seed quality. PICS bags were first developed and disseminated in West and Central Africa for cowpea (black-eyed pea) storage (Baributsa et al., 2010). Later, the bags were tested and disseminated for the storage of other crops including maize, rice, common beans, peanuts, wheat, sorghum, pigeon peas, mung beans, and other dry grains (Murdock and Baributsa, 2014). The use of PICS bags to store grains provides an opportunity to: (i) improve food security by allowing farmers to store grain or other foodstuffs through the lean season when supply is low, (ii) increase incomes of millions of smallholder farmers by providing the flexibility to store grain until prices are higher, and (iii) improve health by mitigating the impact of aflatoxin while reducing insecticide use. The PICS bags also help farmers to preserve their seed for planting, thus increasing availability and improving affordability.

The PICS technology was developed as a response to cowpea farmers' demand for chemical-free and cost-effective storage technology. During the early development of the PICS technology, farmers raised two major concerns: (i) they lacked knowledge on how to use the bags properly, and (ii) the bags were not available locally. To address these challenges, Purdue University and partners designed a program that had two major thrusts: (i) build awareness and train farmers on the proper use of PICS bags, and (ii) develop the supply chain to make PICS available to farmers in rural areas. In each country, the PICS program implemented several activities to build the demand and make the technology available to farmers (Baributsa et al., 2014a).

Market building activities included a cascade of activities that were implemented in a relatively short time - between 4 and 6 months. These activities included:

- ***Identification of partners***: Both service providers and private-sector actors were identified to lead the demand creation and supply chain development, respectively. These partners were chosen based on their experience and capacity to implement their respective activities in a country or region. Service providers included local and international NGOs, national and international research/extension organizations, and farmers' groups. Service providers led the selection of regions, districts, and communities where awareness and training activities would be implemented. The selection of these areas was based on crop production and the severity of storage losses on smallholder farms. After communities were identified, field agents were selected to implement training. The private-sector actors were involved in producing and building the distribution network to make the PICS bags available to farmers in rural areas.
- ***Training of trainers***: These trainings were implemented for the frontline field agents (government extension agents or field agents) who were already working with farmers in local communities. These 1- to 2-day trainings equipped these field agents with the basic knowledge on grain storage challenges (insect pests, mold, rodents, etc.), the economics of storage, strategies to train farmers, monitoring of field activities, and data collection and reporting.
- ***Implementation of village activities***: Several activities were implemented in communities, markets, and farmers' groups to build awareness and train farmers on the proper use of the PICS bags. These trainings were implemented in several steps: (i) awareness building - engage farmers in recognizing the importance of storage losses and introduce PICS bags, (ii) demonstrations - farmers are trained on how to use PICS bags and five pilot farmers in each community volunteer to store grain for several months to test the efficacy of the bags, (iii) follow up - the stored grain is inspected during the storage duration, and (iv) open-the-bag ceremony - the grain stored by the five pilot farmers is opened during a public event attended by the whole community.
- ***Media activities***: Media activities have included radio, television messages, posters and flyers, cellphone videos, short message services (sms), and Unstructured Supplementary Service Data (USSD) code. These activities have played an important role in reaching out to farmers with PICS information. Media helped to: (i) make farmers aware of PICS bags, (ii) train them on how to use the bags properly, and (iii) advertise retail points and provide PICS bag vendors' information to farmers and other users.

Published by Burleigh Dodds Science Publishing Limited, 2020.

- ***Supply chain development***: This involved identifying potential local or regional manufacturers of PICS bags and distributors, and licensed them to ensure product quality and availability. In addition, the PICS program worked with the private sector to market PICS bags to farmers and other users. It is important to note that the huge demands for PICS bags has incentivized the private sector to invest in the manufacturing and distribution of the technology.

Purdue University and partner organizations have led efforts to commercialize the PICS technology, focusing on low-resource farmers in different regions of the world (Baributsa, 2019). Women play an important role in grain storage (Ibro et al., 2014). To increase their participation in PICS activities, all implementing partners were required to have a target of 30% women's participation among field technicians and village activities (Baributsa et al., 2010). The approach used to commercialize PICS is described in Box 1. For the last 12 years, PICS bags have been commercialized in more than 35 countries in SSA, Asia, and recently in Central America and the Caribbean (Fig. 4). The large-scale dissemination of the PICS bags was supported with more than US$25 million in grants funded by the BMGF, USAID, USDA, and several other donors. The BMGF was the single-largest donor with US$23 million, spanning over 12 years in three project phases (PICS1, PICS2, and PICS3). These funds were used to support training

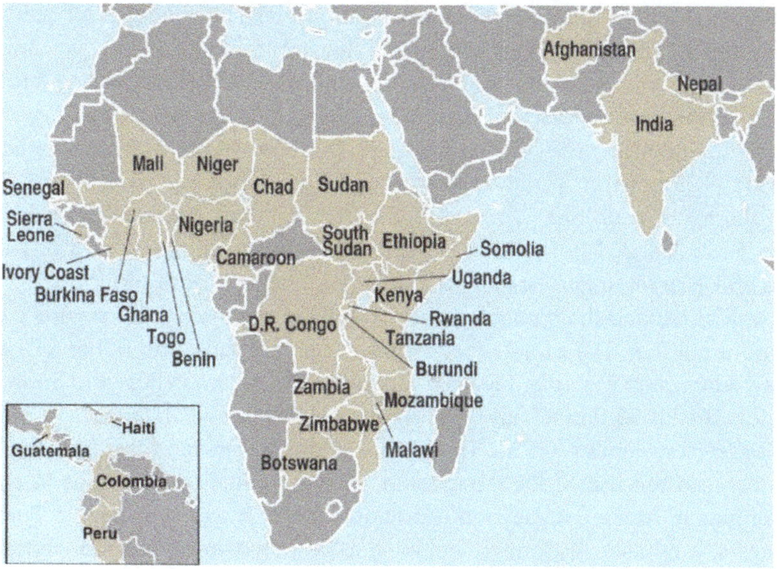


**Figure 4** Countries where the Purdue Improved Crop Storage (PICS) bags are being used/disseminated (2019) available (Source: www.picsnetwork.org).

and media activities led by service providers, and to develop the supply chain led by the private sector.

**Box 1**: PICS's approach to commercialization

- Create the demand by building the capacity of local service providers to raise awareness and train farmers how to use PICS bags
- Target 30% women's (trainers and trainees) participation
- Work with the private sector to manufacture and sell PICS bags to farmers and other users
- Leverage existing efforts by facilitating linkages between the private sector and service providers to increase the demand and availability of PICS bags
- Let the private sector drive the commercialization and service providers continue to provide training on demand.

The PICS1 project (US$12 million) was implemented in ten countries in West and Central Africa (2007–2012) and focused on cowpea storage, cowpea being the most important legume and cash crop for many smallholder farmers in the region (Baributsa et al., 2010, 2014a). Pilot programs were implemented in both Niger and Burkina Faso, with demonstrations in a limited number of villages. This allowed the project to learn and make adjustments for scale-up activities in subsequent years. After the initial pilot effort, the activities were scaled up in the whole region, directly reaching more than 1.6 million farmers in 31 000 villages in West and Central Africa. The PICS2 project (US$1.1 million) was implemented in several countries in SSA and Asia to assess the efficacy of the PICS bags in storing grains other than cowpea including maize, rice, sorghum, common beans, wheat, Bambara nuts, and pigeon peas (Murdock and Baributsa, 2014). The PICS3 project (US$10.1 million), the third phase of the BMGF grant to Purdue, aimed to improve market access and food security as well as expand the commercialization of PICS technology for several crops throughout SSA. The project focused on Nigeria, Ghana, and Burkina Faso in West Africa, and Ethiopia, Tanzania, Uganda, and Malawi in East and Southern Africa. The PICS3 project has directly reached over 2 million farmers in 34 000 villages/communities in SSA. There were efforts to leverage this BMGF grant to increase the use of PICS bags and other hermetic technologies in other countries in Africa and beyond. Additional funding was provided to Purdue University by several donors, including USAID in Kenya, Rwanda, Tanzania, Malawi (more than US$1.2 million) to promote PICS bags and also conduct studies on the impact of credit on storage.

The PICS technology is scalable. Extension activities have scaled the adoption of PICS bags from a few to thousands of villages in several countries in SSA and Asia. Just as PICS bags are a perfectly scalable storage solution for smallholder farmers, they are also very scalable for manufacturing, distribution, and retailing. Several local manufacturers are producing PICS bags around the world, including 12 companies in SSA, two companies in Asia, and one company in Central America. PICS bags are also scalable to the quantity of grain varying from a few kilos to thousands of tons stored by government food security agencies (e.g. the government of Niger purchased 800 000 PICS bags in 2008 to store cowpea). As a result of PICS program activities, millions of farmers have been reached, and millions of bags have been manufactured and sold by the private sector (Box 2). There has been a steady increase in the manufacture and sale of PICS bags for the past 12 years. In the 2018 harvesting season alone, 5.3 million bags were sold globally. This brought the total bags sold globally to 20 million by June 2019. This demonstrates that it is possible to scale up agricultural technologies among smallholder farmers by working with government and development agencies, NGOs, and the private sector. The PICS bag business is entering a transition phase from a project-supported supply chain development to a fully private-sector driven one. PICS Global, a private start-up company, is now working with licensed manufacturers and distributors to expand PICS bag markets in new and existing countries.

**Box 2**: The private sector is driving the commercialization of PICS bags because the business is viable

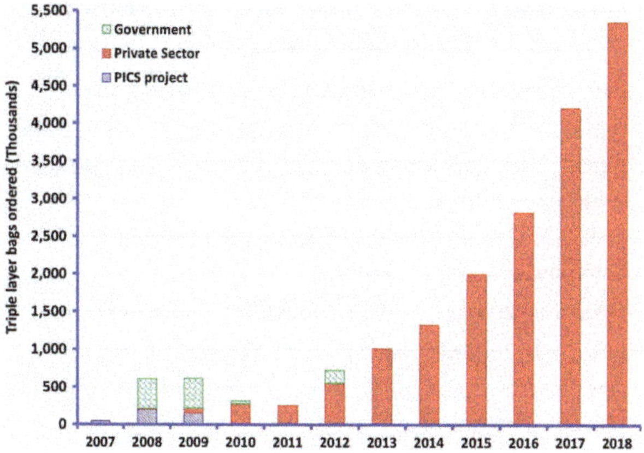


Published by Burleigh Dodds Science Publishing Limited, 2020.

After more than 12 years, the adoption of the PICS technology has resulted in great achievements (Box 3). The key to PICS success is that, since its earliest days, Purdue researchers have worked hand-in-hand with farmers who use the bags every day - making steady and informed improvements. The bags are produced at the cost of about US$1.00 to US$1.50 and are often sold to farmers at US$2.00 to US$3.00. Through investments to reduce storage losses, farmers and the public, in general, have been the biggest beneficiaries of the PICS technology. With sales of 20 million PICS bags, farmers have made or saved more than US$1.5 billion (estimates based on a cash flow of US$25 per 100 kg bag, and each bag is used three times). The private sector, including manufacturers, distributors, and vendors, makes a profit that is sufficient to keep its interest in the business, but small compared to what farmers earn. With a margin of about US$1 for the supply chain, the private sector has made more than US$20 million on the 20 million PICS bags produced and sold to farmers and other users. The PICS program demonstrated that there is a substantial market for hermetic storage technologies among smallholder farmers leading to competition - several brands of hermetic bags were developed and are being sold to farmers in SSA. Other benefits of using PICS and other hermetic bags include the reduction in deaths and illness due to the misuse of insecticides and the mitigation of mold growth (leading to aflatoxin accumulation) that causes stunted growth in children and liver cancer. The later benefits are immeasurable as farmers often tell us, '*There is no price to life*'.

### **Box 3**: Achievement in commercializing PICS bags in sub-Saharan Africa and Asia

- More than 7 million farmers trained to use PICS bags properly.
- More than 65 000 villages reached with training activities.
- More than 10 000 extension/field agents trained.
- Women's participation: 25% of trainers and 40% of trainees.
- 35 countries have PICS activities, and the interest continues to grow around the world.
- 23 licensed local private companies leading PICS commercialization worldwide.
- More than 20 million PICS bags sold from 2007 to June 2019.
- More than 2000 vendors selling PICS bags to farmers.
- US$1.5 billion made or saved by smallholder farmers using PICS bags to store grains.

Published by Burleigh Dodds Science Publishing Limited, 2020.

## 7 Summary and future trends

For the past several years, the use of hermetic bags has proven its impact on food safety and security in developing countries. The demand from smallholder farmers for this technology tremendously increased over the past 12 years, and its distribution has increased to more than 35 countries in SSA (mainly) and Asia. There are different types of hermetic bags (single-, double-, and triple bags) commercially available to smallholder farmers that are chemical-free, of low cost, and easy to use. The dissemination of hermetic bags has been implemented by research institutions, non-government organizations, and government agencies to demonstrate the technology and develop sustainable supply chains. These efforts have generated positive results and responses from farmers, and also generated new issues to be addressed.

One of the major issues raised by farmers is the unavaibility of hermetic bags in villages and rural markets. It is important to note that market penetrations differ among hermetic bag brands. Last-mile distribution remains a challenge, and sales are significantly below the potential in each country. There is a need to explore innovative solutions to improve the availability of hermetic bags in rural areas, including the use of Information and Communication Technology (ICT- e.g. sms and USSD) and youth retail models. Hermetic technology has proven to be effective in preventing insect-caused storage losses to cereal and legume crops. Because hermetic bags are manufactured using raw materials from a variety of sources and of varying quality, involve new manufacturers, are deployed in new geographic regions, are used to store a wide range of crops (some of which have not been tested yet), and are adopted by users from different cultures, we anticipate some issues to arise with respect to the performance of the technology. These issues, if left unresolved, might hinder the acceptance and adoption of hermetic bags. Understanding the material properties of the plastic liner used could provide solutions to improving the durability and usability of the technology. The adoption of national, regional, and international standards will contribute to better product consistency and quality control. It is also important to point out that drying practices and moisture assessment are vital operations in ensuring that grains and other stored products are at a safe moisture content before being placed into hermetic bags. Lastly, studying the behavior of farmers in using hermetic bags could help in the design of extension programs to increase adoption.

The impacts of hermetic bags on food security are well established, but there are still many gaps and questions that need to be addressed. More research is needed to improve and expand the supply chain of the hermetic bag to ensure its availability to farmers in rural areas. With large-scale users getting interested in hermetic bags - Could the sealing of bags be automated?

As more governments are banning the use of plastic - Can biodegradable plastic be used to produce hermetic bags? Understanding and comparing the properties and efficacy of single-, double-, and triple-layer bags will be important in developing standards for hermetic bags. Also, full economic and environmental impact assessments would be useful for hermetic bag technology to help governments make evidence-based decisions on plastic bans.

## 8 Where to look for further information

Several resources are available online:

- *How to properly use hermetic bags for grain storage:* Posters were developed in English and French, and translated into more than 20 local and international languages including Spanish, Swahili, Creole, Djoula, Yoruba, Igbo, Hausa, Chichewa, Tumbuka, Oromiffa, Amharic, Nepali, Kanuri, Hindi, Zarma, Somali, Wolof, Ateso, Kumam, and Twi. These posters are printed and given to farmers, extension agents and vendors. The posters are available online at: https://picsnetwork.org/resources/?tab_id=Posters.
- *How to use hermetic technologies for seed storage:* A poster describing how to use PICS bags and other hermetic containers such as jerrycans and water bottles for seed storage is available online at: https://picsnetwork.org//wp-content/uploads/2018/03/Seed-Storage-Poster-FINAL-reduced-size.pdf.
- *Where to find vendors of PICS bags:* Efforts have been made to make vendors' information available online to facilitate the linkage between potential users and the private sector. A map with pinpoints containing information of manufacturers, distributors and vendors in some of the PICS countries can be found online at: https://picsnetwork.org/where-we-work/?tab_id=distributor.

The following are the resources where information for commercially available hermetic bags in SSA is available:

- AgroZ bag - https://postharvest.nri.org/images/documents/Agroz_bags/AgroZ-bags.pdf.
- ELITE bag - http://eliteinnovations.co.ke/.
- Purdue Improved Crop Storage (PICS) bag - https://picsnetwork.org/.
- SuperGrainbag - https://grainpro.com/what-we-offer/.
- Zerofly bag - https://vestergaard.com/zerofly-hermetic-storage-bag.

## 9 References


Abass, A. B., Fischler, M., Schneider, K., Daudi, S., Gaspar, A., Rüst, J., Kabula, E., Ndunguru, G., Madulu, D. and Msola, D. 2018. On-farm comparison of different postharvest storage technologies in a maize farming system of Tanzania Central Corridor. *Journal of Stored Products Research* 77, 55–65. doi:10.1016/j.jspr.2018.03.002.

ACDI/VOCA 2017. AflaSTOP brief: hermetic storage: save money, safe food. Available at: https://www.acdivoca.org/2017/03/hermetic-storage-save-money-safe-food/.

Afzal, I., Bakhtavar, M. A., Ishfaq, M., Sagheer, M. and Baributsa, D. 2017. Maintaining dryness during storage contributes to higher maize seed quality. *Journal of Stored Products Research* 72, 49–53. doi:10.1016/j.jspr.2017.04.001.

AGRA 2015. Yield Wise: testing strategies to reduce post-harvest losses. Annual report. Available at: http://agra.org/2015AnnualReport/milestones/yieldwise.

AGRA 2017. YieldWise maize program in Tanzania. Available at: https://agra.org/news/agra-yieldwise-maize-program-in-tanzania/.

AgResults 2018a. Kenya on-farm storage challenge project – Airtight (Hermetic) Devices. Summary Report, December 2018.

AgResults 2018b. Kenya randomized rural household survey. Final Report, January 2018.

AgResults 2019. AgResults evaluation: Kenya on-farm storage challenge project. Final Report, December 2019.

Ajao, S. K., Omobowale, M. O., Popoola, K. O., Ala, A. A. and Opit, G. P. 2018. Comparative evaluation of novel low-cost moisture meters suitable for grain moisture measurement. *Arid ZoneJournal of Engineering Technology and the Environment* 14, 219–224.

Alemayehu, S., Abay, F., Ayimut, K. M., Assefa, D., Chala, A., Mahroof, R., Harvey, J. and Subramanyam, B. 2020. Evaluating different hermetic storage technologies to arrest mold growth, prevent mycotoxin accumulation and preserve germination quality of stored chickpea in Ethiopia. *Journal of Stored Products Research* 85, doi:10.1016/j.jspr.2019.101526.

Ameri, S., Deering, A. J. and McNamara, K. T. 2018. The economic benefits of small holder use of hermetic storage for wheat in Afghanistan. *International Journal of Food and Agricultural Economics* 6(3), 33–46. ISSN 2147-8988, E-ISSN, 2149-3766.

Awal, M. A., Ali, M. R., Hossain, M. A., Alam, M. M., Kalita, P. K. and Harvey, J. 2017. Hermetic bag an effective and economic rice storage technology in Bangladesh. Paper presented at the 2017 ASABE Annual International Meeting United States. doi:10.13031/aim.201700329.

Bankole, S., Schollenberger, M. and Drochner, W. 2006. Mycotoxins in food systems in Sub Saharan Africa: a review. *Mycotoxin Research* 22(3), 163–169. doi:10.1007/BF02959270.

Baoua, I. B., Amadou, L., Margam, V. and Murdock, L. L. 2012a. Comparative evaluation of six storage methods for postharvest preservation of cowpea grain. *Journal of Stored Products Research* 49, 171–175. doi:10.1016/j.jspr.2012.01.003.

Baoua, I. B., Margan, V., Amadou, L. and Murdock, L. L. 2012b. Performance of triple bagging hermetic technology for postharvest storage of cowpea grain in Niger. *Journal of Stored Products Research* 51, 81–85. doi:10.1016/j.jspr.2012.07.003.

Baoua, I. B., Amadou, L., Lowenberg-DeBoer, J. D. and Murdock, L. L. 2013. Side by side comparison of GrainPro and PICS bags for postharvest preservation of cowpea grain in Niger. *Journal of Stored Products Research* 54, 13–16. doi:10.1016/j.jspr.2013.03.003.

Baoua, I. B., Amadou, L., Ousmane, B., Baributsa, D.. and Murdock, L. L. 2014a. PICS bags for post-harvest storage of maize grain in West Africa. *Journal of Stored Products Research* 58, 20–28. doi:10.1016/j.jspr.2014.03.001.

Baoua, I. B., Amadou, L., Baributsa, D. and Murdock, L. L. 2014b. Triple bag hermetic technology for post-harvest preservation of Bambara groundnut (*Vigna subterranea* (L.) Verdc.). *Journal of Stored Products Research* 58, 48–52. doi:10.1016/j.jspr.2014.01.005.

Baoua, I. B., Amadou, L., Bakoye, O., Baributsa, D., Murdock, L. L. (2016). Triple bagging hermetic technology for post-harvest preservation of paddy rice *Oryza sativa L.* in the Sahel of West Africa. *Journal of Stored Products Research* 68, 73–79. doi.org/10.1016/j.jspr.2016.04.006.

Baoua, I. B., Bakoye, O. N., Amadou, L., Murdock, L. L. and Baributsa, D. 2018. Performance of PICS bags under extreme conditions in the Sahel zone of Niger. *Journal of Stored Products Research* 76, 96–101. doi:10.1016/j.jspr.2018.01.007.

Baributsa, D., Lowenberg-DeBoer, J., Murdock, L. and Moussa, B. 2010. Profitable chemical-free cowpea storage technology for smallholder farmers in Africa: opportunities and challenges. In: Carvalho, M. O., Fields, P. G., Adler, C. S., Arthur, F. H., Athanassiou, C. G., Campbell, J. F., Fleurat-Lessard, F., Flinn, P. W., Hodges, R. J., Isikber, A. A., Sinha, K. K., Thorpe, G. R., Timlick, B. H., Trematerra, P. and White, N. D. G. (Eds.), *Proceedings of the 10th International Working Conference on Stored Product Protection, 27 June–2 July 2010, Estoril, Portugal*. Julius Kühn-Institut, Berlin, Germany. Available at: http://spiru.cgahr.ksu.edu/proj/iwcspp/iwcspp10.html.

Baributsa, D., Abdoulaye, T., Lowenberg-DeBoer, J., Dabiré, C., Moussa, B., Coulibaly, O. and Baoua, I. 2014a. Market building for post-harvest technology through large-scale extension efforts. *Journal of Stored Products Research* 58, 59–66. doi:10.1016/j.jspr.2014.02.012.

Baributsa, D., Djibo, K., Lowenberg-DeBoer, J., Moussa, B. and Baoua, I. 2014b. The fate of triple- layer plastic bags used for cowpea storage. *Journal of Stored Products Research* 58, 97–102. doi:10.1016/j.jspr.2014.02.011.

Baributsa, D. 2019. Empowering the private sector to reduce hunger - PICS Sales crossed 5 million bags per year in 2018. *PICS. Newsletter* 5(1). Available at: https://picsnetwork.org//wp-content/uploads/2019/09/Volume-5-E-final-sm.pdf.

Baributsa, D. and Njoroge, A. W. 2020. The use and profitability of hermetic technologies for grain storage among smallholder farmers in Eastern Kenya. *Journal of Stored Products Research* 87. doi:10.1016/j.jspr.2020.101618.

Channa, H., Chen, A. Z., Pina, P., Ricker-Gilbert, J. and Stein, D. 2019. What drives smallholder farmers' willingness to pay for a new farm technology? Evidence from an experimental auction in Kenya. *Food Policy* 85, 64–71. doi:10.1016/j.foodpol.2019.03.005.

Chigoverah, A. A. and Mvumi, B. M. 2016. Efficacy of metal silos and hermetic bags against stored-maize insect pests under simulated smallholder farmer conditions. *Journal of Stored Products Research* 69, 179–189. doi:10.1016/j.jspr.2016.08.004.

Chigoverah, A. A. and Mvumi, B. M. 2018. Comparative efficacy of four hermetic bag brands against *Prostephanus truncatus* (Coleoptera: Bostrichidae) in Stored Maize grain. *Journal of Economic Entomology* 111(5), 2467–2475. doi:10.1093/jee/toy217.

Coffi, H., Nyabicha, J. and Ouma, J. O. 2016. The use of hermetic bags for on farm storage of grains and pulses against insect pests. *Outlooks on Pest Management*. Research Information Ltd. All rights reserved. Available at: www.pestoutlook.com.

Coulibaly, J., D'Alessandro, S., Nouhoheflin, T., Aitchedji, C., Damisa, M., Baributsa, D. and Lowenberg-DeBoer, J. 2012. Purdue improved cowpea storage (pics) supply chain study. In: Working Paper #12-4, November 2012. *International Programs in Agriculture*. Purdue University.

De Groote, H., Kimenju, S. C., Likhayo, P., Kanampiu, F., Tefera, T. and Hellin, J. 2013. Effectiveness of hermetic systems in controlling maize storage pests in Kenya. *Journal of Stored Products Research* 53, 27–36. doi:10.1016/j.jspr.2013.01.001.

Donovan, N. K., Foster, K. A. and Salinas, C. A. (2019). Analysis of green coffee quality using hermetic bag storage. *Journal of Stored Products Research* 80, pp. 1–9.

Foy, C. and Wafula, M. 2016. Scaling up of hermetic bag technology (PICS) in Kenya: Review of Successful Scaling of Agricultural Technologies. Available at: https://www.agrilinks.org/library/scaling-hermetic-bag-technology-pics-kenya-review-successful-scaling-agricultural.

Govereh, J., Muchetu, R. G., Mvumi, B. M. and Chuma, T. 2019. Analysis of distribution systems for supply of synthetic grain protectants to maize smallholder farmers in Zimbabwe: implications for hermetic grain storage bag distribution. *Journal of Stored Products Research* 84, 101–520. doi:10.1016/j.jspr.2019.101520.

Guenha, R., Salvador, BdV., Rickman, J., Goulaoc, L. F., Muochaa, I. M. and Carvalho, M. O. 2014. Hermetic storage with plastic sealing to reduce insect infestation and secure paddy seed quality: a powerful strategy for rice farmers in Mozambique. *Journal of Stored Products Research* 59, 275–281. doi:10.1016/j.jspr.2014.06.007.

Hell, K., Edoh Ognakossan, K., Lamboni, Y. (2014). PICS hermetic storage bags ineffective in controlling infestations of Prostephanus truncatus and Dinoderus spp. in traditional cassava chips. *Journal of Stored Products Research* 58, 53–58. doi.org/10.1016/j.jspr.2014.03.003.

Ibro, G., Sorgho, M. C., Idris, A. A., Moussa, B., Baributsa, D. and Lowenberg-DeBoer, J. 2014. Adoption of cowpea hermetic storage by women in Nigeria, Niger and Burkina Faso. *Journal of Stored Products Research* 58, 87–96. doi:10.1016/j.jspr.2014.02.007.

Jones, M., Alexander, C. and Lowenberg-DeBoer, J. 2011a. An initial investigation of the Potential for Hermetic Purdue Improved Crop Storage (PICS) bags to improve incomes for Maize producers in Sub-Saharan Africa. Working Paper #11-3, September 2011. Department of Ag Economics. Purdue University Press, West Lafayette, IN.

Jones, M., Alexander, C. and Lowenberg-DeBoer, J. 2011b. Profitability of hermetic Purdue Improved Crop Storage (PICS) bags for African common bean producers. Working Paper #11-6, November 2011. Department of Ag Economics. Purdue University Press, West Lafayette, IN.

Jonfia-Essien, W., Navarro, S. and Villers, P. 2010. Hermetic storage: a novel approach to the protection of cocoa beans. *African Crop Science Journal* 18(2), 59–68. ISSN 1021-9730. doi:10.4314/acsj.v18i2.65797.

KEBS 2019. Hermetic Storage Bags Specification. Kenya Bureau of Standards, First edition. Available at: http://www.puntofocal.gov.ar/notific_otros_miembros/ken829_t.pdf.

Kitch, L. W. and Ntoukam, G. 1991. Airtight storage of cowpea in triple plastic bags. Technical Bulletin No. 3. Institut de la Recherche Agronomique du Cameroun (IRA) and Bean/Cowpea Collaborative Research Support Program (CRSP).

Kumari, A., Kumar, V., Amitabh, A., Kumar, A., Kumar, M. and Kashyap, V. 2017. Qualitative loss of maize under different bag storage modes. *International Journal of Current Microbiology and Applied Sciences* 6(12), 177–183. ISSN: 2319-7706. Journal homepage. Available at: http://www.ijcmas.com. doi:10.20546/ijcmas.2017.612.023.

Lane, B., Sharma, S., Niu, C., Maina, A. W., Wagacha, J. M., Bluhm, B. H. and Woloshuk, C. P. 2018. Changes in the fungal microbiome of maize during hermetic storage in the United States and Kenya. *Frontiers in Microbiology* 9, 2336. doi:10.3389/fmicb.2018.02336.

Likhayo, P., Bruce, A. Y., Tefera, T. and Mueke, J. 2018. Maize grain stored in hermetic bags: effect of moisture and pest infestation on grain quality. *Journal of Food Quality* 2018, 1–9, Article ID 2515698. doi:10.1155/2018/2515698.

Maina, A. W., Wagacha, J. M., Mwaura, F. B., Muthomi, J. W. and Woloshuk, C. P. 2016. Postharvest practices of maize farmers in Kaiti District, Kenya and the impact of hermetic storage on populations of *Aspergillus* spp. and aflatoxin contamination. *Journal of Food Research* 5(6), 53–66. doi:10.5539/jfr.v5n6p53.

Manandhar, A., Milindi, P. and Shah, A. 2018. An overview of the post-harvest grain storage practices of smallholder farmers in developing countries. *Agriculture* 8(4), 57. doi:10.3390/agriculture8040057. Available at: www.mdpi.com/journal/agriculture.

Martin, D. T., Baributsa, D., Huesing, J. E., Williams, S. B. and Murdock, L. L. 2015. PICS bags protect wheat grain, Triticum aestivum (L.), against rice weevil, Sitophilus oryzae (L.) (Coleoptera: Curculionidae). *Journal of Stored Products Research* 63, 22–30. doi:10.1016/j.jspr.2015.05.001.

Moussa, B., Lowenberg-DeBoer, J. and Baributsa, D. 2010. Adoption of hermetic storage for cowpea in Niger and Burkina Faso in 2010. Available at: http://tinyurl.com/jhvxcy7. Poster presented at the 5th World Cowpea Conference, 27 September–1 October 2010. Saly, Senegal.

Moussa, B., Lowenberg-DeBoer, J., Fulton, J. and Boys, K. 2011. The economic impact of cowpea research in West and Central Africa: A regional impact assessment of improved cowpea storage technologies. *Journal of Stored Products Research* 47(3), 147–156. doi:10.1016/j.jspr.2011.02.001.

Moussa, B., Abdoulaye, T., Coulibaly, O., Baributsa, D. and Lowenberg-DeBoer, J. 2014. Adoption of on-farm hermetic storage for cowpea in West and Central Africa in 2012. *Journal of Stored Products Research* 58, 77–86. doi:10.1016/j.jspr.2014.02.008.

Murdock, L. L., Margam, V., Baoua, I., Balfe, S. and Shade, R. E. 2012. Death by desiccation: effects of hermetic storage on cowpea bruchids. *Journal of Stored Products Research* 49, 166–170. doi:10.1016/j.jspr.2012.01.002.

Murdock, L. L. and Baoua, I. 2014. On Purdue Improved Cowpea Storage (PICS) technology: background, mode of action, future prospects. *Journal of Stored Products Research*, 58, pp. 3–11. https://doi.org/10.1016/j.jspr.2014.02.006

Murdock, L. L. and Baributsa, D. 2014. Hermetic storage for those who need it most -subsistence farmers. In: Arthur, F. H., Kengkanpanich, R., Chayaprasert, W. and Suthisut, D. (Eds). *Proceedings of the 11th International Working Conference on Stored Product Protection*, 24–28 November 2014, Chiang Mai, Thailand, 310–323.

Mutambuki, K., Affognon, H., Likhayo, P. and Baribusta, D. 2019. Evaluation of purdue improved crop storage triple layer hermetic storage bag against *Prostephanus truncates* (Horn) (Coleoptera: Bostrichidae) and *Sitophilus zeamais* (Motsch.) (Coleoptera: Curculionidae). *Insects* 10(7), 204. doi:10.3390/insects10070204.

Mutungi, C. M., Affognon, H. D., Njoroge, A. W., Baributsa, D. and Murdock, L. L. 2014. Storage of mung bean (Vigna radiata [L.] Wilczek) and pigeonpea grains (Cajanus cajan [L.] Millsp) in hermetic triple-layer bags stops losses caused by *Callosobruchus maculatus* (F.) (Coleoptera: Bruchidae). *Journal of Stored Products Research* 58, 39-47. doi:10.1016/j.jspr.2014.03.004.

Mutungi, C. M., Affognon, H. D., Njoroge, A. W., Manono, J., Baributsa, D. and Murdock, L. L. 2015. Triple-layer plastic bags protect dry common beans (Phaseolus vulgaris) against damage by *Acanthoscelides obtectus* (Coleoptera: Chrysomelidae) during storage. *Journal of Economic Entomology* 108(5), 2479–88. doi:10.1093/jee/tov197.

Navarro, S., Donahaye, J. and Varnava, A. 1993. Preservation of grain in hermetically sealed plastic liners with particular reference to storage of barley in Cyprus. In: Navarro, S. and Donahaye, J. (Eds). *Proceedings of the International Conference on Controlled Atmosphere and Fumigation in Grain Storages*, Winnipeg, Canada, June 1992. Caspit Press Ltd., Jerusalem, 223-234.

Navarro, S., Donahaye, E. and Fishman, S. 1994. The future of hermetic storage of dry grains in tropical and subtropical climates. In: Highley, E., Wright, E. J., Banks, H. J. and Champ, B. R. (Eds). *Proceedings of the 6th International Working Conference on Stored-Product Protection*, Canberra, Australia, 17–23 April 1994. CAB International, Wallingford, Oxon, UK, 130–138.

Ndegwa, M. K., De Groote, H., Gitonga, Z. M. and Bruce, A. Y. 2016. Effectiveness and economics of hermetic bags for maize storage: results of a randomized controlled trial in Kenya. *Crop Protection* 90, 17–26. doi:10.1016/j.cropro.2016.08.007.

Ng'ang'a, J., Mutungi, C. and Imathiu, S. M. 2016a. Low permeability triple-layer plastic bags prevent losses of maize caused by insects in rural on-farm stores. *Food Security* 8, 621–633. doi:10.1007/s12571-016-0567-9.

Ng'ang'a, J., Mutungi, C., Imathiu, S. and Affognon, H. 2016b. Effect of triple-layer hermetic bagging on mould infection and aflatoxin contamination of maize during multi-month on-farm storage in Kenya. *Journal of Stored Products Research* 69, 119–128.

Njoroge, A. W., Affognon, H. D., Mutungi, C. M., Manono, J., Lamuka, P. O. and Murdock, L. L. 2014. Triple bag hermetic storage delivers a lethal punch to Prostephanus truncatus (Horn) (Coleoptera: Bostrichidae) in stored maize. *Journal of Stored Products Research* 58, 12–19. doi:10.1016/j.jspr.2014.02.005.

Nkhata, S. G., Ortiz, D., Baributsa, D., Hamaker, B., Rocheford, T. and Ferruzi, M. G. 2019. Assessment of oxygen sequestration on effectiveness of Purdue Improved Crop Storage (PICS) bags in reducing carotenoid degradation during post-harvest storage of two biofortified orange maize genotypes. *Journal of Cereal Science* 87, 68–77. doi:10.1016/j.jcs.2019.02.007.

Nouhoheflin, T., Coulibaly, J. Y., D'Alessandro, S., Aitchédji, C. C., Damisa, M., Baributsa, D. and Lowenberg-DeBoer, J. 2017. Management lessons learned in supply chain development: the experience of PICS bags in West and Central Africa. *International Food and Agribusiness Management Review* 20(3), 427–438. doi:10.22434/IFAMR2016.0167.

Paudyal, S., Opit, G. P., Osekre, E. A., Arthur, F. H., Bingham, G. V., Payton, M. E., Danso, J. K., Manu, N. and Nsiah, E. P. 2017. Field evaluation of the long-lasting treated storage bag, deltamethrin incorporated, (ZeroFly Storage Bag) as a barrier to insect pest infestation. *Journal of Stored Products Research* 70, 44–52. doi:10.1016/j.jspr.2016.11.003.

Rugumamu, C. P. 2004. Insect infestation of maize, Zea mays (L.) in indigenous storage structures in Morogoro region, Tanzania. *Tanzania Journal of Science* 29(2), 1–10. doi:10.4314/tjs.v29i2.18372.

Sanon, A., Dabiré-Binso, L. C. and Ba, N. M. 2011. Triple-bagging of cowpeas within high density polyethylene bags to control the cowpea beetle *Callosobruchus maculatus* F. (Coleoptera: Bruchidae). *Journal of Stored Products Research* 47(3), 210–215. doi:10.1016/j.jspr.2011.02.003.

Sertse, Y. 2016. Radio plus ICT improve the availability of PICS bags in Ethiopia. *PICS Newsletter* 2(3). Available at: https://picsnetwork.org//wp-content/uploads/2016/10/October-2016.pd.

Swathi, Y. and Rajanikanth, P. 2017. Post-harvest losses – use of triple layer plastic bag – an innovative eco-friendly low-cost approach to mitigate the problem in developing countries. *International Journal of Pure and Applied Bioscience* 5(5), 112–116. doi:10.18782/2320-7051.5735.

Thompson, J. F., Reid, M. S., Felix, L., Donis-Gonzalez, I., Mjawa, B. and Ambuko, J. 2017. DryCardTM – A low-cost dryness indicator for dried products. *AIMS Agriculture and Food* 2, 339–344. doi:10.3934/agrfood.2017.4.339.

Tubbs, T., Baributsa, D. and Woloshuk, C. 2016. Impact of opening hermetic storage bags on grain quality, fungal growth and aflatoxin accumulation. *Journal of Stored Products Research* 69, 276–281. doi:10.1016/j.jspr.2016.10.003.

Tubbs, T., Woloshuk, C. and Ileleji, K. E. 2017. A simple low-cost method of determining whether it is safe to store maize. *AIMS Agriculture and Food* 2(1), 43–55. doi:10.3934/agrfood.2017.1.43.

Vales, M. I., Ranga Rao, G. V., Sudini, H., Patil, S. B. and Murdock, L. L. 2014. Effective and economic storage of pigeonpea seed in triple layer plastic bags. *Journal of Stored Products Research* 58, 29–38. doi:10.1016/j.jspr.2014.01.004.

Villers, P., de Bruin, T. and Navarro, S. 2006. *PS7-8* (vol. 6119) -. Development and applications of the hermetic storage technology. In: *9th International Working Conference on Stored Product Protection*, Sao Paulo, Brazil.

Villers, P., Navarro, S. and De Bruin, T. 2008. Development of hermetic storage technology in sealed flexible storage structures. In: *Controlled Atmosphere and Fumigation (CAF) Conference in Chengdu, China*, September 2008, Session 9 (Achievements of CA and Fumigation, and Development Trends).

Walker, S. and Davies, B. 2017. *Feasibility of Up-Scaling the EasyDry M500 Portable Maize Dryer to Kenya*. USAID - Bill & Melinda Gates Foundation.

Walker, S., Jaime, R., Kagot, V. and Probst, C. 2018. Comparative effects of hermetic and traditional storage devices on maize grain: mycotoxin development, insect infestation and grain quality. *Journal of Stored Products Research* 77, 34–44. doi:10.1016/j.jspr.2018.02.002.

Williams, S. B., Baributsa, D. and Woloshuk, C. 2014. Assessing Purdue Improved Crop Storage (PICS) bags to mitigate fungal growth and aflatoxin contamination. *Journal of Stored Products Research* 59, 190–196. doi:10.1016/j.jspr.2014.08.003.

Williams, S. B., Murdock, L. L. and Baributsa, D. 2017. Storage of maize in Purdue Improved Crop Storage (PICS) bags. *PLoS ONE* 12(1), e0168624. doi:10.1371/journal.pone.0168624.

# Chapter 7

# Advances in insect pest management in postharvest storage of cereals: detection and monitoring

*Paul Fields, Agriculture and Agri-Food Canada, Canada; Fuji Jian, University of Manitoba, Canada; and Dianxuan Wang, Henan University of Technology, China*

1 Introduction
2 How to sample
3 Sampling theory
4 Future trends in research
5 Where to look for further information
6 Acknowledgements
7 References

## 1 Introduction

Sampling grain for stored grain insects is expensive in terms of time and resources. There are several reasons to sample: integrated pest management (IPM), regulatory, validation of effectiveness of control measure and understanding insect biology and ecology. Finding stored-product insects is difficult and unreliable as stored-product insects are small, cryptic and often feed within kernels of grain or are found deep within the grain mass. Sampling is done by almost all grain managers with two goals: (1) to identify the species found and (2) to determine if the insect density reaches the economic threshold. Occasionally, there is a third objective to sampling, measuring characteristics of the insects caught, for example, resistance to insecticides (Toon et al., 2018).

As part of an IPM program, many grain managers have a sampling program of their granaries, elevators or bag stacks on an ongoing basis in addition to sampling for insects upon receipt (Anonymous, 2015b). One reason to do this is because stored-product insects have a high rate of population growth (Howe, 1962) and insects move inside grain bulks; populations can go undetected within the grain bulk because the insects are small and cryptic, and finally the grain can be stored for a long period of time which can lead to loss in quality and quantity of grain.

http://dx.doi.org/10.19103/AS.2020.0072.07
Published by Burleigh Dodds Science Publishing Limited, 2020.

Most countries have norms for allowing live insects or dead insects in grains (economic thresholds). Sometimes allowable limits on insect infestation will be stipulated in the contract of sale that is more stringent than the government standards. Several countries have zero tolerance for live and storage insects in grain. In Canada, the Canada Grain Act (Anonymous, 2015a) stipulates that grains must be free of live storage insects in licensed elevators and they cannot receive, store or sell grains that are infested. In Australia and France there are similar controls on the zero tolerance for insects in grain. In the United States and China there is some allowance for live insects in grain. There are a few quarantine insects that infest stored-product insects (Myers and Hagstrum, 2012). The most common one is *Trogoderma granarium* which is found in the Middle East and the Indian subcontinent, but is a quarantine insect in much of the rest of the world (Athanassiou et al., 2019). Commodities are sampled upon arrival into the country to detect quarantine insects.

Sampling is also done to verify if insects have been controlled after a control measure, such as fumigation, contact insecticides or freeze out. With important control measures, for example, controlling an infestation of insects that have resistance to phosphine, bioassays maybe used to verify resistance level. Bioassays are samples of insects with a species of interest with a strain and insect life stage that is tolerant to the insecticide being used (Toon et al., 2018).

## 2 How to sample

There are two broad types of estimates of population that sampling will provide: absolute or relative (Subramanyam and Hagstrum, 1995). Absolute estimates of populations draw the samples for a given unit: grain sample (weight), area of floor (area) or air sample (volume). Insects are extracted from the entire sample, identified and counted. Efforts should be made to get a representative sample (Anonymous, 2015b, 2016a,a) and the implications for sample size, number and distribution of the insects will be discussed in Section 3. There are indirect methods to measure the absolute population density using self-marking traps (Wileyto et al., 2000; Subramanyam and Hagstrum, 1995). These have been used experimentally with moth and beetle species (Campbell and Mullen, 2004), but has not been used commercially.

Relative estimates use traps to capture grain insects. They have different efficiencies depending on several factors outlined below. There have been attempts to estimate absolute population densities from trapping data (Flinn et al., 2009). An Insector system (Fig. 1, an electronic trap) was commercialized and this system can estimate insect densities at the trap location (Jian and Larson, 2006; Jian et al., 2012a). If the purpose is the detection of insects rather than the estimate population densities, trapping can be a more sensitive method of

Published by Burleigh Dodds Science Publishing Limited, 2020.

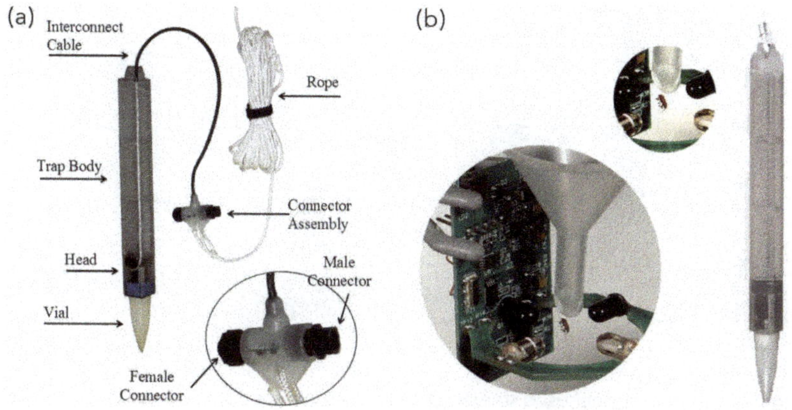


**Figure 1** Insector (electronic grain probe) from OPIsystems. Figure 1a Probes are placed in the grain, secured with rope and connected to an electrical cable to a network of other Insectors and computer. Insects enter the perorated trap body, and then fall past the pair LED sensors (Figure 1b), to be counted, classified and population density estimated using software.

sampling than grain sampling (increasing trapping time increases sample unit and size). For example, using probe traps compared to trier samples, probes traps were 2–10 times more efficient than grain samples at detecting insects (Jian et al., 2014a; White et al., 1990).

## 2.1 Sampling methods for different grain storage types

Almost all commercial operations that deal with grain will sample the grain at receipt, during storage and as it leaves the facility. This includes, producers storing on the farm, elevators, bag stack warehouses and end use buyers (flour mills, food processors). These grain samples are used for grading (quality assessment), in addition to detecting insects.

Detection of insects in grain happens in a number of ways. The simplest method is sieving the grain, collecting the adults and larvae and identifying the insects (Anonymous, 2015b, 2016a,b). Large samples of up to 30 kg can be processed using an inclined sieve (White, 1983). The Carter Dockage Tester commonly used in Canada to process grain samples for grading collects insects into the aspirator discharge pan (Demianyk et al., 1997). However, many insects are inside the kernel and will not be detected upon sieving. One common method for extraction is the Berlese funnel (Fig. 2) (Jian et al., 2016; Smith, 1977). It works well for adults and larvae of *Cryptolestes* spp. and *Tribolium* spp., and mites. It does not work well for larvae of *Sitophilus* spp. and bruchids that cannot escape the seed. There are a number of methods to detect hidden internal infestations. Flotation in various solutions can be used to separate infested and non-infested kernels. The grain is put in an alcohol solution or

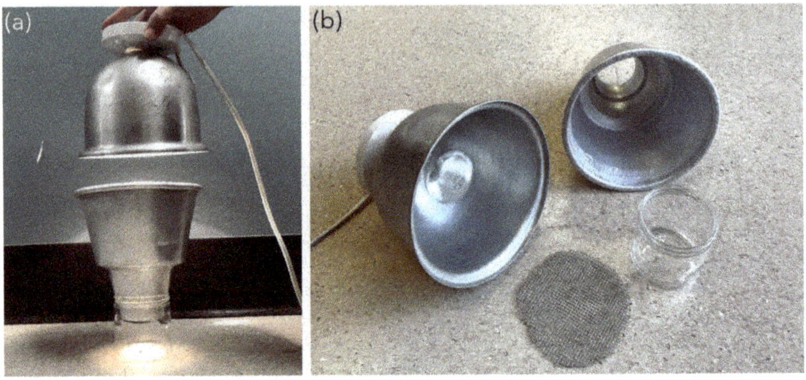


**Figure 2** Berlese funnel used to extract larvae and adults from grain samples (a). These can be set in sets of multiple funnels to run many samples at the same time. The Berlese funnel consists of a metal lid with incandescent light bulb, a funnel with a screen to hold approximately a 1-kg sample of grain and the collection jar (b). As insects are repelled away from the light and heat, they are collected in the jar at the base of funnel.

a salt solution; the infested kernels will float (Brader et al., 2002; Richter and Tchalale, 1994). This method will work only with larger larvae and will not distinguish species. X-rays have been longed used to detect insects, but this has not been used commercially to any great extent (Karunakaran et al., 2003). Internal insects can be detected by crushing the seed and measuring spikes in the conductance across the rolls used to crush the seed (Brabec et al., 2010; Pearson and Brabec, 2007). This has been developed commercially, but has not gained widespread use to date. The presence of insect-damaged kernels is a grading factor in the United States and is used to detect the presence of insect feeding, although all insects could be killed from a previous control effort such as fumigation.

The goal of all grain sampling is to get a representative sample, the 1–10 kg grain sample that is taken from a truck, railcar or bin will represent the entire bulk that can be 5–500 t. For bulk grain that is sampled at the tail gate as the truck unloads into the elevator pit (Fig. 3) (Anonymous, 2016b), the USDA recommends taking at least three samples during the unloading of a truck to avoid the first and last grain to be unloaded in the truck. Larger grain handling facilities may sample trucks manually using trier probes (Fig. 4), or use pneumatic trier probes. These are combined and if the sample unit needs to be reduced, then use a Boerner-type divider that will insure that dockage (Fig. 5), insect-infested grain; different grains are equally divided in the two collection pans (Anonymous, 2015b). There are several sampling devices within an elevator that sample the grain as it moves within the elevator (Figs 6–8). These include the pelican sampler, strip auger sampler, and Woodside sampler (Anonymous, 2015b, 2016a). The grain stream is sampled,

**Figure 3** Sampling grain for a truck as it delivers at country elevator in Canada.

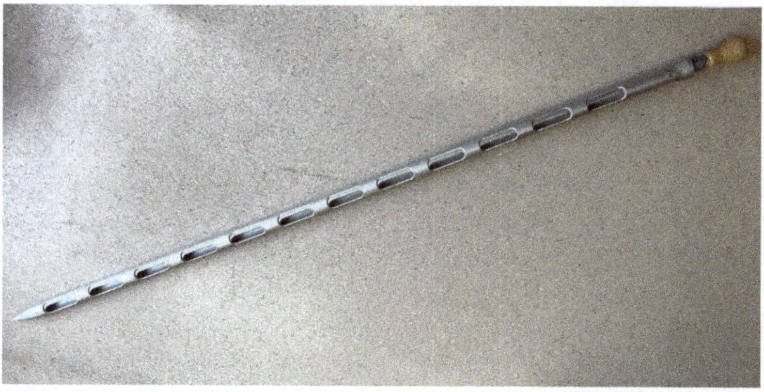

**Figure 4** A 2-m long grain trier probe (open position). The channel that receives the grain can be divided or undivided. The trier is pushed in the grain (closed), the outer sleeve turned to open as shown, grain allowed to flow in, then the outer sleeve closed and the probe pulled from the grain with sample.

divided and reduced and then delivered pneumatically to the grain inspection office. The movement of grain through the pneumatic delivery system is lethal to insects, with mortality ranging from 22% to 73% of adult stored-product insects (Bryan and Elvidge, 1977). The pneumatic delivery systems varied in length from 40 m to 90 m with grain speed reaching 11 m/s. Larvae were not tested, and may be less susceptible to damage as they are often protected inside the kernel (Paliwal et al., 1999). Occasionally, bulk grain will be sampled in grain bins or horizontal storages. Depending on the structure it can be very

**Figure 5** Boerner-type divider that ensures the dockage, insect-infested grain and different seeds are equally divided in the two collection pans.

difficult to access the grain surface, elevators, and farm bins that are half-filled, or it can be very easy to access farm bins that are full or horizontal storage. Grain can be sampled from the surface using a deep-probe trier sampler. Normally this can only sample the top 2 m of the grain. Deeper samples can be taken using a vacuum probe sampler that can go as deep as 12 m (Flinn et al., 2010).

Bag stacks or totes must be sampled manually (Anonymous, 1996) with thin double-sleeved trier probes (Fig. 9, Nobbe trier) that can pierce the fabric of the bag without causing leakage (Fig. 10, (Anonymous, 2015b). Different sized seeds need different diameters of Nobbe trier samplers. Vertical sampling should use a partitioned trier sampler to sample the entire depth of the bag. Ideally the trier sampler will reach the end of the bag. The Canadian Grain Commission recommends all bags (20–50 kg grain) be sampled if 20 or fewer, 6% of bags with 21–1000 bags, and if more than 1000 bags then 3% of the bags should be sampled. Samples can be combined and divided using a Boerner divider (Fig. 5). For totes which can contain 100–500 kg grain,

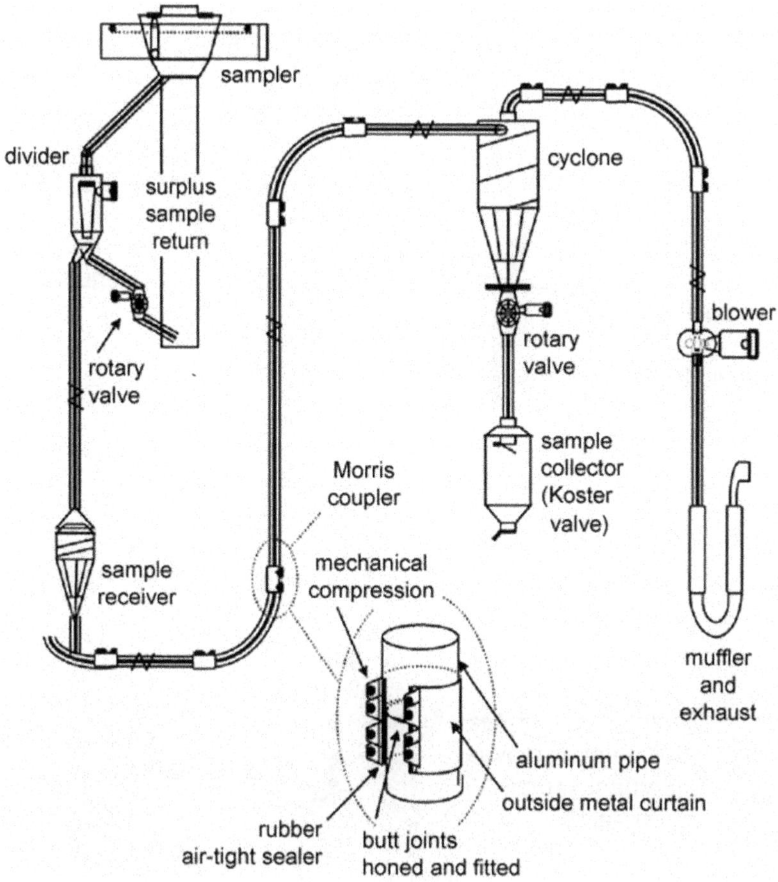


**Figure 6** Automatic pneumatic grain sampling system for grain elevator.

two primary samples should be taken from each tote. Bag stacks in India are typically 9 × 6 m wide and 4 m high with 70–150 t containing a few thousand bags. So sampling the interior bags is difficult (Fig. 10).

## 2.2 Traps

There are a wide variety of traps used to detect stored-product insects (Toews and Nansen, 2012). There are several advantages to traps compared to grain sampling and there are some limitations as well. Some of the advantages of traps are that they can be more sensitive at detecting infestations than grain samples. Traps can be used to estimate the presence or absence of insects. For example, using probe traps compared to trier samples, probes traps are 2–10 times more efficient than grain samples (Jian et al., 2014a; White et al., 1990). Traps work as

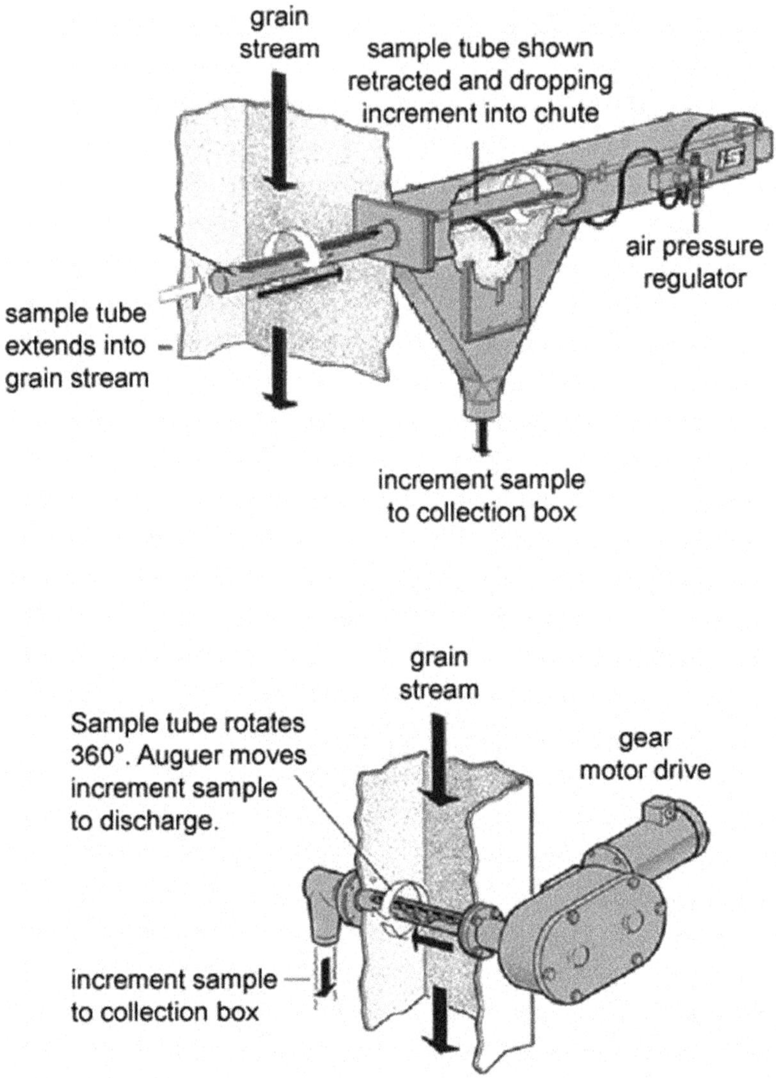


**Figure 7** Strip auger samplers for taking grain samples in grain elevators.

long as they are deployed, whereas in grain samples, the insect must be present in the grain sample to be detected. However, there are several limitations to traps. They are a relative measure of the population density, as many factors affect trap catch, whereas a grain sample gives estimates of population density, insects/kg and so on. Traps will only sample walking or flying insects. Hence immobile internal stages are not detected. Lower temperatures lead to lower trap catches and at less than 5°C most insects cannot crawl and at less than 25°C

**Figure 8** Sample divider in grain elevator. One small pipe goes to the government sampling office, second small pipe goes to company's grain-quality office and the large pipe returns excess grain to grain stream. Grain delivered to offices pneumatically.

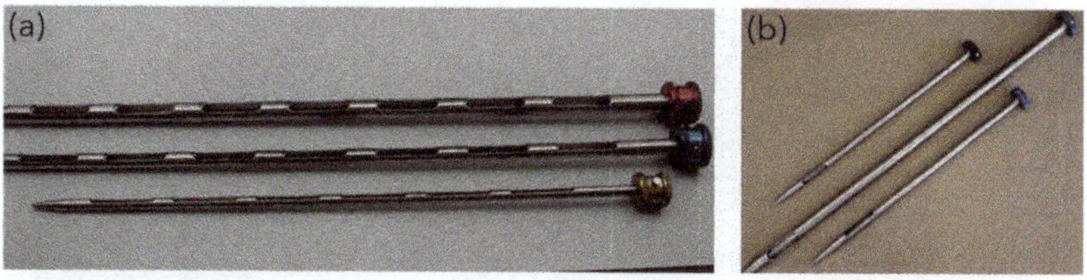


**Figure 9** (a) Double sleeve grain triers for sampling bags. (b) Nobbe grain triers for sampling bags.

most insects cannot fly (Jian et al., 2015). Traps can be biased by differences in insect mobility and trap efficiency for different species, for example, a probe pitfall trap caught 10 *Cryptolestes ferrugineus*/trap, compared to 3 *Sitophilus oryzae*/trap, 2 *Tribolium castaneum*/trap and only 0.4 *Rhyzopertha dominica*/trap, all at the same population density (Fargo et al., 1989).

There are three basic types of traps, grouped on where they are placed: aerial traps, surface traps and pitfall traps. Aerial traps capture flying insects

**Figure 10** Sampling grain in bag stack.

**Figure 11** Delta trap with sticky glue board used with insect pheromones.

and are often combined with pheromones or kairomones. The most common traps used for flying insects are the delta wing traps or diamond traps (Fig. 11) that are used to capture stored-product moths, such as *Plodia interpunctella* or *Ephestia* spp (Phillips et al., 2000; Nansen et al., 2006). These traps have sticky surfaces to capture insects and that can get clogged at high populations or

**Figure 12** Lindgren multiple funnel trap with insect pheromones located outside of grain silo. The white container at the bottom of the jar contains a solution of water and ethylene glycol (slows evaporation) and retains captured insects.

used in dusty environments. They are often used in warehouses, homes and retail stores. Low-profile traps have also been developed (Thinline traps, Trécé Inc., Adair, OK, USA) that catch fewer insects and are easier to overload, but can be placed in places with less room and are more discrete, which can be useful in public areas. Sticky traps have been used extensively to monitor *Lasioderma serricorne* populations in warehouses (Papadopoulou and Buchelos, 2002; Mueller et al., 1990; Miyatake et al., 2016). In areas of high dust the Lindgren Multiple Funnel trap (Fig. 12) or the bucket trap has been used extensively. The Lindgren Multiple Funnel trap has been used extensively to sample outside of grain bins and elevators (Fields et al., 1993; Toews and Nansen, 2012; Holloway et al., 2018). The pheromone for *R. dominica* is very effective outside of the grain to attract flying adults. Light traps using UV or visible light are also effective at catching stored-product moths, *R. dominica, Sitophilus zeamais, T. castaneum* and *L. serricorne* (Miyatake et al., 2016; Mohan et al., 1994; Nualvatna et al.,

**Figure 13** Two Dome traps (one open, one closed) for catching stored-product insects in flour mills and warehouses. The trap has two parts: the top cover with holders for rubber septa with pheromones (figure has one in dark gray) and the bottom with a textured ramp to a central collection well with a paper disk and plant oil. The pheromone and the oil act together to attract insects into the well, and the oil retains and kills the insects.

2003). Certain insects are attracted to specific wavelengths; for example, *P. interpunctella* is attached to UV and blue/violet light (Cowan and Gries, 2009).

There are a variety of surface traps that are placed on the floor of elevators, processing facilities or warehouses. Some are buried into the top of the grain surface. The Dome Trap invented by Mullen et al. (Mullen et al., 1992) and further developed and commercialized by Trécé Inc., (Fig. 13) is a trap that combines pheromones and food attractants and is widely used for monitoring *Tribolium* spp and other crawling insects in flour mills, cereal food production facilities and warehouses. Refuge traps use stored-product insects' thigmotactic behavior to retain insects (Fig. 14) (Barak et al., 1990). Sometimes these traps will have attractants and/or insecticides to retain insects. An example is the corrugated cardboard traps (Fig. 14) used to monitor the quarantine insect, *T. granarium* (Barak and Burkholder, 1976; Barak, 1989).

There are several types of pitfall traps (Figs 1, 15–17) placed on the surface or with the grain mass. They come in various forms. The probe pitfall trap was initially developed by Loschiavo and Atkinson (Loschiavo and Atkinson, 1967; Loschiavo and Atkinson, 1973). The trap is a tube with holes along its length with a removable collection cap at the bottom (Fig. 15). The narrow tube allows the trap to be thrust into the grain for 1 or 2 m. There have been several improvements and modifications to this basic design. It is available

**Figure 14** Corrugated cardboard trap that attracts insects that seek tight spaces.

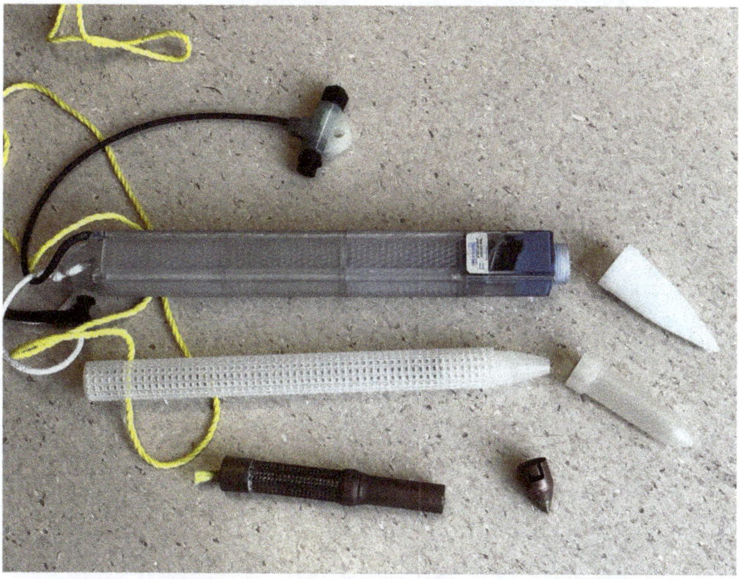

**Figure 15** Three probe pitfall traps: the one on the bottom is made of copper and is one of the original probes, designed by Loschiavo and Atkinson (1973); the one in the middle in white plastic is the WB Probe II trap, sold by Trécé Inc; and the one on the top in clear plastic is the electronic trap Insector designed by OPIsystems (Fig. 1). All traps are placed vertically in the grain and capture insects as they move through the grain.

commercially in plastic (Grain Probe Trap, Trécé Inc., Adair, OK, USA). It has an electronic sensor attached to the collection tip (Fig. 1) (Flinn et al., 2009; Shuman et al., 2005; Toews et al., 2003) (OPIsystems Inc. Calgary AB, Canada). This electronic system can count and classify insects, negating the need to enter the bin to retrieve the counted insects (Jian et al., 2012a). A similar system was developed in China (Fig. 16) (Yao et al., 1999). Considerable effort has been made to correlate the numbers trapped and the corresponding density of insects in the grain (Flinn et al., 2009; Toews et al., 2003). Some traps, such as the PC Trap (AgriSense-BCS Ltd., Pontypridd, UK), have been sunk into the top of the grain surface (Wakefield and Cogan, 2007). Some have used a simple beaker placed into the top of a mound (Wakefield and Cogan, 2007), and are used in the horizontal storages in Berlin. The Chinese have developed a probe pitfall connected to vacuum lines (Fig. 16) and a central collection location (Wang et al., 2018). Pitfall traps have been developed to work in bag stacks (Rajesh et al., 2015), (Fig. 17).

**Figure 16** Grain probe pitfall trap connected to vacuum line to central collection area.

Published by Burleigh Dodds Science Publishing Limited, 2020.

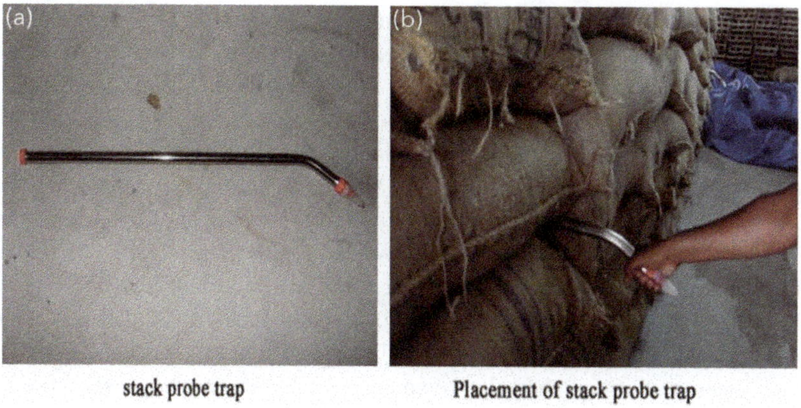


**Figure 17** (a) Probe pitfall trap for sampling in bag stacks, (b) placing probe in bag stacks.

Attractants, pheromones or kairomones are often used in aerial and surface traps, less often in pitfall traps. Pheromones are chemicals produced by insects that elicit a response in the same species. Sex pheromones are usually produced by females to attract males for mating. These have been identified in many of the stored-product moths and in a few of the short-lived adult beetles from the Anobiidae, Bruchidae and Dermestidae species (Phillips et al., 2000). Most of the other stored-product beetles produce aggregation pheromones, which are produced by males and attract both males and females. Currently, there are 11 moth pheromones (Swords and Van Ryckeghem, 2010b) and 17 beetle pheromones available commercially (Swords and Van Ryckeghem, 2010a); the pheromone for *C. ferrugineus* has recently become available (Holloway et al., 2018). Pheromones typically increase trap catch by about 10-fold (Phillips et al., 2000). The pheromones for the moths are very effective, as are the pheromones for the beetles, *R. dominica* in multiple funnel traps and *L. serricorne* and *Tribolium* spp. in surface traps (Hawkin et al., 2011). Kairomones are behavior-modifying chemicals that are perceived by a species but emitted by another. Food volatiles are used in various traps. Bait bags, are a mixture of cracked and whole seeds placed in a mesh bag that attracts a wide range of insects (Collins and Chambers, 2003; Haines et al., 1990). Some traps use extracts of food, such as corn oil with the Dome Trap (Fig. 13) (Trécé Inc., Adair, OK, USA), Indian Meal moth female attractant with sticky traps (Fig. 11) (Insects Limited, Westfield IN, USA) (Sambaraju and Phillips, 2008).

### 2.3 Detection using imaging

Wavelengths from across the electromagnetic spectrum, microwaves, infrared, visible, ultraviolet and X-rays have been used to detect stored-product insects. There is an emerging technology, hyperspectral imaging, that uses images from a few of wavelengths that change with quality (Paliwal et al., 2001) to assess the quality of the grain. Microwaves have been used to detect insects as they move through the grains (Reimer et al., 2018). As insects move through the electrical field generated by the microwave, they can be detected using a sensor with planar active microwave cavity. Near-infrared (NIR) is used to measure protein in cereals, and this technology (Fox and Manley, 2014) has been used to detect insects in flour, as an alternative to the standard floatation method (Toews et al., 2007; Anonymous, 2006), and insects inside kernels (Dowell et al., 1998). The NIR combined with neural networks has been used to identify stored-product insects to genus with over 95% accuracy (Dowell et al., 1999). Hyperspectral imaging, a process that acquires the full range of spectra (1000--1600 nm) for each pixel in an image, has also been used to detect insect or insect damage in grain (Ravikanth et al., 2017). Visible light and form have also been used to automatically detect insects in grain samples (Ravikanth et al., 2017). Some of the challenges are the speed of acquiring the image and processing, and it can only detect external insects. X-rays can detect and identify insects (Karunakaran et al., 2003)

## 2.4 Detection using sound

Insects in stored grain make sound when they move through grain and feed on the grain. As early as 1954, sound was used to detect insects in stored grain (Adams et al., 1954). There are several reviews of using sound to detect insects and the latest gives an overview of the technology and the past reviews (Mankin and Hagstrum, 2012). Different types of microphones have been used: capacitance microphone, microphone, lead zirconate titanate piezoelectric transducer, PVDF piezoelectric film transducer and PZT ultrasonic transducer. The sound waves are often treated with some filtering to remove background noise. There have been a number of commercial systems developed, but none of these are widely used. These systems have been deployed in bulk grain (Mankin and Hagstrum, 2012) bag stacks (Njoroge et al., 2019) and as grain sample in sound-insulated box (Fleurat Lessard et al., 2006). Large insects, (e.g. *Sitophilus* spp or *Tribolium* spp) and late instar larvae make more noise than small insects (e.g. *Cryptolestes* spp, psocids) and early instar larvae. Higher temperatures cause more movement and chewing. Advantages of the system are that it can constantly monitor the grain, and different species may be identified. Constraints of the system are that it gives a relative measure of insect density, false positives due to background noise and attenuation of the sound in the grain. Cell phones have been used to identify mosquito species (Mukundarajan et al., 2017); further work may allow this technology to be used in grains using microphones buried in grains and the data can be seen on mobile phones.

### 2.5 Detection using chemistry

There are several detection methods that detect insects using chemistry to measure uric acid, insect DNA, egg plugs, insect protein or carbon dioxide (Rajendran, 1999). Grain samples are taken and analyzed for various chemicals, the exception being the use of carbon dioxide in a grain bulk. Gas lines are placed into the grain bulk and sampled for carbon dioxide, either using electronic sensor or using colorimetric tubes. Carbon dioxide above ambient levels (0.03%) indicate biological activity either from molds or insects (Muir et al., 1985). Uric acid is found in the feces of insects and has been used to detect insect infestations. However, the sensitivity is such that the infestation needs to be very high in flour (Pixton, 1965) or one kernel/100 g wheat to detect infestation (Wehling et al., 1984). An ELISA test was developed to detect insects in flour (Quinn et al., 1992; Atui et al., 2003) as a replacement for the floatation of insect fragment which is the industry standard (Anonymous, 2006). Egg plugs of *Sitophilus* spp. and eggs of other stored-product insects can be preferentially stained with several different histological stains (Leelaja et al., 2007). It would be possible to combine the staining with some image analyzing software to process large number of samples.

## 3 Sampling theory

### 3.1 Sample unit and size

To know insect density and distribution inside stored-grain bulks, it is always better to collect more samples. However, costs associated with sampling limit the number of samples taken. Therefore, estimations of the optimum (minimum) sample size and sample unit are required in order to meet the basic requirement of a sampling plan with a fixed sample unit and size. Sample unit is defined as the amount of grain taken (weight or volume) or the surface (area) in each single sample. Sample size is the number of samples taken. The optimum sample unit is the ideal amount of grain or area of an observation which is selected to be analyzed. The optimum sample unit is usually changeable with minimum sample unit due to the sample costs in grain industry. A sample unit can be a sample with a few kg or $m^3$ of grain or $m^2$ of the sampled area. The optimum (minimum) sample size is the smallest number of samples that would satisfy the objectives of the sampling plan and achieve the desired precision of estimations of insect number, density or distribution.

There is no recommended method to estimate the optimum (minimum) sample unit. The size of a sampling unit is usually determined by the choice of the sampling device which is based on the insect species being sampled and the facility holding the grain. For example, a small scoop can be used when the grain is transferred inside a belt conveyor. A scoop is usually used to collect samples from the grain stream falling from the end gate of a truck or wagon. In the grain industry, the minimum sample unit for the insect sampling is usually less than 0.5 kg of grain if manual sampling (using probe or grain trier) is conducted, and 2.75 kg of grain if mechanical sampling is conducted. Jian et al. (2011a) recommended using 7 kg sample unit to detect insects at low insect densities. Different countries adopted different sampling protocols and different sample units.

If a large (combined) sample is taken and all of this large sample cannot be measured and/or analyzed, the sampling divider and the nested analysis of variance can be used to sub-divide the large sample. The large sample can be divided into several equal portions (Fig. 5) and one or a few of the subsamples are further analyzed. The nested analysis of variance can be used to estimate the optimum (minimum) size of a sample unit ($S_u$, kg or m³) and sample size in the large (combined) sample (Jian et al., 2011a; Snedecor and Cochran, 1967):

$$S_u = C\sqrt{\frac{C_{bs}}{C_{ws}}\frac{S_{bs}^2}{S_{ws}^2}} \tag{1}$$

where $S_u$ is the weight or volume of a sample unit (kg or m³), $C$ is the weight or volume of the large sample (kg or m³), $S_{bs}^2$ and $S_{ws}^2$ are the variances between and within sample units, respectively, $C_{bs}$ is the cost of locating, sampling and analyzing one sampling unit (usually measured in min), and $C_{ws}$ is the sample inspecting cost to locate, sampling and analyze the large sample (measured in min).

If a fixed number of sample size is conducted, the sample size should be calculated. The method used to estimate the smallest number of sample size requires that the mean should be within a certain range of the mean because the value of the mean variance significantly affects the mean distribution. When the standard error of the mean is defined within a certain range of the mean, the smallest sample size can be calculated as:

$$N = \frac{S^2}{m^2 S_E^2} \tag{2}$$

where $N$ is the smallest sample size, $S^2$ is the sample variance (the average of the squared differences from the mean), $m$ is the expected mean and the $S_E$ is the standard error of the mean expressed as a proportion of $m$. For example, if the sampler wishes to collect enough samples to estimate the mean with a standard error of ± 10% of the mean, then $S_E = 0.1$.

If the distribution of the insect population is known, $N$ can be calculated as:

$$\text{Poisson}(\text{random}):\ N = \frac{1}{m S_E^2} \tag{3}$$

$$\text{Negative binominal (aggregated): } N = \frac{(k+m)}{kmS_E^2} \tag{4}$$

$$\text{If } S^2 = am^2 \; N = \frac{am^{b-2}}{S_E^2} \tag{5}$$

where $k$ is the aggregation factor of the negative binominal, $a$ and $b$ are the constants in the Taylor's power law which can be obtained from Taylor's power law regression by using published data under similar conditions (Jian et al., 2011a). $k$ can be calculated by the method of moments as (Southwood, 1978):

$$k = \frac{\bar{x}^2}{S^2 - \bar{x}} \tag{6}$$

where $\bar{x}$ is the mean of the samples.

When the estimation of the mean is defined within a certain value of the true mean with a given probability, the smallest sample size can be calculated as:

$$N = \frac{(t_{\alpha/2} / d)^2}{S^2 / m^2} \tag{7}$$

where $d$ is the desired fixed proportion of the mean and $t_{\alpha/2}$ is the t value for a given probability ($\alpha$) in the t distribution (Student's distribution). The estimated confidence interval by using Equation 7 is ±dm. In most cases, $\alpha = 0.05$ (95% probability), and $t_{\alpha/2}$ is about 2.0 for a large sample size. For example, if the literature shows that $m = 0.5$ and $S^2 = 5$, then $N = \frac{(2/0.1)^2}{5/0.5^2} = 20$ with a confidence interval ±10% and 95% probability.

If the distribution of the insect population is known and the desired fixed proportion of the mean is specified, $N$ can be calculated as:

$$\text{Poisson (random): } N = \left(\frac{t_{\frac{\alpha}{2}}}{d}\right)^2 \left(\frac{1}{m}\right) \tag{8}$$

$$\text{Negative binominal (aggregated): } N = \left(\frac{t_{\frac{\alpha}{2}}}{d}\right)^2 \left(\frac{k+m}{km}\right) \tag{9}$$

$$\text{If } S^2 = am^2 \; N = \alpha \left(\frac{t_{\frac{\alpha}{2}}}{d}\right)^2 m^{b-2} \tag{10}$$

The distribution pattern of the insect can be evaluated by calculating the dispersion pattern indices including Variance to Mean Ratio, Negative Binominal Distribution Parameter, Liod's Mean Crowding, Morisita's index, Green's Index, Iwao's Patchiness Regression and Taylor's power Law. Subramanyam and Hagstrum (1995) extensively described these indices and methods of calculation.

Jian et al. (2011b) found that the distribution of the sampling count frequency was at a random or near random pattern, and the count frequency of the adults of *C. ferrugineus* fit less than 23% for the negative binomial, positive binomial or Poisson distribution. Therefore, the count frequency of stored-grain insects could not be fitted to either of these distributions except the clumped distribution. If no distribution model can fit the distribution of insect population, a normal approximation (like Taylor's power law and method of Iwao) can be used, provided a minimum sample size is used.

Both methods mentioned above depend on the insect distribution and how precision is defined. Precision is the closeness of the estimations in repeated measurements (sampling), while accuracy refers to the nearness of a measurement to the true value of a parameter being measured. Obtaining an accuracy of measuring is impossible because the true value of insect number and density is not known. However, we can use precision to estimate the accuracy. For example, if 10 samples are drawn from a grain bin, the standard error (division) of the mean of the insect density associated with the 10 samples is 0. This indicates that the true insect density will be close to the mean of the 10 samples. The measured precision can be described by using standard error, standard deviation, relative variation, proportion of the mean or confidence level (interval). Therefore, precision can be measured from the sampling analysis, not accuracy. However, in the sampling literature, the term of accuracy is used, so we will also use the term accuracy in this chapter.

Precision can be justified by the sampler. Without knowing insect distributions and true densities, the samplers usually use statistical tools and based on literatures and/or experience to estimate the sampling accuracy with an assumption that the sample size and unit could provide a representative statistical population. This assumption may be true when the sample size and unit are big enough (e.g., the sample size is equal to the minimum sample size) or insects have a uniform distribution (Jian et al., 2014b).

### 3.2 Sequential sampling

It is difficult to estimate the insect density in stored grain bulks from insect number in each sample unit. Sequential sampling uses variable number of sample size and can take advantage of the sequential order of the sampling to avoid taking too many or too few. The essential step of the sequential sampling method is to check whether a conclusion can be made after every single observation or a group of observations is analyzed. The conclusion can be that a null hypothesis is rejected, insect number reaches the economic threshold or the mean of the insect density reaches a defined level of precision. For example, if the insect numbers found in a previous sample reaches the economic thresholds, further sampling is not needed. Therefore, sequential sampling in general requires fewer sample units than a fixed sample size, thereby reducing costs (Subramanyam et al., 1997).

There are different sequential sampling protocols used in entomology (Binns, 1993). The equation developed by Green (1970) is usually used to calculate the critical stop line for stored-product insect sampling, the cumulative number of insects after which sampling can be terminated:

$$\ln(T_n) = \frac{\ln\left(\frac{d^2}{a}\right)}{b-2} + \frac{b-1}{b-2}\ln(N) \qquad (11)$$

where $N$ is the number of sample units, $T_n$ is the cumulative total insects in the total $N$ samples, and $d$ is the precision expressed as the ratio of the standard error of the mean to the mean. The constants $a$ and $b$ are the regressed values in the Taylor's power law by using published data under similar sampling conditions (Jian et al., 2011a) or the samples in the previous sampling.

Equation 11 is a linear line on the natural log scale and this line can be drawn before sampling (referred to as stop line). After each sampling observation and $T_n$ calculation, the calculated $T_n$ and the corresponding $n$ can be also drawn on the same graph of the stop line. At the time when the plotting of $T_n$ intersect the stop line, the corresponding $n$ value is the sample size of the sequential sampling plan and further sampling is not needed. Sequential sampling plan for *C. ferrugineus* was developed (Subramanyam et al., 1997) and verified (Jian et al., 2011a). This developed sequential sampling plan can be used to estimate whether the density of *C. ferrugineus* is near or above the economic threshold and pest management is needed.

### 3.3 Representative sample

A representative sample is a group of samples with a suitable sample unit and size that accurately represents the distribution of the insect population inside a grain bulk studied. For example, when a group of samples are drawn from a bin with 0.5 insects/kg, the data from the samples should characterize the insect density of 0.5 insects/kg. Using the developed sample plans with correct sample size and correct procedure to collect each sample unit is a correct practice to retrieve the representative samples. Without losing the required accuracy and with minimized cost, there are two approaches to achieve representative samples: a small sampling unit with large sample size or a large sample unit with small sample size. Using either one of the approaches, Jian et al. (2014a and b) concluded: (1) 'less than 20 sampling size would have 90% chance to predict insect density as 0 insects/kg when insect density was ≤ 1.0 adults/kg'; (2) 'using a sampling method to estimate insect densities with ≤60% of relative variance might not be practical when insect densities was ≤1.0 adults/kg because more than 0.09% grain should be sampled' (detecting insects inside a 1000 t wheat bin will need 8889 samples with 0.5 kg sample unit); (3) 'when insect densities were between 1.0 and 2.5 adults/kg, the characterized distribution could be different from the true clumped distribution if sampling units were not large enough (i.e. ≥10 kg)'; (4) 'sample unit with less than 0.5 kg underestimate insect density and cannot be used to characterize insect distribution'; and (5) 'if the random sampling technique was used and 15% of the stored wheat bulk was sampled, 72% of determined means of insect densities of the sampling sets were lower than the introduced insect densities'.

Most of the grain in stored grain bulks will not be infested by insects, and hence there is an absence of insects in most samples. A sampling plan based on presence/absence of insects at low insect densities has been developed (Hagstrum et al., 1985; Subramanyam et al., 1997). However, this developed sampling plan required too many samples at low insect densities (roughly 24.5 t in 82–122 t bins). At high insect densities, the fraction of samples with insects rises progressively more slowly, resulting in poor resolution above one insect per 0.5 kg of wheat. Insects usually have clumped distribution inside stored grain bins (Athanassiou et al., 2003; Jian et al., 2011a,b, 2012b,c,d). Therefore, a slight difference in the sampling location will induce a significant difference of the detected insect numbers. Hagstrum et al. (1985) concluded that one probe sample did not predict very well the insect density in another grain sample taken at a distance of 30 cm. Different sampling methods may result in different characterizations of insect distribution and abundance (Stejskal et al., 2008; Jian et al., 2014a). These conclusions are consistent with the other reports in the literature (Jian et al., 2014b; White and Loschiavo, 1986; Wilkin and Fleurat-Lessard, 1990; Wright and Mills, 1984). Therefore, taking a representative sample from grain bulks is difficult.

The main purpose of the sampling is to infer the information of the entire population by using a simple and inexpensive method without interfering in the grain bulk. Considering the challenge of retrieving a representative sample, sampling methods in grain storage practice are mostly developed based on the demand, experience and cost under storage conditions. These developed methods and plans usually did not consider the sampling theories (the correct sample unit and size) because it is too expensive or not feasible to sample grain bulks if the statistical theory is followed. Different methods and plans are developed in different countries and there is no international accepted standard. In the international trade, the contract with a specified sample size and unit is used as the protocol of sampling plan. In the national trade, each country has developed its own protocol. The most used sampling plan in the international or national trade is the fixed sample unit and size with random sampling. Grain storage managers or farmers occasionally use the sequential sampling or presence/absence sampling plan. The following sampling techniques are also occasionally used: cluster, nested, hierarchical, multistage, double and variable-intensity sampling. There is no published study on these sampling techniques for stored-grain insects. There has been some work done for field crop insects (Cochran, 1977; Southwood, 1978). Regardless of the sample protocol, the data associated with different protocols in different countries under any condition should be analyzed with the consideration of the grain conditions such as the grain storage history, moisture content and temperature. Therefore, sampling the locations with higher temperatures and moisture contents should be prioritized. With nonrandom sampling, the first sample unit can be selected at random or nonrandomly, and the remaining units are usually selected in a systematic fashion (Cochran, 1977). The data collected by these bias sampling plans should be analyzed by using bias statistical methods (Legg, 1993).

Developing a sampling program includes choosing the proper sampling unit and size, determining the sampling locations and the sampling schedule. Sampling schedule should be developed and followed because stored-product insects are cryptic; given the right conditions, they can multiply rapidly. Therefore, a correct sampling location is the basic requirement to achieve a representative sample. The sampling schedule should also be flexible which is based on the previous sampling results such as insect density and distribution. The application of insect control methods should also be considered in the sampling schedule development. Graphing the trends of insect density and distribution over time is a good tool for understanding insect problems in grain storage.

## 4 Future trends in research

The future of sampling in grain storage is more automated with rapid nondestructive samples that can be taken, as the grain is moved or monitored by sensors in the bin (Jayas and Jian, 2012). The information from these sensors will be coupled with computer systems that will estimate insect populations and predict degradation according to the grain type, temperature and moisture.

## 5 Where to look for further information

There have been several reviews of sampling for stored-product insects (Hagstrum and Flinn, 2012; Campbell et al., 2012; Hagstrum and Subramanyam, 2006; Subramanyam and Hagstrum, 1995; Neethirajan et al., 2007; Banga et al., 2018; Rajendran and Steve, 2005). All reviews cover methods, theory and examples. Subramanyam and Hagstrum (1995) have conducted an excellent review of sampling statistics. Campbell et al. (2012) review sampling in food processing facilities, such as flour mills and breakfast food manufactures. There are several reviews just focusing on traps and the interpretation of trap results (Wang et al., 2018; Toews and Nansen, 2012; White et al., 1990).

There are two international conferences that focus primarily on stored-product entomology research. Both publish proceedings with many papers on sampling of stored-product insects: The International Working Conference on Stored-Product Protection has been meeting since 1974 (http://bru.gmprc.ksu.edu/proj/iwcspp/) and the International Conference of Controlled Atmosphere and Fumigation in Stored Products has been meeting since 1980 (http://ftic.co.il/caf-en.php). The *Journal of Stored Product Research* has been published since 1965 and is one of the best places to find information on stored product entomology. The *Journal of Economic Entomology* has a section on Stored-Products and one on Sampling and Biostatistics.

## 6 Acknowledgements

We would like to thank the following people and organizations for providing figures: OPIsystems Inc. for Fig. 1; Canadian Grain Commission for Figs 6, 7, 8 and 9; Insects Limited Inc. for Fig. 11; and S. Mohan for Fig. 17.

## 7 References

Adams, R. E., Wolfe, J. E., Milner, M. and Shellenberger, J. A. 1954. Detection of internal infestation in grain by sound amplification. *Cereal Chemistry* 31, 211–76.

Anonymous 1996. Storage of cereals and pulses–part 1. *G: General Recommendations for the Keeping of Cereals; ISO* 6632-1.

Anonymous 2006. AOAC Official Method 993.26 Light Filth in Whole Wheat Flour. *AOAC International*.

Anonymous 2015a. *Canada Grain Act*. Government of Canada.

Anonymous 2015b. Sampling systems handbook and approval guide. *Canadian Grain Commission*.

Anonymous 2016a. *Inspecting Grain: Practical Procedures for Grain Handlers*. United States Department of Agriculture, Washington.

Anonymous 2016b. *Practical Procedures for Sampling Grain at Farm Sites and Remote Locations*. United States Department of Agriculture, Washington.

Athanassiou, C. G., Kavallieratos, N. G., Palyvos, N. E. and Buchelos, C. T. 2003. Three-dimensional distribution and sampling indices of insects and mites in horizontally-stored wheat. *Applied Entomology and Zoology* 38(3), 413–26. doi:10.1303/aez.2003.413.

Athanassiou, C. G., Phillips, T. W. and Wakil, W. 2019. Biology and control of the khapra beetle, *Trogoderma granarium*, a major quarantine threat to global food security. *Annual Review of Entomology* 64, 131–48. doi:10.1146/annurev-ento-011118-111804.

Atui, M. B., Lazzari, S. M. N., Lazzari, F. A. and Flinn, P. W. 2003. Comparison of ELISA and fragment count methods for detection of insects in wheat flour. In: Credland, P. F., Armitage, D., Bell, C. H., Cogan, P. M. and Highley, E. (Eds) *Proceedings of the 8th International Working Conference on Stored-Product Protection*, 22–26 July 2002. CAB International, York, United Kingdom. Wallingford, UK, pp. 135–8.

Banga, K. S., Kotwaliwale, N., Mohapatra, D. and Giri, S. K. 2018. Techniques for insect detection in stored food grains: an overview. *Food Control* 94, 167–76. doi:10.1016/j.foodcont.2018.07.008.

Barak, A. V. and Burkholder, W. E. 1976. Trapping studies with dermestid sex pheromones. *Environmental Entomology* 5(1), 111–14. doi:10.1093/ee/5.1.111.

Barak, A. V. 1989. Development of a new trap to detect and monitor khapra beetle (Coleoptera: Dermestidae). *Journal of Economic Entomology* 82(5), 1470–7. doi:10.1093/jee/82.5.1470.

Barak, A. V., Burkholder, W. E. and Faustini, D. L. 1990. Factors affecting the design of traps for stored-product insects. *Journal of the Kansas Entomological Society* 63, 466–85.

Binns, M. R. 1993. Sequential sampling for classifying pest status. In: Pedigo, L. P. and Buntin, G. D. (Eds) *Handbook of Sampling Methods for Arthropods in Agriculture*. CRC Press Press, Boca Raton, FL.

Brabec, D., Pearson, T., Flinn, P. and Katzke, D. 2010. Detection of internal insects in wheat using a conductive roller mill and estimation of insect fragments in the resulting flour. *Journal of Stored Products Research* 46(3), 180–5. doi:10.1016/j.jspr.2010.04.003.

Brader, B., Lee, R. C., Plarre, R., Burkholder, W., Kitto, G. B., Kao, C., Polston, L., Dorneanu, E., Szabo, I., Mead, B., Rouse, B., Sullins, D. and Denning, R. 2002. A comparison of screening methods for insect contamination in wheat. *Journal of Stored Products Research* 38(1), 75–86. doi:10.1016/S0022-474X(01)00006-6.

Bryan, J. M. and Elvidge, J. 1977. Mortality of adult grain beetles in sample delivery systems used in terminal grain elevators. *The Canadian Entomologist* 109(2), 209–13. doi:10.4039/Ent109209-2.

Campbell, J. F. and Mullen, M. A. 2004. Distribution and dispersal behavior of *Trogoderma variabile* and *Plodia interpunctella* outside a food processing plant. *Journal of Economic Entomology* 97(4), 1455–64. doi:10.1603/0022-0493-97.4.1455.

Campbell, J. F., Perez-Mendoza, J. and Weier, J. 2012. Insect pest management decisions in food processing facilities. In: Hagstrum, D. W., Phillips, T. W. and Cuperus, G. (Eds) *Stored Product Protection*. Kansas State Research and Extension, KS, pp. 219–32.

Cochran, W. 1977. *Sampling Techniques*. Wiley, New York.

Collins, L. E. and Chambers, J. 2003. The I-SPy Insect Indicator: an effective trap for the detection of insect pests in empty stores and on flat surfaces in the cereal and food trades. *Journal of Stored Products Research* 39(3), 277–92. doi:10.1016/S0022-474X(02)00012-7.

Cowan, T. and Gries, G. 2009. Ultraviolet and violet light: attractive orientation cues for the Indian meal moth, *Plodia interpunctella*. *Entomologia Experimentalis et Applicata* 131(2), 148–58. doi:10.1111/j.1570-7458.2009.00838.x.

Demianyk, C. J., White, N. D. G. and Jayas, D. S. 1997. Rapid detection of rusty grain beetles (Coleoptera : Cucujidae) from wheat samples passing through a mechanical dockage tester. *Canadian Journal of Plant Science* 77(4), 717–19. doi:10.4141/P96-177.

Dowell, F. E., Throne, J. E. and Baker, J. E. 1998. Automated nondestructive detection of internal insect infestation of wheat kernels by using near-infrared reflectance spectroscopy. *Journal of Economic Entomology* 91(4), 899–904. doi:10.1093/jee/91.4.899.

Dowell, F. E., Throne, J. E., Wang, D. and Baker, J. E. 1999. Indentifying stored-grain insects using near-infrared spectroscopy. *Journal of Economic Entomology* 92(1), 165–9. doi:10.1093/jee/92.1.165.

Fargo, W. S., Epperly, D., Cuperus, G. W., Clary, B. C. and Noyes, R. 1989. Effect of temperature and duration of trapping on four stored grain insect species. *Journal of Economic Entomology* 82(3), 970–3. doi:10.1093/jee/82.3.970.

Fields, P. G., Loon, J. V., Dolinski, M. G., Harris, J. L., Burkholder, W. E. and Van-Loon, J. 1993. The distribution of *Rhyzopertha dominica* (f.) in western Canada. *The Canadian Entomologist* 125(2), 317–28. doi:10.4039/Ent125317-2.

Fleurat Lessard, F., Tomasini, B., Kostine, L. and Fuzeau, B. 2006. Acoustic detection and automatic identification of insect stages activity in grain bulks by noise spectra processing through classification algorithms. In: Lorini, I., Bacaltchuk, B., Beckel, H., Deckers, D., Sundfeld, E., Dos Santos, J. P., Biagi, J. D., Celaro, J. C. Faroni, L. R. D., Bortolini, L. De O. F., Sartori, M. R., Elias, M. C., Guedes, R. N. C., Da Fonseca, R. G. da and Scussel, V. M. (Eds) *Proceedings of the 9th International Working Conference on Stored-Product Protection*, 15–18 October 2006. Brazilian Post-harvest Association ABRAPOS, Campinas, Brazil. Passo Fundo, Brazil, pp. 476–86.

Flinn, P. W., Opit, G. P. and Throne, J. E. 2009. Predicting stored grain insect population densities using an electronic probe trap. *Journal of Economic Entomology* 102(4), 1696–704. doi:10.1603/029.102.0438.

Flinn, P. W., Hagstrum, D. W., Reed, C. and Phillips, T. W. 2010. Insect population dynamics in commercial grain elevators. *Journal of Stored Products Research* 46(1), 43–7. doi:10.1016/j.jspr.2009.09.001.

Fox, G. and Manley, M. 2014. Applications of single kernel conventional and hyperspectral imaging near infrared spectroscopy in cereals. *Journal of the Science of Food and Agriculture* 94(2), 174–9. doi:10.1002/jsfa.6367.

Green, R. H. 1970. On fixed precision level sequential sampling. *Researches on Population Ecology* 12(2), 249–51. doi:10.1007/BF02511568.

Hagstrum, D. and Flinn, P. 2012. Insect pest management decisions for raw Comoodities during storage. In: Hagstrum, D. W., Phillips, T. W. and Cuperus, G. (Eds) *Stored Product Protection*. Kansas State Research and Extension, KS.

Hagstrum, D. W., Milliken, G. A. and Waddell, M. S. 1985. Insect distribution in bulk-stored wheat in relation to detection or estimation of abundanace. *Environmental Entomology* 14(6), 655–61. doi:10.1093/ee/14.6.655.

Hagstrum, D. W. and Subramanyam, B. 2006. *Sampling Fundamentals of Stored-Product Entomology*. AACC International, St. Paul, MN.

Haines, C., Rees, D., Ryder, K., Sistyanto, S. and Cahyana, Y. 1990. Brown-rice bait-bags for monitoring insect pest populations in bag stacks of milled rice as an aid to pest control decision-making. In: Fleurat Lessard, F. and Ducom, P. (Eds) *Proceedings of the 5th International Working Conference on Stored-Product Protection*. Imprimerie du Médoc, Bordeaux, France, pp. 1351–60.

Hawkin, K., Stanbridge, D. and Fields, P. 2011. Sampling *Tribolium confusum* and *Tribolium castaneum* in mill and laboratory settings: differences between strains and species. *The Canadian Entomologist* 143(5), 504–17. doi:10.4039/n11-026.

Holloway, J. C., Mayer, D. G. and Daglish, G. J. 2018. Flight activity of *Cryptolestes ferrugineus* in southern New South Wales, Australia. *Journal of Pest Science* 91(4), 1353–62. doi:10.1007/s10340-018-0981-1.

Howe, R. W. 1962. The effects of temperature and humidity on the oviposition rate of *Tribolium castaneum* (Hbst.) (Coleoptera, Tenebrionidae). *Bulletin of Entomological Research* 53(2), 301–10. doi:10.1017/S0007485300048148.

Jayas, D. S. and Jian, J. 2012. Integrated commodity management. In: Navarro, S., Banks, H., Jayas, D., Bell, C., Noyes, R., Ferizli, A., Emekci, M., Isikber, A. and Alagusundaram, K. (Eds) *The Proceedings of the 9th International Conference on Controlled Atmosphere and Fumigation in Stored Products (CAF 2012)*. ARBER Professional Congress Services, Antalya, Turkey, pp. 705–14.

Jian, F. and Larson, R. 2006. Relationship between density of adult rusty grain beetle, *Cryptolestes ferrugineus* (Coleoptera: Laemophloeidae) and insector counts in stored wheat at uniform moisture contect and small temperature fluctuations. In: Lorini, I., Bacaltchuk, B., Beckel, H., Deckers, D., Sundfeld, E., Dos Santos, J. P., Biagi, J. D., Celaro, J. C., Faroni, L. R. D., Bortolini, L. de O. F., Sartori, M. R., Elias, M. C., Guedes, R. N. C., Da Fonseca, R. G. and Scussel, V. M. (Eds) *Proceedings of the 9th International Working Conference on Stored-Product Protection*, 15–18 October 2006. Campinas, Brazil. Passo Fundo, Brazil, pp. 456–62.

Jian, F., Larson, R., Jayas, D. S. and White, N. D. G. 2011a. Evaluation of sampling units and sampling plans for adults of *Cryptolestes ferrugineus* (Coleoptea: Laemophloeidae) in stored wheat under different temperatures, moisture contents, and adult densities. *Journal of Stored Products Research* 47(4), 334–40. doi:10.1016/j.jspr.2011.05.006.

Jian, F., Larson, R., Jayas, D. S. and White, N. D. G. 2011b. Three-dimensional spatial distribution of adults of *Cryptolestes ferrugineus* (Coleoptera: Laemophloeidae) in stored wheat under different temperatures, moisture contents, and adult densities. *Journal of Stored Products Research* 47(4), 293–305. doi:10.1016/j.jspr.2011.04.002.

Jian, F., Chelladurai, V., Jayas, D. S. and White, N. D. G. 2012a. Insector® system to monitor insect activity and density during grain storage and fumigation. In: *The Proceedings of the 9th International Conference on Controlled Atmosphere and Fumigation in Stored Products.* ARBER Professional Congress Services, Antalya, Turkey, pp. 396–402.

Jian, F., Larson, R., Jayas, D. S. and White, N. D. G. 2012b. Three-dimensional temporal and spatial distribution of adult *Rhyzopertha dominica* in stored wheat and corn under different temperatures, moisture contents, and adult densities. *Journal of Economic Entomology* 105(4), 1194–204. doi:10.1603/ec12072.

Jian, F., Larson, R., Jayas, D. S. and White, N. D. G. 2012c. Three dimensional temporal and spatial distribution of adult *Oryzaephilus surinamensis* and *Sitophilus oryzae* in stored wheat under different temperatures, moisture contents, and adult densities. *Journal of Stored Products Research* 49, 155-65. doi:10.1016/j.jspr.2012.01.001.

Jian, F., Larson, R., Jayas, D. S. and White, N. D. G. 2012d. Three dimensional temporal and spatial distribution of adults of *Tribolium castaneum* (Coleoptera: Tenebrionidae) in stored wheat under different temperatures and adult densities. *Agricultural Research* 1(2), 165-74. doi:10.1007/s40003-012-0015-9.

Jian, F., Jayas, D. S. and White, N. D. G. 2014a. How many kilograms of grain per sample unit is big enough? part I - Comparison of insect detection and density estimation between manual probe sampling and Insector® system. *Journal of Stored Products Research* 56, 60-6. doi:10.1016/j.jspr.2013.05.007.

Jian, F., Jayas, D. S. and White, N. D. G. 2014b. How many kilograms of grain per sample unit is big enough? part II - Simulation of sampling from grain mass with different insect densities and distribution patterns. *Journal of Stored Products Research* 56, 67-80. doi:10.1016/j.jspr.2013.05.006.

Jian, F., Fields, P. G., Hargreaves, K., Jayas, D. S. and White, N. D. 2015. Chill-coma and minimum movement temperatures of stored-product beetles in stored wheat. *Journal of Economic Entomology* 108(5), 2471-8. doi:10.1093/jee/tov196.

Jian, F., Doak, S., Jayas, D. S., Fields, P. G. and White, N. D. G. 2016. Comparison of insect detection efficiency by different detection methods. *Journal of Stored Products Research* 69, 138-42. doi:10.1016/j.jspr.2016.07.008.

Karunakaran, C., Jayas, D. S. and White, N. D. G. 2003. X-ray image analysis to detect infestations caused by insects in grain. *Cereal Chemistry Journal* 80(5), 553-7. doi:10.1094/CCHEM.2003.80.5.553.

Leelaja, B. C., Rajashekar, Y. and Rajendran, S. 2007. Detection of eggs of stored-product insects in flour with staining techniques. *Journal of Stored Products Research* 43(3), 206-10. doi:10.1016/j.jspr.2006.05.003.

Legg, D. E. 1993. Bias and variablity in statistical estimates. In: Pedigo, L. P. and Buntin, G. D. (Eds) *Handbook of Sampling Methods for Arthropods in Agriculture*. CRC Press Press, Boca Raton, FL.

Loschiavo, S. R. and Atkinson, J. M. 1967. A trap for the detection and recovery of insects in stored grain. *The Canadian Entomologist* 99(11), 1160-3. doi:10.4039/Ent991160-11.

Loschiavo, S. R. and Atkinson, J. M. 1973. An improved trap to detect beetles (Coleoptera) in stored grain. *The Canadian Entomologist* 105(3), 437-40. doi:10.4039/Ent105437-3.

Mankin, R. and Hagstrum, D. 2012. Acoustic monitoring of insects. In: Hagstrum, D. W., Phillips, T. W. and Cuperus, G. (Eds) *Stored Product Protection*.Kansas State Research and Extension, KS.

Miyatake, T., Yokoi, T., Fuchikawa, T., Korehisa, N., Kamura, T., Nanba, K., Ryouji, S., Kamioka, N., Hironaka, M., Osada, M., Hariyama, T., Sasaki, R. and Shinoda, K. 2016. Monitoring and detecting the cigarette beetle (Coleoptera: Anobiidae) using ultraviolet (LED) direct and reflected lights and/or pheromone traps in a laboratory and a storehouse. *Journal of Economic Entomology* 109(6), 2551-60. doi:10.1093/jee/tow225.

Mohan, S., Gopalan, M., Sundarababu, P. C. and Sreenarayanan, V. V. 1994. Practical studies on the use of light traps and bait traps in the management of *Rhyzopertha dominica* (f.) in rice warehouses. *International Journal of Pest Management* 40(2), 148-52. doi:10.1080/09670879409371873.

Mueller, D., Pierce, L., Benezet, H. and Krischik, V. 1990. Practical application of pheromone traps in food and tobacco Industry. *Journal of the Kansas Entomological Society* 63, 548-53.

Muir, W. E., Waterer, D. and Sinha, R. N. 1985. Carbon dioxide as an early indicator of stored cereal and oilseed spoilage. *Transactions of the ASAE. American Society of Agricultural Engineers* 28, 1673–5.

Mukundarajan, H., Hol, F. J. H., Castillo, E. A., Newby, C. and Prakash, M. 2017. Using mobile phones as acoustic sensors for high-throughput mosquito surveillance. *eLife* 6, e27854. doi:10.7554/eLife.27854.

Mullen, M. A., Highland, H. A., Taggart, R. E. and Lingren, B. W. 1992. *Insect Monitoring System*. USA patent application.

Myers, S. and Hagstrum, D. W. 2012. Quarantine. In: Hagstrum, D. W., Phillips, T. W. and Cuperus, G. (Eds) *Stored Preduct Protection*. Kansas State University, KS, pp. 297–304.

Nansen, C., Phillips, T. W., Morton, P. K. and Bonjour, E. L. 2006. Spatial analysis of pheromone-baited trap captures from controlled releases of male Indianmeal moths. *Environmental Entomology* 35(2), 516–23. doi:10.1603/0046-225X-35.2.516.

Neethirajan, S., Karunakaran, C., Jayas, D. S. and White, N. D. G. 2007. Detection techniques for stored-product insects in grain. *Food Control* 18(2), 157–62. doi:10.1016/j.foodcont.2005.09.008.

Njoroge, A., Affognon, H., Richter, U., Hensel, O., Rohde, B., Chen, D. and Mankin, R. 2019. Acoustic, pitfall trap, and visual surveys of stored product insect pests in Kenyan warehouses. *Insects* 10(4). doi:10.3390/insects10040105.

Nualvatna, K., Makathan, N., Chavapradit, C., Kitkuandee, K. and Uraichuan, J. 2003. The use of light traps for attracting stored-product insects in a rice mill and paddy seed stores. In: Credland, P. F., Armitage, D., Bell, C. H., Cogan, P. M. and Highley, E. (Eds) *Proceedings of the 8th International Working Conference on Stored-Product Protection*, 22–26 July 2002. CAB International, York, United Kingdom. Wallingford, UK, pp. 244–7.

Paliwal, J., Jayas, D. S., White, N. D. G. and Muir, W. E. 1999. Effect of pneumatic conveying of wheat on mortality of insects. *Applied Engineering in Agriculture* 15(1), 65–8. doi:10.13031/2013.5834.

Paliwal, J., Visen, N. S. and Jayas, D. S. 2001. AE--automation and emerging technologies: evaluation of neural network architectures for cereal grain classification using morphological features. *Journal of Agricultural Engineering Research* 79(4), 361–70. doi:10.1006/jaer.2001.0724.

Papadopoulou, S. C. and Buchelos, C. T. 2002. Comparison of trapping efficacy for *Lasioderma serricorne* (f.) adults with electric, pheromone, food attractant and control-adhesive traps. *Journal of Stored Products Research* 38(4), 375–83. doi:10.1016/S0022-474X(01)00039-X.

Pearson, T. and Brabec, D. L. 2007. Detection of wheat kernels with hidden insect infestations with an electrically conductive roller mill. *Applied Engineering in Agriculture* 23(5), 639–45. doi:10.13031/2013.23662.

Phillips, T. W., Cogan, P. M. and Fadamiro, H. Y. 2000. Pheromones. In: Subramanyam, B. and Hagstrum, D. W. (Eds) *Alternatives to Pesticides in Stored-Product IPM*. Kluwer Academic Publishers, Boston, MA.

Pixton, S. W. 1965. Detection of insect infestations in cereals by measurement of uric acid. *Cereal Chemistry* 42, 315–22.

Quinn, F. A., Burkholder, W. and Kitto, B. G. 1992. Immunological technique for measuring insect contamination of grain. *Journal of Economic Entomology* 85(4), 1463-70. doi:10.1093/jee/85.4.1463.

Rajendran, S. 1999. Detection of insect infestation in stored food commodities. *Journal of Food Science and Technology* 36, 283–300.

Rajendran, S. and Steve, L. T. 2005. Detection of insect infestation in stored foods. *Advances in Food and Nutrition Research*. Academic Press, 163–232.

Rajesh, A., Mohan, S. and Nelson, S. 2015. Efficacy of TNAU-stack probe trap in detection, monitoring of pests and validation of fumigation in turmeric bag-stack storage. *Current Biotica* 9, 159–65.

Ravikanth, L., Jayas, D. S., White, N. D. G., Fields, P. G. and Sun, D.-W. 2017. Extraction of spectral information from hyperspectral data and application of hyperspectral imaging for food and agricultural products. *Food and Bioprocess Technology* 10(1), 1–33. doi:10.1007/s11947-016-1817-8.

Reimer, A., Wiebe, K., Rao, J., Yao, B., Gui, Y., Jian, F., Fields, P., Jayas, D. S. and Hu, C.-M. 2018. A compact microwave device for monitoring insect activity in grain samples. *Biosystems Engineering* 175, 27–35. doi:10.1016/j.biosystemseng.2018.08.010.

Richter, K. and Tchalale, P. 1994. Improved early detection of internal infestation by flotation using product-adapted salt solutions. In: Highley, E., Wright, E. J., Banks, H. J. and Champ, B. R. (Eds) *Proceedings of the 6th International Working Conference on Stored-Product Protection*. CAB International, Canberra, Australia, pp. 444–7.

Sambaraju, K. R. and Phillips, T. W. 2008. Responses of adult *Plodia interpunctella* (Hübner) (Lepidoptera: Pyralidae) to light and combinations of attractants and light. *Journal of Insect Behavior* 21(5), 422–39. doi:10.1007/s10905-008-9140-5.

Shuman, D., Weaver, D. K. and Larson, R. G. 2005. Performance of an analytical, dual infrared-beam, stored-product insect monitoring system. *Journal of Economic Entomology* 98(5), 1723–32. doi:10.1603/0022-0493-98.5.1723.

Smith, L. B. 1977. Efficiency of Berlese-Tullegren funnels for removal of the rusty grain beetle, Cryptolestes ferrugineus, from wheat samples. *The Canadian Entomologist* 109(4), 503–9. doi:10.4039/Ent109503-4.

Snedecor, G. and Cochran, W. 1967. *Statistical Methods*. Iowa State University Press, Ames, IA, Section.

Southwood, T. R. E. 1978. *Ecological Methods, with Special Reference to the Study of Insect Populations*. Chapman and Hall, London.

Stejskal, V., Aulicky, R., Kucerova, Z. and Lukas, J. 2008. Method of sampling and laboratory extraction affects interpretation of grain infestation by storage pests. *Journal of Plant Diseases and Protection* 115(3), 129–33. doi:10.1007/BF03356252.

Subramanyam, B. and Hagstrum, D. W. 1995. Sampling. In: Subramanyam, B. and Hagstrum, D. W. (Eds) *Integrated Management of Insects in Stored Products*. Marcel Dekker, Inc., New York.

Subramanyam, B., Hagstrum, D. W., Meagher, R. L., Burkness, E. C., Hutchison, W. D. and Naranjo, S. E. 1997. Development and evaluation of sequential sampling plans for *Cryptolestes ferrugineus* (Stephens) (Coleoptera: Cucujidae) infesting farm-stored wheat. *Journal of Stored Products Research* 33(4), 321–9. doi:10.1016/S0022-474X(97)00014-3.

Swords, P. and Van Ryckeghem, A. 2010a. Summary of commercially available pheromones of common stored-product beetles. In: Carvalho, M. O., Fields, P. G., Adler, C. S., Arthur, F. H., Athanassiou, C. G., Campbell, J. F., Fleurat-Lessard, F., Flinn, P. W., Hodges, R. J., Isikber, A. A., et al. (Eds) *Proceedings of the 10th International Working Conference on Stored-Product Protection*, 27 June–2 July 2010, Estoril, Portugal. Julius Kuhn-Institut, Berlin, Germany, pp. 1004–7.

Swords, P. and Van Ryckeghem, A. 2010b. Summary of commercially available pheromones of common stored product moths. In: Carvalho, M. O., Fields, P. G., Adler, C. S., Arthur, F. H., Athanassiou, C. G., Campbell, J. F., Fleurat-Lessard, F., Flinn, P. W., Hodges, R. J., Isikber, A. A., et al. (Eds) *Proceedings of the 10th International Working Conference on Stored-Product Protection*, 27 June–2 July 2010, Estoril, Portugal. Julius Kuhn-Institut, Berlin and Germany, pp. 1008–10.

Toews, M. D., Phillips, T. W. and Shuman, D. 2003. Electronic and manual monitoring of *Cryptolestes ferrugineus* (Coleoptera: Laemophloeidae) in stored wheat. *Journal of Stored Products Research* 39(5), 541–54. doi:10.1016/S0022-474X(02)00059-0.

Toews, M. D., Perez-Mendoza, J., Throne, J. E., Dowell, F. E., Maghirang, E., Arthur, F. H. and Campbell, J. F. 2007. Rapid assessment of insect fragments in flour milled from wheat infested with known densities of immature and adult *Sitophilus oryzae* (Coleoptera: Curculionidae). *Journal of Economic Entomology* 100(5), 1714–23. doi:10.1603/0022-0493(2007)100[1714:raoifi]2.0.co;2.

Toews, M. D. and Nansen, C. 2012. Trapping and interpreting captures of stored grain insects. In: Hagstrum, D. W., Phillips, T. W. and Cuperus, G. (Eds) *Stored Product Protection*. Kansas State Research and Extension, Manhattan, KS, USA, pp. 243–61.

Toon, A., Daglish, G. J., Ridley, A. W., Emery, R. N., Holloway, J. C. and Walter, G. H. 2018. Significant population structure in Australian *Cryptolestes ferrugineus* and interpreting the potential spread of phosphine resistance. *Journal of Stored Products Research* 77, 219–24. doi:10.1016/j.jspr.2018.05.004.

Wakefield, M. E. and Cogan, P. M. 2007. Laboratory evaluation of traps for the detection of beetle pests in bulk grain. *Journal of Stored Products Research* 43(4), 546–9. doi:10.1016/j.jspr.2007.03.004.

Wang, D., Bai, C., Li, H., Lu, Y. and Guo, X. 2018. Remote monitoring of stored grain insect pests. In: Adler, C. S., Opit, G., Fürstenau, B., Müller-Blenkle, C., Kern, P., Arthur, F. H., Athanassiou, C. G., Bartosik, R., Campbell, J., Carvalho, M. O., Chayaprasert, W., Fields, P., Li, Z., Maier, D., Nayak, M., Nukenine, E., Obeng-Ofori, D., Phillips, T., Riudavets, J., Throne, J., Schöller, M., Stejskal, V., Talwana, H., Timlick, B. and Trematerra P. (Eds) *Proceedings of the 12th International Working Conference on Stored Product Protection*. Julius Kühn Institut, Berlin.

Wehling, R. L., Wetzel, D. L. and Pedersen, J. R. 1984. Stored wheat insect infestation related to uric acid as determined by liquid chromatography. *Journal - Association of Official Analytical Chemists* 67(3), 644–7. doi:10.1093/jaoac/67.3.644.

White, G. G. 1983. A modified inclined sieve for separation of insects from wheat. *Journal of Stored Products Research* 19(2), 89–91. doi:10.1016/0022-474X(83)90031-0.

White, N. D. G. and Loschiavo, S. R. 1986. Effects of insect density, trap depth, and attractants on the capture of *Tribolium castaneum* (Coleoptera: Tenebrionidae) and *Cryptolestes ferrugineus* (Coleoptera: Cucujidae) in stored wheat. *Journal of Economic Entomology* 79(4), 1111–7. doi:10.1093/jee/79.4.1111.

White, N. D. G., Arbogast, R. T., Fields, P. G., Hillmann, R. C., Loschiavo, S. R., Subramanyam, B., Throne, J. E. and Wright, V. F. 1990. The development and use of pitfall and probe traps for capturing insects in stored grain. *Journal of the Kansas Entomological Society* 63, 506–25.

Wileyto, E. P., Norris, J. L., Weaver, D. K. and Arbogast, R. T. 2000. Self-marking recapture models for estimating closed insect populations. *Journal of Agricultural, Biological, and Environmental Statistics* 5(4), 456–74. doi:10.2307/1400660.

Wilkin, D. R. and Fleurat-Lessard, F. 1990. The detection of insects in grain using conventional sampling spears. In: Fleurat Lessard, F. and Ducom, P. (Eds) *Proceedings of the 5th International Working Conference Stored-Product Protection*. Imprimerie du Médoc, Bordeaux, France, pp. 1445–54.

Wright, V. F. and Mills, R. B. 1984. Estimation of stored-product insect regulations in small bins using two sampling techniques. In: Mills, R. B., Wright, V. F., Pedersen, J. R., Mcgaughey, W. H., Beeman, R. W., Kramer, K. J., Speirs, R. D. and Storey, C. L. (Eds) *Proceedings of the 3rd International Working Conference on Stored-Product Entomology*, 23-28 October 1983. Manhattan, KS, pp. 672-9.

Yao, W., Fu, J. and Zhang, Y. 1999. The sensor-based probe traps for monitoring stored-product insects. In: Jin, Z., Liang, Q., Liang, Y., Tan, X. and Guan, L. (Eds) *Proceedings of the 7th International Working Conference on Stored-Product Protection*, 1998. Sichuan Publishing House of Science and Technology, Beijing, China. Chengdu, China, pp. 1437–45.

# Chapter 8

## Advances in insect pest management in postharvest storage of cereals: use of controlled atmosphere and temperature control

*Shlomo Navarro and Hagit Navarro, Green Storage Ltd., Israel*

1 Introduction
2 Controlled and modified atmospheres
3 Temperature manipulation
4 Conclusions and future trends
5 Acknowledgement
6 References

### 1 Introduction

Two technologies that contribute to the increasing demand for cereals uncontaminated by molds, insects, and insecticide residues are modified atmospheres and temperature manipulation. In developed countries, the loss of quality is particularly important. Although in some developing countries, improved storage methods have been implemented, in some others still poor handling and storage methods under warm and humid climatic conditions promote rapid deterioration of stored cereals. Increased public concern over the adverse effects of pesticide residues in food and the environment has led to the partial substitution of the use of contact pesticides (typically organophosphates and pyrethroids) and fumigants by alternative control methods. In this regard, it should be noted that, particularly in developed countries, there is a dramatic decline in the direct application of contact pesticides on grain commodities and a significant increase of insect pest management (IPM) measures.

It is worth noting that of the 14 fumigants listed some 36 years ago by Bond (1984), only two remain today in regular worldwide use, namely, phosphine and methyl bromide, which is used mainly for QPS (quarantine and pre-shipment) conditions. Methyl bromide kills insects relatively quickly, but because of its contribution to stratospheric ozone depletion (UNEP, 2002), it was phased out

http://dx.doi.org/10.19103/AS.2020.0072.09

in developed countries by 2005 (UNEP, 2006). In contrast, phosphine remains popular because it is environmentally benign and easier to apply than methyl bromide. However, its application requires long exposure time and many insects have developed resistance to phosphine over the last two decades.

Cereals can be stored for extended periods, provided that there is no insect infestation and that their water activity is low enough to prevent microbial growth. However, quantitative or qualitative losses still occur. Qualitative losses, for example, may consist of changes in physical appearance, changes in nutritional degradation due to oxidation, an increase in free fatty acids (FFAs), the presence of insects or their fragments, or contamination by mold or the presence of mycotoxins. Some of these are difficult to detect visually. Quantitative postharvest losses of food grain in developing countries have been conservatively estimated during the 1980s at 10–15% by the Food and Agriculture Organization's (FAO) Special Action Programme for the Prevention of Food Losses. For example, losses of corn due only to insects in farmers' stores in Nigeria, Swaziland, and Kenya were of the order of 6–10% (Schulten, 1988).

If the moisture content is maintained sufficiently low, insects remain the main concern for the quality preservation of durable agricultural commodities. Therefore, in this chapter, the major emphasis is placed on technologies based on controlled/modified atmospheres (CAs/MAs) and the manipulation of temperature for the quality preservation of stored grain and the control of insect pests. MAs and temperature manipulation offer safe and environmentally benign alternatives to the use of conventional residue-producing chemical fumigants and contact insecticides for controlling insect pests attacking stored grain. The environmental benefit is simply from the alteration of the physical parameters as opposed to the use of chemical contaminants.

## 2 Controlled and modified atmospheres

Modified atmospheres (MA) is proposed to serve as the general term, including all cases in which the composition of atmospheric gases or their partial pressures in the treatment enclosure have been modified to create in it the conditions favorable for the control of storage insects and preserve the quality of the commodity. In a MA treatment, the atmospheric composition within the treated enclosure may change during the treatment period. This term will comprise all the following designations mentioned in this chapter.

The objective of MA treatments is to attain a composition of atmospheric gases rich in carbon dioxide ($CO_2$) and low in oxygen ($O_2$), or a combination of these two gases at normal or altered atmospheric pressure within the treatment enclosure, for the exposure time necessary to control the storage pests and preserve the quality of the commodity. Terms that define the same method of treatment, namely, MA for the control of storage insect pests or for the preservation of food, include controlled atmosphere (CA), sealed storage, hermetic storage, or atmospheres used at high or low pressures. Therefore, an attempt is made here to propose definitions that will add clarity to the available storage insect control methods, whether at normal atmospheric pressure or under altered atmospheric pressure.

The protection of the beneficial qualities of cereals during storage depends on many factors. Among the detrimental factors that reduce the quality of cereals are insects and microflora (Navarro and Donahaye, 2005). Although in MAs/CAs, the major emphasis is placed on the control of insect pests, for the quality preservation of the product volatiles, just maintaining the vapor pressure in the sealed structure is sufficient (Navarro and Navarro, 2018).

MA/CA offers an alternative that is safe and environmentally benign to the use of conventional residue-producing chemical fumigants for controlling insect pests attacking stored cereal grain. These atmospheres prevent fungal growth and also maintain product quality. An important development stimulating further work on MA took place in the United States between 1980 and 1981. The Environmental Protection Agency (EPA) approved an exemption from tolerance for $CO_2$, nitrogen ($N_2$) and products from an 'inert' gas generator when used to control insects in raw (Federal Register 45, pp. 75663 64, Nov. 1980) and processed (Federal Register 46, pp. 32865 66, June 1981) agricultural products.

The development of MA/CA technology has come about mostly over public concern for the adverse effects of pesticide residues in food and the environment. Although this method has become well established for the control of storage pests, its commercial use is still limited to a few countries. Investigations that are more recent have attempted to integrate MA application into the 21st-century version of raw products and manufactured food storage and transportation (Navarro, 2006).

Atmospheric manipulation for the protection of stored products such as wheat grains has been researched extensively for more than 30 years (Adler et al., 2000; Calderon and Barkai-Golan, 1990; Jay, 1984; Navarro, 2006). The purpose of this chapter is to discuss the concepts and variations of MA and CA, their impact on pests and on the quality of cereals being treated, the structures where they may be considered for use, and their compatibility in commercial settings.

## 2.1 Controlled atmosphere (CA) under normal atmospheric pressure

### 2.1.1 Gas supply from pressurized cylinders

CA is a modified gas composition, usually produced artificially, and maintained unchanged by adding desired gases ($CO_2$ or $N_2$), supplied from pressurized cylinders or otherwise. This supplementary introduction of gases is carried out when their concentration in the sealed container drops to below the desired level. The objective of CA treatment is to attain a composition of atmospheric gases rich in $CO_2$ and low in $O_2$, or a combination of these two gases within the storage enclosure or treatment chamber. These set concentrations are maintained for the time period necessary to control the storage pests.

A widely used source for the production of such atmospheric gas compositions is tanker-delivered liquefied $CO_2$ or $N_2$ when the target CA gas composition is <1% $O_2$ or high $CO_2$ concentration. For the large-scale application of $N_2$ or $CO_2$, vaporizers are essential. These vaporizers consist of a suitably designed receptacle with a heating medium (electricity, steam, diesel fuel, or propane), a super-heated coil with a hot-water jacket, and forced or natural draught.

### 2.1.2 Combustible gases

For the on-site generation of combustible gases (CAs) by the combustion of hydrocarbon fuel to produce a low $O_2$ atmosphere containing some $CO_2$, commercial installations - termed exothermic gas generators or gas burners - are available. Their CA composition is designed to allow the presence of approximately 2–3% $O_2$, with $CO_2$ removed through scrubbers.

Several adaptations are required for their use in the grain industry, that is, tuning the equipment to obtain an $O_2$ level of <1%, utilizing to full advantage the $CO_2$ generated, and removing excess humidity from the atmosphere generated. Combustion of propane and butane yields approximately 13% and 15% $CO_2$, respectively. The CA generated is more toxic than an $N_2$ atmosphere deficient in $O_2$ due to the presence of $CO_2$ in the MA, causing hypercarbia, which, together with hypoxia, is synergistic in its effect on insect mortality.

Bell et al. (1993) conducted tests on a well-sealed, 100-tonne capacity silo to compare the use of two different modified atmospheres: $CO_2$ and the exhaust gas produced by propane combustion. Both treatments were effective in killing grain pests. For both gases, the efficiency of purge increased with the flow rate over the range tested, and atmosphere maintenance was accomplished more efficiently by adding gas to the headspace rather than the hopper bottom. The advantage of head-space maintenance was greater for burner gas (< 1% oxygen) than for 60% $CO_2$ in the air. Costs for MA treatment of the experimental silo were up to 10 times greater for $CO_2$ than for burner gas.

### 2.1.3 On-site $N_2$ generators

Commercial equipment, also termed 'pressure-swing absorption' systems, uses the process of $O_2$ absorption from compressed air passed through a molecular sieve bed. For continuous operation, a pair of absorbers are provided that operate sequentially for $O_2$ absorption and regeneration. $N_2$ at a purity of 99.9% can be obtained through the regulation of inlet airflow; this method of $N_2$ generation is an expanding new approach in CA generation technology. Equipment is now being manufactured that is rated to supply an outlet flow of 120 $m^3$/h at an outlet purity of 98% $N_2$ compared to less-efficient equipment with lower rate levels in the past.

## 2.2 Effects of controlled atmosphere (CA) on insects

### 2.2.1 Effects of low oxygen levels

Insects can tolerate low levels of $O_2$ for prolonged periods. Using $N_2$ to replace $O_2$ must result in $O_2$ being below 2%, preferably 1% for rapid death. This effect is reversed below 1% $O_2$ in $N_2$ where adult rice weevils, *Sitophilus oryzae* (L.) (Navarro, 1978), showed tolerance, increasing the lethal exposure time by apparently closing their spiracles. In particular, *S. oryzae* adults are killed more quickly at 1.0% $O_2$ rather than at 0.1% or 2% $O_2$ under the same conditions.

*Tribolium castaneum* (Hbst.) in $N_2$ showed significant differences in adult mortality between 0.1% and 1.0% $O_2$ (Navarro, 1978). Adults are generally most susceptible to treatment, and *S. oryzae* or *Rhyzopertha dominica* (F.) was found to be more tolerant than *Tribolium* spp. The lowest level of tolerance to the lack of $O_2$ was attained around the 1% concentration level. Therefore, Annis (1987) concluded that $O_2$ levels of 1% are needed to kill insects in 20 days (Table 1).

**Table 1** Suggested provisional dosage regimes for control of all stages of the 12 most common insect species of stored grain, using modified atmospheres at temperatures between 20°C and 29°C* (Navarro and Donahaye, 1990).

| Atmospheric gas concentration | Controls most common grain insects including *Trogoderma granarium* (yes/no) | Exposure period (days) |
|---|---|---|
| <1% $O_2$ (in nitrogen) | Yes | 20 |
| Constant % of $CO_2$ in air | | |
| 40 | No | 17 |
| 60 | No | 11 |
| 80 | No | 8.5 |
| 80 | Yes | 16 |
| $CO_2$ decay in the air from | | |
| >70–35% | No | 15 |
| Pressurized $CO_2$ | | |
| at >20 bar | ** | <0.35 |

* Data, except those on pressurized $CO_2$, compiled from Annis (1987).

Radek et al. (2017) tested how quickly the mobile adults of six species of storage beetles (*Oryzaephilus surinamensis (L.), Cryptolestes ferrugineus, (Stephens) Tribolium confusum J. du Val, T. castaneum, Sitophilus granarius (L.),* and *S. oryzae*) are killed after the introduction of the infested commodity to prevent their further spread to the surrounding storage silos. The trials were conducted in a metal silo containing 25 tonnes of seeds using the system of continual top-down $N_2$ filling to replace the oxygen. The target concentration ($\leq$ 1% $O_2$ and 99% $N_2$) was reached at the bottom of the silo after 12 h of purging $N_2$. Seven days of exposure led to 100% mortality of all tested species except for *S. granarius* (96.7% mortality at the bottom), while 10 days of exposure led to 100% mortality of all adults located at both the bottom and the top of the silo. The commodity temperatures ranged from 15.9°C to 23.9°C through the grain profile and trials. This experiment showed that 10 days of exposure completely prevented the adult mobile pest stages of all tested species from spreading from the treated silo.

### 2.2.2 Effects of high carbon dioxide levels

Elevated $CO_2$ levels cause the spiracles to open, resulting in insect death from water loss. Above 10% $CO_2$ spiracles remain permanently open. Toxic effects are entirely through the tracheae, not the hemolymph; $CO_2$ has direct toxic effects on the nervous system. In some cases, $CO_2$ can acidify the hemolymph, leading to membrane failure in some tissues (Nicolas and Sillans, 1989).

Elevated, but sublethal, $CO_2$ levels for prolonged periods can have deleterious effects on insect development, growth, and reproduction (White et al., 1995; Nicolas and Sillans, 1989). Atmospheres containing about 60% $CO_2$ rapidly kill stored-product insects. At 26°C, about 4 days of exposure would be sufficient to kill all stages (including eggs) of most stored-product insects.

Conyers and Bell (2007) described the feasibility of using $O_2$ levels that disrupt the life cycles of the target beetle species. The aim was to identify MAs that would kill the most susceptible stage and prevent population growth. Simulated burner gas and $N_2$ atmospheres with $O_2$ contents between 3% and 6% were tested, along with an $N_2$-based MA with elevated $CO_2$ (10–20%). The tests were carried out at 20°C and 25°C with 75% and 85% r.h. at each temperature. Prevention of population growth for burner gas or $N_2$ was about 4% $O_2$ for *O. surinamensis*, *S. granarius*, and *S. oryzae*, and about 3% $O_2$ for *C. ferrugineus* and *T. castaneum* at 25°C. At 20°C it was about 3% $O_2$ for all species tested. When $CO_2$ was increased to 10% or 20%, reducing $O_2$ to 5% was sufficient to eliminate the emergence of *S. granarius* at 20°C, but a few individuals emerged at 25°C. For *C. ferrugineus* there was a 95% reduction with 5% $O_2$ plus 20% $CO_2$ at 20°C, but not at 25°C.

### 2.2.3 Combination of high carbon dioxide and low oxygen levels

Atmospheres with 60% $CO_2$ and 8% $O_2$ are very effective at killing internal kernel-feeding insects, while low $O_2$ atmospheres are more rapid in killing external-feeding insects (Banks and Annis, 1990). High $CO_2$ levels, even with 20% $O_2$, rapidly kill insects because of $CO_2$ toxicity. $CO_2$ levels must be at 40% for 17 days, 60% for 11 days, 80% for 8.5 days at temperatures above 20°C, or 70% declining to 35% in 15 days at 20°C (Annis, 1987). Higher temperatures accelerate $CO_2$ toxicity as insect metabolism is elevated. Even low levels of $CO_2$ (7.5–19.2%) for prolonged periods sharply increase immature and adult mortality (White et al., 1995).

### 2.2.4 Effects of temperature and relative humidity on controlled atmosphere fumigation

Insect mortality increases more rapidly as temperatures rise, and their metabolism speeds up. Cool temperatures slow down the rates of mortality, while lower relative humidity hastens toxic effects, notably in high $CO_2$ atmospheres because of the desiccation of insects (Banks and Fields, 1995).

## 2.3 Effects of controlled atmosphere (CA) on product quality

### 2.3.1 Germination of seeds

Seed below the critical moisture content is not significantly affected at high $CO_2$ or low $O_2$ atmospheres. However, with increasing grain moisture contents, $CO_2$-rich atmospheres could reduce the physiological quality of grain by interfering with the enzymatic activity of glutamine-decarboxylase. The adverse effect of $CO_2$ on the germination of rice, maize, and wheat becomes more pronounced at temperatures higher than 47°C and, from observations carried out so far, this adverse effect may not be detectable at all below 30°C. Therefore, if the preservation of germination is of primary importance, the use of $CO_2$-free low $O_2$ atmospheres is preferred if expected temperatures are significantly above 30°C.

Viability of corn stored under hermetic (148 days storage) and non-hermetic (120 days storage) conditions in the Philippines did not indicate significant changes between the initial and final samples (Navarro and Caliboso, 1996; Navarro et al., 1998). In the same trials, the viability of paddy stored under hermetic conditions did not change significantly. To test the viability of wheat stored under hermetic conditions in Israel, two trials were carried out with storage periods of 1440 and 450 days only under hermetic conditions. The viability of wheat changed slightly from an initial 99% to 97% after 1440 days, and from 97% to 91% after 450 days, respectively. In both trials, insect populations were successfully controlled and, the average $CO_2$ concentrations ranged between 10% and 15%.

### 2.3.2 Product quality preservation

Donahaye et al. (2001) reported on quality preservation of 13.4–31.9 tonnes lots of paddy, stacked in flexible enclosures and stored outdoors for 78–183 days. The quality of the paddy was compared with that of three control stacks (5.3–5.6 tonnes capacity) held under tarpaulins in the open for 78–117 days. Percent milling recovery and levels of yellowing in the gastight stacks showed no significant change.

In a study on the quality preservation of stored cocoa beans by bio-generated modified atmospheres, respiration rates of fermented cocoa beans were tested at equilibrium relative humidity (ERH) of 73% at 26°C in hermetically sealed containers. The $O_2$ concentration was reduced to <0.3%, and $CO_2$ concentration increased to 23% within 5.5 days. The FFA content of cocoa beans at 7.0%, 7.5%, and 8.0% moisture content under hermetic conditions of 30°C remained below or close to 1.0% after 90–160 days of storage (Navarro et al., 2010).

## 2.4 Types of structures in which controlled atmosphere (CA) and modified atmosphere (MA) have been used

Controlled atmospheres have been used in a wide array of grain storage structures. The most important consideration is that they must be airtight for long-term storage or relatively airtight for $CO_2$ or $N_2$ fumigation. Acceptable airtightness for $CO_2$ fumigation is determined by negative pressure testing and should at most hold a negative pressure from 500 Pa to 250 Pa in 10 min (Annis and Graver, 1990). Attempts have been made to predict gas tightness relative to leakage areas (Mann et al., 1999; Lukasiewicz et al., 1999). Provisional guidelines based on best estimates from comparative variable pressure tests are presented in Table 2 (Navarro, 1999). The suggested times given in Table 2 were doubled for empty storages as an approximation to the intergranular airspace.

### 2.4.1 In-ground storage

Historically, in-ground storage was widely used worldwide to create hermetic storage where $CO_2$ was produced and $O_2$ was consumed by respiration of grain and microflora. Its use was recorded from Spain to India and China, East Africa, and North America west of the Mississippi River (Sigaut, 1988).

### 2.4.2 Bolted steel silos

Bolted steel silos are not airtight, but they can be sealed for partially successful fumigation with $CO_2$ (Fig. 1). Navarro et al. (1985) reported the treatment of a

**Table 2** Provisional recommended ranges for variable pressure tests carried out in structures destined for gaseous treatments to control storage insects (Navarro 1999).

| Type of gaseous treatment | Structure volume in cubic meters | Variable pressure test decay time (min.) 250–125 Pa | |
|---|---|---|---|
| | | Empty structure | 95% Full |
| Fumigants | Up to 500 | 3 | 1.5 |
| | 500–2000 | 4 | 2 |
| | 2000–15 000 | 6 | 3 |
| CA | Up to 500 | 6 | 3 |
| | 500–2000 | 7 | 4 |
| | 2000–15,000 | 11 | 6 |
| MA, including airtight storage | Up to 500 | 10 | 5 |
| | 500–2000 | 12 | 6 |
| | 2000–15 000 | 18 | 9 |

welded-steel silo of 11 278 m$^3$ containing 6 881 tonnes of soft red winter wheat with $CO_2$. The silo was purged from the aeration duct at the bottom of the silo for 39 h at a rate of 427 kg $CO_2$/h, and during the maintenance phase, $CO_2$ was added from the top for an additional 106 h at a rate of 79 kg $CO_2$/h (Fig. 2). The temperature of the grain ranged from 18°C to 20°C on the surface to 30–33°C in the center of the bulk. After 30 days grain samples were examined from the center midway to the bottom of the bulk, the percent reduction in insect population for *T. castaneum* larvae, *R. dominica* adults, and *C. ferrugineus* was 100%, while for *S. oryzae* it was 98.5%.

Alagusundaram et al. (1995) placed dry ice in insulated coolers under a $CO_2$ impervious plastic sheet above wheat at 2.5 m deep in a 5.6 m diameter silo. $CO_2$ levels were 30% at 0.55 m above the floor where 90% of rusty grain

**Figure 1** Application of $CO_2$ based CA in a 6800 tonnes metal silo bin in Kingston, North Carolina, USA.

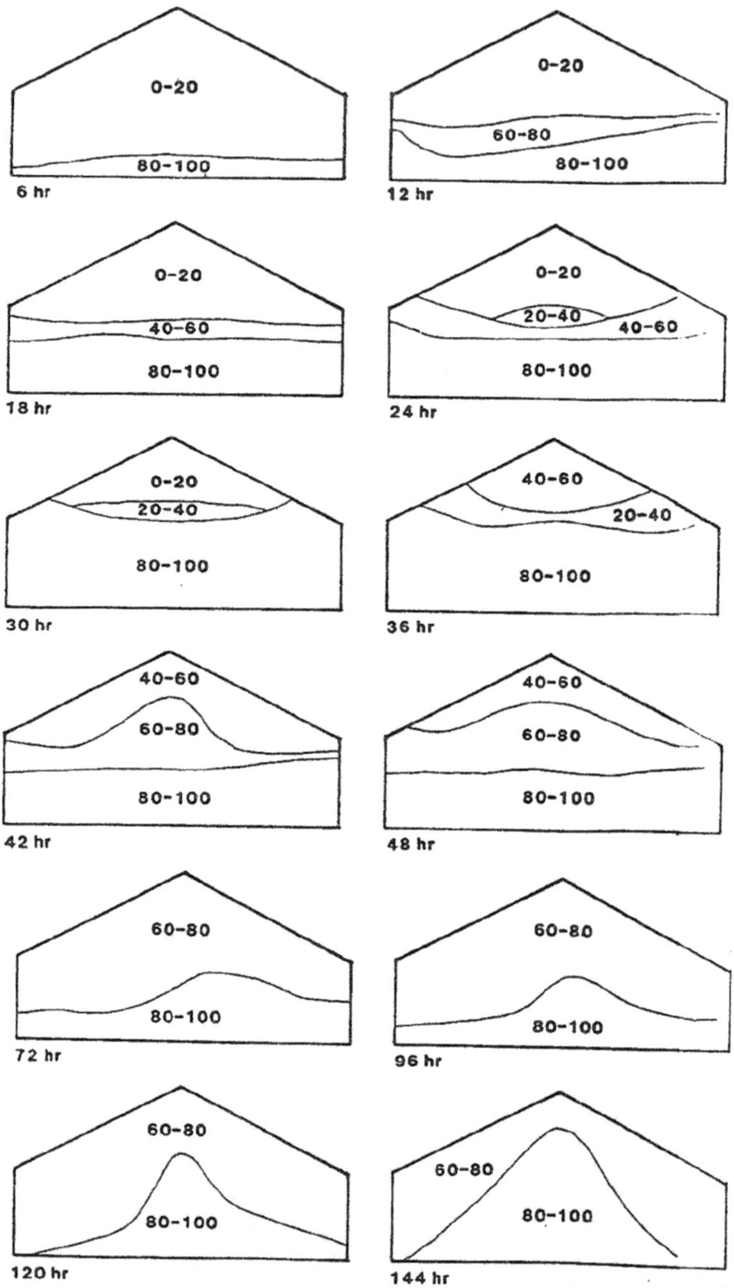


**Figure 2** Iso-concentration lines recorded during the introduction of $CO_2$ into 11 278 $m^3$ welded steel bin partially filled with 6881 tonnes of soft red winter wheat. The initial purge was through the main aeration duct and lasted 39 h. The maintenance of $CO_2$ was from the top of the bin that lasted 106 h (compiled from Navarro et al., 1985).

beetles, *C. ferrugineus*, were killed; $CO_2$ levels of 15% at 2.0 m above the floor resulted in 30% mortality.

A bolted, galvanized-iron silo (21.5 tonnes) was sealed using a polyvinyl resin formulation sprayed onto joints from the inside. The silo was loaded with wheat into which cages of insect-infested wheat were introduced. Conditions were monitored with thermocouples and gas sampling lines. $O_2$ levels were reduced to <1% by purging with $N_2$, and similar levels were then maintained by a slow $N_2$ bleed for 35 days, after which the silo was emptied. All adult insects were dead, but, as expected, some immatures survived. This was because the exposure period was too short of ensuring a complete kill at the observed grain temperatures of <15°C (Williams et al., 1980).

### 2.4.3 Sealed steel silos

Airtight, galvanized-steel silos have been manufactured in Australia for the past 30 years and are commercially available (Moylan Silos, 2011). Welded-steel hopper silos can be modified for $CO_2$ fumigation for a few hundred dollars. $CO_2$ from dry ice must be recirculated through the grain, and a pressure relief valve must be installed at the top of the silo. The top and bottom hatches must be gasket-sealed. After 10 days at 20°C, 75% of applied $CO_2$ was retained, while 99% of the caged *C. ferrugineus* were killed (Mann et al., 1999).

### 2.4.4 Concrete grain elevators

$CO_2$ fumigation of grain has been successful in concrete elevators holding 209 tonnes of wheat. The bottom hopper was sealed, and the grain purged with $CO_2$ for 4 h (1 metric tonne of $CO_2$), and additional gas is added as needed. All caged-test insects were killed (White and Jayas, 2003). A large installation for the application of $CO_2$-based CA was installed to treat more than 200 000 tonnes of rice annually in flat storage structures, each of 5000 tonnes capacity in Mianyang, China (Fig. 3).

$N_2$ was applied in three concrete silos in Cyprus, each containing 2400 tonnes of grain for controlling stored-grain insects (Navarro et al., 2012a). Structural sealing was carried out, and airtight valves were installed to improve the airtightness of the silos. Pressure decay tests (250–125 Pa) carried out in full silos showed decay time from 120 s to 290 s. Using a PSA-$N_2$ generator, the $O_2$ concentration was reduced below 0.9% in 44–56 h, and after that it was maintained between 0.2% and 0.9% up to 23.8 days. Treatment for 18.7 days and 23.8 days on grain temperature at 26°C and 22°C, respectively, was effective for the control of the adults of important stored-grain insects *O. surinamensis*, *T. confusum*, and *R. dominica* placed above and inside the grain mass. In separate bioassays, which were placed in different zones right above the grain mass, complete (100%) mortality of *T. confusum* (all

**Figure 3** Application of $CO_2$-based CA on a 5000 tonnes concrete horizontal silo bin in Mianyang, Sichuan, China.

life stages), *O. surinamensis* (larvae and adults), *S. granarius* (adults), and *R. dominica* (adults) was achieved in 18.7 days and 23.8 days treatment as well.

### 2.5 Commercial use

Numerous MA and CA systems have been developed over the years to manage insect pests and microflora associated with stored products; however, their general commercial use remains somewhat limited (Adler et al., 2000). Exceptions are for organic products where the use of fumigants is not possible because of residues; hermetic storage in plastic structures with the application of MA is the preferred choice (Navarro, 2006).

#### 2.5.1 Hermetic storage

When placed in sealed airtight storage, commodities, and the insects and aerobic microflora that exist within them respire, consuming $O_2$ and producing $CO_2$. This modified atmosphere technology has been utilized to a great extent for durables such as wheat grains. Hermetic grain bags (Africa, Argentina, Asia, Australia, North and South America, Middle East) and sealed bunker storage (Australia, the United States, Middle East) have been implemented into a commercial application to various extents.

Bunker storage, having designed storage capacities to over 10 000 tonnes, is established in permanent locations with a prepared base (usually asphalt or compacted soil with a convex profile) and an airtight cover. This type of storage has been used extensively in Australia, Argentina, Israel, and Cyprus (Adler et al., 2000). While low-moisture content, high-temperature grain supports this type of storage, condensation can remain problematic if the grain is stored with cones or ridges (Navarro et al., 1994). Sealed bunker storage has also been demonstrated as an effective means for utilizing CA or conventional fumigation where the bunker is sealed and flushed with $N_2$ or $CO_2$.

A major challenge that South America is facing is to minimize quality and quantity losses and improve food safety in view of the shortage of permanent storage capacity. As a result, the silo bag system for the temporary storage of dry grain and oilseeds has been adopted. During the 2008 and 2010 harvest seasons, more than 33 million and 43 million tonnes of grain were stored, respectively, in these plastic bags in Argentina. Commodities included corn, soybean, wheat, sunflower, malting barley, canola, cotton seed, rice, lentils, sorghum, beans, and even fertilizers. The silo bag technology is also being adopted in other countries such as the United States, Australia, Bolivia, Brazil, Canada, Chile, Italy, Kazakhstan, Mexico, Paraguay, Russia, South Africa, Sudan, Ukraine and Uruguay (Bartosik, 2011 personal communication). Dry grain can be stored in silo bags for more than 6 months without losing quality (Bartosik, 2012).

## 2.5.2 Controlled atmospheres

$N_2$ and $CO_2$ have been used as agents for controlled atmospheric storage for many years. $CO_2$ has been considered to be more efficient than $N_2$ due to the concentrations necessary for control and the level of gas tightness of the structure being used. A $CO_2$ concentration of about 60% can provide 95% control of most stored-product insect pests at 27°C (Jay, 1971), while $N_2$ use requires interstitial $O_2$ levels to be reduced to 1% or less. Considerable efforts to improve the sealing of storage silos have been made (Mann et al., 1999), which in turn facilitates ease in gas application and retention.

Mann et al. (1999) demonstrated that $CO_2$ generated from dry ice and circulated with a vacuum pump at a concentration of 51% caused 100% mortality of *C. ferrugenius* after 10 days at 20°C. $CO_2$ can also be added to bulk-stored products as compressed gas. White and Jayas (1991) demonstrated that by circulating $CO_2$ released from compressed cylinders, high mortality of several stored-product arthropod pests could be achieved within 14 days. They found that silo sealing was crucial to maintain efficacy, especially when the commodity temperature fell below 20°C, and that utilizing pressure testing techniques (Banks and Annis, 1980) is a useful means of determining the gas tightness of a seal.

$N_2$ production has also changed considerably over the years. Pressure-swing absorption systems have proven successful where a 13 660 m$^3$ silo can be purged to <1.0% $O_2$ in 7 days. Appropriate sealing allows for an accurate calculation for the additional gas application required to compensate for the gas loss due to sorption and pressure cycling caused by pressure change (Cassels et al., 2000); gas concentration can be maintained for appropriate times.

Liquid $N_2$ can be used for topping up the CA but can cost twice that of other sources. Although the CA treatment of grain is an old and proven technology, its commercial application has remained limited. A recent development has been reported by Clamp and Moore (2000), in which $N_2$ supplied as a bulk liquid under pressure was used to treat 1800 tonne silos. Since the $N_2$ treatment was commissioned in 1993, more than 300 000 tonnes were treated in the Newcastle facilities as of 2000 (Clamp and Moore, 2000).

$N_2$ can also be easily generated using molecular membrane generators. These are capable of purging vertical grain storages of 120 tonnes capacity within 3 h (Timlick et al., 2002). By maintaining a slight positive silo pressure, concentrations within a sealed commercial storage could be maintained (compensation for leakage), and insect mortality was significant after 14 days at 17°C.

In terms of efficacy and efficiency, there is not much difference between using $CO_2$ over traditional fumigants such as phosphine. However, a major component that should also be considered is the cost of $CO_2$ treatment as an alternative, especially in developing countries, which are low-income countries, and where phosphine tablets are readily available at low cost. $N_2$ has been considered unsuitable for bulk commodity treatment at export locations because the length of time required for the significant mortality of the pests in question is too long. However, effective management procedures can allow for $N_2$ use when temperatures are appropriate. All require effective sealing and monitoring, and efficiency is directly correlated to temperature. While caution is necessary when utilizing any product as an atmospheric control, there are no residues of concern when utilizing MA. Aeration after treatment is of less concern, thereby allowing for outturn of product in export position, minimizing concerns for worker safety.

Product availability is not an issue. Flexible liners and their associated loading/unloading equipment and $N_2$ generators are all available commercially and can be set up and maintained with product replenished on-site.

Maintenance of the sealing of hermetic storage has proven a challenge at times. Large bunkers and grain bags in Australia often have sealing breached by birds pecking holes in the liner. In Canada, deer often break the seals of hermetically stored grain in bags. Consequently, some focused research on liner integrity may be of use in these types of situations. The discovery of breached seals during CA treatments can be difficult to remedy, underlining the necessity of performing pressure testing before application.

### 2.6 Recent developments

A significant aspect of MA/CA treatment is the ability to monitor gas concentrations in the gastight structures. Similar to fumigation, $O_2$, and/or $CO_2$ concentrations should be monitored to ensure a successful application. As a common practice, gas sampling lines have been installed for monitoring gas concentrations. Such monitoring is performed by technical personnel on-site at predetermined regular times, particularly during gas purging and during the exposure time to MA/CA treatment. The conventional method of such monitoring is the visit of the technical personnel on a regular basis to monitor changes in gas concentrations. This method necessitates precious time and travel of technical personnel, which makes monitoring costly.

Wireless sensors for monitoring $O_2$ and/or $CO_2$ concentrations, relative humidity, and temperature have been developed for enabling technical personnel record data (Centaur Analytics, 2018; GrainPro Ecowise, 2019). Such technology facilitates the analysis of MA/CA treated enclosures without the presence of technical personnel on-site. Since the exposure time of MA/CA treatments lasts many days and the treated structures may be located at distant sites from the operator, such development offers an excellent tool that should be incorporated with MA/CA treatments. Wireless monitoring assists the operator in decision making on the necessity of intervention during the treatment.

## 3 Temperature manipulation

### 3.1 Aeration

An acceptable practice to reduce the commodity temperature is to use mechanical aeration by means of fans. It can be easily applied to stored grain or granular commodities. Aeration may be defined as 'the forced movement of ambient air of suitable quality or of suitably conditioned air through a grain bulk for improvement of grain storability' (Calderon, 1972). Aeration is also called 'active', 'mechanical', 'low volume', or 'forced' ventilation since fan power is used to deliver the air.

Aeration should be distinguished from 'passive' or 'natural' ventilation, which takes place in grain silos with an open manhole, or in granaries with open doors or windows, and is due to natural or convective air currents. Passive aeration takes place also in corn cribs, used traditionally in tropical and subtropical climates. Wind forces ambient air to flow through corn (maize) cribs, causing the slow drying of damp unshelled corn and other grains.

Aeration is a widely used method for the preservation of stored grain. It is used to modify the grain bulk microclimate and to render it unfavorable for the development of damaging organisms in the grain, and at the same time to create favorable conditions for the sustained preservation of grain quality. Within the concept of the storage ecosystem, the role of aeration is to modify one or more of the abiotic factors (temperature, humidity, atmospheric composition) and thereby to 'condition' the stored grain to improve existing conditions in the grain bulk by moving the air of suitable quality through the grain mass.

Forced aeration is an effectively applied method in commercial-scale bulk storage of grain and takes advantage of two important physical properties of the grain bulk:

1. Porosity of the grain bulk: For most cereal grain, the intergranular void volume is 35–55% of the grain bulk volume. The porous nature of bulk grain permits forced air to pass through and come into contact with almost all grain kernels in bulk.
2. Thermal insulation property of the grain bulk: Due to low thermal conductivity, the grain mass is self-insulating. This enables maintenance of a 'modified microclimate' long after the grain bulk has been aerated.

Although the role of temperature has long been recognized, manipulation of temperature by aeration techniques was first brought into focus in the early 1950s in stored-grain management programs in the United States. Pioneering engineering work of US researchers such as Foster (1953), Robinson et al. (1951), Shedd (1953), and Holman (1966) and research on technological aspects of aeration by Hukill (1953), and more recently by Cuperus et al. (1986), Arthur and Casada (2005, 2010), and Reed (2006), form the basis of modern grain aeration systems.

Aeration technology is used to modify the grain bulk microclimate to reduce or eliminate the development of harmful or damaging organisms in the grain by reducing and maintaining grain temperatures at safe levels below humidity levels which support microflora activity. Aeration helps sustain favorable storage conditions for the safe preservation of grain quality. Since then, several authors have reported their findings on aeration carried out in temperate climates. The knowledge accumulated over the last four decades has formed the basis for the present-day aeration technology reviewed by Navarro et al. (2012b). At present, forced aeration of grain is one of the most effective sustainable, economical, non-chemical methods in use for the control of stored-grain conditions, biological activity, and grain quality losses in the world.

### 3.2 Objectives of aeration

The objective of aeration is to maintain the quality of bulk grain in storage. Although aeration can improve storage conditions, aeration does not improve the intrinsic quality attributes of grain.

#### 3.2.1 Cooling the grain bulk for pest suppression

Cooling grain is the primary objective of grain aeration (Reed and Arthur, 2000; Reed and Harner, 1998) when discussing pest suppression. Stored-grain insects are of tropical or subtropical origin and require fairly high temperatures, typically 24–32°C for development. Grain-infesting insects are sensitive to low temperatures. Stored-product insect development is generally stopped below 16°C; there is little insect survival above 43°C. In the southwestern and south-central United States, temperatures of wheat, rice, and sorghum at harvest can range from 32°C to 43°C, depending on the specific crop and location. During fall harvest in the northern United States, grain temperatures from 10°C to 18°C are typical.

At temperatures below 21°C, the population growth of most storage insects is significantly suppressed. Grain temperatures from 16°C to 21°C are considered 'safe' for insect management because feeding and breeding are slow. Complete life cycles at these temperatures take 3 months or more, so insect population growth remains insignificant. Insect damage caused by these low-temperature conditions is minimal (Flinn et al., 1997; Stejskal et al., 2019).

Ileleji et al. (2007) conducted 3 years of experimental trials (2001–2003) in 12.7 tonnes capacity pilot-scale silos to determine the survival, reproduction, and suppression of *Sitophilus zeamais* Motschulsky under three temperature management strategies: no aeration (NA, control), ambient aeration (AA ≤ 23.9°C) and chilled aeration (CA ≤ 18.3°C) from May to November in Indiana, USA. There were no differences in progeny numbers between the NA and AA strategies for most of the storage periods. Their results suggest that maintaining stored maize at temperatures ≤15.0°C for longer periods suppressed *S. zeamais* progeny more effectively than at ≤18.3°C.

Arthur and Casada (2016) conducted a 3-year study to compare temperature profiles in the headspace and in the bulk mass of wheat aerated through updraft aeration and downdraft aeration. In general, downdraft aeration cooled the top portion of the grain mass more quickly than updraft aeration. Populations of the rusty grain beetle, and the red flour beetle, were generally at least 2× to 4× greater in updraft versus downdraft aeration, depending on the specific sampling date, during the first 2 years of the study. An extensive infestation that developed during the third year may have diminished the effects of aeration. The other two insect species caught in abundance were the hairy fungus beetle, and the foreign grain beetle. Results show that downdraft aeration may be preferable over updraft aeration for summer harvested wheat in the US central plains, but afternoon temperatures lead to a buildup of heat in the headspace that affected the temperature of the upper layer of the grain mass.

The crucial control parameter for mite pests is not only temperature but establishing an ERH below about 65%. About 12.5% moisture content (MC) for wheat at 25°C suppresses mite development (Cunnington, 1984; Navarro et al., 2002). Temperatures required to suppress mite development in damp grain (14–16% MC wet-basis) are obtainable in temperate climates. Maintaining low uniform grain temperatures is too expensive at the bulk periphery when mean ambient temperatures are favorable for mite development. Although cooling moist grain is unlikely to prevent moderate mite infestation, aeration is expected to minimize 'hot spots' and heavy mite populations associated with them.

### 3.2.2 Suppression of microflora growth

Low temperatures are required to prevent mold damage in moist grain. Temperatures below 5°C are needed for the suppression of most mold development. For the suppression of *Penicillium* molds, temperatures must be below 0°C. Most fungi do not grow at a relative humidity below 70%, which is equivalent to roughly 13% MC for cereal grains at typical storage temperatures. The moisture content threshold is lower for oilseeds. In practice, mold growth is dependent mainly upon interstitial air humidity. Although cooling grain may not seem like an efficient method for controlling mold, at lower grain temperatures, mold damage is reduced.

### 3.2.3 Maintenance of seed and grain quality

Low kernel temperatures are desirable for better maintenance of seed and grain quality. Studies have shown that the lower the temperature (within certain limits), the longer seeds maintain full viability. A rule of thumb (Harrington, 1973) states that a seed's life span in storage is doubled for each 5°C decrease in temperature (within the range of 0–50°C) and for each 1% decrease in seed MC (within the range of 5–14%). Seeds are commonly stored with ERH from 30% to 40% with good results. For extended storage times of seeds, Vertucci and Roos (1990) recommend the best storage MC is between 19% and 27% ERH.

Rates of chemical deterioration, such as oxidation of fats and vitamin loss occurring in grain during storage, are very slow and sometimes insignificant at low temperatures. The rate of the chemical reaction taking place in stored food is halved with each decrease of 10°C in temperature. Therefore, cool storage is important for the prevention of deteriorative changes in stored grain. Converse et al. (1977) reported on the changes in the quality of wheat stored with and without aeration in concrete silos. Aeration of wheat significantly reduced losses in germination and development of FFAs. Wheat from aerated silos had a better physical appearance and a more desirable aroma than wheat from non-aerated silos.

Grain in storage, even though a living component of the grain bulk ecosystem, is actually in a dormant state in which all biotic activities of the grain are imperceptibly slow. This state of inactivity should be maintained for as long a period as possible since the activation of life processes in grain leads to loss of viability, followed by deterioration in quality. Thus, low ambient temperatures, introduced into the grain bulk by aeration, are very beneficial in keeping the grain in a state of quiescence, needed to maintain grain quality over long periods of storage.

### 3.2.4 Equalization of temperature throughout the grain bulk

Because of self-insulating properties, grain placed in storage during summer harvest retains initial harvest temperatures for a long time before cool weather arrives in the fall (except for grain near bin walls, at exposed conical base, or at the surface). It is recommended that harvest heat be removed by night time downdraft aeration as soon as ambient temperatures are 8–11°C below internal grain mass temperatures to prevent condensation and minimize insect activity at or near the grain surface. The initial cooling should be followed by additional aeration when generally lower ambient temperatures will allow cooling the entire grain mass below 21°C. Aside from pest suppression in freshly harvested wheat, updraft aeration is preferred, especially for cooling fall-harvested maize and soybeans.

### 3.2.5 Prevention of moisture migration in the grain bulk

As the ambient temperature drops during the cool season, the surface (and peripheral) layers of the grain become considerably cooler than the internal grain mass that is left unaerated. Temperature gradients are established in the non-aerated grain bulk that can lead to convection currents that circulate air through the intergranular spaces. In large non-aerated bulks, cold, dense air settles along the outer walls. Warmer air (which contains more moisture than cool air) moves toward the colder upper surface of the grain bulk.

When the warm air reaches the cool layers of the non-aerated grain bulk, moisture condenses and creates wet layers or spots in the grain. Studies (Montross et al., 2002; Montross and Maier, 2001) suggest a moisture equilibration theory for the mechanisms involved in this moisture movement in a non-aerated grain mass. Using the finite-element model, they developed and validated with large-scale trials to demonstrate the effect of moisture condensation due to favorable headspace conditions that deposit moisture into the cool layers of the top of the grain bulk. On the other hand, the traditional natural convection hypothesis suggests that the natural convection currents in an unaerated grain bulk are sufficient to cause large amounts of moisture to 'migrate' to cooler layers or the cooler surface grain, where the air cools to 'dew point' and deposits excess moisture, slowly increasing the grain moisture content in the upper parts of the grain bulk. Equalizing grain bulk temperature by aeration prevents moisture migration because convection currents are slowed, especially when aeration fans are sealed and bulk grain is cored and un-peaked.

### 3.2.6 Prevention of headspace and downspout condensation

Under-roof condensation is a different natural process than moisture migration within the grain bulk. Condensate that drips on the grain involves moisture in humid air, which accumulates in the headspace above the grain bulk and condenses on the under-surface of the bin roof. Silos with sufficient roof vent and open eave gaps (spacing of 1/2 to 1 inch) between sidewall and roof, generally have enough natural ventilation to avoid under-roof condensate. In the United States, roof exhaust fans are routinely installed as silos have become larger in diameter. Condensate is especially problematic in silos with eave gaps that are permanently sealed to prevent fumigant gas losses and easy grain access for insects.

### 3.2.7 Prevention of biological heating of dry grain

In grain bulks where the infestation is localized, insect populations develop in small pockets of grain. The lesser grain borer and the three primary weevil species found in grains in the United States - the rice weevil, the maize weevil, and the granary weevil - are characteristic species that develop localized infestations in bulk grains, creating hot spots. Temperatures of heavily infested grain undergoing widespread heating are typically about 38–43°C. When heavy infestations are discovered, the grain should be fumigated immediately to stop insect activity. Then aeration should be used to re-cool the grain bulk.

### 3.2.8 Prevention of spontaneous heating of moist grain

In warm moist grain (ERH greater than 70%), respiration can become very intensive due to mold development. High levels of respiration produce a phenomenon called 'spontaneous heating'. The heating of the grain bulk is detrimental to grain quality. In spontaneous heating, hot spot temperatures can easily reach 57-60°C creating steep temperature gradients between heated and surrounding cool grain. In bulks containing oil-rich seeds such as cottonseeds, soybeans, and sunflower seeds at sufficiently high-moisture conditions, very high temperatures are generated, and 'spontaneous combustion' can occur, starting a fire. Do not operate aeration fans if fire is detected (by the smell of smoke or burning grain in the exhaust air stream) in a grain bulk. Aeration promotes fire by supplying more oxygen to the heating spot.

### 3.2.9 Limited grain drying by aeration

A small but significant drying effect (from 1/4% to 1/2% moisture loss per aeration cooling cycle) is typically experienced, and during long-term aeration (multiple cooling cycles), up to 2% moisture reduction may occur while cooling large grain bulks. Because of the very low flow rates during aeration, the drying front moves slowly, and this small drying effect is usually limited to the grain near the entrance of the aeration air. This grain moisture loss is reflected in a corresponding shrinkage or market weight loss in the grain bulk. This must be considered in grain management as a cost for keeping grain safe for marketing. Aeration moisture shrinkage as well as 'invisible' handling loss will affect facility records significantly and should be considered when grain receipt and delivery records from storage facilities or sites do not tally.

### 3.2.10 Removal of fumigant residues and odors

The release of or desorption of fumigants at the end of fumigation can be achieved with relatively low airflow rates. The aeration system can be operated intermittently (in pulses) to flush gas vapors from the grain bulk and storage. Aeration systems can be operated for 10–15 min every 2–3 h to allow interstitial air space to reach equilibrium with the concentration of the fumigant in the grain. Thus, the aeration system can be operated several times to ventilate the storage.

Storage odors also can develop in a grain bulk due to hot spots containing insects or moldy grain. Sour odors result from anaerobic activity in the process of fermentation at high MCs (above 18% for cereals). At moderate moisture levels (14–18% MC for cereals), musty odors in grain are usually caused by the growth of certain molds. Other odors occasionally found in grain are considered commercially objectionable foreign odors (COFO) because they are odors that are foreign to grain and render it unfit for normal commercial usage. Most odors can be reduced using aeration; however, residual odors may linger after repeated aeration cycles. Commercial applications based on pilot laboratory studies have used aeration combined with ozone treatment to reduce off-odors in grain.

### 3.3 Preservation of grain by chilled aeration with refrigerated air

Aeration using ambient air may not be sufficient to control fungi on moist grain, or protect grain against mites and insects, control self-heating of grain, or preserve the germination capacity and quality of stored grain in warm climates, or when warm grain is stored immediately after harvest. In answer to these situations, refrigerated-air units for chilling grain have been developed for commodities that can justify the added expense of refrigerated aeration cooling. In this type of aeration process, ambient air is conditioned by passing it through the evaporator coil and a secondary reheat coil of the refrigeration unit, and then blowing the chilled air into the grain bulk via the existing aeration system. Passage through the secondary reheating coil is designed to adjust the air relative humidity to 60–75% to match the target MC of the dry grain. The amount of reheating and the final air temperature are adjustable by the operator to achieve the desired aeration conditions.

Refrigerated aeration has been used for cooling dry grain in subtropical climates when ambient temperatures are too high for successful insect control by aeration with untreated air (Hunter and Taylor, 1980; Navarro et al., 1973). Refrigeration involves a considerable investment, but together with the dehumidified air method, it provides answers to the practicability of aeration for safe commercial storage in tropical climates.

In many grain-growing areas of the world, the MC of grain at harvest is too high for safe storage. This damp grain is readily attacked by molds unless it is protected in some way. The most widely used method of preventing mold growth is to dry the grain to a safe level. Burrell (1982) argued that there is a balance between 'safe moisture content' and 'safe temperature': the lower the temperature of bulk grain, the damper it can be safely stored. This leads to the conclusion that drying need not be so stringently applied since the power required to evaporate moisture from a bulk of grain is far greater than that required to cool the same bulk. For example, the energy required to evaporate 6 percentage points of moisture from a grain mass is at least six times greater than that required to cool the same bulk from 25°C to 5°C by using refrigerated air. Therefore, chilled storage may have an economic advantage over drying. This advantage is even retained when storing excessively damp grain, which requires frequent re-chilling, including the high capital investment costs for equipment.

In the major study 'Enhancing the Quality of U.S. Grain for International Trade' (U.S. Congress, 1989), the maintenance of low-temperature and moisture levels in bulk-stored grain was identified as the primary way to preserve grain quality and prevent damage from molds and insects. Grain chilling is a technology that can be successfully applied under many climatic conditions to preserve grain quality during storage when commodity value and profit margins are sufficient. It permits the short- to long-term storage of grain independent of the ambient conditions. Grain chilling has been applied commercially in over 50 countries during the past 40 years but has gained recognition in the United States only in 1994 (Maier, 1994).

It is estimated that over 80 million tonnes of grain is cooled annually worldwide with grain-chilling systems. Grain chilling is accepted as a grain conditioning technology in much of Western Europe; currently, most new units appear to be marketed in Southeast Asia. In the 1960s grain chillers were primarily used as a means of preserving high-moisture (moist, damp) grain. Later, grain chilling was applied to improve the storability of sensitive commodities subject to the development of heat foci (hot spots), that is, for soybeans and maize, and preserving the quality of high-value dry grain, seeds, and edible beans, primarily against mites and insects (Navarro et al., 2002).

## 3.4 Simulation models in cooling grain

Simulation models were used principally for drying processes for cereals. For cooling dry grain, such simulation models were developed over the last 20 years. One of the early simulation models was utilized by Maier (1994) in field tests and simulation techniques for an in-depth analysis of the chilled aeration and storage of low-moisture maize and wheat in the midwestern United States. Alagusundaram (1994) studied the airflow patterns during drying through wheat, barley, and canola stored in a 4.2-m diameter cylindrical steel bin with a fully perforated floor that was modified to give three different partially perforated floors (straight, cross, and square floor openings). Sinicio et al. (1997) determined the relative importance of several variables in a non-equilibrium mathematical model for simulating aeration of stored wheat with uniform and non-uniform airflow distributions. Results of simulations using weather data from Curitiba, Brazil, indicated that the most important variables in decreasing order, were the fan temperature rise, thin-layer wetting equation, and thin-layer drying equation.

Thorpe (1997) modeled ecosystems in ventilated conical-bottomed farm grain silos differential equations that govern distributions of velocity, moisture, and temperature in conical-bottomed silos that contain peaked bulks of grain. The equations were solved by transforming the shape of the silo into a right cylinder with a flat top and base. Results of Thorpe (1997) work show that aeration of a small farm silo with ambient air results in improved storage conditions compared with those that obtain in unaerated silos. Arthur and Flinn (2000) conducted simulation studies to determine temperature accumulations below defined thresholds and to show the impact of controlled aeration on populations of the rusty grain beetle, *C. ferrigineus*. Recorded data from weather stations in Texas, Oklahoma, Kansas, eastern New Mexico, and eastern Colorado were used to determine hours of temperature accumulation below

23.9°C in June and July, 15.6°C in September and October, and 7.2°C in December.

At an airflow rate of 4.68 $m^3/h/m^3$, which requires 120 h of temperatures below the specified threshold to complete an aeration cycle, summer cooling at 23.9°C in bulk-stored wheat could be completed throughout the hard red winter wheat zone except for extreme southern Texas. Five geographic regions were delineated and the times required for completion of the summer, early-autumn, and late-autumn cooling cycles within each region were estimated. The population growth of the rusty grain beetle was modeled for San Antonio, TX; Abilene, TX; Tulsa, OK; Topeka KS; and Goodland, KS, by predicting the numbers of adults in the top, outer middle, outer periphery, and the center of the bin during a 1-year storage season.

A 'family of curves' to describe several variations of temperature change from 25°, 30°, 35° and 40°C to ambient temperatures of 10°, 15° and 20°C at 64% relative humidity were developed by Navarro and Noyes (2002) (Fig. 4). This family of curves clearly indicates that by reducing or increasing the airflow rate beyond certain limits, the aeration time needed to cool grain satisfactorily may exceed practical limits. At a low airflow rate, below 1.0 (m3/h)/tonne, the aeration time will exceed 600-700 h, which is not practical for grain cooling, especially in geographical regions with marginal ambient temperature conditions.

Faulkner (2004) investigated the aeration characteristics of a grain silo using a computational fluid dynamics package and developed an understanding of the complexities involved with the modeling of grain ecosystems with regard to air movement, moisture migration, and temperature distribution within the grain.

Liu et al. (2016) used a finite volume method to simulate temperature variations and distributions in stored paddy rice in a warehouse-type facility during cooling aeration. A software package was used to develop a model to solve coupled nonlinear heat, mass, and momentum transfer equations. An experiment was conducted in a large-scale (commercial) warehouse storage facility to measure temperature distributions in the stored paddy rice to validate the model. The stored grain was aerated for 30 h, and the temperature was measured at five depths within the grain bulk and at 24 locations for each depth. The simulated temperatures were in good agreement with the measured values, with relative differences less than 6.3%.

Binelo et al. (2019) developed a mathematical model to simulate the distribution of 3D airflow in horizontal bulk storage bins during the aeration process under nonhomogeneous and anisotropic conditions. The criterion of evaluation of the specific airflow and the criterion of optimization of the aeration system were incorporated in the model. With the results, it was possible to perform the analysis of the air pressure, global and local specific airflow rate, and air velocity throughout the grain mass domain. Pressure data were obtained

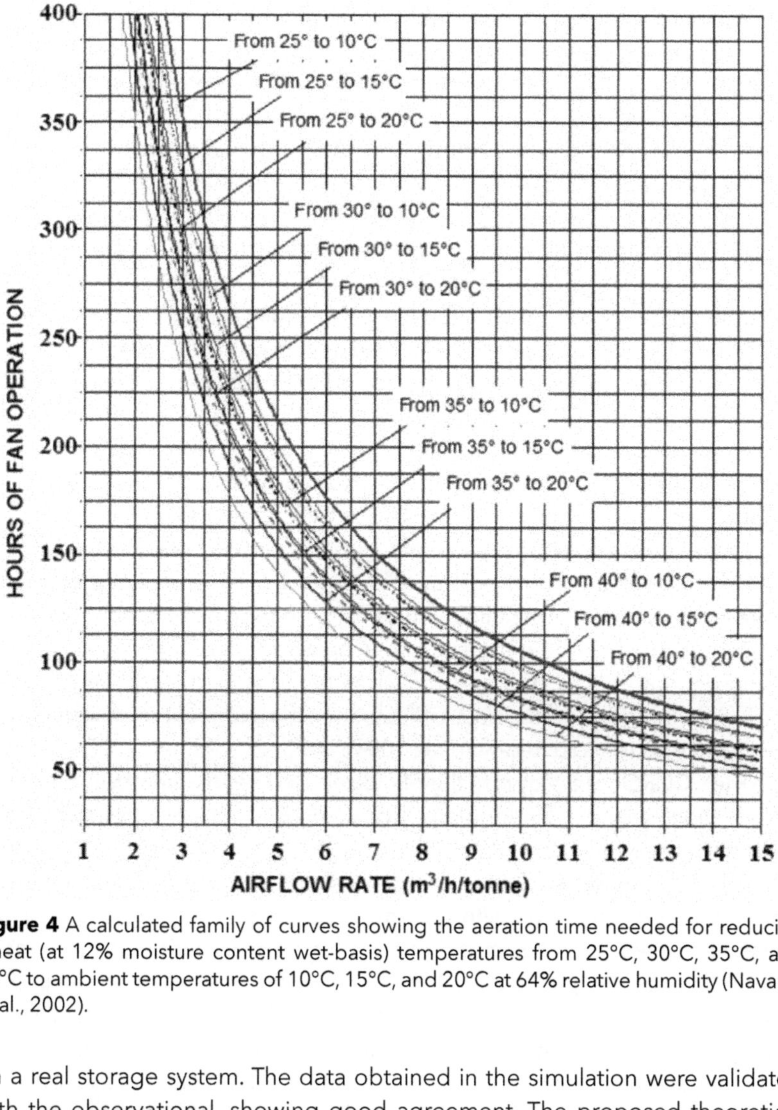


**Figure 4** A calculated family of curves showing the aeration time needed for reducing wheat (at 12% moisture content wet-basis) temperatures from 25°C, 30°C, 35°C, and 40°C to ambient temperatures of 10°C, 15°C, and 20°C at 64% relative humidity (Navarro et al., 2002).

on a real storage system. The data obtained in the simulation were validated with the observational, showing good agreement. The proposed theoretical air inlet profile allowed an 85.3% decrease in the airflow variation, but a 150% increase in maximum inlet air pressure at the center of the storage bin was necessary. It could be used as a guideline for a new aeration system design.

### 3.5 High temperatures

The lethal effect of temperatures from 42°C to 50°C has been the subject for a study by Gonen (1977a,b) and Menon et al. (2000). There is significant survival (< 95% mortality) of several species when exposed for some hours at this range of temperatures. Fields (1992) has listed the results of studies on the susceptibility of stored-product pests to temperatures greater than 50°C. The information suggests that exposure for 2 h at 50°C should eliminate all pests, but studies on heat-tolerant species and stages (e.g. *R. dominica* pupae) are lacking (Banks and Fields, 1995). To control the most heat-tolerant common pest, *R. dominica*, Evans (1987) concluded that a grain temperature of 65°C must be attained briefly. Such temperatures can damage the baking quality of wheat, the malting quality of barley, and the germination of most seeds. Thus the temperature of the grain must be carefully measured and controlled. The effects of high temperatures have been concentrated on obtaining rapid disinfestation without damage to the treated grain (Banks and Fields, 1995).

Navarro et al. (2011) reported on the emerging potential and promising application methods of heat treatment for use in the food industry. The heat sterilization technique in flour mills and food processing facilities uses high-target temperatures ranging from 54°C to 60°C to kill stored-product insects by inducing dehydration or protein coagulation or enzyme destruction. The length of time required to destroy stored-product insects varies from a few hours within a piece of machinery to up to 24 h for a large facility. The key to successfully eradicate the insects is in achieving the target temperatures for a long enough period of time to reach lethal temperatures in areas such as machinery interiors, voids, dust collectors and piping where stored-product insects may be found (Dosland et al. 2006).

The CSIRO developed the fluidized-bed and the falling column (conduction) systems (Dermott and Evans, 1978). These two systems had no industrial acceptance.

Qaisrani and Beckett (2003) described a cross-flow dryer in which the grain flows by gravity from a holding bin into the drying zone consisting of one or several screened grain columns; hot air is forced perpendicularly from the air plenum through the grain in the columns. The width of the grain columns in cross-flow dryers is between 0.25 m and 0.45 m, and the length of grain columns in the drying section is between 3 m and 30 m. Two trials were carried out with batches of 25 tonnes of barley, using an inlet air temperature of 82°C and 72°C for drying. In the third trial, 25 tonnes of wheat were dried from 14% to 10.5% moisture content using an inlet air temperature of 82°C. The aim was to explore the possibility of using this type of dryer for heat disinfestation of grain. The average final grain temperature and temperature distribution across the grain batch were measured during the trials and compared with grain temperatures that are necessary to kill stored-grain pests. Conditions were shown to be favorable for heat disinfesting grain in a cross-flow dryer. No practical data is available on the implementation of this technology for the disinfestation of grains.

### 3.6 Recent developments

A significant aspect of effective aeration control is our ability to monitor the temperature in the aerated silos. Similar to fumigation, $O_2$, or $CO_2$, temperature and humidity should be monitored to ensure successful aeration. As a common practice, temperature sensors have been installed to ensure timely cooling of grain in silos. Such monitoring is performed by technical personnel on-site at predetermined regular times, particularly during aeration. The conventional method of such monitoring is the visit of the technical personnel on a regular basis to monitor the changes in temperature. This method necessitates precious time and travel of the technical personnel, which makes the monitoring expensive.

Wireless sensors for monitoring temperature and relative humidity have been developed to enable the technical personnel record of the data (Centaur Analytics, 2018; Amber Agriculture, 2019; *i*grain $CO_2$ Sniffer Spoilage Detection, 2019). Such development facilitates the analysis of the cooling pattern of the grain bulk without the presence of the technical personnel on-site. Since the aeration to cool the grain mass lasts many days and the aerated structures may be located at distant sites from the operator, this development offers an excellent tool that can be incorporated into the package of the aeration equipment. Such wireless monitoring assists the operator in decision making on the necessity of intervention during the treatment.

## 4 Conclusions and future trends

### 4.1 Conclusions

#### 4.1.1 Controlled and modified atmospheres

1. MA and CA treatments comprise atmospheric gases rich in $CO_2$ and low in $O_2$, or a combination of these two gases that control storage pests and preserve the quality of the commodity.
2. MA/CA offers an alternative that is safe and environmentally benign, to the use of conventional residue-producing chemical fumigants for controlling insect pests.
3. $N_2$ and $CO_2$ have been used as agents for MA/CA storage.
4. Insects can tolerate low levels of oxygen for prolonged periods. Using $N_2$ to replace $O_2$ must result in $O_2$ being below 2%, preferably 1% for considerably rapid death.
5. About 60% $CO_2$ at 26°C about 4 days of exposure would be sufficient to kill all stages (including eggs) of most stored-product insects.
6. $CO_2$ has been considered to be more efficient than $N_2$.
7. The successful application of MA/CA depends on the gas tightness of the treated structure.

8. The commercial use of MA/CA is still limited to a few countries.
9. Insect mortality increases more rapidly as temperatures rise.
10. Seed below the critical moisture content is not significantly affected at high $CO_2$ or low $O_2$ atmospheres.
11. Hermetic storage technology has been utilized to a great extent in developing countries for storage in grain bags, in big bags, in cocoons, and in sealed bunkers.
12. Silo bag system for the temporary storage of dry grain and oilseeds has been adopted.

### 4.1.2 Temperature manipulation

1. Aeration is a widely used method for the preservation of stored grain.
2. Aeration is used to modify the grain bulk microclimate and to render it unfavorable for the development of damaging organisms
3. Aeration creates favorable conditions for the sustained preservation of grain quality.
4. It can be easily applied to stored-grain or granular commodities.
5. Aeration takes advantage of two important physical properties of the grain bulk: (a) porosity of the grain bulk and (b) thermal insulation property of the grain bulk.
6. Cooling grain is the primary objective of grain aeration.
7. At temperatures below 21°C, the population growth of most storage insects is significantly suppressed.
8. Grain temperatures of 16–21°C are considered 'safe' for insect management.
9. Because grain placed in storage during summer harvest retains its initial harvest temperatures, harvest heat should be removed as soon as ambient temperatures are 8–11°C below internal grain mass temperatures to prevent condensation and minimize insect activity at or near the grain surface.
10. In warm climates, refrigerated-air units for chilling grain have been developed for commodities that can justify the added expense of refrigerated aeration cooling.
11. Elevated grain temperatures for the control of insects in silos have not gained favorable support.

## 4.2 Future trends

### 4.2.1 Controlled and modified atmospheres

1. Insect mortality under the influence of MA or CA has been a subject of serious consideration. A road map that will indicate the exposure time

needed in the presence of the specific grain conditions that consider moisture content, temperature, and pest species is lacking in the literature.

2. Since insect mortality increases more rapidly as temperatures rise, insect mortality at elevated $CO_2$ and low in $O_2$, or a combination of these two gases that control storage pests, is of particular interest. Such information is needed, particularly for determining the reduced exposure times to control storage pests where short exposure times are preferred such as in organic products.
3. Insect mortality increases more rapidly as relative humidity lowers. Although in stored products such changes are very limited, exposure times that combine, relative humidity, and temperature of various insect development stages remain to be investigated.
4. The potential use of MA and CA for specific applications like organic products and seeds is of major importance that needs further investigations.
5. Because the successful application of MA and CA depends on the gas tightness of the treated structure, standardized, simple, and accurate gas tightness level determination methods are lacking.
6. Hermetic storage has been demonstrated as an effective method of storage in flexible liners implemented in developing countries. However, the implementation of hermetic storage in rigid structures (concrete or metal) is missing in the literature.
7. Remote sensing to monitor gas concentrations in the gastight structures is an important development that needs further exploration. Although technologies are available for such monitoring, data that determine conditions for successful application of MA and CA have not been reported yet.
8. Simulation models that describe gas movement ($CO_2$ or $O_2$) in relation of various MA and CA application technologies are missing in the literature. Such models may be used to determine the formation of low $CO_2$ or high $O_2$ pockets in the treated grain bulk.

### 4.2.2 Temperature manipulation

1. The efficacy of aeration for cooling grain in temperate and subtropical climates has been demonstrated. Many silos installed in tropical climates were equipped with aeration systems. However, the benefits of such aeration systems remain unclear. Research that will clarify the benefits or lack of benefits of such aeration systems is needed.
2. Although the use of refrigerated air aeration for chilling grain has been implemented in several tropical countries, information on the successful

application of chilled air in relation to arresting development insect populations is lacking in the literature.
3. Cross-flow aeration has been reported as an effective aeration system. However, trials on the effective use of this system and its benefits in controlling grain temperatures are missing in the literature.
4. Although the technological ability to monitor grain temperature in the aerated silos using remotely monitored sensors exists, the value of such technology in relation of time saving and its benefits have not been demonstrated yet.
5. Coring has been a practice implemented in combination with aeration. The value of coring has not been assayed to explore its full benefits in terms of reducing grain temperature and insect development.
6. Spontaneous heating is common in grain storage. However, experimental work to reduce the temperature of grain undergoing heating still has room for studying the airflow rates needed to arrest the deleterious effect of heat on grain.

## 5 Acknowledgement

The authors thank Mr. Nadav Inbari for his technical assistance in the preparation of this manuscript.

## 6 References

Adler, C., Corinth, H. G. and Reichmuth, C. 2000. Ch. 5: Modified atmospheres. In: Subramanyam, Bh. and Hagstrum, D. W. (Eds), *Alternatives to Pesticides in Stored-Product IPM*. Kluwer Academic Publishing, Norwell, MA, pp. 105–46.

Alagusundaram, K., Jayas, D. S., Friesen, O. H. and White, N. D. G. 1994. Airflow patterns through wheat, barley, and canola in bins with partially perforated floors–an experimental investigation. *Applied Engineering in Agriculture* 10(6), 791–6. doi:10.13031/2013.25914.

Alagusundaram, K., Jayas, D. S., White, N. D. G., Muir, W. E. and Sinha, R. N. 1995. Controlling *Cryptolestes ferrugineus* (Stephens) adults in wheat stored in bolted-metal bins using elevated carbon dioxide. *Canadian Agricultural Engineering* 37, 217–23.

Amber Agriculture. 2019. Available at: https://www.amber.ag.com (Accessed on March 2020).

Annis, P. C. 1987. Towards rational controlled atmosphere dosage schedules: a review of current knowledge. In: Donahaye, E. and Navarro, S. (Eds), *Proceedings of the 4th International Working Conference on Stored Product Protection*. Tel Aviv, Israel, pp. 128–48.

Annis, P. C. and Graver, J. van S. 1990. *Suggested Recommendations for the Fumigation of Grain in the ASEAN Regions. Part 2. Carbon Dioxide Fumigation of Bag Stacks Sealed in Plastic Enclosures: An Operations Manual*. Australian Centre for International Agricultural Research, Canberra, Australia, 58pp.

Arthur, F. H. and Flinn, P. W. 2000. Aeration management for stored hard red winter wheat: simulated impact on rusty grain beetle (Coleoptera: Cucujidae) populations. *Journal of Economic Entomology* 93(4), 1364–72. doi:10.1603/0022-0493-93.4.1364.

Arthur, F. H. and Casada, M. E. 2016. Temperature stratification and insect pest populations in stored wheat with suction versus pressure aeration. *Applied Engineering in Agriculture* 32(6), 849–60. doi:10.13031/aea.32.11757.

Arthur, F. H. and Casada, M. E. 2005. Feasibility of summer aeration for management of wheat stored in Kansas. *Applied Engineering in Agriculture* 21, 1027–38.

Arthur, F. H. and Casada, M. E. 2010. Directional flow of summer aeration to manage insect pests in stored wheat. *Applied Engineering in Agriculture* 26(1), 115–22. doi:10.13031/2013.29462.

Banks, H. J. and Annis, P. C. 1990. Comparative advantages of high $CO_2$ or low $O_2$ types of controlled atmospheres for grain storage. In: Calderon, M. and Barkai Golan, R. (Eds), *Food Preservation by Modified Atmospheres*. CRC Press, Boca Raton, FL, pp. 93–122.

Banks, H. J. and Annis, P. C. 1980. Conversion of existing grain storage structures for modified atmosphere use. In: *Controlled Atmosphere Storage of Grains*. Elsevier, Castelgandolfo (Rome), Italy. pp. 461–73.

Banks, J. and Fields, P. 1995. Ch. 11: Physical methods for insect control in stored-grain ecosystems. In: Jayas, D. S., White, N. D. G. and Muir, W. E. (Eds), *Stored Grain Ecosystems*. Marcel Dekker, New York, pp. 353–409.

Bartosik, R. 2012. An inside look at the silo-bag system. In: Navarro, S., Banks, H. J., Jayas, D. S., Bell, C. H., Noyes, R. T., Ferizli, A. G., Emekci, M., Isikber, A. A. and Alagusundaram, K. (Eds), *Proceedings of the 9th International Conference on Controlled Atmosphere and Fumigation in Stored Products*, Antalya, Turkey, 15–19 October 2012. ARBER Professional Congress Services, Turkey, pp, pp. 117–28.

Bell, C. H., Chakrabarti, B., Llewellin, B. E. and Wontner-Smith, T. J. 1993. A comparison of carbon dioxide and burner gas as replacement atmospheres for control of grain pests in a welded steel silo. *Postharvest Biology and Technology* 2(3), 241–53. doi:10.1016/0925-5214(93)90051-4.

Binelo, M. O., Faoro, V., Kathatourian, O. A. and Ziganshin, B. 2019. Airflow simulation and inlet pressure profile optimization of a grain storage bin aeration system. *Computers and Electronics in Agriculture* 164, 104923. (doi:10.1016/j.compag.2019.104923).

Bond, E. J. 1984. *Manual of Fumigation for Insect Control*. FAO Plant Production and Protection Paper No. 54, 432pp.

Burrell, N. J. 1982. Storage of cereal grains and their products. In: Christensen, C. M. (Ed.), *Refrigeration*. American Association of Cereal Chemists, Inc., St. Paul, MN, pp. 407–41.

Calderon, M. 1972. Aeration of grain – benefits and limitations. *EPPO Bulletin* 2(6), 83–94. (doi:10.1111/j.1365-2338.1972.tb02137.x).

Calderon, M. and Barkai-Golan, R. 1990. *Food Preservation by Modified Atmospheres*. CRC Press, Boca Raton, FL.

Cassels, J., Banks, J. and Allanson, R. 2000. Applications of pressure swing absorption (PSA) and liquid nitrogen as methods for providing controlled atmospheres in grain terminals. In: *Proceedings of the 6th International Working Conference on Stored-Product Protection*, Canberra, Australia, pp. 57–63.

Centaur Analytics. 2018. Available at: https://centaur.ag/ (Last visited February 2019).

Clamp, P. and Moore, D. 2000. Nitrogen treatment of grain, Newcastle grain terminal. *Australian Postharvest Technical Conference*, pp. 184–6.

Conyers, S. T. and Bell, C. H. 2007. A novel use of modified atmospheres: storage insect population control. *Journal of Stored Products Research* 43(4), 367–74. (doi:10.1016/j.jspr.2006.09.003).

Converse, H. H., Miller, B. S. and Graves, A. H. 1977. Quality changes of hard red winter wheat in concrete silos. Agr. Res. Serv. USDA. ARS-NC-58, 7p.

Cunnington, A. M. 1984. Resistance of the grain mite, *Acarus siro* L. (Acarina, Acaridae) to unfavourable physical conditions beyond the limit of the development. *Agriculture, Ecosystems and Environment* 11(4), 319–39. doi:10.1016/0167-8809(84)90005-7.

Cuperus, G. W., Prickett, C. K., Bloome, P. D. and Pitts, J. T. 1986. Insect populations in aerated and unaerated stored wheat in Oklahoma. *Journal of the Kansas Entomological Society* 59, 620–7.

Dermott, T. and Evans, D. E. 1978. An evaluation of fluidized-bed heating as a means of disinfesting wheal. *Journal of Stored Products Research* 14(1), 1–12. doi:10.1016/0022-474X(78)90012-7.

Donahaye, J., Navarro, S., Andales, S., del Mundo, A., Caliboso, F., Sabio, G. and Felix, A. 2001. Quality preservation of moist paddy under hermetic conditions. In: *International Conference on Controlled Atmosphere and Fumigation in Stored Products*, Radisson Hotel, Fresno, CA, 29 October – 03 November 2000, Abstract 3.2, p. 75.

Dosland, O. Subramanyam, B., Sheppard, K. and Mahroof, R. 2006. Chapter 10: Temperature modification for insect control. In: Heaps, J. W. (Ed.), *Insect Management for Food Storage and Processing* (2nd edn.). AACC International, St. Paul, MN, pp. 89–103.

Evans, D. E. 1987. Some biological and physical constraints to the use of heat and cold for disinfesting and preserving stored products. In: Donahaye, E. and Navarro, S. (Eds), *Proceedings of the Fourth International Working Conference on Stored-Product Protection*, Tel-Aviv, Israel, pp. 149–64.

Faulkner, G. 2004. Numerical investigation into the aeration of grain silos. University of Southern Queensland. Faculty of Engineering and Surveying, ENG4111 and 4112 Research Project, 144p.

Federal Register. 1981. Carbon dioxide, nitrogen, and combustion product gas; tolerances for pesticides in food administered by the Environmental Protection Agency. *Federal Register* 46(122), 32865–6.

Federal Register. 1980. Tolerances and exemptions from tolerances for pesticides chemicals in or on raw agricultural commodities; carbon dioxide, nitrogen, and combustion product gas. *Federal Register* 45(223), 75663–4.

Fields, P. G. 1992. The control of stored-product insects and mites with extreme temperatures. *Journal of Stored Products Research* 28(2), 89–118. doi:10.1016/0022-474X(92)90018-L.

Flinn, P. W., Hagstrum, D. W.and Muir, W. E. 1997. Effects of time of aeration, bin size, and latitude on insect populations in stored wheat: a simulation study. *Journal of Economic Entomology* 90(2), 646–51. doi:10.1093/jee/90.2.646.

Foster, G. W. 1953. Minimum airflow requirements for drying grain with unheated air. *Agricultural Engineering* 34, 681–4.

Gonen, M. 1977a. Susceptibility of *Sitophilus granarius* and *S. oryzae* (Coleoptera: Curculionidae) to high temperature after exposure to supra-optimal temperature.

*Entomologia Experimentalis et Applicata* 21(3), 243–8. doi:10.1111/j.1570-7458.1977.tb02678.x.

Gonen, M. 1977b. Survival and reproduction of heat-acclimated *Sitophilus granarius* (Coleoptera: Curculionidae) at a moderately high temperature. *Entomologia Experimentalis et Applicata* 21(3), 249–53. doi:10.1111/j.1570-7458.1977.tb02679.x.

GrainPro EcoWise Plus. 2019. Available at: https://grainpro.com/grainpro-ecowise-plus/ (Accessed on November 2019).

Harrington, J. F. 1973. Problems of seed storage. In: Heydecker, W. (Ed.), *Seed Ecology*. Pennsylvania State University Press, University Park, PA and London, UK, pp. 251–63.

Holman, L. E. 1966. Aeration of grain in commercial storages. Marketing Research Report No. 178. Agricultural Marketing Service, Res. Div. United States Department of Agriculture, Washington DC, 46p.

Hukill, W. V. 1953. Grain cooling by air. *Agricultural Engineering* 34(7), 456–8.

Hunter, A. J. and Taylor, P. A. 1980. Refrigerated aeration for the preservation of bulk grain. *Journal of Stored Products Research* 16(3–4), 123–31. doi:10.1016/0022-474X(80)90009-0.

*i*grain $CO_2$ Sniffer Spoilage Detection. 2019. Available at: https://crop-protector.com/products/product/co2-sniffer-spoilage-detection/ (Accessed on March 2020).

Ileleji, K. E., Maier, D. E. and Woloshuk, C. P. 2007. Evaluation of different temperature management strategies for suppression of Sitophilus zeamais (Motschulsky) in stored maize. *Journal of Stored Products Research* 43(4), 480–8. doi:10.1016/j.jspr.2007.01.002.

Jay, E. 1971. *Suggested Conditions and Procedures for Using Carbon Dioxide to Control Insects in Grain Storage Facilities.* US Agric Res Serv. (Rep) ARS pub. 51-46.

Jay, E. 1984. Recent advances in the use of modified atmospheres for the control of stored product insects. In: Baur, F. (Ed.), *Insect Management for Food Storage and Processing.* Am. Assoc. Cereal Chem., St. Paul, MN, pp. 241–54.

Liu, Q., Yang, G., Zhang, Q. and Ding, C. 2016. CFD simulations of aeration for cooling paddy rice in a warehouse-type storage facility. *Transactions of the ASABE* 59(6), 1873–82. doi:10.13031/trans.59.11478.

Lukasiewicz, M., Jayas, D. S., Muir, W. E. and White, N. D. G. 1999. Gas leakage through samples of wall seams of bolted-metal bins. *Canadian Agricultural Engineering* 41, 65–71.

Maier, D. E. 1994. Chilled aeration and storage of U.S. crops – a review, In Proceedings. 6th International Working Conference on Stored-Product Protection (Vol. 1, pp. 300–311). Wallingford: CAB International.

Mann, D. D., Jayas, D. S., Muir, W. E. and White, N. D. G. 1999. Efficient carbon dioxide fumigation of wheat in welded-steel hopper bins. *Applied Engineering in Agriculture* 15(1), 57–63. doi:10.13031/2013.5820.

Menon, A., Subramanyam, Bh., Dowdy, A. and Roesli, R. 2000. Heat treatment: a viable alternative to methyl bromide for managing insects. *World Grain (March)*, 68–69.

Montross, M. D. and Maier, D. E. 2001. A new moisture equilibration theory to predict moisture movement in a non- aerated grain mass. ASAE Paper no. 016110, ASABE, St. Joseph, MI. Available at: www.asabe.org.

Montross, M. D., Maier, D. E. and Haghighi, K. 2002. Validation of a finite-element stored grain ecosystem model. *Transactions of the ASAE* 45(5), 1465–74.

Moylan Silos. 2011. Moylangrainsilos.com. Available at: moylanproducts.htm (Accesesd on 09 February 2011).

Navarro, S. 1978. The effects of low oxygen tensions on three stored-product insect pests. *Phytoparasitica* 6(2), 51–8. doi:10.1007/BF02980382.

Navarro, S. 1999. Pressure tests for gaseous applications in sealed storages: theory and practice. In: Zuxun, J., Quan, L., Yongsheng, L., Xianchang, T. and Lianghua, G. (Eds), *Proc. 7th Int. wkg. Conf. Stored-Product Protection*, 14–19 October 1998, Beijing China, Sichuan Publishing House of Science and Technology, Chengdu, Sichuan Province, pp. 385–90.

Navarro, S. 2006. Chapter 11: Modified atmospheres for the control of stored product insects and mites. In: Heaps, J. W. (Ed.), *Insect Management for Food Storage and Processing* (2nd edn.). AACC International, St. Paul, MN.

Navarro, S. and Navarro, H. 2018. The significance of vapor pressure in quality preservation of stored commodities under gastight conditions. *Julius-Kühn-Archiv* 463, 549–56.

Navarro, S., Athanassiou, C., Varnava, A., Vroom, N., Yilasoumis, D., Leandrou, I. and Hadjioannou, S. 2012a. Control of stored grain insects by using nitrogen in large concrete silos in Cyprus. In: Navarro, S., Banks, H. J., Jayas, D. S., Bell, C. H., Noyes, R. T., Ferizli, A. G., Emekci, M., Isikber, A. A. and Alagusundaram, K. (Eds), *Proceedings of the 9th International Conference on Controlled Atmosphere and Fumigation in Stored Products*, Antalya, Turkey, 15–19 October 2012. ARBER Professional Congress Services, Turkey, , pp. 478–87.

Navarro, S. and Caliboso, F. M. 1996. *Application of Modified Atmospheres under Plastic Covers for Prevention of Losses in Stored Grain*. Final Report for CDR Project No. C7-053, Submitted to U.S. Agency for International Development. U.S. AID, Washington DC, 32pp.

Navarro, S. and Donahaye, E. 1990. Generation and application of modified atmospheres and fumigants for the control of storage insects. In: Proc. Int. Conf. on Fumigation and Controlled Atmosphere Storage of Grain. B. R. Champ, E. Highley, and H. J. Banks, Eds. ACIAR Proc. 25. Brown Prior Anderson Pty. Ltd., Burwood, Victoria, Australia. pp. 152-65..

Navarro, S. and Donahaye, E. 2005. Innovative environmentally friendly technologies to maintain quality of durable agricultural produce. In: Ben-Yehoshua, S. (Ed.), *Environmentally Friendly Technologies for Agricultural Produce Quality*. CRC Press, Taylor & Francis Group, Boca Raton, FL, pp. 205–62.

Navarro, S., Donahaye, E. J., Caliboso, F. M. and Sabio, G. C. 1998. Outdoor storage of corn and paddy using sealed-stacks in the Philippines. In: Villapando, I., Ramos, C. L.. and Salcedo, B. G. A. (Eds), *Proc. 18th ASEAN Seminar on Grains Postharvest Technology*. National Postharvest Institute for Research and Extension, Manila, Philippines, pp. 225–36.

Navarro, S., Donahaye, E. and Calderon, M. 1973. Studies on aeration with refrigerated air. Chilling of wheat in a concrete elevator. *Journal of Stored Products Research* 9(4), 253–9. doi:10.1016/0022-474X(73)90008-8.

Navarro, S., Donahaye, E. and Fishman, S. 1994. The future of hermetic storage in dry grains in tropical and subtropical climates. In: Highley, E., Wright, E. J., Banks, H. J. and Champ, B. R. (Eds), *Proceedings of the 6th International Working Conference on Stored-Product Protection*. Canberra, Australia, CAB International, Wallingford, UK (pp. 130–138).

Navarro, S., Donahaye, E. and Jay, E. 1985. *Modified Atmospheres for Controlling Stored-Product Insects*. Final Scientific Rep. presented to Binational Agric. Res. Dev. Fund US-Israel (BARD) No. I-303-80, 184p.

Navarro, S., Finkelman, S. and Navarro, H. 2011. Application of thermal disinfestation in the food industry. In: Athanassiou, C. G., Adler, C. and Trematerra, P. (Eds), *Proceedings of the Conference of the International Organization for Biological and Integrated Control of Noxious Animals and Plants (IOBC)*. West Palaearctic Regional Section (WPRS) (OILB SROP) Working Group on Integrated Protection of Stored Products Bulletin (vol. 69). Campobasso, Italy, 29 July - 02 June 2009, pp. 7–18.

Navarro, S., Navarro, H., Finkelman, S. and Jonfia-Essien, W. A. 2010. A novel approach to the protection of cocoa beans by preventing free fatty acid formation under hermetic storage. In: Navarro, S. and Riudavets, J. (Eds), *Fumigation, Modified Atmospheres and Hermetic Storage. Proceedings of the 10th International Working Conference on Stored Product Protection*, 27 June to 2 July 2010, Estoril, Portugal Julius-Kühn-Archiv, 425, Bundesforschungsinstitut für Kulturpflanzen, Berlin, 1077p.

Navarro, S. and Noyes, R. 2002. (Eds), *The Mechanics and Physics of Modern Grain Aeration Management*. CRC Press, Boca Raton, FL, 647p.

Navarro, S., Noyes, R. and Jayas, D. S. 2002. Stored grain eco-system and heat, and moisture transfer in grain bulks. In: Navarro, S. and Noyes, R. (Eds), *The Mechanics and Physics of Modern Grain Aeration Management*. CRC Press, Boca Raton, FL, pp. 35–78.

Navarro, S., Noyes, R. T., Cassada, M. and Arthur, F. H. 2012b. Chapter 11: Aeration of grain. In: Hagstrum, D. W., Phillips, T. W. and Cuperus, G. (Eds), *Stored Product Protection*. Kansas State Research and Extension as Publication S156, pp. 121–34. Available at: http://www.ksre.ksu.edu/bookstore/pubs/S156.pdf.

Nicolas, G. and Sillans, D. 1989. Immediate and latent effects of carbon dioxide on insects. *Annual Review of Entomology* 34(1), 97–116. doi:10.1146/annurev.en.34.010189.000525.

Qaisrani, R. and Beckett, S. 2003. Possible use of cross-flow dryers for heat disinfestation of grain. In: Wright, E. J., Webb, M. C. and Highley, E. (Eds), *Stored Grain in Australia 2003. Proceedings of the Australian Postharvest Technical Conference* (vol. 2003), Canberra, 25–27 June 2003. CSIRO Stored Grain Research Laboratory, Canberra, pp. 30–5.

Radek, A., Vlastimil, K., Jan, P. and Vaclav, S. 2017. Field efficacy of brief exposure of adults of six storage pests to nitrogen-controlled atmospheres. *Plant Protection Science* 53(3), 169–76. doi:10.17221/136/2016-PPS.

Reed, C. R. 2006. Aeration. In: *Managing Stored Grain to Preserve Quality and Value*. AACC International, St. Paul, MN, pp. 141–80.

Reed, C. and Arthur, F. H. 2000. Aeration. In: Subramanyam, Bh. and Hagstrum, D. W. (Eds), *Alternatives to Pesticides in Stored-Product IPM*. Kluwer Academic Publishers, New York, NY, pp. 51–72.

Reed, C. and Harner, J. 1998. Cooling of stored wheat in multiple or single cycles using automatic aeration controllers. *Applied Engineering in Agriculture* 14, 495–500.

Robinson, R. N., Hukill, W. V. and Foster, G. H. 1951. Mechanical ventilation of stored grain. *Agricultural Engineering* 32(11), 606–8.

Shedd, C. K. 1953. Resistance of grain and seeds to airflow. *Transactions of the ASAE* 34, 616–9.

Sigaut, F. 1988. A method for identifying grain storage techniques and its application for European agricultural history. *Tools and Tillage* VI, 3–32.

Sinicio, R., Muir, W. E. and Jayas, D. S. 1997. Sensitivity analysis of a mathematical model to simulate aeration of wheat stored in Brazil. *Postharvest Biology and Technology* 11(2), 107–22. doi:10.1016/S0925-5214(97)00017-3.

Schulten, G. G. M. 1988. FAO's experiences with crop loss assessment. *International Journal of Tropical Insect Science* 9(6), 763–7. doi:10.1017/S1742758400005658.

Stejskal, V., Vendl, T., Li, Z. and Aulicky, R. 2019. Minimal thermal requirements for development and activity of stored product and food industry pests (Acari, Coleoptera, Lepidoptera, Psocoptera, Diptera and Blattodea): a review. *Insects* 10(5), 149. doi:10.3390/insects10050149.

Thorpe, G. R. 1997. Modelling ecosystems in entilated conical bottomed farm grain silos. *Ecological Modelling* 94(2–3), 255–86.

Timlick, B., Dickie, G. and McKinnon, D. 2002. Nitrogen as a major component of a controlled atmosphere to manage stored product insect pests in large vertical storage. *Integrated Protection of Stored Products, IOBC Bulletin* 25(3), 193–7.

UNEP. 2002. Montreal Protocol on Substances that Deplete the Ozone Layer, 2002 Assessment, Methyl Bromide Technical Options Committee. United Nations Environment Programme, Nairobi, Kenya.

UNEP. 2006. *Handbook for the Montreal Protocol on Substances that Deplete the Ozone Layer* (7th Edn). Ozone Secretariat United Nations Environment Programme, Nairobi, Kenya. Available at: http://ozone.unep.org/Publications/MP_Handbook/index.shtml.

U.S. Congress. 1989. *Enhancing the Quality of U.S.* Grain for International Trade, Office of Technology Assessment, Washington, DC.

Vertucci, C. W. and Roos, E. E. 1990. Theoretical basis of protocols for seed storage. *Plant Physiology* 94(3), 1019–23. doi:10.1104/pp.94.3.1019.

White, N. and Jayas, D. 1991. Control of Insects and Mites with carbon dioxide in wheat stored at cool temperatures in non-airtight bins. *Journal of Economic Entomology* 86, 1933–42.

White, N. D. G. and Jayas, D. S. 2003. Chapter 10: Controlled atmosphere storage of grain. In: Chakraverty, A., Majumdar, A. S., Raghavan, G. S. W. and Ramaswamy, H. S. (Eds), *Handbook of Postharvest Technology: Cereals, Fruits, Vegetables, Tea, and Spices*. Marcel Dekker Inc., New York, pp. 235–51.

White, N. D. G., Jayas, D. S. and Muir, W. E. 1995. Toxicity of carbon dioxide at biologically producible levels to stored-product beetles. *Environmental Entomology* 24(3), 640–7. doi:10.1093/ee/24.3.640.

Williams, P., Minett, W., Navarro, S. and Amos, T. G. 1980. Sealing a farm silo for insect control by nitrogen swamping for fumigation. *Australian Journal of Experimental Agriculture and Animal Husbandry* 20, 108–14.

# Chapter 9

# Biologically based control strategies for managing stored-product insect pests

*Benjamin Fürstenau and Garnet Marlen Kroos, Julius Kühn-Institut (JKI), Germany*

1 Introduction
2 Semiochemicals and their use in stored-product protection
3 Biological control of stored-product pests
4 Acknowledgements
5 Where to look for further information
6 References

## 1 Introduction

After harvest, many agricultural plant and food products, for example, cereals, oilseeds, legumes, or fruits, are dried and stored for a year-round supply. In the course of transportation and particularly during storage, even these non-perishables are at permanent risk of being attacked and damaged. Post-harvest losses amount to 5–10% worldwide, reaching up to 30% in tropical areas or developing countries (Kumar and Kalita, 2017; Pimentel, 2011; Saad and Abdelgaleil, 2018). Besides physical impacts such as heat or humidity, major threats are biological factors such as rodents, birds, mites, fungi, microorganisms and especially insects (Hagstrum and Phillips, 2017; Hengsdijk and de Boer, 2017; Hubert et al., 2018; Stejskal et al., 2015). Feeding damage and contamination of food and feed products by these pest organisms (e.g. by feces and chemical excretions, live or dead insects and body fragments) as well as mold growth and mycotoxins result in quantitative losses, spoilage or reduction in quality and imply health risks for humans and animals (Gram et al., 2002; Hagstrum et al., 2012; Keskin and Ozkaya, 2013; Stejskal et al., 2014). Therefore, pest insects contribute significantly to the global food loss estimated by the FAO at more than one billion tons per year (Gustavsson et al., 2011; Nayak and Daglish, 2018).

Since the 1940s, conventional chemical pesticides have been successfully adopted (Bale et al., 2008) and seemed to be the best instrument for industrialized agriculture to ensure a stable supply of affordable food and feed

http://dx.doi.org/10.19103/AS.2020.0072.12

in sufficient amounts and quality. Although plant protection products (PPPs) need to be evaluated before they are placed on the market, today chemical pest control strategies to protect stored products from insect infestation are considered critically due to the intrinsic hazard of these compounds and the risk they could pose to users, consumers and the environment (e.g. accumulation of chemical residues in food and human exposure to dangerous chemicals) (Barratt et al., 2018; Hagstrum and Flinn, 2014). As the safeguarding of the environment, soil and biodiversity plays a crucial role in modern society, agricultural policies have been reshaped to be fairer, greener and more innovative. These aspects will be recognized, for example, in the 2020 strategy of the European Common Agricultural Policy (CAP) and issues like biodiversity and sustainability play an important role in the European Union and worldwide (EU, 2019a).

Besides the greening aspects, there are two other developments influencing the general view on conventional chemical control. First, long-term use and misapplication may cause notable resistance against pesticides within pest populations (Chen et al., 2015; Daglish et al., 2013; Freitas et al., 2016; Nayak et al., 2017). Second, the number of authorized PPPs has decreased in Europe over recent years due to stricter laws and regulations. That, in turn, requires a steady optimization of existing plant protection methods in integrated pest management (IPM) strategies. More importantly, it has led to an increased demand for prospective alternatives to conventional synthetic pesticides and effective but also more environment-friendly products to suppress or kill pest populations. Although the continued sustainable use of PPPs is necessary to meet the global goals for food production by 2050 (FAO, 2006; Keulemans et al., 2019), the need for low-risk substances and products is evident in due consideration to agricultural, environmental and socioeconomic objectives.

## 1.1 Biologically based control strategies

As part of an ongoing effort to improve the safety of substances which are released into the environment and used on food and feed or for public hygiene, many countries create action plans, stewardship programs and regulations laying down a framework to achieve the sustainable use of pesticides (EC, 2009). These approaches promote an IPM with priority on the use of low-risk PPPs and other non-chemical techniques that have no or only minimal side effects on human health, non-target and beneficial organisms and on the environment. In addition, these regulations aim to minimize the risks of conventional pesticides, preserve their efficacy and invest in innovative and more favorable products such as biopesticides, biological control tools and next-generation technologies (e.g. RNA-interference). Wherever possible, sustainable biological, physical and other non-chemical alternatives should be preferred over chemical methods if they provide satisfactory pest control. In this context, weighting factors and harmonized risk indicators are useful for comparing conventional and low-risk pesticides (EU, 2019b).

Also in the protection of stored plant products, IPM concepts aim to reduce synthetic pesticides by replacing them with novel and environment-friendly approaches. These include biologically based control strategies (Fig. 1), which are based on the use of biopesticides such as semiochemicals (e.g. pheromones), botanical insecticides (e.g. pyrethrum), plant extract repellents and insect growth regulators (e.g. methoprene) or biological control agents such as entomophagous invertebrates (e.g. parasitoids or predators) and microbial insecticides including bacteria (e.g. *Bacillus thuringiensis* Berliner) and fungi (e.g. *Beauveria bassiana* Vuellemin and *Metarhizium anisopliae* Sorokin) (Hagstrum and Phillips, 2017; Phillips and Throne, 2010). The present chapter focuses on the use of semiochemicals and that of natural enemies of pest insects (i.e. parasitoids). Both alternative management tools are generally considered as relatively safe and environmentally more acceptable than conventional PPPs (Riudavets, 2018; Trematerra, 2012).

Semiochemicals and especially pheromones act by modifying the behavior of a target (pest) organism rather than killing it. In general, they are used at concentrations close to those in nature and increasingly efficient at low population densities. Since pheromones are more species-specific than conventional insecticides, they do not adversely affect other organisms such as natural enemies of the pest species. Moreover, pheromones possess low acute

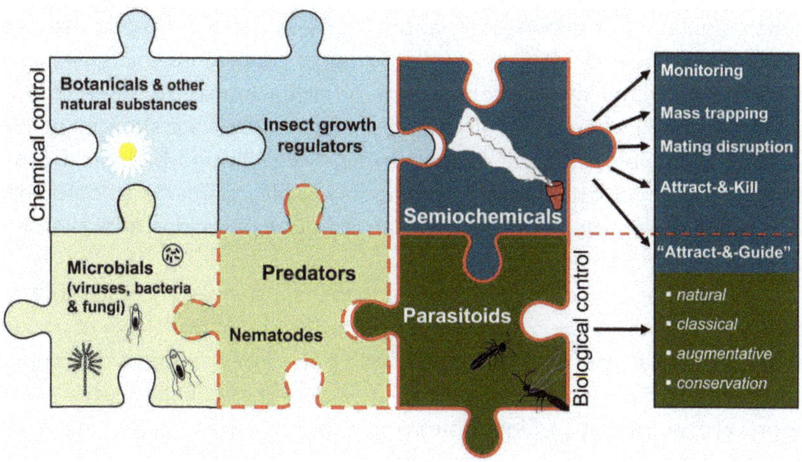


**Figure 1** Biologically based control strategies for the management of stored-product pests based on the use of biopesticides and biological control agents. Puzzle pieces framed in red represent the two alternative control tools which are discussed in detail in this chapter (i.e. semiochemicals and parasitoids).

mammalian toxicity and are usually volatile chemicals that dissipate rapidly and do not cause harmful residues (Savoldelli and Trematerra, 2011; Witzgall et al., 2010). For all these reasons, it is assumed that most semiochemical products present a lower potential risk to human health and the environment than conventional pesticides (ENV/JM/MONO, 2017).

Biological control has proven to be a technically and economically feasible alternative to chemical control and has been adopted successfully for managing agricultural pests on more than 30 million ha worldwide (van Lenteren et al., 2018). Therefore, the introduction of natural enemies such as parasitoids into the stored-product environment is considered another promising alternative to the use of pesticides (Solà, 2017). Parasitoids feed exclusively on their hosts and are capable of locating host individuals in hiding places, even at low (pest) population densities. They continue to reproduce as long as hosts are available and leave no noxious chemical residues (Flinn and Schöller, 2012). Further advantages over conventional chemical control measures are, for example, that work does not have to be interrupted in the treated facility since neither the production process nor employees are affected by the presence of biological control agents.

In this chapter, we will summarize the state of research of these two biologically based approaches for managing stored-product pest insects and discuss how to improve the protection of post-harvest food and feed products in the future by combining several (alternative) control strategies.

## 1.2 Definitions of biopesticides and related terms

The emergence of natural resistance and withdrawal of several products for either regulatory or commercial reasons have led to a reduced assortment of chemical pesticides on the market for crop protection. This is especially a problem for the minor crops sector or niche areas like stored-product protection (SPP) where alternative crop protection tools are strongly needed. In this context, biopesticides such as semiochemicals and biological control agents have an important and increasing role to play, primarily in combination with other tools including conventional chemical pesticides and alternative techniques as part of IPM.

In principle, not only conventional active substances and PPPs but also biopesticides, which are intended to be placed on the market, have to undergo an intensive evaluation, risk assessment and peer review before a decision can be made on national country-specific or wider cross-country and interzonal approval or authorization. Since both pesticides and biopesticides have the same purpose in preventing, destroying, repelling or mitigating any harmful organism, they shall follow the same uniform principles for evaluation and authorization (Leahy et al., 2014). If the active substances of biopesticides are classified as low-risk products, derogations from general regulations and options for fast-track procedures, specified new criteria and longer approval periods will be possible (e.g. up to 15 years in the EU). Such authorization processes, data requirements and evaluation criteria often differ from country to country. Actual information on agricultural pesticides and biocides, low-risk status and biopesticides, as well as an overview of national authorities and international organizations, is made available by the Organization for Economic Co-operation and Development (OECD). In an international context and recognizing the global trade, it would be necessary and beneficial to harmonize definitions and related assessment procedures in order to strengthen cross-nation reliability in food and feed safety.

However, terms and categories related to biopesticides (Fig. 1) are still under discussion and no formally agreed definition exists (Chandler et al., 2011). Therefore, the following descriptions just show four different classifications (I-IV) which consider the European and US American authorization process. The overview is limited to the main existing criteria and wording of the original terms and statements, especially those which are laid down in various legal regulations and guidance documents of the official authorities (as at August 2019), even knowing that they are only binding in the scope of the respective document.

I) The *Environmental Protection Agency (EPA)* uses a general descriptive term, which is not legally defined but forms the basis for the categorization within the registration process (Leahy et al., 2014).

*Biopesticides* (= biological pesticides) are pesticides if any active substance or mixture of substances is intended for preventing, destroying, repelling or mitigating any pest. Pesticides are also plant regulators, defoliants, desiccants or nitrogen stabilizers. They are derived or contain active ingredients from certain natural materials such as animals, plants, bacteria and certain minerals. The list of adopted biopesticidal substances is published on https://www.epa.gov/ingredients-used-pesticide-products/biopesticide-active-ingredients.

Biopesticides are divided into three main classes:

1. *Biochemical pesticides* are naturally occurring chemicals or synthetically derived equivalents that have a history of exposure to humans and the environment demonstrating minimal toxicity, or in the case of synthetically derived biochemical pesticides, are equivalent to a naturally occurring chemical that has such a history (CFR, 2018a). Biochemical pesticides have a non-toxic mode of action to the target pest(s) and include, but are not limited to (a) semiochemicals (insect pheromones and kairomones), (b) natural plant and insect regulators, (c) naturally occurring repellents and attractants and (d) enzymes.

   A *pheromone* is a compound produced by a living organism or is a synthetically derived substance that is structurally similar and functionally identical to a naturally occurring pheromone which, alone or in combination with other such compounds, modifies the behavior of other individuals of the same species (Fig. 2).

2. *Microbial pesticides* consist of a microorganism (e.g. bacterium, fungus, virus or protozoan) as the active ingredient (CFR, 2018b). Microbial pesticides can control many different kinds of pests, although each separate active ingredient is relatively specific for its target pest(s). The most widely used microbial pesticides are subspecies and strains of *B. thuringiensis*. Microbial pesticide is a microbial agent including (a) eucaryotic microorganism (e.g. protozoa, algae and fungi), (b) procaryotic microorganism (e.g. Eubacteria and Archaebacteria) and (c) parasitically replicating microscopic element (e.g. viruses).
3. *Plant-incorporated protectants (PIPs)* are pesticidal substances that plants produce from genetic material that has been added to the plant. According to the US Code of Federal Regulations, PIP means a pesticidal substance that is intended to be produced and used in a living plant, or in the produce thereof, and the genetic material necessary for the production of such a pesticidal substance. It also includes any inert ingredient contained in the plant or produce thereof (CFR, 2018c).

*Pest control organisms* such as insect predators, nematodes and macroscopic parasites are exempt from the requirements of Federal Insecticide, Fungicide, and Rodenticide Act (FIFRA) (CFR, 2019) as long they are adequately regulated by

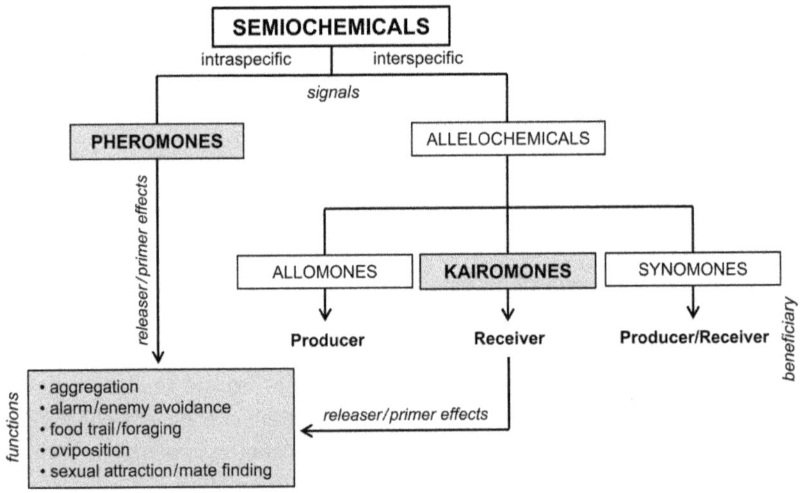


**Figure 2** Classification and some possible functions of semiochemicals modified after Mori (2010) and Ruther et al. (2002). Effects on the receiving (benefiting) organism: releaser effect = inducing (immediate) behavioral responses; primer effect = inducing physiological changes.

the Plant Protection Act. The above-mentioned control agents are not classified as a pesticide. Therefore, their movement and release are regulated. The use of pheromones in traps in conjunction with conventional pesticides, or in other application methods (other than traps), is subject to regulation under FIFRA.

II) The *European Union's (EU) Plant Protection Regulation 1107/2009* does not recognize biopesticides as a special regulatory category of plant protection active ingredients or products but considers microorganisms, semiochemicals and botanicals.

A *biopesticde* that prevents, destroys or controls a harmful organism ('pest') or disease, or protects goods during production, storage and transport, is a pesticide. The term includes, among others, herbicides, fungicides, insecticides, acaricides, nematicides, molluscicides, rodenticides, growth regulators, repellents and rodenticides. The EU legislation does not regulate pesticides as a whole but distinguishes two regulatory systems for PPPs (protect plants, parts of plants and plant products) and biocides, including treated articles (protect food, feed and materials and is used for human and animal health care).

The categories of products related to biopesticidal PPPs are the following:

- *Microorganism* means any microbiological entity (including lower fungi and viruses), cellular or non-cellular, capable of replication or of transferring genetic material (EU, 2013). The term 'microorganism' as defined in Article 3 of Regulation (EC) No 1107/2009 applies to but is not limited to bacteria, fungi, protozoa, viruses and viroids.
- *Semiochemical active substance* refers to active substances that are emitted by plants, animals and other organisms and are used by these organisms for communication and evoke a behavioral or physiological response in individuals of the same or other species (Fig. 2), as well as substances referred to as natural-identical synthesized molecules (SANTE, 2016). Semiochemicals are not considered as active substances when they are used only to attract arthropods, which subsequently receive a lethal dose of an insecticide or are killed by other means. In general, semiochemicals used in traps to attract arthropods only for monitoring purposes are exempt from registration.
- The term 'botanical active substance' covers a very heterogeneous group of substances, ranging from simple plant powders to unprocessed and processed plant extracts. Furthermore, plant extracts may be highly refined (i.e. one single active substance) or represent a complex mixture of components, of which all or only some are biologically active (SANCO, 2014). They refer to active substances obtained from plant material. Therefore, substances referred to as analogs, mimics, natural-identical synthesized molecules and biosimilars are not covered by this guidance document.

A botanical active substance consists of one or more components found in plants and is obtained by subjecting plants or parts of plants of the same species to a process such as pressing, milling, crushing, distillation or extractions. The process may include further concentration, purification and blending, provided that the chemical nature of the components is not intentionally modified by chemical and/or microbial processes. Botanical active substances derived from genetically modified organisms (GMOs) are not evaluated under the assumption of biopesticides.

*Macroorganisms* are not addressed under the European regulation of PPPs because this regulation applies just to substances, including microorganisms. However, biological control agents may be regulated on the national level, for example, in Austria and Hungary. If the biological control agents are not considered as a PPP, its introduction, release and movement have to be regulated to avoid hazards for the agricultural production and ecosystem. Especially non-indigenous biological control agents have to undergo a risk assessment. The EPPO Standard PM 6/3 lists biological control agents widely used in the EPPO region (EPPO, 2016).

III) The *Food and Agriculture Organization of the United Nations* (FAO) has published its definitions of biopesticides and biological control agents in the scope of the 'International Code of Conduct on Pesticide Management' and the glossary of phytosanitary terms.

There is no globally agreed definition of *biological pest control agents* or so-called *biopesticides*. 'Biopesticide' is a generic term generally applied to active substances derived from nature, such as a microorganism, botanical or semiochemical, that may be formulated and applied in a manner similar to a conventional chemical pesticide and that is normally used for short-term pest control.

For the purposes of the above-mentioned guideline for the registration of microbial, botanical and semiochemical pest control agents for plant protection and public health uses, these terms will include products with active substances that are based on microbials, botanicals or semiochemicals. These substances are distinguished from conventional chemical pesticides by a combination of their active substance material and/or nature and their use (WHO and FAO, 2017).

Biological control agent means a natural enemy, antagonist or competitor, or other organisms, used for pest control (FAO, 2019).

IV) The *OECD Program on Biological Pesticides* differentiates four categories of biopesticides (biological pesticides) (ENV/JM/MONO, 2003, 2004, 2017):

1. *Microbials*: bacteria, algae, protozoa, viruses, fungi.

2. *Semiochemicals* are chemicals produced by plants and animals (and synthetic analogs of such substances) which influence the behavior of individuals of the same or other species (Fig. 2). Pheromones modify the behavior of individuals of the same species. Allelochemicals modify the behavior of individuals of a different species.
3. *Macrobials/invertebrates* or *invertebrate biological control agents* include insects, mites and nematodes that are natural enemies, antagonists or competitors of a pest.
4. *Plant extracts/botanicals.*

All the above-mentioned terms are defined in the scope of special regulations. It has to be underlined that the terms may be used in similar, but not identical, ways and differ from country to country or between member states. Until further harmonization, applicants, administrators and stakeholders have to recognize carefully which pesticides are regulated and how products are defined in the relevant authorization process and plant protection legislation.

## 2 Semiochemicals and their use in stored-product protection

Every living organism receives information from its habitat and simultaneously emits information into the environment, which, in turn, may be perceived and used by others. This intra- and interspecific information exchange among organisms within and between different trophic levels (= *multitrophic interactions*) proceeds via semiochemicals, which are often volatile organic compounds (VOCs) of low to medium molecular weight but can also be less- or non-volatile (contact) chemicals (Law and Regnier, 1971; Nordlund and Lewis, 1976). These behavior-modifying substances can be further divided into pheromones, which act between organisms of the same species, and allelochemicals, which act between organisms of different species (e.g. Dicke and Sabelis, 1988; Mori, 2010; Ruther et al., 2002) (Fig. 2).

Semiochemicals are present in all environments, including man-made habitats such as food stores, silos or warehouses, and are released from many different sources. As such, and since they hold considerable potential for pest management, semiochemicals also provide various opportunities for the control of stored-product pests (Anand et al., 2015; Bell, 2014; Cox, 2004; Hagstrum and Phillips, 2017; Trematerra, 2012). Due to their ability to attract a wide range of insect species, semiochemicals can be used to monitor specific pest populations as well as are assumed to reduce infestations by mass trapping and methods based on attract-and-kill or by mating disruption (MD) (Fig. 1) (Cox, 2004; Phillips, 1997; Savoldelli and Trematerra, 2011). While detection of a specific pest gives information only about its presence, monitoring is the collection of all relevant data regarding trap catches over time in a defined place (e.g. number, sex, flight curves).

Population monitoring by traps baited with aggregation or sex pheromones is already a standard today for a multitude of important agricultural and stored-product pests. It enables the user to determine initial insect infestations earlier than by visual inspections or direct sampling. Food odors which have been shown to attract stored-product insects (Fouad et al., 2013; Mikolajczak et al., 1984; Ndomo-Moualeu et al., 2016; Phillips et al., 1993) can be combined with pheromones in order to increase the attractiveness of the lure (Athanassiou et al., 2006; Campbell, 2012; Stevenson et al., 2017).

However, although a great number of semiochemicals of several pest species have been identified and the advantages of using these compounds alternatively or additionally for SPP have been highlighted many times, the existing potential is not exhausted. Most literature on the use of semiochemicals deals mainly with only two parties and their mutual effects on each other without, in many cases, looking at the multitrophic context. Actually, very few of the known compounds have been launched to the market and have been used in SPP so far. Here we will provide some possible explanations and discuss the existing knowledge about bioactive VOCs present in stored-product environments with a special focus on pheromones and its use in biologically based SPP.

## 2.1 Pheromones of stored-product pests

Pheromones are substances which promote communication between individuals of the same species eliciting (immediate) behavioral responses in the receiver (= *releaser or signaler pheromone*) or causing physiological changes such as sexual maturation (= *primer pheromone*) (Fig. 2) (Cardé and Millar, 2009; Hartlieb and Anderson, 1999; Karlson and Lüscher, 1959; Regnier and Law, 1968). According to the type of behavioral change induced, releaser pheromones can be further classified, for example, into sex, aggregation or trail pheromones (Mori, 2010). In general, pheromones consist of more than a single chemical component and are produced by a variety of glands located in different parts of the insects' body. They are released in minute amounts and with the exception of most short-range or contact pheromones they are highly volatile chemicals commonly used as long-range cues (Kaissling, 2014; Levinson and Levinson, 1995). Depending on species and function, both males and females can be the producer of pheromones (Bell, 2014; Burkholder and Ma, 1985; Silverstein, 1981; Trematerra, 1997; Witzgall et al., 2010).

In the 1960s the newly reemerging field of chemical ecology began to prosper (Hartmann, 2008), and practical applications were envisioned with special emphasis on interfering in the chemical communication of pest organisms. Because of their ability to alter insect behavior (e.g. by attracting, repelling or confusing), pheromones are considered as an additional tool for IPM to reduce or control pest populations and, thus, minimize the risk of further damage. In agriculture, under field conditions, several biologically active compounds are already used as biopesticides to control herbivorous pests and vectors of diseases and, thus, represent a successful alternative to conventional PPPs (Allison and Cardé, 2016; Gut and Miller, 2015; Witzgall et al., 2010).

Pheromones of many important pest organisms infesting stored products have also been identified over the last decades, and their use in SPP is still a very promising, albeit only partially applied, approach (Anand et al., 2015; Burkholder and Ma, 1985; Losey et al., 2019; Phillips, 1997; Phillips and Throne, 2010; Trematerra, 1997, 2012). Actually, pheromones of more than 30 species of stored-product pest insects are commercially available (Bell, 2014; Swords and Van Ryckeghem, 2010a,b). But, as mentioned earlier, the potential of these bioactive compounds to alternatively protect stored products from insect infestations has not yet been fully exploited.

Most insect pheromones are highly specific to a target species and represent a very sensitive and reliable tool for insect detection. This is also a disadvantage, if a different chemical or pheromone trap must be deployed for pest control and monitoring for each pest species. However, among some stored-product pest beetle families (e.g. Bostrichidae, Dermestidae and Tenebrionidae), cross-species attraction has been shown. In these cases, the involved pheromone components can attract more species than only the producing one (Cox, 2004; Cox and Collins, 2002). Traps baited with lures containing pheromones of various insects are another usual feature of IPM strategies as they attract simultaneously different pest species (e.g. Barak and Burkholder, 1985).

There are two main pheromone systems in stored-product insects (Levinson and Levinson, 1995; Trematerra, 2012). One refers to those species whose adults usually live for less than a month and do not need feed to reproduce while their larval stages require feeding for growth and development. These insects are mostly moths in which generally the adult females produce a volatile pheromone that attracts individuals of the opposite sex. The second pheromone system describes those insects whose adults live longer than a month and have to feed to reproduce. These species belong exclusively to the order Coleoptera, in which mostly males produce an aggregation pheromone that attracts both sexes (Barak and Burkholder, 1985; Benelli et al., 2017; Holloway et al., 2018; McKay et al., 2019; Yang et al., 2017).

### *2.1.1 Sex pheromones in species with short-lived adults*

Primarily, stored-product moths and particularly pyralid species of the subfamily Phycitinae rely on female-produced sex pheromones for mate finding. The chemical structures of Lepidopteran sex pheromones are simple but diverse. The components are usually straight-chain compounds, that is, unsaturated C10–C18 fatty alcohols and their derivatives (acetates and aldehydes) with one or more double bonds or C17–C23 polyenyl hydrocarbons and their epoxides (Ando et al., 2004; Ando and Yamakawa, 2011). The sex pheromone of the Angoumois grain moth *Sitotroga cerealella* Olivier (Lepidoptera: Gelichiidae), for example, is (*Z*,*E*)-7,11-hexadecadienyl acetate (Vick et al., 1974), and its efficacy to capture males in stored rice has been shown (Akter and Ali, 2016).

Different moth species from one family may produce the same pheromone components for the attraction of sex mates, but the odor cocktail usually differs. This, for instance, is the case in several pyralids. The Indian meal moth *Plodia interpunctella* Hübner shares (*Z*,*E*)-9,12-tetradecadienyl acetate (Z9,*E*12-14:Ac, commercially known as TDA or ZETA) as the main component of its sex pheromone, among others, with the almond moth *Cadra cautella* Walker, an important pest of a wide range of tropical stored products (Brady, 1973; Brady et al., 1971; Kuwahara and Casida, 1973; Kuwahara et al., 1971), and with the Mediterranean flour moth *Ephestia kuehniella* Zeller, one of the major pests in European food-processing industries (Trematerra and Gentile, 2010). Further components of the *P. interpunctella* female-produced sex pheromone are (*Z*,*E*)-9,12-tetradecadienal (*Z*9,*E*12-14:Ald), (*Z*,*E*)-9,12-tetradecadienol (*Z*9,*E*12-14:OH) and (*Z*)-9-tetradecenyl acetate (*Z*9-14:OAc) (Teal et al., 1995; Zhu et al., 1999).

Not only females but also males of stored-product moths produce compounds with pheromonal activity. Males of the warehouse moth *E. elutella* Hübner (Lepidoptera: Pyralidae), *C. cautella* and *P. interpunctella* emit a courtship pheromone that attracts females (Grant, 1976; Grant and Brady, 1975; Krasnoff and Vick, 1984). In the case of *E. elutella* it consists of (*E*)-phytol and saturated γ-lactones (Phelan et al., 1986). However, any potential application has not yet been studied, which should be done in future investigations.

### 2.1.2 Aggregation and sex pheromones in species with long-lived adults

It is assumed that the release of pheromones by (male) beetles is closely related to feeding or contact with food (Phillips and Throne, 2010). In most cases, these are aggregation pheromones, which mostly consist of long-chain unsaturated acids, alcohols, aldehydes, acetates and esters with one or more double bonds in *E* or *Z* configuration, forming multiple isomers. Over the last decades, several pheromones of pest beetles have been identified. The commercially available pheromone components, as well as the suppliers of formulated lures for beetles infesting stored products, are summarized in Swords and Van Ryckeghem (2010a).

The main component of the aggregation pheromone of the confused flour beetle *Tribolium confusum* Jacqueline du Val (Coleoptera: Tenebrionidae) is 4,8-dimethyldecanal (DMD) (*tribolure*) (Suzuki, 1980; Suzuki and Sugawara, 1979). Verheggen et al. (2007) identified further compounds with electrophysiological and behavioral activity (i.e. benzoquinones and unsaturated hydrocarbons), which may also be components of the aggregation pheromone of flour beetles.

Males of the granary weevil *Sitophilus granarius* Linnaeus (Coleoptera: Dryophtoridae) produce an aggregation pheromone, (2S,3R)-1-ethylpropyl-2-methyl-3-hydroxypentanoate (*sitophilate*), and a second component not yet identified (Chambers et al., 1996; Phillips et al., 1989). Yang et al. (2017) described squalene as the male-produced aggregation pheromone in the coffee bean weevil *Araecerus fasciculatus* De Geer (Coleoptera: Anthribidae).

Nevertheless, and although not belonging to the first pheromone system of short-lived species, there are also quite a number of known sex pheromones in long-lived stored-product pest beetles. Rodríguez (2018) recently reviewed the results on the chemical identification of pheromones of bruchids (Chrysomelidae), most of them involved in mate finding. The adzuki bean weevil *Callosobruchus chinensis* Linnae, for example, is a field-to-storage pest of legumes, whose females produce a sex pheromone known as *homofarnesal* with two isomeric components, 2*E*- and 2*Z*-homofarnesal, (2*E*,6*E*)-7-ethyl-3,11-dimethyl-2,6,10-dodecatrienal and (2*Z*,6*E*)-7-ethyl-3,11-dimethyl-2,6,10-dodecatrienal (Chiluwal et al., 2018; Shimomura et al., 2008).

The male-produced pheromone of the dried bean beetle *Acanthoscelides obtectus* Say (Coleoptera: Chrysomelidae) had been described as methyl (2*E*)-2,4,5-tetradecatrienoate (Horler, 1970; Mori, 2012), but no successful attraction of females in the laboratory could be demonstrated. However, a study published 50 years later showed that the volatile blend released by males, which proved to be active in olfactometer bioassays and attracted virgin females, comprises a total of six components: methyl (*E*,*R*)-2,4,5-tetradecatrienoate, methyl (2*E*,4*Z*,7*Z*)-2,4,7-decatrienoate, methyl (2*E*,4*Z*)-2,4-decadienoate, octadecanal and the sesquiterpenes (3*Z*,6*E*)- and (3*E*,6*E*)-*α*-farnesene (Vuts et al., 2015). These newly identified components of the *A. obtecus* sex pheromone are supposed to help developing an effective semiochemical-based control strategy against this important pest in the future. This example shows that in-depth research and follow-up studies, even many years after the first discovery of a certain (bioactive) compound, are highly important in order to understand the exact role of pheromones and verify their efficacy for commercialization and subsequent application in specific control programs.

The main sex pheromone component of *Trogoderma* spp. females, (*Z*)-14-methyl-8-hexadecenal (*trogodermal*) (Mori et al., 1978), can be used to capture the harmful Khapra beetle *T. granarium* Everts (Coleoptera: Dermestidae)

(Upadhyay and Ahmad, 2011). In the yellow mealworm beetle *Tenebrio molitor* Linnaeus (Coleoptera: Tenebrionidae), both a female-produced (4-methyl-1-nonanol) and a male-produced ((*Z*)-3-dodecenyl acetate) sex pheromone could be identified, each attracting the opposite sex (Bryning et al., 2005).

## 2.2 Effects of semiochemicals on the stored-product environment

The use of semiochemicals in biologically based SPP may also imply miscellaneous side effects. For example, the presence of compounds with pheromonal activity can influence the stored-product environment and community, apart from mediating intraspecific communication. Stored goods may be contaminated or directly affected, and other insects than the emitter can perceive and use this information for their own benefit.

The male-produced aggregation pheromones of the lesser grain borer *Rhyzopertha dominica* Fabricius (Coleoptera: Bostrichidae), (*S*)-(+)-1-methylbutyl (*E*)-2-methyl-2-pentenoate and (*S*)-(+)-1-methylbutyl (*E*)-2,4-dimethyl-2-pentenoate (*dominicalure 1 and 2*) (Williams et al., 1981), as well as 2-pentanol, may lead to severe changes in odor and baking properties of wheat kernels (Khorramshahi and Burkholder, 1981; Sánchez-Mariñez et al., 1997; Seitz and Sauer, 1996). The malodorous exudates of *Tribolium* spp. can turn flour pink and render food toxic and unsuitable for human or animal consumption (El-Mofty et al., 1989; Işikber and Öztekin, 2009). Quinonic secretions, with predominantly para-benzoquinones (e.g. 2-methyl-1,4-benzoquinone and 2-ethyl-1,4-benzoquinone), are widespread within the beetle family Tenebrionidae, and in *Tribolium* spp. they are produced by the adults in defense against predators and pathogens (Attygalle et al., 1991; Geiselhardt et al., 2009; Hassemer et al., 2015). Despite their carcinogenic and enterotoxic activity (Lis et al., 2011), it has recently been shown that benzoquinones can be successfully applied in a push-and-pull strategy (Hassemer et al., 2019).

Pheromones can be eavesdropped by natural enemies of the releasing insect and used as foraging kairomones for host finding (Belda and Riudavets, 2010; Godfray, 1994; Quicke, 1997; Steidle and Van Loon, 2003) (see section 3.4). In this context, it has been demonstrated that the predatory beetle *Teretrius* (synonym *Teretriosoma*) *nigrescens* Lewis (Coleoptera: Histeridae) is strongly attracted to the aggregation pheromone of the larger grain borer *Prostephanus truncatus* Horn (Coleoptera: Bostrichidae), denominated *trunc-call 1+2* (1-methylethyl (*E*)-2-methyl-2-pentenoate and 1-methylethyl (*E,E*)-2,4-dimethyl-2,4-heptadienoate) (Birkinshaw et al., 2004; Cork et al., 1991).

Several egg parasitoids in the genus *Trichogramma* (Hymenoptera: Trichogrammatidae) rely on sex pheromones of their hosts for host finding such as *T. evanescens* Westwood, which responds to the pyralid pheromone main component TDA (Noldus et al., 1991; Schöller and Prozell, 2002). Females of the larval ectoparasitoid *Bracon* (synonym *Habrobracon*) *hebetor* Say (Hymenoptera: Braconidae) use the male-produced sex pheromone of its host, the greater wax moth *Galleria mellonella* Linnaeus (Lepidoptera: Pyralidae) as an indirect cue for the localization of possible oviposition sites (Dweck et al., 2010). Furthermore, compounds from mandibular glands of host larvae (2-acylcyclohexane-1,3-diones) have been shown to induce oviposition movements, arrestment and trail following in *B. hebetor* (Corbet, 1971; Strand et al., 1989).

## 2.3 Application of semiochemicals

As mentioned before, approved and commercially available bioactive compounds may be utilized for the implementation of four main methodologies in semiochemical-based management of stored-product pests: (i) monitoring, (ii) mass trapping, (iii) mating disruption (MD) and (iv) attracticide methods (Fig. 1) (Bell, 2014; Cox, 2004; Savoldelli and Trematerra, 2011). In any case, the success of trapping a target pest species strongly depends on the degree of insect attraction to a specific lure (Stevenson et al., 2017).

In order to attract insects, synthetic sex and aggregation pheromones alone or in combination with food attractants are usually impregnated in rubber septa placed into a specific trap but can also be mixed directly into the insect glue of sticky traps. Since more than 30 years, semiochemical traps or slow-release formulations of lures baited with pheromones, kairomones and food-based VOCs for detecting and monitoring pest populations have been commercially available. Nowadays, they are an essential and important part of routine surveillance or control measures of stored-product pests (Campbell et al., 2002; Hagstrum et al., 2012; Sambaraju and Phillips, 2008; Swords and Van Ryckeghem, 2010a,b).

Apart from detecting the presence of pest individuals or estimating the degree of the infestation, semiochemical-baited traps can help to get information about temporal and special patterns in flight activities and seasonal abundance of stored-product insects (Athanassiou et al., 2018; Campbell et al., 2002; Campbell and Mullen, 2004; Daglish et al., 2017; McKay et al., 2019). Different trap types, its application and the corresponding literature sources were reviewed by Anand et al. (2015). However, it should be taken into account that behavioral variations due to changing environmental conditions or within different populations of stored-product insects, as well as changing behaviors to trapping devices due to age, sex, type of lure and the presence of other insects, may significantly reduce the effectiveness of many of these approaches (Campbell, 2012; Gerken et al., 2018; Semeao et al., 2011).

In some cases, pheromones have shown their potential to be used for MD of stored-product moths. In commercial field settings, the main component of the pyralid sex pheromone, TDA, demonstrated to disrupt the mating of *P. interpunctella, E. kuehniella* and *C. cautella* (Athanassiou et al., 2016; Burks et al., 2011; Ryne et al., 2006; Sieminska et al., 2009; Trematerra et al., 2011, 2017; Trematerra and Savoidelli, 2013). The principle of MD is that numerous pheromone dispensers are equally distributed in a defined space, in order to produce a large number of pheromone plumes (Arthur, 2018). In this way, storage facilities can be treated under controlled conditions with huge amounts of pheromones to saturate the environment with unnaturally high concentrations of these synthetic compounds so that finally males are confused and not (or only hardly) able to find their mating partner. Further pheromone research with regard to MD of stored-product pests has been done, for example, on the cigarette beetle *Lasioderma serricorne* Fabricius (Coleoptera: Ptinidae) using the synthetic pheromone *serricornin* (7-hydroxy-4,6-dimethylnonan-3-one) (Mahroof and Phillips, 2014).

Moreover, TDA has been described to suppress pest populations through mass trapping or attracticide methods and to act as a repellent and behavioral stimulant or deterrent (Campos and Phillips, 2014; Savoldelli and Trematerra, 2011; Trematerra and Gentile, 2010). However, complete elimination of moth infestations was never achieved. This efficacy gap between biopesticides and conventional chemical insecticides (Chandler et al., 2011) often discourages farmers and retailers from using these alternative (low-risk) compounds as a common control tool during storage and trade of post-harvested food and feed products. For this reason, and despite several studies on successful application, pheromones are mainly used only as a monitoring tool for stored-product pests (Athanassiou and Arthur, 2018; Cox, 2004; Mohandass et al., 2007). Besides, many recent publications with regard to semiochemical research primarily deal with the monitoring of pest insects and not their control (Athanassiou et al., 2018; Baek et al., 2015; Belda and Riudavets, 2013; Holloway et al., 2018; Losey et al., 2019; Miyatake et al., 2016), so that new findings and further knowledge about the aforementioned alternatives are rather sparse.

A major issue is the difficulty of correctly assessing the effectiveness of semiochemical-based methods such as mass trapping, MD or attract-and-kill. This is principally due to methodological problems, as, for example, the lack of identical control facilities and a sufficient number of replicates (Trematerra, 2018), as well as missing harmonized test guidelines. One promising trial design for the efficacy evaluation of MD has been described in the EPPO standard 'mating disruption pheromones' (EPPO, 2008), but, in general, and as described earlier, there are few existing field studies with regard to MD of stored-product pests (e.g. Trematerra et al., 2017). The main problem and possible obstacle in the (routine) implementation of MD into practice is the simultaneous monitoring of the target pest population, which is necessary for evaluating the success of this alternative control method. For this, it is crucial to find or select the right lures for both of the approaches, MD and monitoring, without one method influencing the other. Relating thereto, it has already been assumed that the monitoring of pyralid moths using pheromone traps may be affected by MD (Trematerra et al., 2017). Alternatively, oviposition traps have been tested to evaluate the offspring production of *P. interpunctella* females as well as water traps, which, however, also exercised control effects (Trematerra and Savoldelli, 2013).

The attractant generally used for MD and monitoring of stored-product moths is mainly limited to TDA, although the composition of pheromones is highly diverse in most cases. In addition to the main component, the sex pheromone of *P. interpunctella* consists of at least three further active compounds (Zhu et al., 1999). In field trapping experiments, the four-component blend attracted significantly more males of *P. interpunctella* than traps baited with Z9,*E*12-14:OAc alone. In contrast, the highest number of *E. kuehniella* males was trapped with the major component, suggesting that minor pheromone components may be responsible for the species specificity of the blend.

Referring to this, the sex pheromone of the leafminer, *Phyllocnistis citrella* Stainton (Lepidoptera: Gracillariidae), an important citrus field pest in the USA, contains three components of which only two compounds (a triene and a diene in a 3:1 ratio) are sufficient for attraction in the field. MD experiments revealed that the triene component alone worked as well as or even better than the 3:1 blend, which is still used for monitoring (Lapointe et al., 2015, 2011). For future applications and studies dealing with the use of semiochemicals in SPP, this must be kept in mind and tested extensively.

## 2.4 Future trends in semiochemical research

Generally, more bioactive compounds must be found and integrated into modern SPP to meet the IPM objectives as well as the greening aspects. VOCs from (infested) grains or further processed plant products can be used to find alternative attractants or repellents for the monitoring and control of stored-product insects. For example, laboratory tests have shown that the red flour beetle *Tribolium castaneum* Herbst (Coleoptera: Tenebrionidae) is attracted to grains infested with *R. dominica* and flour infested with conspecifics but is repelled by grains infested with *S. granarium* (Stevenson et al., 2017). In this regard, identification of new semiochemicals and determination of their role in intraspecific interactions can help protect stored products from pests, for example, through push-pull tactics achieved by combining repellents and attractants. This approach has been recently shown for the lesser mealworm *Alphitobius diaperinus* Panzer (Coleoptera: Tenebrionidae), an important pest in the poultry production. Para-benzoquinones, which are generally produced by the adults for defense or enemy avoidance, were combined with the aggregation pheromone, and significant numbers of pest beetles were trapped (Hassemer et al., 2019).

In contrast to laboratory studies that mostly analyze grain odor when the sample is artificially heated or damaged, it is necessary to assess VOCs that naturally occur in the stored-product environment, for example, those compounds which are released from stored products. Since odor profiles of stored goods can alter upon the presence of (pest) insects (e.g. pest-derived or feeding-induced odors and volatiles originating from beneficial insects or fungi/mold), it would be highly beneficial to monitor this change of scent for early detection of an infestation.

Ubiquitous VOCs, as well as specific feeding-induced odors, could play an important role as general indicators for product quality, insect infestation or fungal spoilage of stored goods (Morath et al., 2012). Fungal odors may also be used as monitoring tools to attract pest beetles (Pierce et al., 1991) or to attract parasitoids for biological control (Steiner et al., 2007). Whether feeding or egg deposition by stored-product insects actually induces odor profile changes in stored products, equally to herbivore-infested plants in the field, needs to be studied more extensively. Also, already-known bioactive VOCs, which are normally not part of the stored-product odor environment such as citronellal (*rhodinal*, 3,7-dimethyl-6-en-1-al) or citral (*lemonal*, 3,7-dimethyl-2,6-octadienal), have shown its potential to be used in SPP by attracting *L. serricorne* adults (Lü and Liu, 2016).

## 2.5 Conclusions

Although a multitude of semiochemicals are already known, just a few are commercially available or registered. Moreover, most of them, in particular, pheromones, are primarily applied for the monitoring of stored-product pests and not their control. One of the main reasons for this is the existing efficacy gap between semiochemicals and conventional chemical pesticides. A better understanding of the mode of action and higher efficacy of these biopesticides may help raise their acceptance among farmers and policymakers and, hence, point out their part in an IPM and sustainable SPP. Adapted data requirements, globally harmonized test standards and a prioritized fast track authorization for low-risk substances within the existing system may strengthen the development, commercialization and use of biopesticides. Major objectives for future semiochemical research shall be the identification of new bioactive VOCs present in the stored-product environment, verification of their potential to control pest insects under natural (stored-product) conditions as well as the development of novel biopesticides (e.g. effective mixtures of attractants/repellents) for the great number of species-specific uses. Once placed on the market, all products must comply with the control level in force, which may vary in commercial standards of cereal quality, legal regulations of food hygiene and requirements of the retailers. To attain these claims, a wide expert knowledge of the biology of the harmful organism, the active substance and pest management is required. Overall, due to consumer demands and political trends, it can be assumed that biopesticides (i.e. semiochemicals) will play a key part in future IPM strategies. In this context, it is very likely that their role in agriculture in general, but also in the protection of stored-product pest insects will increase significantly.

## 3 Biological control of stored-product pests

Biological control of insect pests is a strategy that, along with good practice and different non-chemical approaches, forms part of IPM which aims at minimizing the use of conventional synthetic pesticides. The basic concept of biological control is that living organisms of one (non-pest) population are used to reduce the population of a specific pest organism. There are four different biological control strategies (Fig. 1) (Eilenberg et al., 2001; Hajek, 2004; van Lenteren et al., 2018).

- *Natural biological control* is achieved by reducing a pest population through naturally occurring antagonists.
- In *classical biological control* natural enemies of a specific pest species are collected at their original habitat, introduced to a new environment and then released into the area of infestation.
- For *augmentative biological control* (*inundative* or *inoculative*) large numbers of laboratory-reared natural enemies have to be released repeatedly into the infested environment (Parnell et al., 2016; van Lenteren, 2012).
- *Conservation biological control* is based on the manipulation or preservation of the (stored-product) environment, making conditions most suitable for the survival of naturally occurring enemies of the pest (Begg et al., 2017).

In the field, the application of entomophagous invertebrates against crop pests is widely spread and well-established. As a result, the industry specialized in biological control or rather rearing, mass production and distribution of biological control agents is growing (van Lenteren, 2012). The import, release and use of living species in agricultural environments or protected production have to consider special regulations which are summarized under the EPPO standards on the safe use of biological control (EPPO, 2019). To date approximately 350 species of natural enemies of pest insects, in particular, parasitoids, predatory insects and mites, as well as entomopathogenic nematodes, are commercially available worldwide but mainly in the EU, Brazil, USA, Canada, Australia, China and New Zealand (Solà, 2017; van Lenteren et al., 1997).

Biological control has several advantages over conventional chemical control measures and represents an environment-friendly alternative (Flinn and Schöller, 2012; van Lenteren et al., 2018). Therefore, and because the stored-product environment provides excellent conditions for the occurrence and release of beneficial insects due to mostly stable temperatures and humidity (Niedermayer et al., 2016b; Schöller and Flinn, 2000; Schöller et al., 2006), biological control also constitutes a promising strategy in the protection of post-harvest plant and food products (Belda and Riudavets, 2013; Droby et al., 2009; Riudavets, 2018), which is constantly increasing in Central Europe (Schöller, 2015). Over the last 30 years, biological control of stored-product insects has been an integral part of many laboratory experiments and also practical applications in grain stores and processing companies (Schöller et al., 2018, and references therein). Due to this research, nowadays, for a large number of stored-product insect pests, their natural enemies are known (468 species according to Hagstrum and Subramanyam, 2009), and several examples for successful use have been shown (Hagstrum and Phillips, 2017; Hagstrum et al., 2012; Prozell and Schöller, 2018).

The most effective biological control agents against stored-product pest insects are, on the one hand, generally less host-specific insect predators that usually prey on more than one host species, mainly because they require a lot of food for development. The predacious reduviid *Amphibolus venator* Klug (Hemiptera: Reduviidae), for example, is a natural enemy of many stored-product pests (Abd-Elgayed and Youssef, 2015), just like the warehouse pirate bug *Xylocoris flavipes* Reuter (Hemiptera: Anthocoridae), which uses immature stages of several beetle and moth species as hosts (Adarkwah et al., 2019; Berger et al., 2017; Murata et al., 2016). Another predatory bug of stored-product beetles is *Alloeocranum biannulipes* Montrouzier (Hemiptera: Reduviidae), which was shown to prey on *Dinoderus porcellus* Lesne (Coleoptera: Bostrichidae), a common pest of dried cassava chips in West Africa (Loko et al., 2017, 2019). *Teretrius nigrescens,* a predatory histerid beetle, preys on eggs, larvae and adults of most stored-product insect species and particularly on *P. truncatus* (Holst and Meikle, 2003; Nboyine et al., 2015; Schöller et al., 1997; Stewart-Jones et al., 2004).

The other and at the same time dominant group of natural enemies of stored-product pest insects is that of parasitoid Hymenoptera. Parasitoids lay their eggs on (*ecto*parasitoid) or in (*endo*parasitoid) the bodies or eggs of other insect species, and after hatching, the parasitoid larvae develop by feeding on the host, causing its eventual death (Godfray, 1994). These parasitic wasps are often host specific and complete their development on or in only one host. *Koinobiont* parasitoids allow the host to continue its development while feeding upon it, whereas *idiobont* wasp species develop on non-growing hosts which have been initially injected a venom that may induce (permanent) paralysis, arrested development or death (Heimpel and Casas, 2008; Heimpel and Collier, 1996; Jervis et al., 2008; Quicke, 1997).

### 3.1 Parasitoids of stored-product moths

There are several examples of parasitoids attacking stored-product moths. The chalcid wasp *Antrocephalus mitys* Walker (Hymenoptera: Chalcididae), for example, has been described to parasitize pupae of *C. cautella* (Pereira et al., 2013). Among the most studied species of natural enemies for the biological control of stored-product moths is the braconid wasp *Habrobracon hebetor*, which paralyzes the host larva before parasitizing it (Eliopoulos and Stathas, 2008; Mbata and Warsi, 2019). Preferred hosts of this cosmopolitan gregarious ectoparasitoid are the wandering last instars of different pyralid species (Dweck et al., 2010; Ghimire and Phillips, 2014; Prozell and Schöller, 2018). In field experiments, *H. hebetor* proved his efficiency in suppressing *P. interpunctella* infestations in different storage facilities, especially when released in combination with *Trichogramma* spp. (Hymenoptera: Trichogrammatidae) (Flinn and Schöller, 2012; Grieshop et al., 2006).

The genus *Trichogramma* comprises a group of polyphagous egg parasitoids, which are frequently used for the biological control of Lepidopteran pest species (Hansen and Jensen, 2002). In SPP they are typically released to prevent packaged products from moth infestation (Grieshop et al., 2007). Another well-studied parasitoid is the ichneumonid wasp *Venturia canescens* Gravenhorst (Hymenoptera: Ichneumonidae), which has shown to control populations of *E. kuehniella* (Belda and Riudavets, 2012a; Phillips and Throne, 2010). This parasitoid co-inhabits bakeries and mills in Central and Southern Europe, together with *H. hebetor* (Schöller et al., 2018). Further-described pyralid host species of this solitary koinobiont endoparasitoid (typically a single egg is laid inside a host) are *P. interpunctella* and the rice moth *Corcyra cephalonica* Stainton (Eliopoulos and Stathas, 2003; Harvey and Thompson, 1995; Jones et al., 2015).

### 3.2 Parasitoids of stored-product beetles

Among the prominent representatives of parasitoids of stored-product beetles is *Anisopteromalus calandrae* Howard (Hymenoptera: Pteromalidae), a primary generalist idiobiont ectoparasitoid that parasitizes late instars and early pupae of a wide variety of species developing inside grains or rather those concealed within the host food or a cocoon such as *R. dominica*, *Sitophilus* spp., *L.*

*serricorne*, *S. cerealella* and the bean weevils *A. obtectus*, *Callosobrochus maculatus* Fabricius and *C. chinensis* (Coleoptera: Chrysomelidae) (Belda and Riudavets, 2010; Belda and Riudavets, 2012b; Bender et al., 2019; Berger et al., 2017; Ghimire and Phillips, 2007; Kasamatsu and Abe, 2015; Niedermayer et al., 2016a).

Moreover, there is a series of bethylid wasps, a hymenopteran family with more than 2000 species, that also attacks stored-product beetles (Amante et al., 2017c). To these idiobiont ectoparasitoids belongs *Cephalonomia tarsalis* Ashmead, which occurs frequently in stored-product environments and represents the most common natural enemy of the saw-toothed grain beetle *Oryzaephilus surinamensis* Linnaeus and the merchant grain beetle *O. mercator* Fauvel (Coleoptera: Silvanidae) (Collatz et al., 2009). Another species of the genus *Cephalonomia* is *C. waterstoni* Gahan, a natural enemy of the rusty grain beetle *Cryptolestes ferrugineus* Stephens, *C. pusillus* Schonherr and *C. turcicus* Grouvelle (Coleoptera: Cucujidae) (Amante et al., 2017a,c).

A further member of the Bethylidae family is *Holepyris sylvanidis* Brèthes, a larval ectoparasitoid whose females preferably attack fourth instars of *Tribolium* spp. (Amante et al., 2017b; Fürstenau et al., 2016). Although to date a commercial rearing and distribution of this parasitic wasp does not exist and the knowledge about its biology and behavior is still rather low, Adler et al. (2012) could show that the release of 300 *H. sylvanidis* adults (ca. 50% females) at a two-week interval was sufficient to control a local population of *T. confusum* in a grain mill in Germany. As a result, a second heat treatment in fall, which had to be adopted in the past, was not necessary during the current and following year. Therefore, and based on the results of subsequent lab studies (see later in this chapter), *H. sylvanidis* represents a potential candidate for biological control of beetle larvae infesting stored products, in particular, of the genus *Tribolium* (Amante et al., 2017b; Fürstenau et al., 2016; Fürstenau and Hilker, 2017).

Another group of parasitoids which uses stored-product beetles as hosts belongs to the family Pteromalidae. *Lariophagus distinguendus* Förster, a solitary ectoparasitoid of larvae and prepupae, is one of the most studied species (e.g. Steidle and Schöller, 1997; Steidle et al., 2001, 2003) which has been successfully applied in the biological control of different pest beetles that develop endophytically in seeds (Niedermayer et al., 2016a,b; Niedermayer and Steidle, 2013). The solitary ectoparasitoid *Theocolax elegans* Westwood is a pteromalid wasp species, which parasitizes mainly immature stages of stored-product beetles that develop inside cereal and legume grains such as *R. dominica*, *C. maculatus*, *Sitophilus* spp. and *L. serricorne* but also *S. cerealella*. *Theocolax elegans* has proven its capacity to successfully reduce populations of several of these cereal pests (Flinn et al., 2006; Germinara et al., 2016; Namikawa and Amornsak, 2016; Schöller et al., 2018; Sitthichaiyakul et al., 2018).

### 3.3 Application of parasitoids

The successful application of parasitoids against storage pests, especially against pyralid moths, has been demonstrated in many cases (Athanassiou and Arthur, 2018; Flinn and Schöller, 2012). However, the number of known species which are commercially available and which are suitable for the biological control of stored-product pests, in general, is still limited. In this context, a literature search by Solà and Riudavets (in preparation) on the biological control of insects infesting stored products revealed that in the last 10 years a total of just 14 parasitoid species were investigated more in detail (Solà, 2017). What needs to be noted here is that the majority of these studies were performed in the laboratory, and results of field applications or efficacy measurements are the exception. Worldwide, 182 parasitoid species are commercialized, but less than 5% of them (only seven species) are natural enemies related to stored-product pests (van Lenteren, 2012; van Lenteren et al., 2018). In contrast, nine species which have also been studied and proven to effectively control their respective hosts, are not yet commercially available (e.g. *A. mitys, C. waterstoni, H. sylvanidis, V. canescens*) (Solà, 2017).

Another indication that research and progress on the biological control of stored-product pests are somehow stuck is that in the course of the 12th International Working Conference on Stored Product Protection (IWCSPP2018) from 237 submitted contributions which subsequently have been published in the corresponding proceedings (Adler et al., 2018), only 11 (<5%) dealt with natural enemies of stored-product pests. Two of these papers were on biological control and its progress in SPP in general (Prozell and Schöller, 2018; Trdan and Bohinc, 2018), whereas the other studies investigated seven different parasitoid species and their possible application against stored-product pests. In contrast, more than 10% of the submitted contributions were on chemical control and the use of phosphine ($PH_3$). This goes in line with the assumption of van Lenteren (2012) that possible reasons for the slow uptake of augmentative biological control, for example, are the posture of the pesticide industry, which considers biological control as laborious and of limited use, as well as many users and farmers who may prefer conventional and proven control methods. In addition, it appears that industries specialized in biological pest control, although steadily expanding their markets, show little interest in developing new biological solutions for protecting stored products, which, on the other hand, results in a lack of commercially available natural enemies (Riudavets, 2018). All this, in turn, may have led to relatively few (new) parasitoid species being scientifically studied and/or placed on the market in recent years.

Biological control in stored products is commercialized since 1998, mostly against moths in bakeries, food-processing industries, retail trade and private households (Schöller et al., 2018). Since 2009, the EU has been promoting the use of biological pest control in its *Sustainable Use of Pesticides Directive* (EC, 2009; van Lenteren et al., 2018). But there are still only very few commercial breeders and suppliers of natural enemies for stored-product pest management. In Italy, for example, the application of parasitoids for biological control in food facilities constitutes more the exception (Trematerra et al., 2017). However, there is considerable potential for a wide use of beneficial insects against stored-product pests in Europe, and there are several examples of successful distribution of parasitoids in the organic food trade as well as in commercial food-processing facilities, particularly in Germany (Flinn and Schöller, 2012; Hansen, 2007; Schöller et al., 2006).

Prozell and Schöller (2018) recently postulated the question as to whether biological control in Germany works. They assume that this is the case and concluded that, after its introduction into the organic market, biological control had been adopted by the conventional sector as well. Many pest control operators now use natural enemies as customers are increasingly demanding pesticide-free products. In the USA, natural enemies of insects in stored-product environments received governmental approval as a pest mitigation practice exempting them for their use in stored raw commodities and processed food (Leahy et al., 2014). As a consequence, several large commercial insectaries were initiated to mass-rear parasitoids and predators for use in large-scale release in stored-product environments in North America (Ghimire and Phillips, 2014; Hagstrum and Phillips, 2017), which, however, has not yet led to commercial availability.

In stored-product environments, commercial use of natural enemies is principally against pyralid moths (ca. 50% of the types of application) and stored-product pest beetles (Prozell and Schöller, 2018). Practical applications are predominantly empty room treatment, treatments in rooms with stored products in packages and bulk grain treatments (Schöller, 2015). An overview of application sites for biological control in Europe was recently given by Schöller et al. (2018). Following the augmentative release strategy, usually, large numbers of laboratory-reared parasitoids are repeatedly released into the area to be treated (inundative release) as adults from small plastic containers or they are stuck as pupae on cardboard strips from where they emerge (Flinn and Schöller, 2012; Schöller et al., 2006). Once they become established in the infested facility and are able to maintain a pest population below damaging thresholds, natural enemies can play a key role in the control of target species (Belda and Riudavets, 2013).

But as mentioned before, the high potential of natural enemies as an alternative control tool against stored-product pests is not exhausted, or, rather, their possible key role in sustainable SPP is only partially exploited. Altogether, there are many different reasons for the still limited use of natural enemies in the control of stored-product pests and the lack of commercially available parasitoids. Overall, the basic knowledge about the correct use of biological control strategies in the protection of stored goods is still insufficient. This may be because the effective application of natural enemies and its release into stored-product environments is rather difficult because of its complexity and the interactions of changing or diverse physical, chemical and biological factors (Hagstrum and Subramanyam, 2009). The restraining factor for use in food facilities is the potential contamination of the products by the beneficial organisms themselves (e.g. by dead bodies or body parts), which always requires a final cleaning step before further processing and placing the food on the market (Ghimire and Phillips, 2014).

In this context, the lack of thresholds for the presence of arthropod filth fragments in stored food and the need to improve regulations for their use are reasons why biological control is not yet implemented by the food industry (Stejskal et al., 2015). Furthermore, and as mentioned before, the effectiveness of most natural enemies of stored-product pest insects has been investigated almost exclusively under laboratory conditions; that is why there is very little information on their performance and efficacy in practical or natural situations.

Altogether, it is of great importance to have detailed information on the ecological functions, biology, host range, behavior, life history and performance of the pest and its antagonist under different conditions. This information is necessary to determine exactly how, when and to what extent natural enemies have to be released in order to ultimately increase their control potential (Droby et al., 2009; Riudavets, 2018; Schöller et al., 1997). Also, knowledge of whether and how insect pests and their natural enemies respond to climate change (e.g. increased temperatures) is of general relevance to adapt existing strategies for future biological insect pest management (Chidawanyika et al., 2012).

## 3.4 Parasitoids and the role of semiochemicals

As stated previously, the discrepancy between the number of known parasitoids using stored-product insects as hosts (468 species) (Hagstrum and Subramanyam, 2009) and that of commercially available species (seven species) (van Lenteren, 2012; van Lenteren et al., 2018) is in part also because information on the biology and foraging behavior and particularly, the odor-mediated host and mate finding the process of parasitoids is relatively scarce. But it is precisely this knowledge that is needed to better understand the linkage between host location behavior of beneficial insects and the population dynamics of their hosts (Mills and Heimpel, 2018). This, in turn, could help to advance commercial antagonist rearing and promote the effective use of parasitoids, with the main aim of improving the biological control of insects infesting stored goods (Barratt et al., 2018). Bioactive VOCs responsible for intra- and interspecific information transfer among organisms as well as between organisms and their environment may not only be used for early detection, monitoring, and control (e.g. attract-and-kill, MD) of stored-product pests as described previously in this chapter (Phillips and Throne, 2010; Trematerra et al., 2017) but also for manipulating the behavior of their natural enemies. The parasitoids' search for suitable hosts (Fig. 3) and interactions with conspecifics (e.g. location of mating partners) is mainly mediated by olfactory cues (Fors et al., 2018; Metzger et al., 2010; Quicke, 1997; van Oudenhove et al., 2017).

One important weak point in the application of parasitoids in classical biological control programs is the lack of specific traps and attractants (Schöller, 1998), which would help to monitor naturally occurring beneficials as well as to recover additionally released ones. The development of suitable lures and traps for the detection of populations of natural enemies of pest insects requires much more research on the odor-mediated host search of the respective species (Trematerra, 2012). In this context, a promising strategy constitutes the use of semiochemicals to attract (and recover) parasitoids after their release into the storage environment and to possibly enhance their performance and host-finding efficiency (Cox, 2004; Mbata et al., 2017). Therefore, it is necessary to identify and synthesize new behavior-modifying compounds which could lead parasitoids to infestation sites (*'attract-and-guide'*) (Fig. 1) and, at best, increase parasitism rates (Eliopoulos et al., 2017; Hötling et al., 2014).

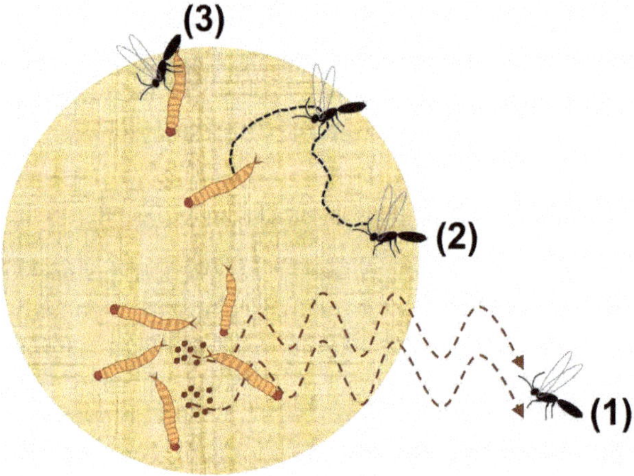


**Figure 3** Scheme of the chemically mediated host-finding process of a parasitic wasp by the example of the larval ectoparasitoid *H. sylvanidis*: (1) host habitat and host location from a distance by highly volatile (*directly* or *indirectly host-associated*) long-range attractants (i.e. host fecal odors from *T. confusum* larvae and habitat volatiles); (2) short-range host location by trail following of less volatile contact cues deposited by host larva on the substrate (i.e. CHC-trails laid by *T. confusum* larvae on wheat grist); (3) host recognition and acceptance by host-specific contact kairomones (i.e. host larval CHC from *T. confusum*).

The origin of bioactive compounds, involved in the orientation and communication of beneficial insects, is diverse and includes the host (*directly host-associated chemical cues*) and the host's habitat or food plant (*indirectly host-associated chemical cues*) (Steidle and Van Loon, 2003; Wajnberg and Colazza, 2013). Some current studies demonstrated the presence of pheromonal components in stored-product parasitoids, including hydrocarbons, aldehydes, indoles and more complex compounds such as polydeoxypropionates (Collatz et al., 2009; Goubault et al., 2008; König et al., 2015; Kühbandner et al., 2012; Tröger et al., 2014).

Nevertheless, to date, most work has been done on the chemistry of the parasitoid's host species and the habitat where the host is located. Whether or to which degree these host-associated odors may influence the parasitoid's behavior under storage conditions in most cases still needs to be verified. The effect of host-related chemical cues, as, for example, host pheromones or VOCs released from host feces on the host-finding behavior has been described previously for several parasitic wasps (Weiss, 2006) and also for parasitoid species attacking insects developing inside seeds (Chiu-Alvarado et al., 2009; Steidle and Van Loon, 2003). Fürstenau et al. (2016) showed that females of the larval ectoparasitoid *H. sylvanidis* are strongly attracted to a mix of specifically host-associated components of the fecal odor from *T. confusum* (Fig. 3). More interestingly, addition of volatiles ubiquitously present in the host's environment enriched with wheat grist (which individually have shown not to be active) enhanced significantly the attractiveness of two host-specific key compounds, (*E*)-2-nonenal and 1-pentadecene, identified from the larval feces.

Especially at low host population densities, habitat odor released from (non-)infested stored food prevail the storage environment. The effect of these background volatiles on the olfactory response of parasitoids to host-specific volatile cues emitted by pest insects is fairly unknown. Some experiments were done on the attraction of parasitoids (i.e. *A. calandrae* and *T. elegans*) to odors from healthy wheat grains as well as to grains which have been infested by the parasitoid's host (Belda and Riudavets, 2010; Germinara et al., 2009, 2016). Females of the granary weevil parasitoid *L. distinguendus* use volatiles emitted from non-infested and infested wheat grains, which are supposed to be induced by some protein substances originated from the weevil larvae saliva (Steidle et al., 2005; Tang et al., 2016). While *H. sylvanidis* females are highly attracted to specifically host-associated volatiles from larval feces but not directly to the host, different studies revealed that *H. hebetor* uses odors emanated from their host larvae for host location (Mbata et al., 2017; Strand et al., 1989).

After finding the host's (micro)-habitat and localization of a host (population) by using highly volatile host-indicating cues, for close-range host search parasitoids may follow specific trails composed of less-volatile compounds laid by the host (Fig. 3). Trail following has been widely investigated in parasitoids of heteropteran pests which leave specific walking trails on the leaf surface of field crops (Conti and Colazza, 2012; Peri et al., 2016). However, some bethylids, such as *C. tarsalis* and *C. waterstoni*, which are commonly associated with beetle larvae living in processed, stored products, also demonstrated to follow host-specific trails (Howard et al., 1998; Howard and Flinn, 1990). Female *H. sylvanidis* proved to orient along trails left by its preferred host *T. confusum* on processed, stored products. Chemical analysis and different bioassays revealed that host larval cuticular hydrocarbons (CHCs) are the active compounds eliciting trail-following behavior (Fürstenau and Hilker, 2017).

The CHCs which are exploited by parasitic wasps as contact kairomones not only elicit trail following but also mediate host recognition by several natural enemies of herbivorous insects (e.g. Colazza et al., 2014; Howard, 1992). Host recognition bioassays with *H. sylvanidis* revealed that parasitoid females equally recognize live and freshly-killed larvae of *Tribolium* spp. as hosts by their CHC pattern (Fig. 3) (Fürstenau and Hilker, 2017). Therefore, and in contrast to other parasitoid species using stored-product pests as hosts, *H. sylvanidis* most probably rely especially on chemical cues rather than on visual or rather respiratory cues of the moving host to locate and recognize their resource within the substrate (Collatz et al., 2010; Howard et al., 1998). Wang et al. (2019) recently demonstrated that two cytochrome p450 genes (*TmCYP4G122* or/and *CYP4G123*) are responsible for the CHC biosynthesis in *T. molitor* and thus, indirectly influence the olfactory response of the pupal ectoparasitoid *Scleroderma guani* Xiao et Wu Sulzer (Hymenoptera: Bethylidae). Mealworm pupae, which lacked one or both of these genes, were no longer accepted as hosts by its natural enemy.

### 3.5 Future trends in parasitoid research

Some parasitoid species have demonstrated to be able to locate their hosts in stored-product environments and, therefore, represent a useful tool for controlling different stored-product pests (e.g. Adarkwah et al., 2014; Ghimire and Phillips, 2014; Niedermayer et al., 2016a). Nevertheless, there are still many possibilities to sustainably optimize the existing control methods, especially with regard to the additional use of semiochemicals.

All parasitoids, whether naturally occurring or artificially released into storage facilities, have to deal with the challenging task during host location to distinguish between relevant, resource-specific or host-indicating odors and unspecific habitat background noise (Beyaert et al., 2010; Bruce and Pickett, 2011; Schröder and Hilker, 2008). The application or slow release of specifically host-associated semiochemicals into a storage facility might possibly help to improve the searching ability of the parasitoids by attracting-and-guiding or rather encouraging them to remain close to infestation sites without diverting from the host (Cox, 2004; Phillips and Throne, 2010). Hereby, it is important to ensure that higher densities of parasitoids do not lead to increased mutual interferences, which would most likely be disruptive to biocontrol (Eliopoulos et al., 2017).

It has already been shown that the application of host-indicating kairomones originated from host moth scales can lead to increased parasitism rates of parasitoids (i.e. *Trichogramma pretoisum* Riley) under field conditions (Gross et al., 1984). Whether this effect occurs in stored-product environments has still to be tested. In any case, it should be noted that the saturation of the air with specific bioactive VOCs may also reduce the efficiency of beneficial organisms if they are released inundatively (Schöller, 1998). However, first results of an ongoing semi-field study on the olfactory host search of *H. sylvanidis* showed that the females' host-finding success is positively influenced by the additional release of specifically host-associated volatiles which previously have demonstrated to attract the parasitoid in olfactometer bioassays (Awater-Salendo et al., in preparation). The importance of semiochemicals for effective host location varies over the life span of a parasitoid. Environmental factors and previous encounters with direct host stimuli or indirect host-indicating cues can modify their reactions (e.g. Peri et al., 2016; Turlings et al., 1993; Vet et al., 1995). The experience of an adult parasitic wasp with such compounds revealing the presence of a host species may increase its sensitivity to these cues. In addition to non-associative learning mechanisms (*sensitization* or *habituation*), the experience of a host-indicating kairomone may also need to be associated with a reward in order to obtain increased responsiveness, and this is referred to as *associative learning* (Smith, 1993; van Alphen and Vet, 1986).

The parasitoid's ability to learn different odors may not only increase the host-finding success but also the number of eggs laid and, thus, may have a general positive effect on the total fitness of the parasitoid (e.g. Steidle, 1998; Steidle and Van Loon, 2003). This, in turn, may lead to improved biological control of the target pest organism. However, a recent study on the host-finding behavior of the parasitoid *Cotesia glomerata* Linnaeus (Hymenoptera: Braconidae), natural enemy of a wide range of *Pieris* butterfly species, showed that the foraging efficiency of oviposition-experienced females is reduced when non-host-infested plants which release similar herbivore-induced plant volatiles are abundant (Vosteen et al., 2019). Therefore, it is necessary to investigate more in detail whether and how prior olfactory experience with or the learning of specific odor-cues (e.g. host-indicating kairomones) in the laboratory (Bender et al., 2019) affects the parasitoid's host search under natural conditions. In this regard, it would be important to verify whether laboratory-reared parasitoids, which have been 'trained' with specific odors before being released into an infested storage environment, reduce or control the pest population in a different way from inexperienced individuals.

Another future approach to improving biological control of stored-product pest insects constitutes the development of specially designed rearing boxes for the controlled release of parasitic wasps and, thereby, to avoid or prevent food contamination with live and dead parasitoids as already shown for *L. distinguendus* (Niedermayer and Steidle, 2013) and for *H. hebetor* (Solà et al., 2018). Moreover, Trematerra et al. (2017) performed field tests in a chocolate factory and demonstrated that the combination of MD and the release of parasitoids might represent a further promising technique in SPP. The application of parasitoids in combination with microbial insecticides (e.g. Bt) or entomopathogenic nematodes also showed high potential in the management of stored-product moths and beetles (Mbata and Warsi, 2019). An innovative project by Müller-Blenkle et al. (2018) deals with the automatic acoustic detection and specific identification of infestations in grain silos with the help of a 'Beetle Sound Tube'-System. After the early detection of pest insects, parasitoids shall be introduced into the silo via the tube system in order to nip the infestation in the bud and, finally, to increase the effectiveness of biological control.

### 3.6 Conclusions

Overall, biological control represents an environment-friendly strategy to protect stored products from insect infestations. Nevertheless, for a large number of important stored-product pest species, the (best) parasitoid candidates still have to be found and/or commercialized. Moreover, the potential of known natural enemies must be better exploited, principally by evaluating their efficacy to reduce a pest population under storage conditions. Also, ambitious research on the olfactory host search of natural enemies of stored-product pests is strongly needed in order to identify new semiochemicals, which could help to influence or improve the parasitoids' performance and increase the host location success. That includes the finding of the most effective odor blends as well as the best methods for their correct application and release. All these approaches could contribute significantly to bringing new species of natural enemies on the market and thus, promoting the use of biological control of insects infesting stored products. Combined with effective IPM strategies, this, in turn, could help to ensure the security of post-harvest stored food.

## 4 Acknowledgements

The authors are grateful to Nadine Austel, Tina Gasch, Torsten Meiners and Matthias Schöller for their helpful input and useful comments on an earlier version of this chapter.

## 5 Where to look for further information

A very good and comprehensive introduction to the role of semiochemicals and especially pheromones in modern insect pest control is Witzgall et al. (2010), *Sex pheromones and their impact on pest management* (see in the References and the corresponding section in this chapter for full details).

The following review article provides detailed information on the biorational approaches to managing stored-product insects:

- Phillips, T. W. and Throne, J. E. 2010, Biorational approaches to managing stored-product insects. *Annual Review of Entomology* 55: 375–97.

The following two in-depth reviews deal exclusively with the potential of using semiochemicals or rather pheromones for stored-product protection:

- Cox, P. D. 2004, Potential for using semiochemicals to protect stored products from insect infestation. *Journal of Stored Products Research* 40: 1–25.
- Trematerra P. 2012, Advances in the use of pheromones for stored-product protection. *Journal of Pest Science* 85: 285–99.

The book chapters and articles below provide a very good overview of biological control in general, the biology and behavior of natural enemies (e.g. parasitoids) of stored-product pests and their application as alternative control measure for stored-product protection:

- Flinn, P. W. and Schöller, M. 2012, Biological control: Insect pathogens, parasitoids, and predators. In: Hagstrum, D. W., Phillips, T. W. and Cuperus, G. (Eds), *Stored Product Protection*. Kansas State University, Manhatten, KS, pp. 203–12.
- Schöller, M., Prozell, S., Suma, P., and Russo, A. 2018, Biological control of stored-product insects. In: Athanassiou, C. and Arthur, F. H. (Eds), *Recent Advances in Stored Product Protection*. Springer, Berlin, Germany, 183–209.
- Steidle, J. L. M. and Van Loon, J. J. A. 2003, Dietary specialization and infochemical use in carnivorous arthropods: testing a concept. *Entomologia Experimentalis et Applicata* 108: 133–48.
- Wajnberg, E., Bernstein, C. and van Alphen, J. 2008, *Behavioural Ecology of Insect Parasitoids: From Theoretical Approaches to Field Applications*. Blackwell Publishing Ltd., Oxford.

Scientific institutions or organizations with high expertise in the areas covered in this chapter include:

- International Working Conference on Stored-Product Protection (IWCSPP) held every four years (http://bru.gmprc.ksu.edu/proj/iwcspp/).

- Julius Kühn-Institut (JKI), Federal Research Centre for Cultivated Plants, Institute for Ecological Chemistry, Plant Analysis and Stored Product Protection, Berlin, Germany (https://www.julius-kuehn.de/en/ecological-chemistry-plant-analysis-and-stored-product-protection/fields-of-activity/stored-product-protection/).
- The Crop Research Institute (CRI), Division of Crop Protection and Plant Health, Stored Product Pest Control, Prague, Czech Republic (https://www.vurv.cz/index.php?p=ochrana_zasob&site=vyzkum_en).
- United States Department of Agriculture (USDA), Stored Product Insect and Engineering Research, Manhattan, KS, USA (https://www.ars.usda.gov/plains-area/mhk/cgahr/spieru/).

## 6 References

Abd-Elgayed, A. A. and Youssef, N. A. 2015. Effect of some stored insect pest species on biological aspects of the predator, *Amphibolus venator* Klug (Hemiptera: Reduviidae). *Annals of Agricultural Sciences* 60(1), 47–51. doi:10.1016/j.aoas.2015.01.003.

Adarkwah, C., Ulrichs, C., Schaarschmidt, S., Badii, B. K., Addai, I. K., Obeng-Ofori, D. and Schöller, M. 2014. Potential of Hymenopteran larval and egg parasitoids to control stored-product beetle and moth infestation in jute bags. *Bulletin of Entomological Research* 104(4), 534–42. doi:10.1017/S0007485314000285.

Adarkwah, C., Obeng-Ofori, D., Opuni-Frimpong, E., Ulrichs, C. and Schöller, M. 2019. Predator-parasitoid-host interaction: biological control of *Rhyzopertha dominica* and *Sitophilus oryzae* by a combination of *Xylocoris flavipes* and *Theocolax elegans* in stored cereals. *Entomologia Experimentalis et Applicata* 167, 118–28.

Adler, C., Schöller, M. and Beier, S. 2012. Development of insects in a flour mill treated with *Holepyris sylvanidis* (Hym., Bethylidae) for biological control of the confused flour beetle *Tribolium confusum* (Col., Tenebrionidae). In: Athanassiou, C. G., Kavallieratos, N. G. and Weintraub, P. G. (Eds), *Integrated Protection of Stored Products: Proceedings of the meeting at Volos (Greece)*, 4–7 July 2011 (IOBC WPRS bulletin 81). Montfavet, pp. 169–70.

Adler, C. S., Opit, G., Fürstenau, B., Müller-Blenkle, C., Kern, P., Arthur, F. H., Athanassiou, C. G., Bartosik, R., Campbell, J., Carvalho, M. O., Chayaprasert, W., Fields, P., Li, Z., Maier, D., Nayak, M., Nukenine, E., Obeng-Ofori, D., Phillips, T., Riudavets, J., Throne, J., Schöller, M., Stejskal, V., Talwana, H., Timlick, B. and Trematerra, P. 2018. *Proceedings of the 12th International Working Conference on Stored Product Protection (IWCSPP)*. Julius-Kühn-Archiv 463, Berlin, Germany.

Adler, C. S., Opit, G., Fürstenau, B., Müller-Blenkle, C., Kern, P., Arthur, F. H., Athanassiou, C. G., Bartosik, R., Campbell, J., Carvalho, M. O., Chayaprasert, W., Fields, P., Li, Z., Maier, D., Nayak, M., Nukenine, E., Obeng-Ofori, D., Phillips, T., Riudavets, J., Throne, J., Schöller, M., Stejskal, V., Talwana, H., Timlick, B. and Trematerra, P. 2018. *Proceedings of the 12th International Working Conference on Stored Product Protection (IWCSPP)*. Julius-Kühn-Archiv 463, Berlin, Germany.

Akter, T. and Ali, M. 2016. Efficacy of pheromone on capturing Angoumois grain moth (*Sitotroga cerealella*) in stored rice (*Oryza sativa* L.) grain. *The Agriculturists* 13(2), 74–80. doi:10.3329/agric.v13i2.26590.

Allison, J. and Cardé, R. 2016. *Pheromone Communication in Moths - Evolution, Behavior, and Application*. University of California Press, Oakland, CA.

Amante, M., Russo, A., Schöller, M. and Steidle, J. L. M. 2017a. Olfactory host location in the rusty grain beetle parasitoid *Cephalonomia waterstoni* (Gahan) (Hymenoptera: Bethylidae). *Journal of Stored Products Research* 71, 1–4. doi:10.1016/j.jspr.2016.12.006.

Amante, M., Schöller, M., Hardy, I. C. W. and Russo, A. 2017b. Reproductive biology of *Holepyris sylvanidis* (Hymenoptera: Bethylidae). *Biological Control* 106, 1–8. doi:10.1016/j.biocontrol.2016.12.004.

Amante, M., Schöller, M., Suma, P. and Russo, A. 2017c. Bethylids attacking stored-product pests: an overview. *Entomologia Experimentalis et Applicata* 163(3), 251–64. doi:10.1111/eea.12587.

Anand, P., Nandagopal, V., Prasad, T. V., Jagadiswari, R. and Korada, R. R. 2015. Pheromones for the management of insect pests of stored products: a review. *Journal of Applied Zoological Researches* 26, 11–36.

Ando, T. and Yamakawa, R. 2011. Analyses of lepidopteran sex pheromones by mass spectrometry. *Trends in Analytical Chemistry* 30(7), 990–1002. doi:10.1016/j.trac.2011.03.010.

Ando, T., Inomata, S.-I. and Yamamoto, M. 2004. Lepidopteran sex pheromones. In: Schulz, S. (Ed.), *The Chemistry of Pheromones and Other Semiochemicals I*. Springer, Berlin, Heidelberg, pp. 51–96.

Arthur, F. H. 2018. Structural pest management for stored product insects. In: Athanassiou, C. and Arthur, F. H. (Eds), *Recent Advances in Stored Product Protection*. Springer, Berlin, Germany, pp. 65–81.

Athanassiou, C. and Arthur, F. H. 2018. *Recent Advances in Stored Product Protection*. Springer, Berlin.

Athanassiou, C., Bray, D. P., Hall, D. R., Phillips, C. and Vassilakos, T. N. 2018. Factors affecting field performance of pheromone traps for tobacco beetle, *Lasioderma serricorne*, and tobacco moth, *Ephestia elutella*. *Journal of Pest Science* 91(4), 1381–91. doi:10.1007/s10340-018-0987-8.

Athanassiou, C. G., Kavallieratos, N. G. and Trematerra, P. 2006. Responses of *Sitophilus oryzae* (Coleoptera: Curculionidae) and *Tribolium confusum* (Coleoptera: Tenebrionidae) to traps baited with pheromones and food volatiles. *European Journal of Entomology* 103(2), 371–8. doi:10.14411/eje.2006.050.

Athanassiou, C. G., Kavallieratos, N. G., Sciarretta, A. and Trematerra, P. 2016. Mating disruption of *Ephestia kuehniella* (Zeller) (Lepidoptera: Pyralidae) in a storage facility: spatio-temporal distribution changed after long-term application. *Journal of Stored Products Research* 67, 1–12. doi:10.1016/j.jspr.2016.01.001.

Attygalle, A. B., Blankespoor, C. L., Meinwald, J. and Eisner, T. 1991. Defensive secretion of *Tenebrio molitor* (Coleoptera: Tenebrionidae). *Journal of Chemical Ecology* 17(4), 805–9. doi:10.1007/BF00994202.

Baek, S., Perez, A. E., Turcotte, R. M., White, J. B., Adedipe, F. and Park, Y. 2015. Response of *Tenebrio molitor* (Coleoptera: Tenebrionidae) adults to potato: implications for monitoring and sampling. *Journal of Stored Products Research* 60, 5–10. doi:10.1016/j.jspr.2014.11.002.

Bale, J. S., van Lenteren, J. C. and Bigler, F. 2008. Biological control and sustainable food production. *Philosophical Transactions of the Royal Society of London. Series B, Biological Sciences* 363(1492), 761–76. doi:10.1098/rstb.2007.2182.

Barak, A. V. and Burkholder, W. E. 1985. A versatile and effective trap for detecting and monitoring stored-product Coleoptera. *Agriculture, Ecosystems and Environment* 12(3), 207–18. doi:10.1016/0167-8809(85)90112-4.

Barratt, B. I. P., Moran, V. C., Bigler, F. and van Lenteren, J. C. 2018. The status of biological control and recommendations for improving uptake for the future. *BioControl* 63(1), 155–67. doi:10.1007/s10526-017-9831-y.

Begg, G. S., Cook, S. M., Dye, R., Ferrante, M., Franck, P., Lavigne, C., Lövei, G. L., Mansion-Vaquie, A., Pell, J. K., Petit, S., Quesada, N., Ricci, B., Wratten, S. D. and Birch, A. N. E. 2017. A functional overview of conservation biological control. *Crop Protection* 97, 145–58. doi:10.1016/j.cropro.2016.11.008.

Belda, C. and Riudavets, J. 2010. Attraction of the parasitoid *Anisopteromalus calandrae* (Howard) (Hymenoptera : Pteromalidae ) to odors from grain and stored product pests in a Y-tube olfactometer. *Biological Control* 54(1), 29–34. doi:10.1016/j.biocontrol.2010.02.005.

Belda, C. and Riudavets, J. 2012a. The influence of the rearing host on the response of the parasitoid *Venturia canescens* (Gravenhorst) (Hymenoptera: Ichneumonidae) to odours from *Ephestia kuehniella* and *Plodia interpunctella* in a Y-tube olfactometer. *BioControl* 57(6), 801–8. doi:10.1007/s10526-012-9461-3.

Belda, C. and Riudavets, J. 2012b. Reproduction of the parasitoids *Anisopteromalus calandrae* (Howard) and *Lariophagus distinguendus* (Förster) on arenas containing a mixed population of the coleopteran pests *Sitophilus oryzae* and *Rhyzopertha dominica*. *Journal of Pest Science* 85(3), 381–5. doi:10.1007/s10340-011-0401-2.

Belda, C. and Riudavets, J. 2013. Natural enemies associated with lepidopteran pests in food and feed processing companies. *Journal of Stored Products Research* 53, 54–60. doi:10.1016/j.jspr.2013.02.006.

Bell, C. H. 2014. Pest management. In: Motarjemi, Y. and Lelieveld, H. (Eds), *Food Safety Management*. Academic Press, San Diego, pp. 799–820.

Bender, E., Jahnke, S. M. and Köhler, A. 2019. Chemotaxic responses of *Anisopteromalus calandrae* (Howard) (Hymenoptera: Pteromalidae) to odors of larvae, pupae, and the diet of *Lasioderma serricorne* (Fabricius) (Coleoptera: Ptinidae). *Neotropical Entomology* 49(2), 171–8. doi:10.1007/s13744-019-00742-0.

Benelli, G., Romano, D., Stefanini, C., Kavallieratos, N. G., Athanassiou, C. G. and Canale, A. 2017. Asymmetry of mating behaviour affects copulation success in two stored-product beetles. *Journal of Pest Science* 90(2), 547–56. doi:10.1007/s10340-016-0794-z.

Berger, A., Degenkolb, T., Vilcinskas, A. and Schöller, M. 2017. Evaluating the combination of a parasitoid and a predator for biological control of seed beetles (Chrysomelidae: Bruchinae) in stored beans. *Journal of Stored Products Research* 74, 22–6. doi:10.1016/j.jspr.2017.08.009.

Beyaert, I., Wäschke, N., Scholz, A., Varama, M., Reinecke, A. and Hilker, M. 2010. Relevance of resource-indicating key volatiles and habitat odour for insect orientation. *Animal Behaviour* 79(5), 1077–86. doi:10.1016/j.anbehav.2010.02.001.

Birkinshaw, L. A., Hodges, R. J. and Addo, S. 2004. Flight behaviour of *Prostephanus truncatus* and *Teretrius nigrescens* demonstrated by a cheap and simple pheromone-baited trap designed to segregate catches with time. *Journal of Stored Products Research* 40(2), 227–32. doi:10.1016/S0022-474X(02)00094-2.

Brady, U. E. 1973. Isolation, identification and stimulatory activity of a second component of the sex pheremone system (complex) of the female almond moth, *Cadra cautella* (Walker). *Life Sciences* 13(3), 227–35. doi:10.1016/0024-3205(73)90021-0.

Brady, U. E., Tumlinson, J. H., Brownlee, R. G. and Silverstein, R. M. 1971. Sex stimulant and attractant in the Indian meal moth and in the almond moth. *Science* 171(3973), 802–4. doi:10.1126/science.171.3973.802.

Bruce, T. J. A. and Pickett, J. A. 2011. Perception of plant volatile blends by herbivorous insects – finding the right mix. *Phytochemistry* 72(13), 1605–11. doi:10.1016/j.phytochem.2011.04.011.

Bryning, G. P., Chambers, J. and Wakefield, M. E. 2005. Identification of a sex pheromone from male yellow mealworm beetles, *Tenebrio molitor*. *Journal of Chemical Ecology* 31(11), 2721–30. doi:10.1007/s10886-005-7622-x.

Burkholder, W. E. and Ma, M. 1985. Pheromones for monitoring and control of stored-product insects. *Annual Review of Entomology* 30(1), 257–72. doi:10.1146/annurev.en.30.010185.001353.

Burks, C. S., McLaughlin, J. R., Miller, J. R. and Brandl, D. G. 2011. Mating disruption for control of *Plodia interpunctella* (Hübner) (Lepidoptera: Pyralidae) in dried beans. *Journal of Stored Products Research* 47(3), 216–21. doi:10.1016/j.jspr.2011.03.001.

Campbell, J. F. 2012. Attraction of walking *Tribolium castaneum* adults to traps. *Journal of Stored Products Research* 51, 11–22. doi:10.1016/j.jspr.2012.06.002.

Campbell, J. F. and Mullen, M. A. 2004. Distribution and dispersal behavior of *Trogoderma variabile* and *Plodia interpunctella* outside a food processing plant. *Journal of Economic Entomology* 97(4), 1455–64. doi:10.1603/0022-0493-97.4.1455.

Campbell, J. F., Dowdy, A. K. and Mullen, M. A. 2002. Monitoring stored-product pests in food processing plants with pheromone trapping, contour mapping, and mark-recapture. *Journal of Economic Entomology* 95(5), 1089–101. doi:10.1603/0022-0493-95.5.1089.

Campos, M. and Phillips, T. W. 2014. Attract-and-kill and other pheromone-based methods to suppress populations of the Indianmeal moth (Lepidoptera: Pyralidae). *Journal of Economic Entomology* 107(1), 473–80. doi:10.1603/ec13451.

Cardé, R. T. and Millar, J. G. 2009. Pheromones. In: Resh, V. H. and Cardé, R. T. (Eds), *Encyclopedia of Insects* (2nd edn.). Academic Press, San Diego, pp. 766–72.

CFR. 2018a. Title 40: Protection of environment, chapter I – Environmental Protection Agency (continued). Subchapter E – pesticide programs. Part 158 – data requirements for pesticides. Subpart U – Biochemical pesticides, §158.2000. *CFR/Code of Federal Regulations 26*.

CFR. 2018b. Title 40: Protection of environment, chapter I – Environmental Protection Agency (continued). Subchapter E – pesticide programs. Part 158 – data requirements for pesticides. Subpart V – Microbial Pesticides, §158.2100. *CFR/Code of Federal Regulations 26*.

CFR. 2018c. Title 40: Protection of environment, chapter I – Environmental Protection Agency (continued). Subchapter E – pesticide programs. Part 174 – procedures and requirements for plant-incorporated protectants. Subpart A – General provisions; §40 CFR 174.3. *CFR/Code of Federal Regulations 26*.

CFR. 2019. Title 40: Protection of environment, chapter I – Environmental Protection Agency (continued). Subchapter E – pesticide programs. Part 152 – pesticide registration and classification procedures. Subpart B - Exemptions, §152.20. *CFR/Code of Federal Regulations 25*.

Chambers, J., Van Wyk, C. B., White, P. R., Gerrard, C. M. and Mori, K. 1996. Grain weevil, *Sitophilus granarius* (L.): antennal and behavioral responses to male-produced volatiles. *Journal of Chemical Ecology* 22(9), 1639–54. doi:10.1007/BF02272404.

Chandler, D., Bailey, A. S., Tatchell, G. M., Davidson, G., Greaves, J. and Grant, W. P. 2011. The development, regulation and use of biopesticides for integrated pest management. *Philosophical Transactions of the Royal Society of London: Series B, Biological Sciences* 366(1573), 1987–98. doi:10.1098/rstb.2010.0390.

Chen, Z., Schlipalius, D., Opit, G., Subramanyam, B. and Phillips, T. W. 2015. Diagnostic molecular markers for phosphine resistance in U.S. populations of *Tribolium castaneum* and *Rhyzopertha dominica*. *PLoS ONE* 10(3), e0121343. doi:10.1371/journal.pone.0121343.

Chidawanyika, F., Mudavanhu, P. and Nyamukondiwa, C. 2012. Biologically based methods for pest management in agriculture under changing climates: challenges and future directions. *Insects* 3(4), 1171–89. doi:10.3390/insects3041171.

Chiluwal, K., Kim, J., Bae, S. D., Roh, G. H. and Park, C. G. 2018. Methyl salicylate and trans-anethole affect the pheromonal activity of homofarnesal, the female sex pheromone of azuki bean beetle. *Entomological Research* 48, 354–61. doi:10.1111/1748-5967.12294.

Chiu-Alvarado, P., Barrera, J. F. and Rojas, J. C. 2009. Attraction of *Prorops nasuta* (Hymenoptera: Bethylidae), a parasitoid of the coffee berry borer (Coleoptera: Curculionidae), to host-associated olfactory cues. *Annals of the Entomological Society of America* 102(1), 166–71. doi:10.1603/008.102.0119.

Colazza, S., Cusumano, A., Lo Giudice, D. and Peri, E. 2014. Chemo-orientation responses in hymenopteran parasitoids induced by substrate-borne semiochemicals. *BioControl* 59(1), 1–17. doi:10.1007/s10526-013-9546-7.

Collatz, J., Tolasch, T. and Steidle, J. L. M. 2009. Mate finding in the parasitic wasp *Cephalonomia tarsalis* (Ashmead): more than one way to a female's heart. *Journal of Chemical Ecology* 35(7), 761–8. doi:10.1007/s10886-009-9659-8.

Collatz, J., Fuhrmann, A., Selzer, P., Oehme, R. M., Hartelt, K., Kimmig, P., Meiners, T., Mackenstedt, U. and Steidle, J. L. M. 2010. Being a parasitoid of parasites: host finding in the tick wasp *Ixodiphagus hookeri* by odours from mammals. *Entomologia Experimentalis et Applicata* 134(2), 131–7. doi:10.1111/j.1570-7458.2009.00943.x.

Conti, E. and Colazza, S. 2012. Chemical ecology of egg parasitoids associated with true bugs. *Psyche: A Journal of Entomology* 2012, 1–11, Article ID 651015. doi:10.1155/2012/651015.

Corbet, S. A. 1971. Mandibular gland secretion of larvae of the flour moth, *Anagasta kuehniella*, contains an epideictic pheromone and elicits oviposition movements in a Hymenopteran parasite. *Nature* 232(5311), 481–4. doi:10.1038/232481b0.

Cork, A., Hall, D. R., Hodges, R. J. and Pickett, J. A. 1991. Identification of major component of male-produced aggregation pheromone of larger grain borer, *Prostephanus truncatus* (Horn) (Coleoptera: Bostrichidae). *Journal of Chemical Ecology* 17(4), 789–803. doi:10.1007/BF00994201.

Cox, P. D. 2004. Potential for using semiochemicals to protect stored products from insect infestation. *Journal of Stored Products Research* 40(1), 1–25. doi:10.1016/S0022-474X(02)00078-4.

Cox, P. D. and Collins, L. E. 2002. Factors affecting the behaviour of beetle pests in stored grain, with particular reference to the development of lures. *Journal of Stored Products Research* 38(2), 95–115. doi:10.1016/S0022-474X(01)00010-8.

Daglish, G. J., Holloway, J. C. and Nayak, M. K. 2013. Implications of Methoprene resistance for managing *Rhyzopertha dominica* (f.) in stored grain. *Journal of Stored Products Research* 54, 8–12. doi:10.1016/j.jspr.2013.03.006.

Daglish, G. J., Ridley, A. W., Reid, R. and Walter, G. H. 2017. Testing the consistency of spatio-temporal patterns of flight activity in the stored grain beetles *Tribolium castaneum* (Herbst) and *Rhyzopertha dominica* (f.). *Journal of Stored Products Research* 72, 68–74. doi:10.1016/j.jspr.2017.03.005.

Dicke, M. and Sabelis, M. W. 1988. Infochemical terminology: based on cost-benefit analysis rather than origin of compounds? *Functional Ecology* 2(2), 131-9. doi:10.2307/2389687.

Droby, S., Wisniewski, M., Macarisin, D. and Wilson, C. 2009. Twenty years of postharvest biocontrol research: is it time for a new paradigm? *Postharvest Biology and Technology* 52(2), 137–45. doi:10.1016/j.postharvbio.2008.11.009.

Dweck, H. K. M., Svensson, G. P., Gündüz, E. A. and Anderbrant, O. 2010. Kairomonal response of the parasitoid, *Bracon hebetor* Say, to the male-produced sex pheromone of its host, the greater waxmoth, *Galleria mellonella* (L.). *Journal of Chemical Ecology* 36(2), 171–8. doi:10.1007/s10886-010-9746-x.

EC. 2009. Sustainable use directive. *Official Journal of the European Union* L309, 71–86.

Eilenberg, J., Hajek, A. and Lomer, C. 2001. Suggestions for unifying the terminology in biological control. *BioControl* 46(4), 387–400. doi:10.1023/A:1014193329979.

El-Mofty, M. M., Sakr, S. A., Osman, S. I. and Toulan, B. A. 1989. Carcinogenic effect of biscuits made of flour infested with *Tribolium castaneum* in *Bufo regularis*. *Oncology* 46(1), 63–5. doi:10.1159/000226684.

Eliopoulos, P. A. and Stathas, G. J. 2003. Temperature-dependent development of the koinobiont endoparasitoid *Venturia canescens* (Gravenhorst) (Hymenoptera: Ichneumonidae): effect of host instar. *Environmental Entomology* 32(5), 1049–55. doi:10.1603/0046-225X-32.5.1049.

Eliopoulos, P. A. and Stathas, G. J. 2008. Life tables of *Habrobracon hebetor* (Hymenoptera: Braconidae) parasitizing *Anagasta kuehniella* and *Plodia interpunctella* (Lepidoptera: Pyralidae): effect of host density. *Journal of Economic Entomology* 101(3), 982–8. doi:10.1603/0022-0493(2008)101[982:LTOHHH]2.0.CO;2.

Eliopoulos, P. A., Kapranas, A., Givropoulou, E. G. and Hardy, I. C. 2017. Reproductive efficiency of the bethylid wasp *Cephalonomia tarsalis*: the influences of spatial structure and host density. *Bulletin of Entomological Research* 107(2), 139–47. doi:10.1017/S0007485316000651.

ENV./JM/MONO. 2003. Guidance for registration requirements for microbial pesticides. *OECD Series on Pesticides* 18.

ENV./JM/MONO. 2004. Guidance for information requirements for regulation of invertebrates as biological control agents (IBCAs). *OECD Series on Pesticides* 21.

ENV./JM/MONO. 2017. Guidance document on semiochemical active substances and plant protection products. *OECD Series on Pesticides* 93.

EPPO. 2008. Mating disruption pheromones, *PP* 38, 322-325.

EPPO. 2016. Safe use of biological control. List of biological control agents widely used in the EPPO Region. *EPPO Standards* PM 6/3 English: 39.

EPPO. 2019. PM6 – safe use of biological control. Available at: https://gd.eppo.int/standards/PM6/.

EU. 2013. Commission regulation (EU) No 283/2013 setting out the data requirements for active substances, in accordance with Regulation (EC) No 1107/2009 of the European Parliament and of the Council concerning the placing of plant protection products on the market Part B. *Official Journal of the European Union* L 93/1.

EU. 2019a. The post-2020 common agricultural policy: environmental benefits and simplification. *DG Agriculture and Rural Development*, 1–19.

EU. 2019b. Amending Directive 2009/128/EC of the European Parliament and of the Council as regards the establishment of harmonised risk indicators. *Official Journal of the European Union* L 127/4.

FAO. 2006. Guidelines on efficacy evaluation for the registration of plant protection products. *International Code of Conduct on the Distribution and Use of Pesticides* 61.

FAO. 2019. ISPM 5 – glossary of phytosanitary terms. *International Standards for Phytosanitary Measures* 34.

Flinn, P. W. and Schöller, M. 2012. Biological control: insect pathogens, parasitoids, and predators. In: Hagstrum, D. W., Phillips, T. W. and Cuperus, G. (Eds), *Stored Product Protection*. Kansas State University, Manhatten, KS, pp. 203–12.

Flinn, P. W., Kramer, K. J., Throne, J. E. and Morgan, T. D. 2006. Protection of stored maize from insect pests using a two-component biological control method consisting of a hymenopteran parasitoid, *Theocolax elegans*, and transgenic avidin maize powder. *Journal of Stored Products Research* 42(2), 218–25. doi:10.1016/j.jspr.2005.02.002.

Fors, L., Mozuraitis, R., Blazyte-Cereskiene, L., Verschut, T. A. and Hambäck, P. A. 2018. Selection by parasitoid females among closely related hosts based on volatiles: identifying relevant chemical cues. *Ecology and Evolution* 8(6), 3219–28. doi:10.1002/ece3.3877.

Fouad, H. A., Faroni, L. R. D. A., Vilela, E. F. and de Lima, E. R. 2013. Flight responses of *Sitotroga cerealella* (Lepidoptera: Gelechiidae) to corn kernel volatiles in a wind tunnel. *Arthropod-Plant Interactions* 7(6), 651–8. doi:10.1007/s11829-013-9275-y.

Freitas, R. C. P., Faroni, L. R. D. A., Haddi, K., Viteri Jumbo, L. O. and Oliveira, E. E. 2016. Allyl isothiocyanate actions on populations of *Sitophilus zeamais* resistant to phosphine: toxicity, emergence inhibition and repellency. *Journal of Stored Products Research* 69, 257–64. doi:10.1016/j.jspr.2016.09.006.

Fürstenau, B. and Hilker, M. 2017. Cuticular hydrocarbons of *Tribolium confusum* larvae mediate trail following and host recognition in the ectoparasitoid *Holepyris sylvanidis*. *Journal of Chemical Ecology* 43(9), 858–68. doi:10.1007/s10886-017-0885-1.

Fürstenau, B., Adler, C., Schulz, H. and Hilker, M. 2016. Host habitat volatiles enhance the olfactory response of the larval parasitoid *Holepyris sylvanidis* to specifically host-associated cues. *Chemical Senses* 41(7), 611–21. doi:10.1093/chemse/bjw065.

Geiselhardt, S., Schmitt, T. and Peschke, K. 2009. Chemical composition and pheromonal function of the defensive secretions in the subtribe Stizopina (Coleptera, Tenebrionidae, Opatrini). *Chemoecology* 19(1), 1-6. doi:10.1007/s00049-008-0001-7.

Gerken, A. R., Scully, E. D. and Campbell, J. F. 2018. Red flour beetle (Coleoptera: Tenebrionidae) response to volatile cues varies with strain and behavioral assay. *Environmental Entomology* 47(5), 1252–65. doi:10.1093/ee/nvy107.

Germinara, G. S., De Cristofaro, A. and Rotundo, G. 2009. Antennal olfactory responses to individual cereal volatiles in *Theocolax elegans* (Westwood) (Hymenoptera: Pteromalidae). *Journal of Stored Products Research* 45(3), 195–200. doi:10.1016/j.jspr.2009.02.002.

Germinara, G. S., De Cristofaro, A. and Rotundo, G. 2016. Electrophysiological and behavioral responses of *Theocolax elegans* (Westwood) (Hymenoptera:

Pteromalidae) to cereal grain volatiles. *BioMed Research International* 2016, 5460819. doi:10.1155/2016/5460819.

Ghimire, M. N. and Phillips, T. W. 2007. Suitability of five species of stored-product insects as hosts for development and reproduction of the parasitoid *Anisopteromalus calandrae* (Hymenoptera: Pteromalidae). *Journal of Economic Entomology* 100(5), 1732–9. doi:10.1093/jee/100.5.1732.

Ghimire, M. N. and Phillips, T. W. 2014. Oviposition and reproductive performance of *Habrobracon hebetor* (Hymenoptera: Braconidae) on six different Pyralid host species. *Annals of the Entomological Society of America* 107(4), 809–17. doi:10.1603/AN14046.

Godfray, H. 1994. *Parasitoids: Behavioral and Evolutionary Ecology*. Princeton University Press, Princeton, NJ.

Goubault, M., Batchelor, T. P., Romani, R., Linforth, R. S. T., Fritzsche, M., Francke, W. and Hardy, I. C. W. 2008. Volatile chemical release by bethylid wasps: identity, phylogeny, anatomy and behaviour. *Biological Journal of the Linnean Society* 94(4), 837-52. doi:10.1111/j.1095-8312.2008.01022.x.

Gram, L., Ravn, L., Rasch, M., Bruhn, J. B., Christensen, A. B. and Givskov, M. 2002. Food spoilage–interactions between food spoilage bacteria. *International Journal of Food Microbiology* 78(1-2), 79–97. doi:10.1016/s0168-1605(02)00233-7.

Grant, G. G. 1976. Female coyness and reciptivity during courtship in *Plodia interpunctella* (Lepidoptera: Pyralidae). *The Canadian Entomologist* 108(9), 975–9. doi:10.4039/Ent108975-9.

Grant, G. G. and Brady, U. E. 1975. Courtship behavior of phycitid moths. I. Comparison of *Plodia interpunctella* and *Cadra cautella* and role of male scent glands. *Canadian Journal of Zoology* 53(6), 813–26. doi:10.1139/z75-095.

Grieshop, M. J., Flinn, P. W. and Nechols, J. R. 2006. Biological control of Indianmeal moth (Lepidoptera: Pyralidae) on finished stored products using egg and larval parasitoids. *Journal of Economic Entomology* 99(4), 1080–4. doi:10.1603/0022-0493-99.4.1080.

Grieshop, M. J., Flinn, P. W., Nechols, J. R. and Scholler, M. 2007. Host-foraging success of three species of Trichogramma (Hymenoptera: Trichogrammatidae) in a simulated retail environment. *Journal of Economic Entomology* 100(2), 591–8. doi:10.1093/jee/100.2.591.

Gross, H. R., Jr., Lewis, W. J., Beevers, M. and Nordlund, D. A. 1984. *Trichogramma pretiosum* (Hymenoptera: Trichogrammatidae): effects of augmented densities and distributions of *Heliothis zea* (Lepidoptera: Noctuidae). *Environmental Entomology* 13(4), 981–5. doi:10.1093/ee/13.4.981.

Gustavsson, J., Cederberg, C., Sonesson, U., et al. 2011. *Global Food Losses and Food Waste: Extent, Causes and Prevention, Food and Agriculture Organisation of the United Nations*. FAO, Rome.

Gut, L. J. and Miller, J. R. 2015. Mating disruption for the 21st century: matching technology with mechanism. *Environmental Entomology* 44(3), 427–53. doi:10.1093/ee/nvv052.

Hagstrum, D. W. and Flinn, P. W. 2014. Modern stored-product insect pest management. *Journal of Plant Protection Research* 54(3), 205–10. doi:10.2478/jppr-2014-0031.

Hagstrum, D. W. and Phillips, T. W. 2017. Evolution of stored-product entomology: protecting the world food supply. *Annual Review of Entomology* 62, 379–97. doi:10.1146/annurev-ento-031616-035146.

Hagstrum, D. W. and Subramanyam, B. 2009. *Stored Product Insect Resources*. AACC International, St. Paul, MN.

Hagstrum, D. W., Phillips, T. W. and Cuperus, G. 2012. *Stored Product Protection*. Kansas State University, Manhatten, KS.

Hajek, A. E. 2004. *Natural Enemies: An Introduction to Biological Control*. Cambridge University Press, Cambridge.

Hansen, L. S. 2007. Potential for widespread application of biological control of stored-product pests–the European perspective. *Journal of Stored Products Research* 43(3), 312–3. doi:10.1016/j.jspr.2006.08.001.

Hansen, L. S. and Jensen, K. M. V. 2002. Effect of temperature on parasitism and host-feeding of *Trichogramma turkestanica* (Hymenoptera: Trichogrammatidae) on *Ephestia kuehniella* (Lepidoptera: Pyralidae). *Journal of Economic Entomology* 95(1), 50–6. doi:10.1603/0022-0493-95.1.50.

Hartlieb, E. and Anderson, P. 1999. Olfactory-released behaviours. In: Hansson, B. (Ed.), *Insect Olfaction*. Springer, Berlin Heidelberg, New York, pp. 315–50.

Hartmann, T. 2008. The lost origin of chemical ecology in the late 19th century. *Proceedings of the National Academy of Sciences of the United States of America* 105(12), 4541–6. doi:10.1073/pnas.0709231105.

Harvey, J. A. and Thompson, D. J. 1995. Host behaviour and its influence on foraging and acceptance by the solitary parasitoid wasp,*Venturia canescens* (Hym: Ichneumonidae). *Entomophaga* 40(2), 193–210. doi:10.1007/BF02373068.

Hassemer, M. J., Sant'Ana, J., de Oliveira, M. W. M., Borges, M., Laumann, R. A., Caumo, M. and Blassioli-Moraes, M. C. 2015. Chemical composition of *Alphitobius diaperinus* (Coleoptera: Tenebrionidae) abdominal glands and the influence of 1,4-benzoquinones on its behavior. *Journal of Economic Entomology* 108(4), 2107–16. doi:10.1093/jee/tov147.

Hassemer, M. J., Borges, M., Withall, D. M., Pickett, J. A., Laumann, R. A., Birkett, M. A. and Blassioli-Moraes, M. C. 2019. Development of pull and push-pull systems for management of lesser mealworm, *Alphitobius diaperinus*, in poultry houses using alarm and aggregation pheromones. *Pest Management Science* 75(4), 1107–14. doi:10.1002/ps.5225.

Heimpel, G. E. and Casas, J. 2008. Parasitoid foraging and oviposition behaviour in the field. In: Wajnberg, E., Bernstein, C. and Van Alphen, J. (Eds), *Behavioural Ecology of Insect Parasitoids: from Theoretical Approaches to Field Applications*. Blackwell Publishing Ltd., Oxford, pp. 51-70.

Heimpel, G. E. and Collier, T. R. 1996. The evolution of host-feeding behaviour in insect parastoids. *Biological Reviews* 71(3), 373-400. doi:10.1111/j.1469-185X.1996.tb01279.x.

Hengsdijk, H. and de Boer, W. J. 2017. Post-harvest management and post-harvest losses of cereals in Ethiopia. *Food Security* 9(5), 945–58. doi:10.1007/s12571-017-0714-y.

Holloway, J. C., Mayer, D. G. and Daglish, G. J. 2018. Flight activity of *Cryptolestes ferrugineus* in southern New South Wales, Australia. *Journal of Pest Science* 91(4), 1353–62. doi:10.1007/s10340-018-0981-1.

Holst, N. and Meikle, W. G. 2003. *Teretrius nigrescens* against larger grain borer *Prostephanus truncatus* in African maize stores: biological control at work? *Journal of Applied Ecology* 40(2), 307–19. doi:10.1046/j.1365-2664.2003.00805.x.

Horler, D. F. 1970. (-) Methyl n-tetradeca-trans-2,4,5-trienoate, an allenic ester produced by the male dried bean beetle, *Acanthoscelides obtectus* (Say). *Journal of the Chemical Society. Perkin Transactions* 6, 859–62. doi:10.1039/j39700000859.

Hötling, S., Haberlag, B., Tamm, M., Collatz, J., Mack, P., Steidle, J. L., Vences, M. and Schulz, S. 2014. Identification and synthesis of macrolide pheromones of the grain beetle *Oryzaephilus surinamensis* and the frog *Spinomantis aglavei*. *Chemistry* 20(11), 3183–91. doi:10.1002/chem.201304414.

Howard, R. W. 1992. Comparative analysis of cuticular hydrocarbons from the ectoparasitoids *Cephalonomia waterstoni* and *Laelius utilis* (Hymenoptera: Bethylidae) and their respective hosts, *Cryptolestes ferrugineus* (Coleoptera: Cucujidae) and *Trogoderma variabile* (Coleoptera:Dermestidae). *Annals of the Entomological Society of America* 85(3), 317–25. doi:10.1093/aesa/85.3.317.

Howard, R. W. and Flinn, P. W. 1990. Larval trails of *Cryptolestes ferrugineus* (Coleoptera: Cucujidae) as kairomonal host-finding cues for the parasitoid *Cephalonomia waterstoni* (Hymenoptera: Bethylidae). *Annals of the Entomological Society of America* 83(2), 239–45. doi:10.1093/aesa/83.2.239.

Howard, R. W., Charlton, M. and Charlton, R. E. 1998. Host-finding, host-recognition, and host-acceptance behavior of *Cephalonomia tarsalis* (Hymenoptera: Bethylidae). *Annals of the Entomological Society of America* 91(6), 879–89. doi:10.1093/aesa/91.6.879.

Hubert, J., Stejskal, V., Athanassiou, C. G. and Throne, J. E. 2018. Health hazards associated with arthropod infestation of stored products. *Annual Review of Entomology* 63, 553-73. doi:10.1146/annurev-ento-020117-043218.

Işikber, A. A. and Öztekin, S. 2009. Comparison of susceptibility of two stored-product insects, *Ephestia kuehniella* Zeller and *Tribolium confusum* du Val to gaseous ozone. *Journal of Stored Products Research* 45(3), 159–64. doi:10.1016/j.jspr.2008.12.003.

Jervis, M. A., Ellers, J. and Harvey, J. A. 2008. Resource acquisition, allocation, and utilization in parasitoid reproductive strategies. *Annual Review of Entomology* 53, 361–85. doi:10.1146/annurev.ento.53.103106.093433.

Jones, T. S., Bilton, A. R., Mak, L. and Sait, S. M. 2015. Host switching in a generalist parasitoid: contrasting transient and transgenerational costs associated with novel and original host species. *Ecology and Evolution* 5(2), 459–65. doi:10.1002/ece3.1333.

Kaissling, K. 2014. Pheromone reception in insects (the example of silk moths). In: Mucignat-Caretta, C. (Ed.), *Neurobiology of Chemical Communication*. CRC Press Press/Taylor & Francis, Boca Raton, London, New York, pp. 99–146.

Karlson, P. and Lüscher, M. 1959. 'Pheromones': a new term for a class of biologically active substances. *Nature* 183(4653), 55–6. doi:10.1038/183055a0.

Kasamatsu, E. and Abe, J. 2015. Influence of body size on fecundity and sperm management in the parasitoid wasp *Anisopteromalus calandrae*. *Physiological Entomology* 40(3), 223–31. doi:10.1111/phen.12106.

Keskin, S. and Ozkaya, H. 2013. Effect of storage and insect infestation on the mineral and vitamin contents of wheat grain and flour. *Journal of Economic Entomology* 106(2), 1058–63. doi:10.1603/ec12391.

Keulemans, W., Bylemans, D. and De Coninck, B. 2019. Farming without plant protection products. Can we grow without using herbicides, fungicides and insecticides? *European Parliamentary Research Service*, 44.

Khorramshahi, A. and Burkholder, W. E. 1981. Behavior of the lesser grain borer *Rhyzopertha dominica* (Coleoptera: Bostrichidae). *Journal of Chemical Ecology* 7(1), 33–8. doi:10.1007/BF00988633.

König, K., Seeger, L. and Steidle, J. L. 2015. Sexy mouth odour? Male oral gland pheromone in the grain beetle parasitoid *Lariophagus distinguendus* (Forster) (Hymenoptera: Pteromalidae). *BioMed Research International* 2015, 216952. doi:10.1155/2015/216952.

Krasnoff, S. B. and Vick, K. W. 1984. Male wing-gland pheromone of *Ephestia elutella*. *Journal of Chemical Ecology* 10(4), 667–79. doi:10.1007/BF00994227.

Kühbandner, S., Sperling, S., Mori, K. and Ruther, J. 2012. Deciphering the signature of cuticular lipids with contact sex pheromone function in a parasitic wasp. *The Journal of Experimental Biology* 215(14), 2471–8. doi:10.1242/jeb.071217.

Kumar, D. and Kalita, P. 2017. Reducing postharvest losses during storage of grain crops to strengthen food security in developing countries. *Foods* 6(1), 1–22. doi:10.3390/foods6010008.

Kuwahara, Y. and Casida, J. E. 1973. Quantitative analysis of the sex pheromone of several phycitid moths by electron-capture gas chromatography. *Agricultural and Biological Chemistry* 37(3), 681–4. doi:10.1080/00021369.1973.10860730.

Kuwahara, Y., Kitamura, C., Takashi, S., Hara, H., Ishii, S. and Fukami, H. 1971. Sex pheromone of the almond moth and the Indian meal moth: cis-9, trans-12-tetradecadienyl acetate. *Science* 171(3973), 801–2. doi:10.1126/science.171.3973.801.

Lapointe, S. L., Stelinski, L. L. and Robinson, R. D. 2011. A Novel pheromone dispenser for mating disruption of the leafminer *Phyllocnistis citrella* (Lepidoptera: Gracillariidae). *Journal of Economic Entomology* 104(2), 540–7. doi:10.1603/ec10267.

Lapointe, S. L., Keathley, C. P., Stelinski, L. L., Urrutia, W. H. and Mafra-Neto, A. 2015. Disruption of the leafminer *Phyllocnistis citrella* (Lepidoptera: Gracillariidae) in citrus: effect of blend and placement height, longevity of disruption and emission profile of a new dispenser. *The Florida Entomologist* 98(2), 742–8. doi:10.1653/024.098.0251.

Law, J. H. and Regnier, F. E. 1971. Pheromones. *Annual Review of Biochemistry* 40, 533–48. doi:10.1146/annurev.bi.40.070171.002533.

Leahy, J., Mendelsohn, M., Kough, J., Jones, R. and Berckes, N. 2014. Biopesticide oversight and registration at the U.S. Environmental Protection Agency. In: Gross, A. D., Coats, J. R., Duke, S. O., et al. (Eds), *Biopesticides: State of the Art and Future Opportunities*. American Chemical Society, pp. 3–18.

Levinson, A. and Levinson, H. 1995. Reflections on structure and function of pheromone glands in storage insect species. *Anzeiger für Schädlingskunde Pflanzenschutz Umweltschutz* 68(5), 99–118. doi:10.1007/BF01906539.

Lis, Ł. B., Bakuła, T., Baranowski, M. and Czarnewicz, A. 2011. The carcinogenic effects of benzoquinones produced by the flour beetle. *Polish Journal of Veterinary Sciences* 14(1), 159–64. doi:10.2478/v10181-011-0025-8.

Loko, Y. L., Djagoun, A. D., Dannon, E. A., Datinon, B., Dansi, A., Thomas-Odjo, A. A. and Tamo, M. 2017. Functional response of the predators *Alloeocranum biannulipes* (Hemiptera: Reduviidae) and *Teretrius nigrescens* (Coleoptera: Histeridae) feeding on *Dinoderus porcellus* (Coleoptera: Bostrichidae) infesting yam chips. *Environmental Entomology* 46(1), 84–91. doi:10.1093/ee/nvw156.

Loko, Y. L. E., Gavoedo, D. M., Toffa, J., Orobiyi, A., Thomas-Odjo, A. and Tamò, M. 2019. Life table of the predator *Alloeocranum biannulipes* Montrouzier and Signoret (Hemiptera: Reduviidae) and a test of its ability to suppress populations of *Dinoderus porcellus* Lesne (Coleoptera: Bostrichidae) in stored yam chips. *Biological Control* 130, 60–9. doi:10.1016/j.biocontrol.2018.12.011.

Losey, S. M., Daglish, G. J. and Phillips, T. W. 2019. Orientation of rusty grain beetles, *Cryptolestes ferrugineus* (Coleoptera: Laemophloeidae), to semiochemicals in field and laboratory experiments. *Journal of Stored Products Research* 84, 101513. doi:10.1016/j.jspr.2019.101513.

Lü, J. and Liu, S. 2016. The behavioral response of *Lasioderma serricorne* (Coleoptera: Anobiidae) to citronellal, citral, and rutin. *SpringerPlus* 5(1), 798. doi:10.1186/s40064-016-2553-2.

Mahroof, R. M. and Phillips, T. W. 2014. Mating disruption of *Lasioderma serricorne* (Coleoptera: Anobiidae) in stored product habitats using the synthetic pheromone serricornin. *Journal of Applied Entomology* 138(5), 378–86. doi:10.1111/jen.12097.

Mbata, G. N. and Warsi, S. 2019. *Habrobracon hebetor* and *Pteromalus cerealellae* as tools in post-harvest integrated pest management. *Insects* 10(4), 85. doi:10.3390/insects10040085.

Mbata, G. N., Eason, J., Payton, M. E. and Davis, M. F. 2017. Putative host volatiles used by *Habrobracon hebetor* (Hymenoptera: Braconidae) to locate larvae of *Plodia interpunctella* (Lepidoptera: Pyralidae). *Journal of Insect Behavior* 30(3), 287–99. doi:10.1007/s10905-017-9619-z.

McKay, T., Bowombe-Toko, M. P., Starkus, L. A., Arthur, F. H. and Campbell, J. F. 2019. Monitoring of *Tribolium castaneum* (Coleoptera: Tenebrionidae) in rice mills using pheromone-baited traps. *Journal of Economic Entomology* 112(3), 1454–62. doi:10.1093/jee/toy422.

Metzger, M., Fischbein, D., Auguste, A., Fauvergue, X., Bernstein, C. and Desouhant, E. 2010. Synergy in information use for mate finding: demonstration in a parasitoid wasp. *Animal Behaviour* 79(6), 1307–15. doi:10.1016/j.anbehav.2010.03.003.

Mikolajczak, K. L., Zilkowski, B. W., Smith, C. R. and Burkholder, W. E. 1984. Volatile food attractants for *Oryzaephilus surinamensis* (L.) from oats. *Journal of Chemical Ecology* 10(2), 301–9. doi:10.1007/BF00987858.

Mills, N. J. and Heimpel, G. E. 2018. Could increased understanding of foraging behavior help to predict the success of biological control? *Current Opinion in Insect Science* 27, 26–31. doi:10.1016/j.cois.2018.02.013.

Miyatake, T., Yokoi, T., Fuchikawa, T., Korehisa, N., Kamura, T., Nanba, K., Ryouji, S., Kamioka, N., Hironaka, M., Osada, M., Hariyama, T., Sasaki, R. and Shinoda, K. 2016. Monitoring and detecting the cigarette beetle (Coleoptera: Anobiidae) using ultraviolet (LED) direct and reflected lights and/or pheromone traps in a laboratory and a storehouse. *Journal of Economic Entomology* 109(6), 2551–60. doi:10.1093/jee/tow225.

Mohandass, S., Arthur, F. H., Zhu, K. Y. and Throne, J. E. 2007. Biology and management of *Plodia interpunctella* (Lepidoptera: Pyralidae) in stored products. *Journal of Stored Products Research* 43(3), 302–11. doi:10.1016/j.jspr.2006.08.002.

Morath, S. U., Hung, R. and Bennett, J. W. 2012. Fungal volatile organic compounds: a review with emphasis on their biotechnological potential. *Fungal Biology Reviews* 26(2–3), 73–83. doi:10.1016/j.fbr.2012.07.001.

Mori, K. 2010. Pheromones in chemical communication. In: Herrmann, A. (Ed.), *The Chemistry and Biology of Volatiles*. John Wiley and Sons Ltd., West Sussex, UK, pp. 123–49.

Mori, K. 2012. Pheromone synthesis. Part 249: Syntheses of methyl (R,E)-2,4,5-tetradecatrienoate and methyl (2E,4Z)-2,4-decadienoate, the pheromone components of the male dried bean beetle, Acantoscelides obtectus (Say). *Tetrahedron* 68, 1936–1946.

Mori, K., Suguro, T. and Uchida, M. 1978. Synthesis of optically active forms of (Z)-14-methyl-hexadec-8-enal. The pheromone of female dermestid beetle. *Tetrahedron* 34(20), 3119–23. doi:10.1016/0040-4020(78)87008-2.

Müller-Blenkle, C., Kirchner, S., Szallies, I., Adler, C. 2018. A new approach to acoustic insect detection in grain storage. In: Adler, C. S., Opit, G., Fürstenau, B., et al. (Eds), *Proceedings of the 12th International Working Conference on Stored Product Protection* (vol. 463). Julius-Kühn-Archiv, Berlin, Germany, pp. 328–37.

Murata, M., Imamura, T. and Miyanoshita, A. 2016. Suppression of the stored-product pest *Sitophilus zeamais* (Coleoptera: Curculionidae) by simultaneous release of a predatory bug and a parasitoid wasp. *Biocontrol Science and Technology* 26(6), 872–6. doi:10.1080/09583157.2016.1159283.

Namikawa, S. K. and Amornsak, W. 2016. Antennal sensilla morphology of *Theocolax elegans* (Westwood) (Hymenoptera: Pteromalidae), a larval parasitoid of the maize weevil, *Sitophilus zeamais* (Motschulsky) (Coleoptera: Curculionidae). *Agriculture and Natural Resources* 50(5), 374–9. doi:10.1016/j.anres.2015.12.003.

Nayak, M. K. and Daglish, G. J. 2018. Importance of stored product insects. In: Hagstrum, D. W., Phillips, T. W. and Cuperus, G. (Eds), *Recent Advances in Stored Product Protection*. Springer, Berlin, Germany, pp. 1–17.

Nayak, M. K., Falk, M. G., Emery, R. N., Collins, P. J. and Holloway, J. C. 2017. An analysis of trends, frequencies and factors influencing the development of resistance to phosphine in the red flour beetle *Tribolium castaneum* (Herbst) in Australia. *Journal of Stored Products Research* 72, 35–48. doi:10.1016/j.jspr.2017.03.004.

Nboyine, J. A., Asante, S. K., Nutsugah, S. K., Abudulai, M., Ansaah-Agyapong, F., Luke, B. and Clottey, V. 2015. Biological control of the larger grain borer, *Prostephanus truncatus* (Horn) in stored maize using the fungal pathogen, *Beauveria bassiana* and the predator *Teretrius nigrescens* Lewis. *Journal of Stored Products and Postharvest Research* 6, 30–7.

Ndomo-Moualeu, A. F., Ulrichs, C. and Adler, C. 2016. Behavioral responses of *Callosobruchus maculatus* to volatile organic compounds found in the headspace of dried green pea seeds. *Journal of Pest Science* 89(1), 107–16. doi:10.1007/s10340-015-0652-4.

Niedermayer, S. and Steidle, J. L. M. 2013. The Hohenheimer Box – a new way to rear and release *Lariophagus distinguendus* to control stored product pest insects. *Biological Control* 64(3), 263–9. doi:10.1016/j.biocontrol.2012.12.005].

Niedermayer, S., Krogmann, L. and Steidle, J. L. M. 2016a. Lost in space? Host-finding ability of the parasitoids *Lariophagus distinguendus* and *Anisopteromalus calandrae* in empty grain storage facilities to control residual pest populations. *BioControl* 61(4), 379–86. doi:10.1007/s10526-016-9717-4.

Niedermayer, S., Pollmann, M. and Steidle, J. L. M. 2016b. *Lariophagus distinguendus* (Hymenoptera: Pteromalidae) (Förster)-past, present, and future: the history of a biological control method using L. *distinguendus* against different storage pests. *Insects* 7(3), 39. doi:10.3390/insects7030039.

Noldus, L. P. J. J., Van Lenteren, J. C. V. and Lewis, W. J. 1991. How Trichogramma parasitoids use moth sex pheromones as kairomones: orientation behaviour in a wind tunnel. *Physiological Entomology* 16(3), 313–27. doi:10.1111/j.1365-3032.1991.tb00570.x.

Nordlund, D. A. and Lewis, W. J. 1976. Terminology of chemical releasing stimuli in intraspecific and interspecific interactions. *Journal of Chemical Ecology* 2(2), 211–20. doi:10.1007/BF00987744.

Parnell, J. J., Berka, R., Young, H. A., Sturino, J. M., Kang, Y., Barnhart, D. M. and DiLeo, M. V. 2016. From the lab to the farm: an industrial perspective of plant beneficial microorganisms. *Frontiers in Plant Science* 7, 1110. doi:10.3389/fpls.2016.01110.

Pereira, A. I. A., Pikart, T. G., Ramalho, F. S., Manickavasagam, S., Serrão, J. E. and Zanuncio, J. C. 2013. *Antrocephalus mitys* (Hymenoptera: Chalcididae) in laboratory cultures of *Tenebrio molitor* (Coleoptera: Tenebrionidae), and possible role in biological control of *Ephestia cautella* (Lepidoptera: Pyralidae). *Florida Entomologist* 96(2), 634–7. doi:10.1653/024.096.0233.

Peri, E., Salerno, G., Slimani, T., Frati, F., Conti, E., Colazza, S. and Cusumano, A. 2016. The response of an egg parasitoid to substrate-borne semiochemicals is affected by previous experience. *Scientific Reports* 6, 27098. doi:10.1038/srep27098.

Phelan, P. L., Silk, P. J., Northcott, C. J., Tan, S. H. and Baker, T. C. 1986. Chemical identification and behavioral characterization of male wing pheromone of *Ephestia elutella* (Pyralidae). *Journal of Chemical Ecology* 12(1), 135–46. doi:10.1007/BF01045597.

Phillips, J. K., Chong, J. M., Andersen, J. F. and Burkholder, W. E. 1989. Determination of the enantiomeric composition of (R*, S*)-1-ethylpropyl 2-methyl-3-hydroxypentanoate, the male-produced aggregation pheromone of *Sitophilus granarius*. *Entomologia Experimentalis et Applicata* 51(2), 149–53. doi:10.1111/j.1570-7458.1989.tb01225.x.

Phillips, T. W. 1997. Semiochemicals of stored-product insects: research and applications. *Journal of Stored Products Research* 33(1), 17–30. doi:10.1016/S0022-474X(96)00039-2.

Phillips, T. W. and Throne, J. E. 2010. Biorational approaches to managing stored-product insects. *Annual Review of Entomology* 55, 375–97. doi:10.1146/annurev.ento.54.110807.090451.

Phillips, T. W., Jiang, X. L., Burkholder, W. E., Phillips, J. K. and Tran, H. Q. 1993. Behavioral responses to food volatiles by two species of stored-product Coleoptera, *Sitophilus oryzae* (Curculionidae) and *Tribolium castaneum* (Tenebrionidae). *Journal of Chemical Ecology* 19(4), 723–34. doi:10.1007/BF00985004.

Pierce, A. M., Pierce, H. D., Borden, J. H. and Oehlschlager, A. C. 1991. Fungal volatiles: semiochemicals for stored-product beetles (Coleoptera: Cucujidae). *Journal of Chemical Ecology* 17(3), 581–97. doi:10.1007/BF00982128.

Pimentel, D. 2011. Food for thought: a review of the role of energy in current and evolving agriculture. *Critical Reviews in Plant Sciences* 30(1–2), 35–44. doi:10.1080/07352689.2011.554349.

Prozell, S. and Schöller, M. 2018. Does it really work? 25 years biological control in Germany. In: Adler, C. S., Opit, G., Fürstenau, B., et al. (Eds), *Proceedings of the 12th International Working Conference on Stored Product Protection* (vol. 463). Julius-Kühn-Archiv, Berlin, Germany, pp. 439–41.

Quicke, D. L. J. 1997. *Parasitic Wasps*. Chapman & Hall Ltd, New York, NY.

Regnier, F. E. and Law, J. H. 1968. Insect pheromones. *Journal of Lipid Research* 9(5), 541–51.

Riudavets, J. 2018. Biological control of stored products pests. *Integrated Protection of Stored Products/IOBC-WPRS Bulletin*. Ljubljana, Slovenia, 107–9.

Rodríguez, S. A. 2018. Semiochemistry of the Bruchidae species. *Journal of Stored Products Research* 77, 148–55. doi:10.1016/j.jspr.2018.04.011.

Ruther, J., Meiners, T. and Steidle, J. L. M. 2002. Rich in phenomena-lacking in terms. A classification of kairomones. *Chemoecology* 12(4), 161–7. doi:10.1007/PL00012664.

Ryne, C., Ekeberg, M., Jonzén, N., Oehlschlager, C., Löfstedt, C. and Anderbrant, O. 2006. Reduction in an almond moth *Ephestia cautella* (Lepidoptera: Pyralidae) population by means of mating disruption. *Pest Management Science* 62(10), 912–8. doi:10.1002/ps.1256.

Saad, M. M. G. and Abdelgaleil, S. A. M. 2018. Effectiveness of monoterpenes and phenylpropenes on *Sitophilus oryzae* L. (Coleoptera: Curculionidae) in stored wheat. *Journal of Asia-Pacific Entomology* 21(4), 1153–8. doi:10.1016/j.aspen.2018.08.006.

Sambaraju, K. R. and Phillips, T. W. 2008. Responses of adult *Plodia interpunctella* (Hübner) (Lepidoptera: Pyralidae) to light and combinations of attractants and light. *Journal of Insect Behavior* 21(5), 422–39. doi:10.1007/s10905-008-9140-5.

Sánchez-Mariñez, R. I., Cortez-Rocha, M. O., Ortega-Dorame, F., Morales-Valdes, M. and Silveira, M. I. 1997. End-use quality of flour from *Rhyzopertha dominica* infested wheat. *Cereal Chemistry Journal* 74(4), 481–3. doi:10.1094/CCHEM.1997.74.4.481.

SANCO. 2014. Guidance document on botanical active substances used in plant protection products. *European Commission Health & Food Safety Directorate-General. Safety of the Food Chain Chemicals, Contaminats, Pesticides* 11470/2012- rev 8.

SANTE. 2016. Guidance document on semiochemical active substances and plant protection products. *European Commission Health & Food Safety Directorate-General. Safety of the Food Chain Pesticides and Biocides* 12815/2014 rev. 5.2.

Savoldelli, S. and Trematerra, P. 2011. Mass-trapping, mating-disruption and attracticide methods for managing stored-product insects: success stories and research needs. *Stewart Postharvest Review* 7, 1–8.

Schöller, M. 1998. Integration of biological and non-biological methods for controlling arthropods infesting stored products. *Postharvest News and Information* 9, 15N–20N.

Schöller, M. 2015. Recent advances in the commercial application of beneficials against stored-product and cultural heritage pests. In: Trematerra, P. and Hamel, D. (Eds), *Integrated Protection of Stored Products/IOBC-WPRS Bulletin*. Zagreb, Croatia, pp. 345–8.

Schöller, M. and Flinn, P. W. 2000. Parasitoids and predators. In: Subramanyam, B. and Hagstrum, D. W. (Eds), *Alternatives to Pesticides in Stored-Product IPM*. Kluwer Academic Publishers, Boston, pp. 229–71.

Schöller, M. and Prozell, S. 2002. Response of *Trichogramma evanescens* to the main sex pheromone component of *Ephestia spp.* and *Plodia interpunctella*, (Z,E)-9,12-tetra-decadenyl acetate (ZETA). *Journal of Stored Products Research* 38(2), 177–84. doi:10.1016/S0022-474X(01)00015-7.

Schöller, M., Prozell, S., Al-Kirshi, A. -G. and Reichmuth, C. 1997. Towards biological control as a major component of integrated pest management in stored product protection. *Journal of Stored Products Research* 33(1), 81–97. doi:10.1016/S0022-474X(96)00048-3.

Schöller, M., Flinn, P. W., Grieshop, M. J. and Zdarkova, E. 2006. Biological control of stored product pests. In: Heaps, J. W. (Ed.), *Insect Management for Food Storage and Processing*. American Association of Cereal Chemists International, St. Paul, MN, pp. 67–87.

Schöller, M., Prozell, S., Suma, P. and Russo, A. 2018. Biological control of stored-product insects. In: Athanassiou, C. and Arthur, F. H. (Eds), *Recent Advances in Stored Product Protection*. Springer, Berlin, Germany, pp. 183–209.

Schröder, R. and Hilker, M. 2008. The relevance of background odor in resource location by insects: a behavioral approach. *BioScience* 58(4), 308–16. doi:10.1641/B580406.

Seitz, L. and Sauer, D. 1996. Volatile compounds and odors in grain sorghum infested with common storage insects. *Cereal Chemistry* 73, 744–50.

Semeao, A. A., Campbell, J. F., Whitworth, R. J. and Sloderbeck, P. E. 2011. Response of *Tribolium castaneum* and *Tribolium confusum* adults to vertical black shapes and its potential to improve trap capture. *Journal of Stored Products Research* 47(2), 88–94. doi:10.1016/j.jspr.2011.01.002.

Shimomura, K., Nojima, S., Yajima, S. and Ohsawa, K. 2008. Homofarnesals: female sex attractant pheromone components of the southern cowpea weevil, *Callosobruchus chinensis*. *Journal of Chemical Ecology* 34(4), 467–77. doi:10.1007/s10886-008-9451-1.

Sieminska, E., Ryne, C., Löfstedt, C. and Anderbrant, O. 2009. Long-term pheromone-mediated mating disruption of the Mediterranean flour moth, *Ephestia kuehniella*, in a flourmill. *Entomologia Experimentalis et Applicata* 131(3), 294–9. doi:10.1111/j.1570-7458.2009.00858.x.

Silverstein, R. M. 1981. Pheromones: background and potential for use in insect pest control. *Science* 213(4514), 1326–32. doi:10.1126/science.213.4514.1326.

Sitthichaiyakul, S., Kengkanpanich, R., Noochanapai, P. and Amomsak, W. 2018 Host-age preference of *Theocolax elegans* (Westwood) (Hymenoptera: Pteromalidae), a larval parasitoid of the lesser grain borer, *Rhyzopertha dominica* (Fabricius) (Coleoptera: Bostrichidae) and the cowpea weevil, *Callosobruchus maculatus* (Fabricius) (Coleoptera: Chrysomelidae). In: Adler, C. S., Opit, G., Fürstenau, B., et al. (Eds), *Proceedings of the 12th International Working Conference on Stored Product Protection* (vol. 463). Julius-Kühn-Archiv, Berlin, Germany, pp. 454–8.

Smith, B. H. 1993. Merging mechanisms and adaptation: an ethological approach to learning and generalization. In: Papaj, D. R. and Lewis, A. C. (Eds), *Insect Learning - Ecology and Evolutionary Perspectives*. Chapman & Hall, New York, pp. 126–57.

Solà, M. 2017. *Approaches for the Biological Control of Stored Product Pests*. IRTA, Cabrils, Barcelona. Universitat Autònoma de Barcelona, Spain, p. 222.

Solà, M., Castañé, C., Lucas, E. and Riudavets, J. 2018. Optimization of a banker box system to rear and release the parasitoid *Habrobracon hebetor* (Hymenoptera: Braconidae) for the control of stored-product moths. *Journal of Economic Entomology* 111(5), 2461–6. doi:10.1093/jee/toy219.

Steidle, J. L. M. 1998. Learning pays off: influence of experience on host finding and parasitism in *Lariophagus distinguendus*. *Ecological Entomology* 23(4), 451-. doi:10.1046/j.1365-2311.1998.00144.x.

Steidle, J. L. M. and Schöller, M. 1997. Olfactory host location and learning in the granary weevil parasitoid *Lariophagus distinguendus* (Hymenoptera: Pteromalidae). *Journal of Insect Behavior* 10(3), 331–42. doi:10.1007/BF02765601.

Steidle, J. L. M. and Van Loon, J. J. A. 2003. Dietary specialization and infochemical use in carnivorous arthropods: testing a concept. *Entomologia Experimentalis et Applicata* 108(3), 133–48. doi:10.1046/j.1570-7458.2003.00080.x.

Steidle, J. L. M., Steppuhn, A. and Reinhard, J. 2001. Volatile cues from different host complexes used for host location by the generalist parasitoid *Lariophagus distinguendus* (Hymenoptera: Pteromalidae). *Basic and Applied Ecology* 2(1), 45–51. doi:10.1078/1439-1791-00035.

Steidle, J. L. M., Steppuhn, A. and Ruther, J. 2003. Specific foraging kairomones used by a generalist parasitoid. *Journal of Chemical Ecology* 29(1), 131–43. doi:10.1023/a:1021932731350.

Steidle, J. L. M., Fischer, A. and Gantert, C. 2005. Do grains whisper for help? Evidence for herbivore-induced synomones in wheat grains. *Entomologia Experimentalis et Applicata* 115(1), 239–45. doi:10.1111/j.1570-7458.2005.00295.x.

Steiner, S., Erdmann, D., Steidle, J. L. M. and Ruther, J. 2007 Host habitat assessment by a parasitoid using fungal volatiles. *BMC Frontiers Zoology* 4, 1–10.

Stejskal, V., Aulicky, R. and Kucerova, Z. 2014. Pest Control strategies and damage potential of seed-infesting pests in the Czech stores - a review. *Plant Protection Science* 50, 165–73.

Stejskal, V., Hubert, J., Aulicky, R. and Kucerova, Z. 2015. Overview of present and past and pest-associated risks in stored food and feed products: European perspective. *Journal of Stored Products Research* 64, 122–32. doi:10.1016/j.jspr.2014.12.006.

Stevenson, B. J., Cai, L., Faucher, C., Michie, M., Berna, A., Ren, Y., Anderson, A., Chyb, S. and Xu, W. 2017. Walking responses of *Tribolium castaneum* (Coleoptera: Tenebrionidae) to its aggregation pheromone and odors of wheat infestations. *Journal of Economic Entomology* 110(3), 1351–8. doi:10.1093/jee/tox051.

Stewart-Jones, A., Hodges, R. J., Birkinshaw, L. A. and Hall, D. R. 2004. Responses of *Teretrius nigrescens* toward the dust and frass of its prey, *Prostephanus truncatus*. *Journal of Chemical Ecology* 30(8), 1629–46. doi:10.1023/b:joec.0000042073.92464.7d.

Strand, M. R., Williams, H. J., Vinson, S. B. and Mudd, A. 1989. Kairomonal activities of 2-acylcyclohexane-1,3 diones produced by *Ephestia kuehniella* Zeller in eliciting searching behavior by the parasitoid *Bracon hebetor* (Say). *Journal of Chemical Ecology* 15(5), 1491–500. doi:10.1007/BF01012378.

Suzuki, T. 1980. 4,8-Dimethyldecanal: the aggregation pheromone of the flour beetles, *Tribolium castaneum* and *T. confusum* (Coleoptera: Tenebrionidae). *Agricultural and Biological Chemistry* 44(10), 2519–20. doi:10.1271/bbb1961.44.2519.

Suzuki, T. and Sugawara, R. 1979. Isolation of an aggregation pheromone from the flour beetles, *Tribolium castaneum* and *T. confusum* (Coleoptera: Tenebrionidae). *Applied Entomology and Zoology* 14(2), 228–30. doi:10.1303/aez.14.228.

Swords, P. and Van Ryckeghem, A. 2010a. Summary of commercially available pheromones of common stored-product beetles. In: Carvalho, M. O., Fields, P. G., Adler, C. S., et al. (Eds), *Proceedings of the 10th International Working Conference on Stored Product Protection* (vol. 425). Julius-Kühn-Archiv, Estoril, Portugal, pp. 1004–7.

Swords, P. and Van Ryckeghem, A. 2010b. Summary of commercially available pheromones of common stored product moths. In: Carvalho, M. O., Fields, P. G., Adler, C. S., et al. (Eds), *Proceedings of the 10th International Working Conference on Stored Product Protection* (vol. 425). Julius-Kühn-Archiv, Estoril, Portugal, pp. 1008-10.

Tang, Q., Yang, T. and Jiang, J. 2016. Herbivore-induced rice grain volatiles affect attraction behavior of herbivore enemies. *Interciencia* 41, 319–24.

Teal, P. E. A., Heath, R. R., Dueben, B. D., Coffelt, J. A. and Vick, K. W. 1995. Production and release of (*Z,E*)-9,12-tetradecadienal by sex pheromone glands of females of *Plodia interpunctella* (Lepidoptera: Pyralidae). *Journal of Chemical Ecology* 21(6), 787–99. doi:10.1007/BF02033461.

Trdan, S. and Bohinc, T. 2018. Twelve years (2005–2017) of scientific and professional work in the field of stored products pests protection in Slovenia. In: Adler, C.S., Opit, G., Fürstenau, B., et al. (Eds), Proceedings of the *12th International Working Conference on Stored Product Protection* (vol. 463). Julius-Kühn-Archiv, Berlin, Germany, pp. 522–32.

Trematerra, P. 1997. Integrated pest management of stored-product insects: practical utilization of pheromones. *Anzeiger für Schädlingskunde Pflanzenschutz Umweltschutz* 70(3), 41–4. doi:10.1007/BF01996919.

Trematerra, P. 2012. Advances in the use of pheromones for stored-product protection. *Journal of Pest Science* 85(3), 285–99. doi:10.1007/s10340-011-0407-9.

Trematerra, P. 2018. Efficacy of pheromones for managing of the Mediterranean flour moth, *Ephestia kuehniella* Zeller, in food and feed processing facilities. In: Adler, C. S., Opit, G., Fürstenau, B., et al. (Eds), *Proceedings of the 12th International Working Conference on Stored Product Protection* (vol. 463). Julius-Kühn-Archiv, Berlin, Germany, pp. 485–93.

Trematerra, P. and Gentile, P. 2010. Five years of mass trapping of *Ephestia kuehniella* Zeller: a component of IPM in a flour mill. *Journal of Applied Entomology* 134(2), 149–56. doi:10.1111/j.1439-0418.2009.01470.x.

Trematerra, P. and Savoldelli, S. 2013. The use of water traps and presence of spermatophores to evaluate mating disruption in the almond moth, *Ephestia cautella*, during exposure to synthetic sex pheromone. *Journal of Pest Science* 86(2), 227–33. doi:10.1007/s10340-012-0477-3.

Trematerra, P., Athanassiou, C., Stejskal, V., Sciarretta, A., Kavallieratos, N. and Palyvos, N. 2011. Large-scale mating disruption of *Ephestia* spp. and *Plodia interpunctella* in Czech Republic, Greece and Italy. *Journal of Applied Entomology* 135(10), 749–62. doi:10.1111/j.1439-0418.2011.01632.x.

Trematerra, P., Oliviero, A., Savoldelli, S. and Schöller, M. 2017. Controlling infestation of a chocolate factory by *Plodia interpunctella* by combining mating disruption and the parasitoid *Habrobracon hebetor*. *Insect Science* 24(3), 503–10. doi:10.1111/1744-7917.12319.

Tröger, A., van Beek, T. A., Huigens, M. E., Silva, I. M., Posthumus, M. A. and Francke, W. 2014. Structure elucidation of female-specific volatiles released by the parasitoid wasp *Trichogramma turkestanica* (Hymenoptera: Trichogrammatidae). *Beilstein Journal of Organic Chemistry* 10, 767–73. doi:10.3762/bjoc.10.72.

Turlings, T. C. J., Wäckers, F., Vet, L. E. M., Lewis, W. J. and Tumlinson, J. H. 1993. Learning of host-finding cues by hymenopterous parasitoids. In: Papaj, D. R. and Lewis, A. (Eds), *Insect Learning - Ecology and Evolutionary Perspectives*. Chapman and Hall, New York, pp. 51–78.

Upadhyay, R. K. and Ahmad, S. W. 2011. Management strategies for control of stored grain insect pests in farmer stores and public ware houses. *World Journal of Agricultural Science* 7, 527–49.

van Alphen, J. J. M. and Vet, L. E. M. 1986. An evolutionary approach to host finding and election. In: Waage, J. K. and Greathead, D. (Eds), *Insect Parasitoids*. Aademic Press, London, pp. 23–61.

van Lenteren, J. C. 2012. The state of commercial augmentative biological control: plenty of natural enemies, but a frustrating lack of uptake. *BioControl* 57(1), 1–20. doi:10.1007/s10526-011-9395-1.

van Lenteren, J. C., Roskam, M. M. and Timmer, R. 1997. Commercial mass production and pricing of organisms for biological control of pests in Europe. *Biological Control* 10(2), 143–9. doi:10.1006/bcon.1997.0548.

van Lenteren, J. C., Bolckmans, K., Köhl, J., Ravensberg, W. J. and Urbaneja, A. 2018. Biological control using invertebrates and microorganisms: plenty of new opportunities. *BioControl* 63(1), 39–59. doi:10.1007/s10526-017-9801-4.

van Oudenhove, L., Mailleret, L. and Fauvergue, X. 2017. Infochemical use and dietary specialization in parasitoids: a meta-analysis. *Ecology and Evolution* 7(13), 4804–11. doi:10.1002/ece3.2888.

Verheggen, F., Ryne, C., Olsson, P. O., Arnaud, L., Lognay, G., Högberg, H. E., Persson, D., Haubruge, E. and Löfstedt, C. 2007. Electrophysiological and behavioral activity of secondary metabolites in the confused flour beetle, *Tribolium confusum*. *Journal of Chemical Ecology* 33(3), 525–39. doi:10.1007/s10886-006-9236-3.

Vet, L. E. M., Lewis, W. J. and Cardé, R. T. 1995. Parasitoid foraging and learning. In: Cardé, R. T. and Bell, W. J. (Eds), *Chemical Ecology of Insects 2*. Chapman & Hall, New York, pp. 65–101.

Vick, K. W., Su, H. C. F., Sower, L. L., Mahany, P. G. and Drummond, P. C. 1974. (*Z,E*)-7,11-hexadecadien-1-ol acetate: the sex pheromone of the angoumois grain moth, *Sitotroga cerealella*. *Experientia* 30(1), 17–8. doi:10.1007/BF01921569.

Vosteen, I., van den Meiracker, N. and Poelman, E. H. 2019. Getting confused: learning reduces parasitoid foraging efficiency in some environments with non-host-infested plants. *Oecologia* 189(4), 919–30. doi:10.1007/s00442-019-04384-2.

Vuts, J., Francke, W., Mori, K., Zarbin, P. H. G., Hooper, A. M., Millar, J. G., Pickett, J. A., Tóth, M., Chamberlain, K., Caulfield, J. C., Woodcock, C. M., Tröger, A. G., Csonka, ÉB. and Birkett, M. A. 2015. Pheromone bouquet of the dried bean beetle, *Acanthoscelides obtectus* (Col.: Chrysomelidae), now complete. *European Journal of Organic Chemistry* 2015(22), 4843–6. doi:10.1002/ejoc.201500196.

Wajnberg, E. and Colazza, S. 2013. *Chemical Ecology of Parasitoids*. Wiley-Blackwell, Chichester.

Wang, S. Y., Hackney Price, J. and Zhang, D. 2019. Hydrocarbons catalysed by TmCYP4G122 and TmCYP4G123 in *Tenebrio molitor* modulate the olfactory response of the parasitoid *Scleroderma guani*. *Insect Molecular Biology* 28(5), 637–48. doi:10.1111/imb.12581.

Weiss, M. R. 2006. Defecation behavior and ecology of insects. *Annual Review of Entomology* 51, 635–61. doi:10.1146/annurev.ento.49.061802.123212.

WHO and FAO. 2017. Guidelines for the registration of microbial, botanical and semiochemical pest control agents for plant protection and public health uses. *International Code of Conduct on Pesticide Management* 76.

Williams, H. J., Silverstein, R. M., Burkholder, W. E. and Khorramshahi, A. 1981. Dominicalure 1 and 2: components of aggregation pheromone from male lesser grain borer *Rhyzopertha dominica* (f.) (Coleoptera: Bostrichidae). *Journal of Chemical Ecology* 7(4), 759–80. doi:10.1007/BF00990308.

Witzgall, P., Kirsch, P. and Cork, A. 2010. Sex pheromones and their impact on pest management. *Journal of Chemical Ecology* 36(1), 80–100. doi:10.1007/s10886-009-9737-y.

Yang, S., Zhang, X. F., Gao, Y. L., Chen, D., She, D. M., Zhang, T. and Ning, J. 2017. Male-produced aggregation pheromone of coffee bean weevil, *Araecerus fasciculatus*. *Journal of Chemical Ecology* 43(10), 978–85. doi:10.1007/s10886-017-0894-0.

Zhu, J., Ryne, C., Unelius, C. R., Valeur, P. G. and Lofstedt, C. 1999. Reidentification of the female sex pheromone of the Indian meal moth, *Plodia interpunctella*: evidence for a four-component pheromone blend. *Entomologia Experimentalis et Applicata* 92(2), 137–46. doi:10.1046/j.1570-7458.1999.00533.x.

# Chapter 10

## Advances in insect pest management in postharvest storage of cereals: novel techniques

*Peter Follett, USDA-ARS, USA; Georgios Akepsimaidis and Nicolas Meneses, Bühler AG, Switzerland; and Matthew Murdoch and Heidi Kotilainen, Bühler UK Limited, UK*

1 Introduction
2 Irradiation for control of stored product pests
3 Sources of ionizing radiation
4 Case study 1: low-energy electron beam treatment
5 Case study 2: high-energy X-ray treatment
6 Conclusion
7 Future trends
8 Where to look for further information
9 References

### 1 Introduction

Food irradiation is the process by which food is exposed to ionizing radiation in order to control infestation by insects, kill harmful bacteria and parasites, delay ripening, prevent sprouting of produce, and prevent or delay the growth of molds and bacteria that cause food spoilage. Irradiation has become an important way to ensure that foods marketed internationally meet quality and safety standards, especially since some countries limit or prohibit the use of chemicals due to safety-related issues. Irradiation can also reduce the huge amount of waste resulting from contamination and spoilage. Most countries base their food irradiation regulations on the Codex General Standard for Irradiated Foods (Codex, 2003). This standard recommends that any food may be irradiated to a maximum dose of 10 kGy (10 000 Gy) and can remain wholesome and safe.

According to estimates of the Food and Agriculture Organization (FAO), 5-10% of cereals is lost during storage due to insect infestation (Ahmed, 2001). Most domestic and export markets have zero tolerance for grain contamination

http://dx.doi.org/10.19103/AS.2020.0072.13
Published by Burleigh Dodds Science Publishing Limited, 2020.

by insects. Fumigation treatment with phosphine or methyl bromide ensures that grain can be stored and exported in a clean and uninfested condition. However, resistance to phosphine poses a major challenge (Donahaye, 2000) and methyl bromide is being phased out as an ozone-depleting substance, and therefore alternative control methods are needed. Irradiation is a viable alternative to fumigation as a pre- or post-shipment disinfestation treatment for control of stored food pests. Irradiation is fast, cost competitive, broadly effective against insect and mite pests and pathogens, temperature insensitive, residue free and immune to resistance development (Follett, 2014). Irradiation may be particularly helpful in controlling phosphine and methyl bromide resistant populations and could help manage resistance by preventing the spread of resistant insects in exported grains (Follett et al., 2013).

Ionizing radiation breaks chemical bonds within DNA and other biomolecules, thereby disrupting normal cellular function in insects and other organisms. Radiotolerance in insects can vary among the different life stages and between different taxonomic groups (e.g. insect families and orders) (Follett, 2014). For example, the adult and pupal stages are more tolerant of radiation than the larval and egg stages, and Lepidoptera (moths and butterflies) are generally more tolerant than Coleoptera (beetles) and Diptera (flies). Unlike other disinfestation techniques, irradiation does not need to kill the pest immediately to provide quarantine security, and therefore live (but sterile or not viable) insects may occur with the stored product. The goal of a disinfestation treatment normally is to prevent reproduction, and therefore the desired response for an irradiation treatment is to prevent development to the adult stage or to induce sterility in adult insects.

## 2 Irradiation for control of stored product pests

Most of the regulations adopted worldwide related to insects have focused on regulated pests of horticultural crops, so-called phytosanitary pests (Roberts and Follett, 2018). The United States Food and Drug Administration has approved radiation doses up to 1000 Gy (1 kGy) for preservation and disinfestation of fresh fruits and vegetables (FDA, 1986). Disinfestation means controlling any arthropod pests infesting the commodity, particularly insects. A generic dose of 150 Gy is approved by the United States Department of Agriculture (USDA) and the International Plant Protection Convention (IPPC) for control of tephritid fruit flies in fresh commodities (Follett, 2009; Roberts and Follett, 2018). The USDA, Australia and New Zealand have also approved a 400 Gy for all other insects except pupae and adults of Lepidoptera, which may require a higher dose. Most stored product pests are also controlled at irradiation doses ≤400 Gy (Hallman, 2013). At least 33 countries have approved irradiation for treatment of stored products (Barkai-Golan and Follett, 2017). The largest use of food

Published by Burleigh Dodds Science Publishing Limited, 2020.

irradiation worldwide is the treatment of spices at 5–10 kGy for disease-causing microorganisms. Insects can be controlled at 50–400 Gy, i.e., at less than 1/10 the pathogen dose (Follett, 2014; Barkai-Golan and Follett, 2017), so treatment for pathogens will easily control any stored product insects.

## 2.1 Desired response of irradiation for stored product pests

The goal of irradiation treatment generally is to retard development or sterilize adults (Barkai-Golan and Follett, 2017). With stored products pests it is not always clear what the desired response should be for a disinfestation treatment, e.g. immediate mortality, sterility or arrested feeding. The high doses required for acute mortality in insects (e.g. 1000–3000 Gy) are impractical. Lower radiation doses (50–400 Gy) can prevent development to the adult stage or can sterilize adults, thereby preventing reproduction and propagation. However, post-treatment feeding and frass production by live but non-viable insects may be an unwanted effect of a low sterilizing dose. Sublethal doses of irradiation may cause protracted development and extended longevity (Hasan and Khan, 1998). For commercial application against stored product pests, the ideal irradiation treatment should cause reduced feeding or cessation of feeding in addition to retarding development and preventing reproduction (Follett et al., 2013). Cornwall (1964) reported that *Sitophilus granarius* treated at 160 Gy consumed half the food compared to untreated insects. Watters and McQueen (1967) found that four stored products pests could still cause damage to grain for 5 weeks after being irradiated at 62.5 Gy, but the damage was significantly reduced.

## 2.2 Generic and specific doses

During the past 20 years, irradiation has been used increasingly as a phytosanitary treatment for control of quarantine insect pests in fruits traded internationally as a replacement for methyl bromide (Roberts and Follett, 2018). As with stored cereals and grains, trade in fresh produce can only be conducted when the importing country has confidence that measures are in place to guarantee that pests are not present on the exported produce, or, if present, are non-viable. In addition to the 150 Gy generic dose for the fruit fly family Tephritidae, phytosanitary irradiation doses have also been approved for many specific pests of economic importance such as codling moth (200 Gy), sweetpotato weevil (150 Gy) and the false spider mite (300 Gy) (see complete lists in Barkai-Golan and Follett, 2017). In 2017, the USDA approved a generic dose of 290 Gy for moths in the family Tortricidae. Additional generic treatments would be useful. A generic dose (175 Gy) has been proposed for Curculionidae based on information for about 20 species of weevils, including three species of stored product pests, *Sitophilus oryzae*, *S. granarius* and *S. zeamais* (Follett et al., 2018).

Similar generic radiation doses have not been approved for stored product pests for two reasons which historically have been perceived as problematic: irradiation does not kill insects immediately, and there are no routine tests to determine whether surviving insects are sterile (Donahaye, 2000). The approvals for phytosanitary irradiation to control quarantine insect pests of fresh horticultural commodities for export, and the rapidly expanding use of the technology for this purpose (Follett, 2009), demonstrate that the presence of live but sterile or non-viable insects is acceptable to regulatory authorities, and facility certification and proper documentation can be used to ensure treatment at the required dose. Hence, even though stored product insects may be alive immediately after treatment, their feeding and life span will be greatly reduced, and reproduction will be prevented.

As most stored product insects are cosmopolitan, they are not considered as quarantine pests and are not regulated. The exclusion of stored product insects from the current irradiation approvals also may be due to the high radiation tolerance of several stored products insects. Highly radiotolerant stored product insects include *Sitotroga cerealella*, the Angoumois grain moth which is controlled at 450 Gy but not 400 Gy (Ignatowicz, 2004; Hallman and Phillips, 2008), and the depressed flour beetle, *Palorus subdrepressus*, which may require 400 Gy or more for control (Brower, 1973); the Indianmeal moth, *Plodia interpunctella* (Hubner), may also require radiation doses above 400 Gy to completely sterilize adults (Ignatowicz, 2004). Irradiation studies have been conducted with many stored grain insects, and most are probably controlled at radiation doses well below 400 Gy (Hasan and Khan, 1998; Ahmed, 2001; Ignatowicz, 2004; Hallman, 2013). Few studies have performed large-scale tests with no survivors to confirm a dose that provides quarantine security with high confidence (Follett and Hennessey, 2007). Examples of large-scale confirmatory testing in stored product pests include the testing of 32 025 *Sitophilus oryzae* adults in rice at 120 Gy (Follett et al., 2013), 22 083 *Plodia interpunctella* adults at 388 Gy, and 15 264 *Sitotroga cerealella* adults at 505 Gy (Hallman and Phillips, 2008), all with no survivors. Kapra beetle, *Trogoderma granarium*, is a destructive grain pest with a limited distribution and an example of a regulated stored product pest; the United States restricts the import of rice from countries with known infestations of this pest. An irradiation dose of 200 Gy appears to stop reproduction in *T. granarium* (Gao et al., 2004).

## 2.3 Resistance management using irradiation

Phosphine is one of the most widely used fumigants for stored products. Phosphine is used for control of pests in stored seed and food-grains, high-value commodities such as coffee beans and tobacco, dried fruits, and certain processed foodstuffs. Heavy dependence of phosphine for fumigating stored

Published by Burleigh Dodds Science Publishing Limited, 2020.

products has resulted in resistance in at least 11 species of stored product insect (including *Sitophilus oryzae*) in 45 countries (Donahaye, 2000; Chaudhry, 2000; Benhalima et al., 2004; Nayak et al., 2007; Pimentel et al., 2010). A notable feature of phosphine resistance is that it can be selected relatively quickly from field populations of most stored product insects, which suggests that the frequency of resistance traits is inherently high. Establishing quarantines that restrict the movement of grains from areas with high resistance levels may be a useful resistance management tactic. Irradiation treatment could be used as a pre-shipment mitigation option to control resistant pests in such a quarantine system, thereby allowing movement of grain out of the quarantine area while slowing the rate of resistance development by preventing the transport of resistant insects. An advantage to irradiation is that it can be applied to the commodity in its final packaging - even packaging that does not permit the penetration of fumigants. Furthermore, insects do not develop resistance to irradiation (Brower, 1974; Hasan and Khan, 1998).

## 2.4 Public perception of irradiated food

The 'radiation' part of irradiation scares some people, even though irradiated foods are not radioactive and are safe to consume. Retailers may be reluctant to carry irradiated products because of the possibility of negative customer response. The international symbol that must be on irradiated foods is referred to as the 'Radura', a stylized plant in a circle; while labeling allows consumers the choice to purchase or not, a labeling requirement also implies a warning and may be an impediment to wider adoption by the food industry despite clear benefits in terms of food safety, reduced spoilage and pest control (Follett and Weinert, 2012).

The negative image of irradiation has been perpetuated in the stored products research community. Phillips (2006) concluded that irradiation is not practical for major food industries due to the high initial cost of irradiation facilities and negative public perception. Phillips and Throne (2010) suggested irradiation is not seriously considered for stored products pest control because of public concerns about the safety of facilities using radioactive isotopes, the misperception that irradiated food becomes radioactive, and the zero tolerance for live insects in treated product by importing countries and the public at large. The first concern can be avoided altogether by use of non-radioactive machine sources such as X-ray or electron beam to treat the stored products. The second objection can be overcome through education. Food irradiation does not impart any radioactivity. Bruhn and Noell (1987) and Eustice and Bruhn (2006) showed that factual education about food irradiation increases public acceptance. There is now significant commercial experience showing that when irradiated foods are offered for retail sale, consumers will purchase and re-purchase it, implying that irradiated foods can be marketed profitably and without risk to reputation (Roberts and Henon, 2015). A fraction of the public will not buy irradiated food, but a much larger fraction will. The volume of irradiated fruit that is imported and sold in the United States and New Zealand has increased steadily over the past 10 years without any consumer backlash or incident (Roberts and Follett, 2018).

As far as the third objection, unlike other disinfestation techniques, irradiation does not kill the pest immediately, and therefore shipments of irradiated cereals may contain 'live' but nonviable insect pests. This puts an added level of importance on the certification procedures for irradiation equipment or facilities and proper documentation accompanying irradiated cereals confirming treatment at approved doses (Follett, 2014). It also places responsibility on researchers to ensure that the minimum absorbed dose approved for each stored product pest or pest group has an adequate margin of safety. The desired response for insects is preventing development to the reproductive stage or sterilizing the reproductive stage of the insect. For stored product insects, adults often may occur with the commodity. When the adult stage occurs in the commodity - and it is typically the most tolerant stage - then the measure of treatment efficacy is indicated by the level of sterility rather than mortality (Follett, 2014). Adult females irradiated at a sterilizing dose may oviposit (particularly if they were gravid when irradiated), but eggs will not hatch or hatching neonates do not develop. Large-scale tests at a dose predicted to cause 100% mortality are useful to confirm efficacy. Most irradiation studies with stored products insects have not included large-scale tests to confirm efficacy with high precision, although a few have (Hallman and Phillips, 2008; Follett et al., 2013).

The World Health Organization, the United Nations Food and Agriculture Organization, the American Medical Association and the International Atomic Energy Agency, all endorse food irradiation. The Food and Drug Agency (FDA) has approved irradiation for pork, beef, poultry, shellfish, eggs, fresh fruits and vegetables, spinach and lettuce, sprouting seeds, and spices and vegetable seasonings. A growing number of foods are irradiated in the country where they are produced, particularly fresh fruits (Roberts and Follett, 2018). Recently added to the list are starfruit and mangos from Vietnam, figs and pomegranates from Peru, and lychee and mangos from Australia. The availability of irradiated foods is still quite limited, probably because of the oft-repeated belief that consumers are averse to the technology.

## 3 Sources of ionizing radiation

The source of ionizing radiation can be gamma rays produced by radionuclides ($^{60}$Co or $^{137}$Cs), or electrons or X-rays generated from machine sources operated within certain energy limits (Diehl, 1995). Irradiation by either isotopes or machine sources (e-beam, X-rays) has the same mode of action: the energy absorbed from gamma rays, X-rays or electrons breaks chemical bonds within DNA and other biomolecules, thereby disrupting normal cellular function in the insect or pathogen (Alpen, 1998). The different sources of ionizing radiation are equally effective in controlling insects and pathogens (Mastrangelo et al., 2010; Tallentire and Miller, 2015).

Most food and agricultural products treated by irradiation are processed in facilities using cobalt-60 ($^{60}Co$) as the source of ionizing radiation. The amount of $^{60}Co$ available worldwide is limited and it is proving increasingly difficult and expensive to deliver large amounts of radioisotopes due to heightened security concerns and the logistical complexities associated with trans-boundary shipments. Owing to these uncertainties, the availability of alternative technologies for food irradiation would be beneficial. Electron beam and X-ray machines provide an alternative technology that employs electricity to generate ionizing radiation (Diehl, 1995; Cleland, 2013; Pillai et al., 2013) and avoid the procurement, transport, storage, disposal and safeguard issues associated with radioisotopes. Electron beams and X-ray do not involve the use of any radioactive materials and can be turned off when not in use. E-beam is more efficient and ideal for high-throughput treatment, whereas X-ray systems are ideal for bulk or pallet-scale treatment of cereals and grains.

### 3.1 Low-energy electron beam

Electron penetrability of the product depends on the energy at which they are accelerated. The lower the energies of the electrons, the lower penetrability they have. Increasing the beam power (for a given beam voltage) will increase the number of emitted electrons. The absorbed dose is the result of interaction between the accelerated electrons and the product matrix. Higher beam voltage or beam power will result in higher absorbed dose by the product.

Hayashi et al. (1998) defined electrons with energies of 300 keV or lower as low-energy electrons or soft-electrons. Several studies have shown that homogenous treatment of surfaces with low-energy electrons can decontaminate dry food ingredients such as spices without detrimental effects (Hayashi et al., 1997; Hayashi, 1998; Hayashi et al., 1998; Todoriki et al., 2002). The dose distribution can be controlled by the energies of electrons applied to the food material to the depth where microorganisms reside. Inactivation of pathogenic microorganisms is achieved by damaging their DNA or RNA (Moeller et al., 2008; Nicholson et al., 2000.)

Chalise et al. (2004) and Rahman et al. (2006) showed that low-energy electrons were effective in inactivation of *Escherichia coli* and *Bacillus subtilis* in both the vegetative and spore stages. The doses required for microbial reduction were 20-fold higher than the doses needed to eliminate insect pests. For instance, bacterial spores on dry foods might require 2–3 kGy to cause a 2-log reduction (99% of the population), whereas most of insect pests are controlled at 100–200 Gy. Thus, microbial control would also ensure insect pest control. Since both bacteria and insects are an issue for many cereals, eliminating both with a one-step process would be more efficient and would therefore remove the need for multiple decontamination treatments.

### 3.2 X-ray radiation

X-ray is another type of machine source radiation, but unlike e-beam, it has excellent penetration and is therefore appropriate for treatment of bulk grains. The optimal design and energy requirements of an X-ray machine depend on the geometry and density of the products to be treated and the desired throughput. Small-scale X-ray machines use X-ray tubes with energies <300 keV. For example, a prototype cabinet X-ray system is under development by Applied Energy Devices (Albuquerque, NM) using the 4-pi X-ray tube (165 keV) (Fig. 1). This tube emits X-rays in 360° along a 14-inch cylindrical surface, allowing treatment of the target, such as a box of fruit, from the bottom and then from

**Figure 1** Prototype cabinet X-ray system.

the top using a two-pass conveyor system. Development of small-scale cabinet-style X-ray machines using 150–300 keV X-ray emitters with large transmission window tubes could provide in-line treatment capability for packaged cereals and grains. These systems require limited shielding, are portable, and would provide maximum logistical flexibility for disinfestation treatment.

Large-scale X-ray systems with energies of 5–7.5 meV use a linear accelerator with an X-ray converter plate and can treat pallets of food products. Large accelerator irradiation machines such as the Rhodotron® (IBA, Belgium) can be configured as e-beam or X-ray or dual systems and can be scaled to product size and throughput requirements. The FEERIX research and development irradiator at the Aérial – Technology Resource Centre in Strasbourg, France, uses a Rhodotron® that scans electron across a 2.2 m tantalum target to convert to X-rays (Fig. 2); the large scan horn is suitable for treating pallets of product. High-energy X-ray systems have much higher throughput than low-energy systems but require significant shielding and are therefore less suitable for in-line treatment.

### 3.3 Historical use for stored product disinfestation

Machine sources of irradiation have been used for over 100 years. Hunter (1912) was the first entomologist to conduct radiation sensitivity studies on rice

**Figure 2** Large accelerator irradiator machine.

Published by Burleigh Dodds Science Publishing Limited, 2020.

weevil using X-ray. Runner (1916) demonstrated control of cigarette beetle, *Lasioderma serricorne*, in tobacco using X-rays, and in 1929, the American Tobacco Company used X-ray treatment to control *L. serricorne* in cigars exported to Europe (Diehl, 1995). In 1963, the US FDA approved irradiation of wheat and flour for insect disinfestation. The longest and highest volume use of irradiation to control insects was the electron beam treatment of grain imported into the former Soviet Union through the port of Odessa in the Ukraine from 1980–2007. This system could treat 400 tonnes/h at 200–400 Gy for insect disinfestation, and at its peak, 400 000 tonnes of grain per year was irradiated (Zakladnoy et al., 1989), demonstrating the industrial-scale disinfestation potential of e-beam.

# 4 Case study 1: low-energy electron beam treatment

## 4.1 Introduction

A low-energy electron beam (LEEB) machine was designed for treatment of seeds and grains (Kotilainen et al., 2019). The process works by exposing grains or seeds to low-energy electrons. The electrons are emitted from vacuum-sealed lamps at energies lower than 300 keV. The treatment chamber - in which the grains are exposed to electrons - consists of two lamps facing each other (Fig. 3). The grains are fed by a vibrating conveyor into the treatment zone. The accelerated low-energy electrons emitted from the lamps surround the grains or seeds. The absorbed dose in the product determines the degree of reduction of any insects or microorganisms. Hence, the absorbed dose is a critical parameter to measure. The grains or seeds are in the treatment chamber for less than 30 msec. Within this treatment time, the target dose for insect disinfestation and microbial disinfection must be delivered. For microbial reduction, the target dose is in the range of 8–15 kGy, whereas for insect pest control the target dose is in the range of 100–400 Gy. For grain insects, irradiation must reach the inner part of the grain where the immature life stages (eggs, larvae, pupae) occur. Microbial contamination is located at the surface of the grains and the maximum dose needed for control of most

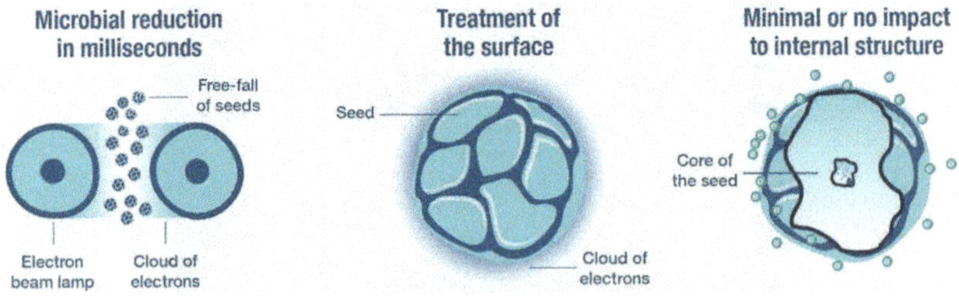


**Figure 3** Schematic of grains exposed to low-energy electron beam (LEEB) treatment.

Published by Burleigh Dodds Science Publishing Limited, 2020.

microorganisms is <10 kGy. Due to the low penetrability of LEEB doses, high doses can be achieved at the surface of the grain (e.g. 10–12 kGy) and lower but insecticidal doses reach the inner parts of the grain (e.g. 400–500 Gy) to control insect pests.

### 4.2 Experimental design

Test insects were maintained in culture on rice at the Biologische Beratung, Berlin, Germany. The LEEB treatment of the infested grains took place at Bühler AG, Switzerland. The principle of LEEB processing at industrial capacities (1 tonne/h) was tested on white rice and using two species of stored product pests of rice. The lesser grain borer *Rhyzopertha dominica* and the rice weevil *Sitophilus oryzae* were inoculated on whole rice kernels (Paella-rice) (Fig. 4).

In each sample, 30 adult beetles were transferred into a plastic Petri-dish (9 cm diameter) filled half-full of rice and placed for 7 days in a climate chamber at 27°C and 65% RH. After 7 days, the adults were removed and the samples containing the beetle eggs and rice were shipped to Bühler, where the treatment took place. Infested rice was held for 3 weeks in order to simulate a real-case scenario of infestation in which eggs, larvae and pupae can be found simultaneously, in order to test the effectiveness of LEEB against immature stages of the selected pests. For each species, there were three untreated travel control samples (effect of travel alone), three process control samples (effect of travel and process handling) and six treated samples (effect of travel and LEEB treatment). The travel control was needed to validate survival during shipment between the two different locations (laboratory and treatment location). A process control was needed to rule out any additional factor that might have an impact on pest reduction during handling at the processing facility.

The absorbed dose was measured using radiochromic B3 film following the guidelines and practices for irradiation dosimetry applications according to ISO/ASTM 51818 (ISO/ATM, 2013a) and ISO/ASTM 51275:2013 (ISO/ATM,

**Figure 4** *Rhyzopertha dominica* and *Sitophilus oryzae* infesting rice.

2013b). B3 film is preferred for low-energy electron beams due to its small thickness (18μm) and the low penetrability of the beams. Due to the small size of grains, standard dosimetry using dosimeters must be modified to accurately measure absorbed dose. A dosimeter that simulated the size of a grain was developed. A sample of these specialized dosimeters were randomly distributed in a product batch and processed under normal treatment conditions. After thermal stabilization, the B3 film was scanned using a calibrated EPSON V-600 perfection scanner. Riso Scan 1.03 software provided by RISO HDRL was used to measure the average absorbed dose of each irradiated dosimeter. The scanner can quantify and relate the absorbance in the optical spectrum of the film to absorbed dose in the simulated grain (Jacob Helt-Hansen, 2004). Statistical analysis of the measured doses was used to estimate the distribution of doses received by the product and to confirm it was within microbial and regulatory limits to a 95% confidence level.

The samples with eggs, larvae and pupae were shipped back to Berlin immediately after LEEB treatment. Upon receipt, the Petri-dishes with infested rice were placed at 27°C and 65% RH to allow pest development. After 45 days, the number of adult beetles in each sample was counted.

The infested rice samples were treated by LEEB according to the set of parameters in Table 1.

Surface doses were significantly higher than the doses needed to achieve pest control; however, the aim of the treatment was to achieve high doses at the surface in order to reduce microbial contamination in addition to insect pest control. The chosen LEEB treatment parameters delivered low doses to the inner part of the grains up to 500–600 μm in depth.

### 4.3 Results and discussion

Holding samples for 45 days was sufficient time to allow all immature stages of the pest to develop to the adult stage. The presence of adults would indicate a failure on the LEEB treatment, whereas no adults would indicate that the LEEB treatment prevented successful development to the adult stage.

**Table 1** Process parameters LEEB treatment

| Variable | Value |
|---|---|
| LEEB acceleration voltage | 225–300 keV |
| LEEB surface dose | 7–12 kGy |
| LEEB surface dose rate | 233–240 kGy/s |
| Throughput | 1 TPH (tonnes/h) |
| Batch size | 0.5 kg |

The LEEB process at 225 and 300 keV effectively controlled both rice pests (Table 2). For *R. dominica* and *S. oryzae*, a significant number of adults developed in the travel and process control samples, whereas in all LEEB-treatment samples there were no adults present. This indicates that the applied doses prevented development in the two target insects developing inside rice grains. These preliminary results suggest that an industrial LEEB treatment of 1 TPH is feasible. However, further validation at lower doses, on more cereal and grain crops and against additional insect pests is needed. Although this case study focuses on insect control, e-beam can also effectively control pathogens found on the surface of cereals and grains, thereby preventing spoilage and ensuring food safety.

LEEB has promise as a non-thermal decontamination technology for low-moisture foods such as cereals and grains. The efficacy of the technology has

**Table 2** Effect of LEEB treated on insect development after 45 days of storage

| Species | Product | LEEB treatment | Number of adult pests |
|---|---|---|---|
| *Rhyzopertha dominica* | White rice | Travel control | 101 |
| | | Travel control | 94 |
| | | Travel control | 88 |
| | | Process Control | 83 |
| | | Process Control | 92 |
| | | Process Control | 97 |
| | | LEEB 225 keV | 0 |
| | | LEEB 225 keV | 0 |
| | | LEEB 225 keV | 0 |
| | | LEEB 300 keV | 0 |
| | | LEEB 300 keV | 0 |
| | | LEEB 300 keV | 0 |
| *Sitophilus oryzae* | White rice | Travel control | 19 |
| | | Travel control | 82 |
| | | Travel control | 110 |
| | | Process Control | 33 |
| | | Process Control | 38 |
| | | Process Control | 37 |
| | | LEEB 225 keV | 0 |
| | | LEEB 225 keV | 0 |
| | | LEEB 225 keV | 0 |
| | | LEEB 300 keV | 0 |
| | | LEEB 300 keV | 0 |
| | | LEEB 300 keV | 0 |

been successfully tested with several dry food products (spices, herbs, nuts, edible seeds, vegetable seeds, agro-seeds and fruits in addition to cereals and grains) with a significant reduction in insects and pathogens. LEEB technology does not use chemicals or introduce water, has a small physical-footprint and consumes less energy than do thermal and conventional irradiation technologies. Currently, the decontamination of foods with irradiation (e.g. spices) is usually outsourced to large multipurpose gamma-facilities (using radioactive cobalt-60), which makes it difficult to integrate into a Hazard Analysis Critical Control Points (HACCP) plan. Also, irradiation might not be easily accessible to food processors due to logistics and transport costs to treatment facilities. The LEEB machine is self-shielded and portable. LEEB irradiation decontamination is largely limited to the surface of the grains, but the relatively low doses required for insect disinfestation can penetrate to the inner parts of food matrixes. Depending on the applied energy levels it may be possible to inactivate bacteria on seeds and nuts without impacting the germination potential or inducement of lipid oxidation, respectively.

## 5 Case study 2: high-energy X-ray treatment

### 5.1 Introduction

*Sitophilus oryzae* (rice weevil) was studied as a model stored product pest to demonstrate development of a disinfestation treatment with high confidence, including large-scale testing. The primary source of storage loss in rice in terms of weight and sensory quality is rice weevil. The adult female rice weevil bores an oblong shallow cavity in the kernel, deposits an egg and then seals the egg cavity with a clear and gluey secretion (Metcalf et al., 1962; Plarre, 2010). Each kernel only hosts one rice weevil egg. The entire larval and pupal period of the beetle is spent hidden inside the kernel, and during the larval period the kernel is completely hollowed out. Once the adult beetle emerges from the pupa, it remains for a short time inside the kernel until it has bored its way to the outside. Adult weevils may crawl back into hollow kernels, where they are difficult to detect. Adult weevils can live for 6–8 months or more, and females may lay 300–400 eggs in their life-time (Metcalf et al., 1962; Longstaff, 1981). Few studies have been conducted with stored product pests using the large-scale tests needed to confirm the efficacy of an irradiation dose with no survivors (Barkai-Golan and Follett, 2017).

### 5.2 Experimental design

Research was conducted to identify an irradiation dose that would prevent development of immature rice weevils and sterilize adults (Follett et al., 2013). Rice infested with adult or immature weevils was treated at radiation doses of

30, 60, 90 or 120 Gy, or left untreated as a control. Irradiation treatment was conducted at a commercial X-ray irradiation facility (Calavo Growers Inc., Keaau, Hawaii) using an electron linear accelerator (5 MeV, model TB-5/15, Titan Corp., San Diego, California). Opti-chromic detectors (FWT-70-83/40M, Far West Technology, Goleta, California) were placed in several containers with rice in each treatment to measure dose variation. Dosimeters were read with an FWT-200 reader at 600-nm absorbance. After irradiation, live and dead beetles were counted weekly for 15–24 weeks and the weight loss of infested (irradiated and unirradiated) and uninfested rice was measured. Weight loss data were subjected to analysis of variance and significant treatment effects were compared by Tukey's test at alpha = 0.05. Large-scale confirmatory testing was conducted with weevil-infested rice at a radiation dose predicted to provide 100% sterility of adult weevils to demonstrate its effectiveness with high confidence.

## 5.3 Results and discussion

Treatment of adult weevils at a radiation dose of 120 Gy resulted in no live adults after 2 weeks, indicating that this dose caused adult mortality and sterility, whereas a total of 1261 adult beetles emerged during 24 weeks in the untreated controls. Treatment of immature life stages (a mixture of eggs, larvae and pupae) with a radiation dose of 90 or 120 Gy resulted in no adults emerging after 5 or 2 weeks, respectively, indicating that these doses reduced development to the adult stage and prevented reproduction, whereas a total of 4275 adults emerged throughout 15 weeks in the untreated controls. Irradiation treatment reduced weight loss in infested rice in addition to controlling *S. oryzae*. Weight loss in weevil-infested rice irradiated at a dose of 120 Gy was not statistically different from uninfested rice ($F_{3,48} = 17.8$, $P < 0.0001$) (Fig. 5), suggesting feeding was arrested. This is important as frass or grain dust may increase after irradiation treatment if the sterilizing dose does not also arrest feeding activity. In large-scale confirmatory testing, a radiation dose of 120 Gy applied to 38 025 adult weevils in rice resulted in no reproduction (Follett et al., 2013).

Australia and Japan have required 99.99% efficacy as a basis for approving treatments as meeting quarantine security requirements, which can be demonstrated by treatment of 29 956 individuals with no survivors (Follett and Neven, 2006). The required response may be mortality, sterility or prevention of maturity. The goal is to demonstrate that insect survival rates are sufficiently low to ensure quarantine security at a specified level of precision. In our study, large-scale validation tests at a radiation dose of 120 Gy resulted in no progeny from 38 025 treated adult rice weevils. The calculated level of confidence is 97.8% that the true survival of rice weevil is less than 0.0001 (Couey and Chew, 1986; Follett et al., 2013).

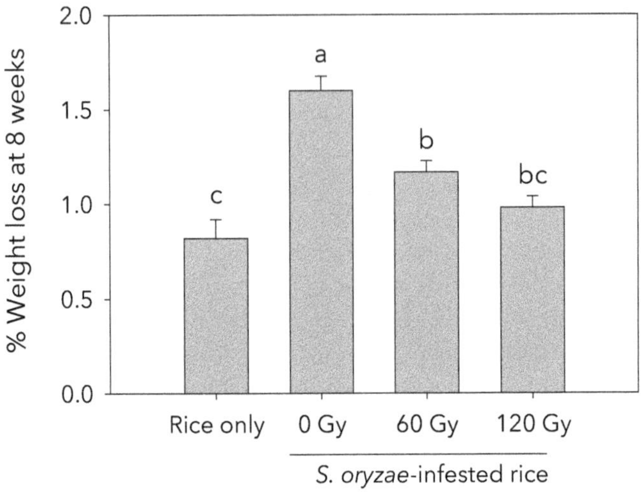


**Figure 5** Comparison of weight loss in uninfested rice ('rice only') and *Sitophilus oryzae*-infested rice irradiated at different doses after 8 weeks post-treatment.

## 6 Conclusion

Irradiation is a viable alternative to fumigation as a pre- or post-shipment quarantine treatment for control of cereal and grain pests. Irradiation is fast, cost competitive, broadly effective against insect and mite pests, and temperature insensitive, and it does not leave residues and has minimal effects on product quality. Due to the massive effects on insect DNA and other biomolecules, irradiation is immune to resistance development. For stored product pests, irradiation may be particularly helpful in controlling phosphine and methyl bromide resistant populations and could help manage resistance by preventing the spread of resistant insects in exported grains. Irradiation can also be used to eliminate human pathogenic bacteria from foods, making them safer, and to prevent or delay the growth of molds and bacteria that cause spoilage of stored grains. Irradiation from machine sources (e-beam and X-ray) offers maximum flexibility and is the technology of the future.

## 7 Future trends

Future research should examine the potential for combining irradiation with other disinfestation and preservation techniques such as modified atmospheres, cold, heat and fumigation treatments which may enhance effectiveness, reduce irradiation (or the combination treatment) time and costs, and preserve product quality (Lacroix and Follett, 2015). Bioactive packaging using plant essential oils and nanotechnology is a particularly interesting approach that provides extended pest control and prevents both reinfestation of insects (Hossain et al., 2019a) and reinfection by molds and bacteria (Hossain et al., 2019b) in packaged cereals and may be an improvement over current bioactive packaging that uses insect growth regulators against insects alone. Studies are also needed to estimate the environmental and economic impacts of irradiation compared with other insect pest infestation control and prevention technologies in grains and cereals.

## 8 Where to look for further information

The following references provide a comprehensive overview of food irradiation research, applications and technology:

- Barkai-Golan, R., Follett, P. A. (2017), *Irradiation for Quality Improvement, Microbial Safety and Phytosanitation of Fresh Produce*, Academic Press, San Diego, California, USA, 301 p.
- Diehl, J. F. (1995), *Safety of Irradiated Foods*, Marcel-Dekker, New York, 464 p.
- Fan, X. F., Sommers, C. H. (2013), *Food Irradiation Research and Technology*, IFT Press, Wiley-Blackwell, Oxford, UK.
- Ferreira, C. F. R., Antonio, A. L. and Cabo Verde, S. (2018), *Food Irradiation Technologies: Concepts, Applications and Outcomes*, Royal Society of Chemistry, Cambridge, UK.

## 9 References

Ahmed, M. 2001. Disinfestation of stored grains, pulses, dried fruits and nuts, and other dried foods. In: Molins, R. (Ed.), *Food Irradiation: Principles and Applications*. John Wiley & Sons, New York, NY, pp. 77–112.

Alpen, E. L. 1998. *Radiation Biophysics* (2nd edn.). Academic Press, New York, NY.

Benhalima, H., Chaudhry, M. Q., Mills, K. A. and Price, N. R. 2004. Phosphine resistance in stored-product insects collected from various grain storage facilities in Morocco. *J. Stored Prod. Res.* 40(3), 241–9. doi:10.1016/S0022-474X(03)00012-2.

Brower, J. H. 1973. Depressed flour beetle: sensitivity to gamma irradiation. *J. Econ. Entomol.* 66(6), 1318–20. doi:10.1093/jee/66.6.1318.

Brower, J. H. 1974. Fitness and radioresistance of the rice weevil, *Sitophilus oryzae* (Coleoptera: Curculionidae), after 25 generations of substerilizing irradiation. *J. Kans. Entomol. Soc.* 47, 437–44.

Bruhn, C. M. and Noell, J. W. 1987. Consumer in-store response to irradiated papayas. *Food Technol.* 41(9), 83–5.

Chalise, P. R., Rahman, M. S., Ghomi, H., Hayashi, Y., Watanabe, M., Okino, A., Ano, T., Shoda, M. and Hotta, E. 2004. Bacterial inactivation using low-energy pulsed-electron beam. *IEEE Trans. Plasma Sci.* 32(4), 1532–9. doi:10.1109/TPS.2004.830967.

Chaudhry, M. Q. 2000. Phosphine resistance. *Pest. Outlook* 11(3), 88–91. doi:10.1039/b006348g.

Cleland, M. R. 2013. Advances in electron beam and x-ray technologies for food irradiation. In: Sommers, C. H. and Fan, X. (Eds), *Food Irradiation Research and Technology*. Blackwell Publishing, Ames, IA, pp. 9–28.

Codex Alimentarius Commission. 2003. General standard for the irradiation of foods, Codex Stan. 106-1983, Revised 2003. Food and Agriculture Organization of the United Nation/World Health Organization, Rome.

Cornwall, P. B. 1964. Insect control. In: Jefferson, S. (Ed.), *Massive Radiation Techniques*. George Newnes, London, UK, pp. 141–98.

Couey, H. M. and Chew, V. 1986. Confidence limits and sample size in quarantine research. *J. Econ. Entomol.* 79(4), 887–90. doi:10.1093/jee/79.4.887.

Diehl, J. 1995. *Safety of Irradiated Foods* (2nd edn.). Marcel-Dekker, New York, NY, 454p.

Donahaye, E. J. 2000. Current status of non-residual control methods against stored product pests. *Crop Prot.* 19(8–10), 571–6. doi:10.1016/S0261-2194(00)00074-0.

Eustice, R. F. and Bruhn, C. M. 2006. Consumer acceptance and marketing of irradiated foods. In: Sommers, C. H. and Fan, X. (Eds), *Food Irradiation Research and Technology*. Blackwell Publishing, Ames, IA, pp. 63–83.

Follett, P. A. 2009. Generic radiation quarantine treatments: the next steps. *J. Econ. Entomol.* 102(4), 1399–406. doi:10.1603/029.102.0401.

Follett, P. A. 2014. Phytosanitary irradiation for fresh horticultural commodities: generic treatment, current issues, and next steps. *Stewart Postharvest Rev.* 3(1), 7.

Follett, P. A. 2018. Irradiation for quarantine control of coffee berry borer, Hypothenemus hampei (Coleoptera: Curculionidae: Scolytinae) in coffee and a proposed generic dose for snout beetles (Coleoptera: Curculionoidea). *J. Econ. Entomol.* 111(4), 1633–7.

Follett, P. A. and Neven, L. G. 2006. Current trends in quarantine entomology. *Annu. Rev. Entomol.* 51, 359–85. doi:10.1146/annurev.ento.49.061802.123314.

Follett, P. A. and Hennessey, M. K. 2007. Confidence limits and sample size for determining nonhost status of fruits and vegetables to tephritid fruit flies as a quarantine measure. *J. Econ. Entomol.* 100(2), 251–7. doi:10.1603/0022-0493(2007)100[251:classf]2.0.co;2.

Follett, P. A. and Weinert, E. D. 2012. Phytosanitary irradiation for tropical commodities in Hawaii: generic treatments, commercial adoption, and current issues. *Radiat. Phys. Chem.* 81(8), 1064–7. doi:10.1016/j.radphyschem.2011.12.007.

Follett, P. A., Snook, K., Janson, A., Antonio, B., Haruki, A., Okamura, M. and Bisel, J. 2013. Irradiation quarantine treatment for control of *Sitophilus oryzae* (Coleoptera: Curculionidae) in rice. *J. Stored Prod. Res.* 52, 63–7. doi:10.1016/j.jspr.2012.09.004.

Food and Drug Administration (FDA). 1986. Irradiation in the production, processing, and handling of food. *Fed. Regist.* 51(75), 13376.

Gao, M. X., Wang, C. Y., Li, S. R. and Zhang, S.-F. 2004. The effect of irradiation on *Trogoderma granarium* in grain and legume. *Acta Phytophylacica Sin.* 31, 377–82.

Hallman, G. J. 2013. Control of stored products pests by ionizing radiation. *J. Stored Prod. Res.* 52, 36–41. doi:10.1016/j.jspr.2012.10.001.

Hallman, G. J. and Phillips, T. W. 2008. Ionizing irradiation of adults of Angoumois grain moth (Lepidoptera: Gelechiidae) and Indianmeal moth (Lepidoptera: Pyralidae) to prevent reproduction, and implications for a generic irradiation treatment for insects. *J. Econ. Entomol.* 101(4), 1051–6. doi:10.1603/0022-0493(2008)101[1051:iioaoa]2.0.co;2.

Hasan, M. and Khan, A. R. 1998. Control of stored product pests by irradiation. *Integr. Pest Manag. Rev.* 3(1), 15–29. doi:10.1023/A:1009621606024.

Hayashi, T., Takahashi, Y. and Todoriki, S. 1997. Low energy electron effects on the sterility and viscosity of grains. *J. Food Sci.* 62(4), 858–60. doi:10.1111/j.1365-2621.1997.tb15472.x.

Hayashi, T. 1998. Decontamination of dry food ingredients with 'soft-electrons' (low-energy electrons). *Jpn Agric. Res. Q.* 32, 293–9.

Hayashi, T., Takahashi, Y. and Torodriki, S. 1998. Sterilization of foods with low- energy electrons ('soft-electrons'). *Radiat. Phys. Chem.* 52(1–6), 73–6. doi:10.1016/S0969-806X(98)00077-2.

Hossain, F., Follett, P., Salmieri, S., Vu, K. D., Harich, M. and Lacroix, M. 2019a. Synergistic effects of nanocomposite films containing essential oil nanoemulsions in combination with ionizing radiation for control of rice weevil *Sitophilus oryzae* in stored grains. *J. Food Sci.* 84(6), 1439–46. doi:10.1111/1750-3841.14603.

Hossain, F., Follett, P. A., Vu, K. D., Harich, M., Salmieri, S. and Lacroix, M. 2019b. Antifungal activity of combined treatments of irradiation and essential oil encapsulated chitosan nanocomposite films in in vitro and in situ conditions. *Inter. J. Food Microbiol.* 295, 33–40.

Hunter, W. D. 1912. Results of experiments to determine the effects of Roentgen rays upon insects. *J. Econ. Entomol.* 5(2), 188–93. doi:10.1093/jee/5.2.188.

Ignatowicz, S. 2004. Irradiation as an alternative to methyl bromide fumigation of agricultural commodities infested with quarantine stored product pests. In: *Irradiation as a Phytosanitary Treatment of Food and Agricultural Commodities*. IAEA-TECDOC-1427; International Atomic Energy Agency, Vienna, Austria, pp. 51–66.

International Plant Protection Convention (IPPC) 2009. Phytosanitary treatments for regulated pests. ISPM 28. IPPC/Food and Agriculture Organization of the United Nations, Rome, Italy.

ISO/ASTM. 2013a. 51818:2013 'Standard practice for dosimetry in an electron beam facility for radiation processing at energies between 80 and 300keV'. Switzerland.

ISO/ASTM. 2013b. 51275:2013 'Practice for use of a radiochromic film dosimetry system'. Switzerland.

Jacob Helt-Hansen. 2004. *RisoScan 1.03 - User Manual and Toolset for Retrospective Validation*. Riso National Laboratory.

Kotilainen, H., Meneses, N., Ibbotson, T. and Conde-Petit, B. 2019. ). Laatu. Non-thermal, in-plant microbial reduction solution for dry foods.. [White paper]. Retrieved from https://digital.buhlergroup.com/laatu/.

Lacroix, M. and Follett, P. A. 2015. Combination irradiation treatments for food safety and phytosanitary uses. *Stewart Postharvest Rev.* 11(3), 10.

Longstaff, B. C. 1981. Biology of the grain pest species of the genus *Sitophilus* (Coleoptera: Curculionidae): a critical review. *Prot. Ecol.* 2, 83–130.

Mastrangelo, T., Parker, A. G., Jessup, A., Pereira, R., Orozco-Davila, D., Islam, A., Dammalage, T. and Walder, J. M. M. 2010. A new generation of x-ray irradiators for insect sterilization. *J. Econ. Entomol.* 103(1), 85–94. doi:10.1603/ec09139.

Metcalf, C. L., Flint, W. P. and Metcalf, R. L. 1962. *Destructive and Useful Insects* (4th edn.). McGraw-Hill Book Co., New York, NY.

Moeller, R., Setlow, P., Horneck, G., Berger, T., Reitz, G., Rettberg, P., Doherty, A. J., Okayasu, R. and Nicholson, W. L. 2008. Roles of major, small, acid-soluble spore proteins and spore-specific and universal DNA repair mechanisms in resistance of *Bacillus subtilis* spores to ionizing radiation from x-rays and high-energy charged-particle bombardment. *J. Bacteriol.* 190(3), 1134–40. doi:10.1128/JB.01644-07.

Nayak, M. K., Collins, P. J. and Pavic, H. 2007. Developing fumigation protocols to manage strongly phosphine-resistant rice weevils, *Sitophilus oryzae* (L.). In: Donahaye, E. J., Navarro, S., Bell, C., Jayas, D., Noyes, R. and Phillips, T. W. (Eds), *Proceedings of the International Conference Controlled Atmosphere and Fumigation in Stored Products Gold-Coast Australia*, 8–13 August 2007. FTIC Ltd. Publishing, Israel, pp. 267–73.

Nicholson, W. L., Munakata, N., Horneck, G., Melosh, H. J. and Setlow, P. 2000. Resistance of *Bacillus* endospores to extreme terrestrial and extraterrestrial environments. *Microbiol. Mol. Biol. Rev.* 64(3), 548–72. doi:10.1128/mmbr.64.3.548-572.2000.

Phillips, T. W. 2006. The science and technology of postharvest insect control: challenges, accomplishments, and future directions. In: Heaps, J. W. (Ed.), *Insect Management for Food Storage and Processing* (2nd edn.). American Association of Cereal Chemists, St. Paul, MN, pp. 211–22.

Phillips, T. W. and Throne, J. E. 2010. Biorational approaches to managing store-product insects. *Annu. Rev. Entomol.* 55, 375–97. doi:10.1146/annurev.ento.54.110807.090451.

Pillai, S. D., Braby, L. and Maxim, J. 2013. Technical challenges and research directions in electronic food pasteurization. In: Fan, X. and Sommers, C. H. (Eds), *Food Irradiation Research and Technology*. Blackwell Publishing, IFT Press, pp. 425–33.

Pimentel, M. A., Faroni, L. R., Silva, F. H., Batista, M. D. and Guedes, R. N. 2010. Spread of phosphine resistance among Brazilian populations of three species of stored product insects. *Neotrop. Entomol.* 39(1), 101–7. doi:10.1590/s1519-566x2010000100014.

Plarre, R. 2010. An attempt to reconstruct the natural and cultural history of the granary weevil, *Sitophilus granarius* (Coleoptera: Curculionidae). *Eur. J. Entomol.* 107(1), 1–11. doi:10.14411/eje.2010.001.

Rahman, M. S., Ghomi, H., Chalise, P. R., Hayashi, Y., Watanabe, M., Okino, A., Ano, T., Shoda, M. and Hotta, E. 2006. Inactivation of cells and spores of *Bacillus subtilis* using low energy pulsed electron beam. *Jpn. J. Appl. Phys.* 45(33), L881–3. doi:10.1143/JJAP.45.L881.

Roberts, P. B. and Henon, Y. 2015. Consumer response to irradiated food: purchase versus perception. *Stewart Postharvest Rev.* 3, 5.

Roberts, P. B. and Follett, P. A. 2018. Food irradiation for phytosanitary and quarantine treatment. In: Ferreira, C. F. R., Antonio, A. L. and Cabo Verde, S. (Eds), *Food Irradiation Technologies: Concepts, Applications and Outcomes*. Royal Society of Chemistry, Cambridge, UK, pp. 169–82.

Runner, G. A. 1916. Effect of roentgen rays on the tobacco or cigarette beetle and the results with a new form of roentgen tube. *J. Agric. Res.* 6, 383–8.

Tallentire, A. and Miller, A. 2015. Microbial effectiveness of X-rays used for sterilization purposes. *Rad. Phys. Chem.* 107, 128–30. doi:10.1016/j.radphyschem.2014.09.012.

Todoriki, S., Kikuchi, O. K., Nakaoka, M., Miike, M. and Hayashi, T. 2002. Soft electron (low energy electron) processing of foods for microbial control. *Radiat. Phys. Chem.* 63(3–6), 349–51. doi:10.1016/S0969-806X(01)00588-6.

United States Department of Agriculture (USDA) - Animal and Plant Health Inspection Service (APHIS) 2006. Treatments for fruits and vegetables. *Fed. Regist.* 71(18), 4451–64. Rules and Regulations.

Watters, F. L. and McQueen, K. F. 1967. Effectiveness of gamma irradiation for control of five species of stored-product insects. *J. Stored Prod. Res.* 3(3), 223–34. doi:10.1016/0022-474X(67)90049-5.

Zakladnoy, G. A., Meshenin, A. I., Pertsofsky, E. S., Salimov, R. A., Cherepkov, V. G., Bogolyubov, B. F. and Stenev, I. S. 1989. Radiation disinfestation at a port elevator with capacity of 400 t/h. *Rad. Phys. Chem.* 34, 991–4.

# Chapter 11

# Advances in post-harvest detection and control of fungal contamination of cereals

*Naresh Magan, Esther Garcia-Cela, Carol Verheecke-Vaessen and Angel Medina, Cranfield University, UK*


1 Introduction
2 Factors affecting the activity of spoilage moulds and mycotoxin contamination
3 Volatiles as an early indicator of spoilage of stored cereals
4 Carbon dioxide monitoring for real-time identification of the risk of spoilage moulds and mycotoxins in stored cereals
5 Control strategies to minimise spoilage and mycotoxin contamination of stored grain: monitoring of relative humidity (RH) and temperature and the use of molecular methods
6 Control strategies: hermetic storage of grain and use of natural compounds
7 Control strategies: biological control and irradiation techniques
8 Control technologies: cold plasma technology and ozone fumigation
9 Conclusions
10 Where to look for further information
11 References


## 1 Introduction

Today, quality assurance systems (QA) based on Hazard Analysis and Critical Control Point (HACCP) principles are recommended in preference to quality control systems (QC) which requires careful monitoring and withholding of contaminated material from the market place. The use of HACCP systems are indeed now the norm in the development of prevention strategies for spoilage and mycotoxin contamination in cereal food/feed chains. This requires effective identification of the critical control points (CCPs) and monitoring of these points, including the drying and post-harvest phases of cereal chains. Thus,

http://dx.doi.org/10.19103/AS.2020.0072.14

reliable and accurate methods are required for enabling management of the CCPs to be implemented (Tothill and Magan, 2003). Consequently, the rapid detection of spoilage microorganisms in the food production and processing chain is critical regardless of the system being used or implemented. There are, however, different levels of detection that may be required in the cereal chain. The diagnostic tool required depends very much on the type and amount of information required. For example: Is a qualitative 'yes' or 'no' information required? Is a quantitative measurement needed? What is the sensitivity required? How often does monitoring need to be done? Today, technological developments are rapid and on-line approaches are often preferred. Of course the economics of implementation in the cereal chain are determined by a range of interacting abiotic and biotic factors.

Post-harvest storage of cereals is a critical and key phase in the cereal food/feed chain because of the potential for significant losses where management is compromised. Poor post-harvest drying and storage can result in significant losses which have been quantified as 5–10% in developed countries and up to 30–35% in lower middle-income countries (LMICs). In addition, many spoilage moulds also produce mycotoxins which can have implications for downstream processing and for both consumers and the animal feed chain. This can lead to significant economic impacts, especially in LMICs. Thus, it is important that effective monitoring and control systems can be utilised to minimise the impact that spoilage moulds and mycotoxins may have on grain quality and facilitate the development of appropriate storage solutions to conserve quality.

It should be borne in mind that cereal grains are usually alive and respiring. The level of respiration is influenced by the prevailing temperature and moisture content of the cereal grain when entering storage. The relationship between cereal grain, abiotic interactions, pest immigration/emigration and spoilage mould activity including toxin contamination was integrated in the stored grain ecosystem approach by Wallace and Sinha in the 1970s (Wallace, 1973). This approach has been used to better understand the role of the different components and the environmental conditions which will allow different groupings of moulds/insects to predominate and indicate the relative risk of spoilage. It should be noted that there is an important link between grain silo hygiene, storage pest infestation and mould spoilage. This is because often pest activity can occur under drier conditions that allow growth, and their activity can produce metabolic heat and water which can facilitate the initiation of a 'hot spot' of mould spoilage in stored cereals rapidly resulting in significant spoilage, dry matter and nutritional losses and toxin contamination. Therefore, in such a hot spot, rapid environmental changes (e.g. temperature, relative humidity, gas and volatile concentrations) occur. These parameters can be potentially monitored in real time to improve post-harvest management and minimise risks of spoilage and mycotoxin contamination.

## 2 Factors affecting the activity of spoilage moulds and mycotoxin contamination

A significant amount of information has now been gathered on the interacting conditions of water availability (water activity, $a_w$) and temperature on germination, growth and also, in some cases, on mycotoxin production (Magan et al., 2010; Sanchis and Magan 2004). These have shown that in many cases there are boundary conditions for different phases of fungal activity. These are more limited for *Fusarium* than for *Penicillium* or *Aspergillus* species (Fig. 1). Indeed, germination occurs over a wider range of conditions than growth or mycotoxin production (Sanchis and Magan, 2004). This has resulted in the development of predictive models of the relationship between storage $a_w$ and temperature and the relative risk of growth or mycotoxin production. This has been developed for *F. graminearum* and *deoxynivalneol, P. verrucosum* and ochratoxin A both in stored wheat, and *A. flavus* and aflatoxin $B_1$ production in maize (Magan et al., 2010; Medina et al., 2017).

Additionally, Garcia et al. (2011) suggested that estimations through linear logistic regression could be an alternative to the development of safe storage predictions. In fact, several studies have used this approach to model the effect of temperature and $a_w$, on the growth boundaries of *Aspergillus* species in general media, in simulated food matrices and in cereal kernels (Astoreca et al., 2012, Marin et al., 2012, 2008; Tassou et al., 2009). Currently, studies at Cranfield are following the same mathematical approach to model the growth boundary conditions for *F. graminearum*. Figure 2 shows different levels of probability of growth under different interacting conditions of temperature

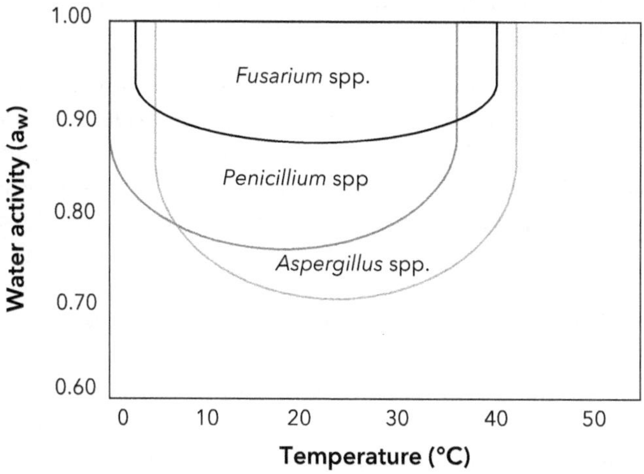


**Figure 1** Diagrammatic representation of the boundary conditions for growth of key spoilage mycotoxigenic fungi in relation to water activity × temperature.

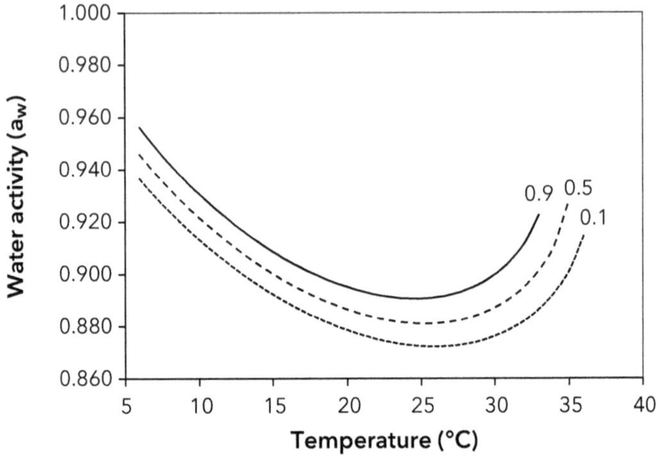


**Figure 2** Predicted growth/no growth boundaries for one month with respect to water activity and temperatures for *F. graminerarum* on wheat-based media. Lines show different levels of probability.

and $a_w$. Probabilistic growth models built with data obtained from wheat-based media have been validated with a set of published data on wheat grain (Ramirez et al., 2006) that showed 83% agreement with other models (Marin et al., 2008; Garcia-Cela et al., 2019). These results confirm that probabilistic models can be applied as accurate predictive tools of fungal spoilage in stored commodities such as grain.

## 3 Volatiles as an early indicator of spoilage of stored cereals

Many moulds which colonise cereal grain in storage produce characteristic volatile fingerprints, including a mixture of alcohols, aldehydes and ketones (Magan and Evans, 2000). For example, using gas chromatography-mass spectrometry (GC-MS) has shown that the volatile profile produced by *P. verrucosum* is very different from those produced by bacteria and yeasts (Olsson et al., 2002; Needham et al., 2005). Magan and Evans (2000) also showed that in cereal grain that is poorly stored, the predominant volatiles were 3-methyl-1-butanol, 1-octen-3-ol and 3-octanone (Magan and Evans, 2000).

Studies on grains with different levels of moulding and natural samples showed that by using a standardised grain amount and a conducting polymer array of sensors and a radial based-neural network it was possible to carry out real-time analyses of grain samples within 10 minutes (Evans et al., 2000). This enabled a decision to be made on whether the grain was acceptable or not. Interestingly, results were very comparable with those produced by an odour panel classification. This approach was based on acceptable or unacceptable criteria. This identified a small group of samples in a so-called zone of uncertainty which suggested that they were in the process of going mouldy and thus needed further investigation (Fig. 3).

Attempts were also made to investigate the potential for differentiation between mould and mite contamination by using both a conducting polymer sensor array combined with a metal oxide array for the pest infestation. This suggested that potential exists to combine real-time analyses of different types of contaminants from a single sampling. This needs further investigation as the volatile organic compounds (VOCs) produced by spoilage moulds and storage pests may be very different in types and relative concentrations.

Other studies using metal oxide sensors have suggested that the presence of 1% contaminated grain could be discriminated (de Lacy Costello et al., 2003). However, in trials with harvested, naturally contaminated grain some false positives were obtained when compared to other criteria used at intake facilities for processing. The results were, importantly, not found to be linked to grain moisture content. These studies all suggested that quality of grain can be effectively managed using this approach.

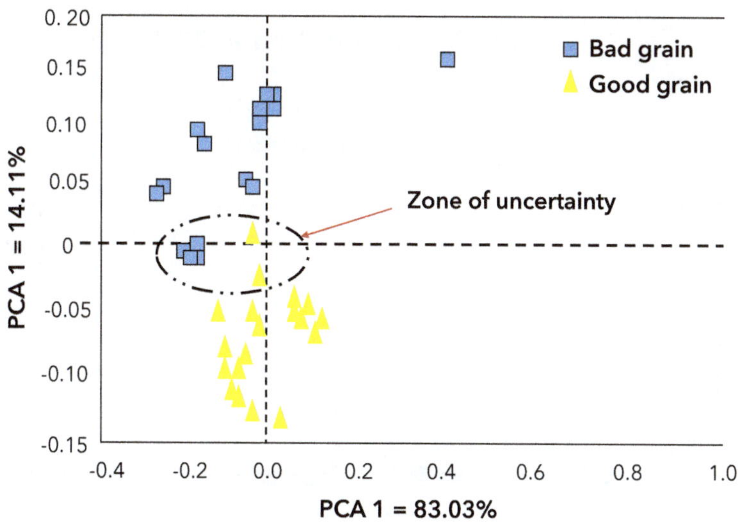


**Figure 3** Principal Component Map (PCA) of commercial wheat samples analysed using the sensor array for volatile fingerprints to differentiate between samples (from Evans et al., 2000).

## 4 Carbon dioxide monitoring for real-time identification of the risk of spoilage moulds and mycotoxins in stored cereals

Harvested and stored cereal grains and nuts are normally all alive and respiring. If effectively dried to safe moisture contents such as 14.5% for wheat, 15% for maize and 7.5–8% for peanuts then they respire at a basal level which can be quantified. These basal levels will of course change with the temperature of storage. Figure 4 shows the effect of changing water availability from the baseline level for safe storage to increasing moisture contents and how this influences the relative respiration rates in different cereals and nuts.

The use of $CO_2$ as a potential indicator of the activity of spoilage moulds and perhaps pests was originally examined by the research group of Dirk Maier at Purdue University. They suggested that $CO_2$ monitoring was a sensitive and early method for detecting hot spots due to water ingress (Bhat et al., 2003; Goncalves, 2009; Maier et al., 2010). Originally, they suggested sensors being placed in the head-space and the exhaust outlet fan.

In parallel with this approach, temperature sensor networks were being developed to place in a detailed matrix configuration within grain silos to monitor in real time as a monitoring system for managing stored grains silos (Uddin et al., 2006). Indeed the use of temperature sensors with wireless connectivity has now become quite common in grain silos, usually in steel-tethered cables

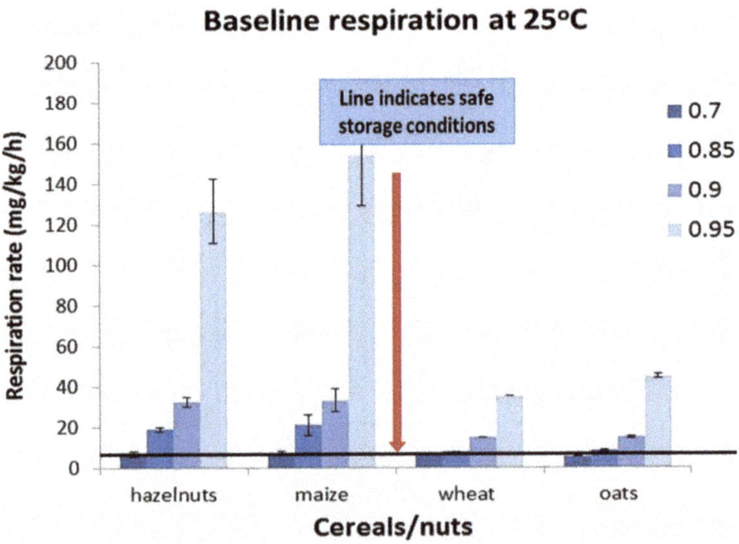


**Figure 4** Effect of different water activity levels of stored cereals and hazelnuts (0.70–0.95 $a_w$) on the respiration rate at 25°C. The line indicates the basal respiration level under safe storage conditions.

in cereal silos. However, as grain is a very good insulator the question arises as to whether temperature reflects earlier changes in $CO_2$, and thus the latter may be a more sensitive indicator of initiation of mould spoilage or pest activity in stored cereals.

The relationship between $CO_2$ production in stored cereals under different temperatures and moisture content storage conditions has been measured in wheat, maize, oats and rice. The temporal and total accumulated $CO_2$ production has been used to quantify the relative dry matter losses (DMLs) under different safe, intermediate and unsafe storage conditions for mould spoilage (Garcia-Cela et al., 2018, 2019; Martín Castaño et al., 2017a,b; Mylona and Magan 2011; Mylona et al., 2012). In addition, the DMLs under different temperatures × moisture contents can be used to correlate with the levels of mycotoxins present in the different storage conditions. This has the advantage of identifying the DML tolerances with regard to meeting the prevailing legislative limits for specific mycotoxins. Figure 5 shows an example of this relationship for wheat and zearalenone.

Real-time monitoring of changes in temperature, relative humidity and $CO_2$ was recently conducted in a pilot silo (2 tonnes) at Barilla Ltd (Parma, Italy). This showed that $CO_2$ changes occurred more rapidly than changes in the temperature. Therefore, $CO_2$ was a sensitive indicator of changes in the grain condition when a wet pocket was introduced. This suggests that biological activity can be detected very sensitively using $CO_2$, while temperature is a reflection of this activity and occurs much later. Indeed, mycotoxins were detected in the wet pocket samples, but not in the areas which remained at a safe moisture content (Garcia-Cela et al., unpublished data).

## 5 Control strategies to minimise spoilage and mycotoxin contamination of stored grain: monitoring of relative humidity (RH) and temperature and the use of molecular methods

There have been a number of approaches used for trying to manage stored commodities especially food/feed grain post-harvest. This has included effective monitoring and management of the RH and temperature during drying, to reach safe moisture contents where spoilage moulds cannot grow, use of natural preservatives, biological control and physical decontamination approaches using irradiation or gaseous fumigation such as the use of ozone ($O_3$).

Depending on the weather conditions at harvest, cereal commodities often have to be effectively dried before going into store to prevent initiation of spoilage mould activity, especially in pockets of wet grain (Fleurat-Lessard, 2017). Maize is often harvested at >18–19% moisture content (m.c.) which

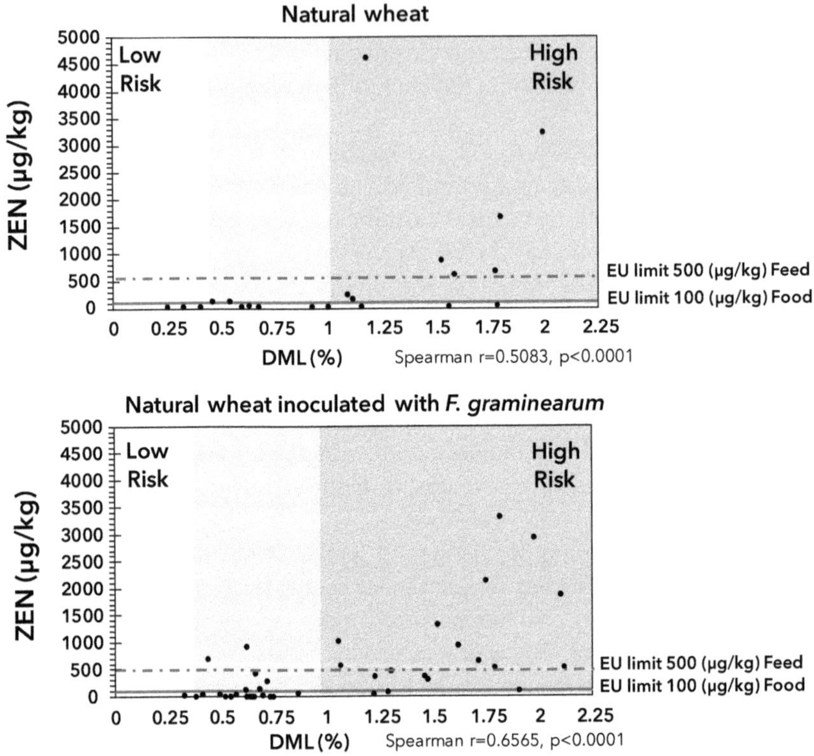


**Figure 5** Scatter-plot of dry matter losses (DMLs) and zearalenone in stored wheat after 15 days of storage under all the environmental conditions examined producing by natural mycobiota (a) and both natural mycobiota and *F. graminearum* (b). The red lines indicate the EU legislative limits for unprocessed wheat for human consumption (solid line) and animal feed (dash line) (EC, 576/2006, 1881/2006) (from Garcia-Cela et al., 2017).

represents ERH values of >85%. This will allow moulds to grow of not effectively dried to 15% m.c. (=70% ERH). Cereals are either dried using heated air streams in the silos or via ambient air in flat-bed stored or in barns. The latter results in the movement of water to the top layers of the cereal that can become rewetted, while the bottom layers can become over-dried. This can result in the initiation of moulding and potentially mycotoxin contamination in the upper layers. Heated air is the most effective treatment if the grain is not being subsequently used for seed, as apposed to processing for food and feed. The monitoring of intergranular RH and temperature can thus be used as an indicator of the presence of any pockets of wetter grain during uneven drying regimes before being placed in store. Today, often, temperature or temperature + RH sensors are used in tethered steel cables in new fabricated grain silos for real-time monitoring of these two abiotic factors during medium-term storage to allow aeration or removal of grain where the risk of mould or pest activity due to poor hygiene may occur. However, in steel silos, where extremes in seasonal temperatures can occur (e.g. North America, Scandinavia), moisture migration due to condensation can have an impact on both mould spoilage and mycotoxin contamination. A study in Canada showed that the key risk factors were (1) grain clumps accumulated around or directly under manhole openings, (2) debris and residue of old grain or grain clumps collected from the bin walls or left on the storage floor and augers and (3) clumps of grain accumulated around side doors. Condensation may lead to hot spots for the growth of *P. verrucosum,* which likes cooler conditions, and for ochratoxin A (OTA) contamination. Moisture migration and inadequate aeration management can exacerbate such events. Other mycotoxins including ochratoxin B and citrinin were also found (Limay-Rios et al., 2017). Temperature and RH measurements in different regions of the silos would allow more effective indication, the possible initiation of mould activity and potential for mycotoxin contamination.

There are now commercial sensors for RH and temperature for *in situ* detection in real time of changes in stored cereal grains. However, the sensors must by ATEX compliant (not catch fire during use) and must have the necessary accuracy. However, as mentioned earlier in Section 4, temperature changes often occur slightly later than $CO_2$ and thus while the industry favours the monitoring of temperature and/or RH, $CO_2$ may provide an earlier indication of changes due to the condensation or other forms of water ingress or poor hygiene practices.

### 5.1 Molecular methods in stored grain

The rapid development of molecular methods based on the polymerase chain reaction (PCR) and the use of quantitative PCR is now being utilised for a better understanding of the kinetics of the fungal biome in stored cereals and other commodities. In particular, for specific mycotoxigenic fungi the development of RT-qPCR approaches for the detection of the activity of specific genes involved in the biosynthesis of mycotoxins has been particularly useful (Schmidt-Heydt et al., 2008). This has allowed the kinetics of invasion of stored cereals by mycotoxigenic fungi to be tracked based on key genes, especially regulatory genes involved in the biosynthesis of the mycotoxins (e.g. *aflR, aflS, A. flavus; FUM* genes, fumonisins; *otaPKS*, ochratoxin A). Indeed, studies by Abdel-Hadi et al. (2010) showed that the temporal colonisation of stored peanuts could be monitored using the structural *aflD* gene involved in aflatoxin production under different storage conditions of $a_w$ conditions.

Multiplex PCR approaches have also been applied to examine the simultaneous amplification of species-specific genes (both structural and regulatory genes) involved in toxin production. For example, Wang et al. (2012) developed a multiplex PCR assay for simultaneous detection of different chemotypes involved in the contamination with type B trichothecene mycotoxins in wheat grain. However, this was focussed on the genetic identification of certain mycotoxin chemotypes of *Fusarium* species only. Recently, Sadhasivam et al. (2017) developed a sensitive multiplex PCR assay for the detection of a wide range of mycotoxigenic fungi in wheat grain samples and correlated this with multi-mycotoxin contamination. While these are sensitive and effective techniques they cannot be done in real time and require central laboratory facilities for the assays to be completed. This is more difficult where a short window exists for deciding on whether the grain meets the regulatory standards or not.

## 6 Control strategies: hermetic storage of grain and use of natural compounds

It has been shown earlier that grains and their associated mycobiota respire at rates determined by both the storage moisture content and the prevailing temperature (see Fig. 4). There have been approaches to store moist cereals destined for animal feed in either bag or silo systems which are entirely sealed to allow a build-up of the $CO_2$ levels which inhibit spoilage moulds and pests from proliferation. The approach used in hermetic storage is to utilise the increased $CO_2$ production by the grain and associated mycobiota of moist grains such that the increased concentration in sealed storage systems results in the inhibition of moulds and pests. The use of polypropylene bag systems has often resulted in condensation or sweating and when re-used can contain subsequent grain batches. More recently, a number of studies have examined the use of the Purdue Improved Crop Storage (PICS) bags consisting of a triple layer to prevent any damage by rodents or pests during storage. This has also been examined for mitigation of fungal growth and aflatoxin production in moist stored maize and groundnuts. Williams et al. (2014, 2017) showed that the PICS bags successfully suppressed the development of *A. flavus* and resulting aflatoxins contamination in maize over a wide range of initial moisture contents when compared with non-hermetic storage bag systems. Njoroge et al. (2014) suggested that grains stored in PICS bags absorbed less moisture than in woven polypropylene bags. This could be partially due to the PICS bags' construction with the two inner layers made of high-density polyethylene (HDPE) and an outer layer made of woven polypropylene. Sealed PICS bags were shown to rapidly increase the $CO_2$ levels and reduce the $O_2$ levels thus facilitating control of pests and most spoilage moulds (Udomkun et al., 2017). Sudini et al. (2015) evaluated the efficacy of PICS bags for protecting groundnuts from quality deterioration and aflatoxin contamination caused by *A. flavus* and found that there was less toxin production in PICS bags, when compared to cloth bags under similar conditions. Tubbs et al. (2016) examined the impact of subsequent opening of hermetically sealed storage bags and found that there was less fungal growth and aflatoxins present when compared with other bag systems. However, these systems do not include integrated preservation systems. It is now possible with nanotechnology systems to immobilise slow release food grade preservatives onto the inner layer of such bags to further develop such systems for the control of both spoilage moulds and mycotoxins. Such approaches need to be considered in the future for storage of cereals destined for human consumption and animal feed.

## 6.1 Natural preservative compounds for controlling moulds and mycotoxins in stored cereals

Traditionally moist cereal grains destined for feed have been treated with the salts of individual aliphatic acids or mixtures to prevent the initiation of spoilage. However, these are fungistatic in nature and require very good mixing and coverage of the feed for efficacy. In addition, under-treated pockets can allow some spoilage or mycotoxigenic fungi to grow and utilise the lower concentrations of such preservative as a carbon source. This often results in a stimulation in growth or mycotoxin production as well (Mutasa and Magan, 1990, Suhr and Nielsen, 2004; Arroyo et al., 2005; Ribes et al., 2018).

In the last decade or so, a significant amount of work has been carried out on the use of natural plant extracts, especially essential oils for controlling spoilage and mycotoxigenic moulds in stored cereal and other commodities (Prakash et al., 2015; Nazzaro et al., 2017). However, while a significant amount of screening has been done, often less attention has been made in simulating storage conditions under which fungal growth and mycotoxin control needs to be achieved. Often only one particular temperature and freely available water conditions have been used. Studies by Aldred et al. (2008) examined the effect of water activity × temperature effects on efficacy of some essential oils and anti-oxidants to *P. verrcuosum* and *A. westerdijkiae* in relation to $ED_{50}$ values for both growth and control of OTA production in wheat-based media and in stored wheat. This showed that under simulated storage conditions for wheat that resveratrol was more effective than essential oils tested and, in addition, at sub-optimal concentrations some of the essential oils resulted in a stimulation in OTA production. Recently, Mylona et al. (2019) examined the effect of two garlic-derived compounds, propyl propane thiosulfonate (PTS) and propyl propane thiosulfinate (PTSO), on growth and mycotoxin production by *Fusarium graminearum* (dexynivalenol, DON, in wheat), *F. langsethiae* (T-2/HT-2 toxin, in oats) and *F. verticillioides* (fumonisins, in maize). The overall conclusions were that in vitro screening for efficacy should include important parameters such as water availability, temperature and perhaps pH stress to identify the most effective candidate compounds for control of colonisation, and more importantly, mycotoxin production. Potential efficacy against Fusaria was demonstrated; however, efficacy *in situ* was not as effective as that observed *in vitro*. In addition, the efficacy of the treatments depended on the specific '*Fusarium* species-toxin' pathosystem. In addition, the essential oil formulations used must be of 'GRAS' quality, allowing their use in food. This requires a favourable safety profile, which many do not have (Prakash et al., 2015). It is possible to use the volatile components of essential oils or other plant-derived compounds, but on a large scale, this can become difficult without accurate and directional release of the necessary concentrations. Many slow release formulations are now being developed to address these issues. However, the cost-benefit analyses of such alternative compounds have often not been calculated. Indeed, the work done with resveratrol, while demonstrating good efficacy (Aldred et al., 2008) was found to be too expensive for use when compared to the aliphatic acid mixtures being used today.

## 7 Control strategies: biological control and irradiation techniques

There has been significant interest in the use of different microorganisms to try and control mycotoxigenic moulds in stored grains. The candidates have included lactic acid bacteria, yeasts and filamentous fungi including non-toxigenic strains (e.g. non-toxigenic strains of *Aspergillus flavus*). Most biocontrol approaches have involved pre-harvest control of spoilage and toxigenic moulds with the hope that some carry over may occur into storage (Bandyopadhya et al., 2016; Bhatnagar et al., 2018). However, for animal feed, which is often stored at a higher moisture content, potential exists for the use of BCAs because the environmental conditions are more conducive for their active growth. For example, a strain of *Pichia anomala* was used for the control of spoilage moulds such as *P. roqueforti* and toxigenic species such as *P. verrucosum* in moist cereals, especially wheat and barley (Mokiou and Magan, 2008). Comparisons of different fluidised bed-dried formulations were more effective than others for controlling total fungal populations and those of *P. verrucosum* and OTA contamination in stored wheat under different $a_w$ levels. More studies are required to examine the potential for post-harvest preservation of stored cereals using BCAs, especially where the grain is destined for animal feed where this may also improve nutritional quality. Recently, studies of *Bacillus* species and their secreted metabolites have been suggested to be effective at reducing *A. flavus* growth and toxin contamination of maize seed (Chalivendra et al., 2018). However, whether this would be effective in stored maize grain needs to be determined. The benefits and limitations in relation to the use of non-toxigenic strains of *A. flavus* and *Fusarium* species for mycotoxin control in developing countries was recently discussed by Kagot et al. (2019).

### *7.1 Irradiation as a method of control*

Historically, gamma irradiation of agricultural commodities was focussed on pest control. Thus, food irradiation with up to 10 kGy has often been considered a safe and effective technology since the 1980s by several international food organizations (FAO/IAEA/WHO, 1981). Later, doses >10 kGy were also considered safe for some niche products and markets (FAO/IAEA/WHO, 1999). However, it is not as widespread as other conventional technologies due to the high costs of irradiation units and because of a negative perception of consumers relative to its safety.

Some studies have been conducted on spices including red chillies and chilli powder and on black pepper and other herbs (see Calado et al., 2014). The control of spoilage moulds required about 4–6 kGys for between log2–3 to log5 reduction. Irradiation with 6 kGy was sufficient to reduce the fungal load by 5 logs. However, usually at least 10 kGy is required for the elimination of most species. Some species such as *Aspergillus, Penicillium* and *Rhizopus* spp. were the most resistant to irradiation doses of 5 kGy.

For cereals and pulses, Wang and Yu (2010) found that *Alternaria, Aspergillus* and *Fusarium* were reduced 10-fold in wheat by 3 kGys irradiation. In rice, *Penicillium* and *Rhizopus* species could not be detected after treatment with 3–5 kGys. During subsequent storage for 6–12 months, the detection of moulds did not significantly change. Studies by Maity et al. (2009) showed that the fungal diversity of rice seeds was significantly reduced during storage of up to 12 months by exposure to 3 kGy gamma irradiation.

Other studies have shown that for wheat, barley, maize and sorghum 5 kGys could completely inhibited fungi; however, bacteria were not completely eliminated. For these contaminants 10 kGy was required to improve the total hygiene of grains. Interestingly, no significant alteration of their nutritional constituents was observed (Aziz and Mattar, 2007).

In contrast, D'Ovidio et al. (2007) found that very high doses of 30 kGy was necessary for complete inhibition of *Aspergillus, Penicillium* and *Fusarium* spp.

For mycotoxin control, results have been more variable. Generally, doses >10 kGy are required to eliminate a significant amount of mycotoxins in food matrices, especially cereals. Patulin is an exception, as it is completely destroyed in apple juice by radiation doses between 2.5 kGy and 5 kGy. Some effects on toxicity of aflatoxins were found after irradiation (Calado et al., 2014).

## 8 Control technologies: cold plasma technology and ozone fumigation

This is a novel non-thermal food processing technology that uses energetic reactive cold gases to disinfect the surfaces of food commodities and packaging. It has been applied to different foodstuffs including grains, cereal flour and perishables. It uses electricity and a carrier gas, such as air or oxygen, nitrogen or helium. The primary mode of action is via UV light and reactive chemical products of the cold plasma ionization process. A wide array of cold plasma systems that operate at atmospheric pressures or in low pressure treatment chambers are under development. However, while this may be an attractive treatment for batches of cereal grain, in silos this technology may be more difficult to utilise, except prior to filling. The other aspects which need to be evaluated for different cereals are the potential impact on nutritional and sensorial impacts of treatment.

A system employing high voltage (HV) (80 kV) dielectric barrier discharge (DBD) closed system was recently used to assess the potential for control of the surface mycobiota of barley and wheat grain. The reductions of barley and wheat mycobiota were 2.1 $\log_{10}$ CFU/g and 2.5 $\log_{10}$ CFU/g, respectively (Los et al., 2018). This required direct treatment for 20 min followed by a 24-hour sealed post-treatment retention time. Treatment was relatively effective against *P.verrucosum*, responsible for OTA contamination of temperate cereals. Some studies have also shown that this technique can be effective for pest disinfestation The insecticidal impact of cold plasma-treated wheat grain against red flour beetles (*Tribolium castaneum*) has been examined. Cold plasma treatment was carried out at 800 V for 1–4 minutes. This showed that mortality and weight gain were increased by the 4 min treatment. It was suggested that this approach could be considered as a practical and effective method for seed protection against such stored grain pests, especially during long-term storage of grain (Afsheen et al., 2019). This strategy needs to be examined now on a cost-benefit basis to examine the economics of this approach.

## 8.1 Ozone ($O_3$) fumigation of stored cereals for spoilage control

Ozone ($O_3$) is a strong germicide and has three oxygen atoms. It has been used in the food industry as an anti-microbial agent to minimise contaminants in raw commodities and is lethal to many bacteria and is thus used in food packaging applications (FDA, 2001; Kim et al., 1999). Natural levels of $O_3$ are between 0.01 ppm and 0.15 ppm but exposure to high levels produces some harmful effects on health such as respiratory and cardiovascular diseases (Frampton et al., 2015; Horstman et al., 1995). Thus occupational health and safety recommendations include legislative limits for human exposure to $O_3$ of 0.1 ppm for 8 h continuously or 0.3 ppm for 15 min (Suslow, 2004). In addition, $O_3$ is very reactive and will degrade rubber tubing and other materials readily. Thus materials which are used practically for exposure of stored cereals to $O_3$ need to be made of materials which cannot be easily corroded (e.g. PTFE tubing, stainless steel silos).

There has been a lot of interest in the use of gaseous fumigation of stored cereals and other food/feed commodities for controlling spoilage and mycotoxin production. The efficacy of such gaseous preservatives depends on the effective sealing of the grain silo to ensure that the target concentration can be reached as well as the relationship between exposure time × concentration used. $O_3$ has been tested in a range of commodities including cereals such as wheat, maize and a range of nuts, especially peanuts and Brazil nuts (Kell et al., 2001; Giordano et al., 2010; Otniel et al., 2013).

The efficacy of $O_3$ on spoilage fungi varies depending on whether control is for fungal spores, mycelial growth or sporulation. Thus the type of spore, spore morphology, moisture content of substrate and $O_3$ concentration and residence time will all influence efficacy (Mylona et al., 2014). In addition, for some commodities, especially those with a high level of fatty acids/lipids the effect can be negative influencing the quality of the food by causing oxidation of lipids, changing colour, modifying some vitamins and phenolic compounds and sometimes causing off-odours (Gabler et al., 2010; Patil et al., 2010).

Studies with maize and other commodities such as nuts have suggested that exposure to $O_3$ gas can certainly influence germinative capacity of mycotoxigenic moulds and in some cases reduce mycotoxin contamination. For example, Sultan (2010) reported that the populations of *A. flavus* in stored peanuts were significantly decreased after exposure to 300 ppm for 1 h. El-Desouky et al. (2012) showed that aflatoxin $B_1$ in wheat was effectively degraded by approximately 55%, 64%, 74% and 77% after exposure to 40 ppm $O_3$ for 5 min, 10 min, 15 min and 20 min. In a study by Sahab et al. (2013) exposure to 40 ppm $O_3$ for 10 min led to 94.6% and 99.5% reduction in aflatoxin $B_1$ and $B_2$, respectively, with degradation rates of 94.6% and 80.8% at 50 ppm with a residence time of 5 min. Chen et al. (2014) suggested that the optimal condition for detoxification of aflatoxins in stored peanuts was using 6 mg/L for 30 min, with degradation rates of the total aflatoxins and aflatoxin $B_1$ being 66%. Luo et al. (2014) reported that $O_3$ treatment of maize samples spiked with 83 ppb aflatoxins at 90mg/L $O_3$ for 20 min and 40 min resulted in reductions of aflatoxins of 18.1 ppb and 9.9 ppb. Mycelial growth of *A. flavus* in maize was decreased by 83% when exposed to 2 g of gaseous $O_3$ for 2 min (Halima et al., 2015). In contrast, other studies have suggested limited effects, both *in vitro* and *in situ* on *Fusarium verticillioides* and *Aspergillus carbonarious* and *Aspergillus westerdijkiae* (Mylona et al., 2014; Akbar, 2015). Indeed, Mylona et al. (2014) showed that conidial germination of *F. verticillioides* was initially inhibited by up to 300 ppm $O_3$ exposure for 1 h. However, subsequent repair mechanisms resulted in germinative capacity to be retained and both germination and indeed some fumonisin production. However, some control was achieved in stored maize using different concentrations of $O_3$ at a flow rate of 4 L/min for 1 h in maize inoculated with *F. verticillioides* and stored for 15 and 30 days at two different $a_w$ levels (Fig. 6).

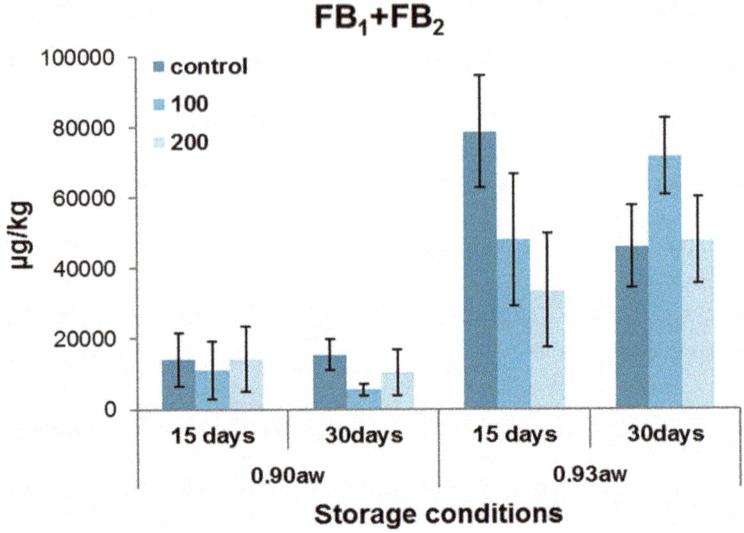


**Figure 6** Effect of $O_3$ on sum of fumonisins ($B_1$+$B_2$) production in maize grain rewetted between 0.90 $a_w$ and 0.93 $a_w$, inoculated with *F. verticillioides* and exposed to 0 ppm, 100 ppm and 200 ppm for 1 h in glass columns and stored at 25°C for 15–30 days (from Mylona et al. 2014).

Practically, $O_3$ can be generated by two methods including the use of ultra-violet rays of the sun and lighting. $O_3$ is also produced artificially by three methods including UV-light, the corona discharge method (CDM) and electrolytic $O_3$ generation (EOG). The most common method for $O_3$ production today is the corona-discharge method, because it has many advantages such as greater sustainability of the unit and $O_3$ generation over a wider range of concentrations (Gonçalves, 2009).

In summary, the effects of gaseous $O_3$ to control spoilage moulds and mycotoxins in stored cereals have been variable. Some very good positive control of mould spoilage and toxin contamination has been achieved as detailed previously. However, in many cases a very high concentration is needed or a low concentration for a lengthy residence time to ensure that efficacy can be achieved. This will influence the economics of its use and the practical aspects that it is not a very user-friendly gas for operators to work with, especially in stored grain silos.

## 9 Conclusions

This chapter has highlighted that a combination of the two most relevant abiotic factors (temperature and $a_W$) can be used as a prediction tool in the risk of growth/mycotoxin contamination. Extensive ecophysiological information about mycotoxigenic fungi has been published. However, most of the work has been carried out on agar media and has not identified the boundary conditions for growth/toxin in the relevant stored commodities, with a few exceptions. Therefore, more information in the so-called zone of uncertainty storage regimes should be produced to develop accurate mathematical models that must be validated on real matrices.

Also, the identification and description of early indicators of biological activity in silos such as VOC changes and $CO_2$ increases have been reviewed in depth. In particular, the $CO_2$ sensor technology is able to develop more sensitive and reliable sensors which can be used to monitor very small changes in the respiratory activity of cereals and nuts and their associated mycobiota in storage. The economic use of $CO_2$ sensors may be a problem as the costs are higher because of the need for having them compliant with ATEX regulations.

Incorporation of the sensor technologies in silos for real-time monitoring combined with mathematical models is a corner stone for an integrated silo management approach. This could be effectively developed as a mobile phone application, for example, to assist with better post-harvest management of stored commodities, especially cereals, in real time, to allow the evaluation of relative risks of mould spoilage and/or mycotoxin contamination and whether this risk is low, intermediate or high. This will allow more rapid remedial actions to be implemented, including aeration regimes or more rapid removal and processing of such batches to minimise rejection based on quality criteria and mycotoxins. This would also reduce waste streams.

Additionally, other methods including molecular methods; biocontrol approaches, which may be more applicable to feed grain; use of natural compounds, which need to be 'GRAS' compatible; irradiation, which has consumer resistance; and the application of $O_3$ in stored grain silos as alternative strategies have all been discussed in this chapter. While some combination treatments/fungi have shown promising results, there are still economical and practical aspects that act as a barrier for the practical transference of these technologies.

## 10 Where to look for further information

Since the cereal grain trade is global in nature it is important to consider storage in relation to producer countries, transport over long distances, especially through different climatic zones which can allow some re-adsorption of moisture by hygroscopic commodities or condensation or sweating of grains which can result in initiation of mould activity and potentially mycotoxin contamination. The impacts of climate-related abiotic factors may also change the mycobiota and the mycotoxins present in specific cereal commodities.

Significant detailed information on drying and post-harvest storage management including links to a wide range of useful information is available via the recently completed EU projects: ***Safe Food and Feed through an Integrated ToolBox for Mycotoxin Management;*** MyToolBox; www.mytoolbox.eu; the E-platform is now available for detailed free access to static information on post-harvest storage of cereals and nuts; **https:\\mytoolbox-platform.com**.

The complementary EU Project: ***Integrated and innovative key actions for mycotoxin management in the food and feed chains***; MycoKey; (www.mycokey.eu) also has very useful links to information and publications.

Bandyopadhyay, R., Ortega-Beltran, A., Akande, A., Mutegi, C., Atehnkeng, J., Kaptoge, L., Senghor, A. L., Adhikari, B. N. and Cotty, P. J. 2016. Biological control of aflatoxins in Africa: current status and potential challenges in the face of climate change. *World Mycotoxin Journal* 9, 771–89.

Medina, A., Mohale, S., Samsudin, N. I. P., Rodriguez-Sixtos, A., Rodriguez, A. and Magan, N. 2017[a]. Biocontrol of mycotoxins: dynamics and mechanisms of action. *Current Opinion in Food Science* 17, 41–8.

Medina, A., Akbar, A., Baazeem, A., Rodriguez, A. and Magan, N. 2017b. Climate change, food security and mycotoxins: do we know enough? *Fungal Biology Review* 31, 143–54.

FAO. 2020. Climate change: Unpacking the burden of food safety. Food Safety Series, No. 8. FAO, Rome, Italy, pp. 176..ISBN: 978-92-5-132293-2.

## 11 References

Abdel-Hadi, A., Carter, D. and Magan, N. 2010. Temporal monitoring of the nor-1 (*aflD*) gene of *Aspergillus flavus* in relation to aflatoxin $B_1$ production during storage of peanuts under different environmental conditions. *Journal of Applied Microbiology* 109(6), 1914–22. doi:10.1111/j.1365-2672.2010.04820.x.

Afsheen, S., Fatima, U., Iqbal, T., Abrar, M., Muhammad, S., Saeed, A., Isa, M., Malik, M. F. and Shamas, S. 2019. Influence of cold plasma treatment on insecticidal properties of wheat seeds against red flour beetles. *Plasma Science and Technology* 21(8). doi:10.1088/2058-6272/ab19ee.

Akbar, A. H. 2015. Growth and ochratoxin production by *Aspergillus* species in Coffee beans: impact of climate change and control using $O_3$ (PhD thesis). Cranfield University, Cranfield, UK.

Aldred, D., Cairns-Fuller, V.. and Magan, N. 2008. Environmental factors affect efficacy of some essential oils and resveratrol to control growth and ochratoxin A production by *Penicillium verrucosum* and *A.westerdijkiae* on wheat grain. *Journal of Stored Products Research* 44(4), 341-6. doi:10.1016/j.jspr.2008.03.004.

Arroyo, M., Aldred, D. and Magan, N. 2005. Environmental factors and weak organic acid interactions have differential effects on control of growth and ochratoxin A production by *Penicillium verrucosum* isolates in bread. *International Journal of Food Microbiology* 98(3), 223-31. doi:10.1016/j.ijfoodmicro.2004.07.004.

Astoreca, A., Vaamonde, G., Dalcero, A., Ramos, A. J. and Marín, S. 2012. Modelling the effect of temperature and water activity of *Aspergillus flavus* isolates from corn. *International Journal of Food Microbiology* 156(1), 60–7. doi:10.1016/j.ijfoodmicro.2012.03.002.

Aziz, N. H. and Mattar, Z. A. 2007. Influence of gamma-radiation on the occurrence of toxigenic *Penicillium* strains and mycotoxins production in different feedstuffs. *Egyptian Journal of Biotechnology* 25, 130–44.

Bandyopadhyay, R., Ortega-Beltran, A., Akande, A., Mutegi, C., Atehnkeng, J., Kaptoge, L., Senghor, A. L., Adhikari, B. N. and Cotty, P. J. 2016. Biological control of aflatoxins in Africa: current status and potential challenges in the face of climate change. *World Mycotoxin Journal* 9(5), 771–89. doi:10.3920/WMJ2016.2130.

Bhat, C., Maier, D. E. and Illeliji, K. E. 2003. Exploratory use of a portable $CO_2$ sensor for early detection of spoilage in a large maize storage tank. Paper no. 03-6149. ASAE, St. Joseph, MI.

Bhatnagar, D., Rajasekaran, K., Gilbert, M., Cary, J. W. and Magan, N. 2018. Research advances to safeguard food and feed supply from aflatoxin contamination. *World Mycotoxin Journal* 11(1), 47–72. doi:10.3920/WMJ2017.2283.

Calado, T., Venancio, A. and Abrunhosa, L. 2014. Irradiation for mold and mycotoxin control: a review. *Comprehensive Reviews in Food Science and Food Safety* 13(5), 1049–61. doi:10.1111/1541-4337.12095.

Chalivendra, S., De Robertis, C., Pineda, J. R., Ham, J. H. and Damann, K. 2018. Rice phyllosphere *Bacillus* species and their secreted metabolites suppress *Aspergillus flavus* growth and aflatoxin production *in vitro* and in maize seeds. *Toxins* 10(4), 159; doi:10.3390/toxins10040159.

Chen, R., Ma, F., Li, P. W., Zhang, W., Ding, X. X., Zhang, Q., Li, M., Wang, Y. R. and Xu, B. C. 2014. Effect of ozone on aflatoxins detoxification and nutritional quality of peanuts. *Food Chemistry* 146, 284–8. doi:10.1016/j.foodchem.2013.09.059.

de Lacy Costello, B. P. JdL., Ewen, R. J., Gunson, H., Ratcliffe, N. M., Sivanand, P. S. and Spencer-Phillips, P. T. N. 2003. A prototype sensor system for the early detection of microbially linked spoilage in stored wheat grain. *Measurement Science and Technology* 14(4), 397–409. doi:10.1088/0957-0233/14/4/301.

D'Ovidio, K. L., Trucksess, M. W., Devries, J. W. and Bean, G. 2007. Effects of irradiation on fungi and fumonisin B1 in corn, and of microwave-popping on fumonisins in popcorn. *Food Additives and Contaminants* 24(7), 735–43. doi:10.1080/02652030701216453.

El-Desouky, T. A., Sharoba, A. M. A., El-Desouky, A. I., El-Mansy, H. A. and Naguib, K. 2012. Effect of ozone gas on degradation of aflatoxin b1 and *Aspergillus flavus* fungal. *Journal of Environmental and Analytical Toxicology* 2(1), 128.

Evans, P., Persaud, K. C., McNeish, A. S., Sneath, R. W., Hobson, N. and Magan, N. 2000. Evaluation of a radial basis function neural network for determination of wheat quality from electronic nose data. *Sensors and Actuators B* 69(3), 348–58. doi:10.1016/S0925-4005(00)00485-8.

FAO/IAEA/WHO. 1981. Wholesomeness of irradiated food. Technical Report Series 659. Joint FAO/IAEA/WHO Expert Committee, Geneva, Switzerland.

FAO/IAEA/WHO. 1999. High-dose irradiation: wholesomeness of food irradiated with doses above 10 kGy. Report of a Joint FAO/IAEA/WHO study group. *World Health Organization Technical Report Series* 890, 1–197.

FDA. 2001. *United States Food and Drug Administration, Code of Federal Regulations 21 CFR part 173* (vol. 368).

Fleurat-Lessard, F. 2017. Integrated management of the risks of stored grain spoilage by seed-borne fungi and contamination by storage mould mycotoxins: an update. *Journal of Stored Products Research* 71, 22–40. doi:10.1016/j.jspr.2016.10.002.

Frampton, M. W., Pietropaoli, A., Dentler, M., Chalupa, D., Little, E. L., Stewart, J., Frasier, L., Oakes, D., Wiltshire, J., Vora, R. and Utell, M. J. 2015. Cardiovascular effects of ozone in healthy subjects with and without deletion of glutathione- transferase M1. *Inhalation Toxicology* 27(2), 113–9. doi:10.3109/08958378.2014.996272.

Gabler, F. M., Smilanick, J. L., Mansour, M. F. and Karaca, H. 2010. Influence of fumigation with high concentrations of ozone gas on postharvest grey mould and fungicide residues on table grapes. *Postharvest Biology and Technology* 55(2), 85–90. doi:10.1016/j.postharvbio.2009.09.004.

Garcia, D., Ramos, A. J., Sanchis, V. and Marín, S. 2011. Modelling the effect of temperature and water activity in the growth boundaries of *Aspergillus ochraceus* and *Aspergillus parasiticus*. *Food Microbiology* 28(3), 406–17. doi:10.1016/j.fm.2010.10.004.

Garcia-Cela, E., Kiaitsi, E., Sulyok, M., Medina, A. and Magan, N. 2018. *Fusarium graminearum* in stored wheat: use of $CO_2$ production to quantify dry matter losses and relate this to relative risks of zearalenone contamination under interacting environmental conditions. *Toxins* 10(2). doi:10.3390/toxins10020086.

Garcia-Cela, E., Kiaitsi, E., Sulyok, M., Krska, R., Medina, A., Petit Damico, I. and Magan, N. 2019. Influence of storage environment on maize grain: $CO_2$ production, dry matter losses and aflatoxins contamination. *Food Additives and Contaminants. Part A, Chemistry, Analysis, Control, Exposure and Risk Assessment* 36(1), 175–85. doi:10.1080/19440049.2018.1556403.

Giordano, B. N. E., Simao, V., Manfio, D., Galvao, S., Scussel, J. N. and Scussel, V. M. 2010. Reduction of in-shell Brazil nut (*Bertholletia excelsa* HBK) aflatoxin contamination by ozone gas application during storage. *Julius-Kühn-Archiv* 425, 566–72.

Gonçalves, A. A. 2009. Ozone: an emerging technology for the seafood industry. *Brazilian Archives of Biology and Technology* 52(6), 1527–39. doi:10.1590/S1516-89132009000600025.

Halima, Z. H., Riyam, H. T. and Amna, M. A. 2015. Study the effect of ozone gas and ultraviolet radiation and microwave on the degradation of aflatoxin $B_1$ produce by *Aspergillus flavus* on stored Maize grains. *Journal of Agriculture and Veterinary Science* 8, 5–12.

Horstman, D. H., Ball, B. A., Brown, J., Gerrity, T. and Folinsbee, L. J. 1995. Comparison of pulmonary responses of asthmatic and no asthmatic subjects performing the light exercise while exposed to a low level of ozone. *Toxicology and Industrial Health* 11(4), 369–85. doi:10.1177/074823379501100401.

Kagot, V., Okoth, S., De Boevre, M. and De Saeger, S. 2019. Biocontrol of *Aspergillus* and *Fusarium* mycotoxins in Africa: benefits and limitations. *Toxins* 11(2), 109. doi:10.3390/toxins11020109.

Kell, S. A., Mason, L. J., Maier, D. E. and Woloshuk, C. P. 2001. Efficacy and fumigation characteristics of ozone in stored maize. *Journal of Stored Products Research* 37(4), 371–82. doi:10.1016/S0022-474X(00)00040-0.

Kim, J. G., Yousef, A. E. and Chism, G. M. 1999. Use of ozone to inactivate microorganisms on lettuce [Cited by Xu, L., 1999. Use of ozone to improve the safety of fresh fruits and vegetables]. *Food Technology* 53, 58–61.

Limay-Rios, V., Miller, J. D. and Schaafsma, A. W. 2017. Occurrence of *Penicillium verrucosum*, ochratoxin A, ochratoxin B and citrinin in on farm stored winter wheat from the Canadian Great Lakes Region. *PLoS ONE* 12(7), e0181239. doi:10.1371/journal.pone.0181239.

Los, A., Ziuzina, D., Akkermans, S., Boehm, D., Cullen, P. J., Van Impe, J. and Bourke, P. 2018. Improving microbiological safety and quality characteristics of wheat and barley by high voltage atmospheric cold plasma closed processing. *Food Research International* 106, 509–21. doi:10.1016/j.foodres.2018.01.009.

Luo, X., Wang, R., Wang, L., Li, Y., Bian, Y. and Chen, Z. 2014. Effect of ozone treatment on aflatoxin b1 and safety evaluation of ozonized corn. *Food Control* 37, 171-6. doi:10.1016/j.foodcont.2013.09.043.

Magan, N. and Evans, P. 2000. Volatiles in grain as an indicator of fungal spoilage, odour descriptors for classifying spoiled grain and the potential for early detection using electronic nose technology: a review. *Journal of Stored Protection* 36, 319–40.

Magan, N., Aldred, D., Mylona, K. and Lambert, R. J. 2010. Limiting mycotoxins in stored wheat. *Food Additives and Contaminants. Part A, Chemistry, Analysis, Control, Exposure and Risk Assessment* 27(5), 644–50. doi:10.1080/19440040903514523.

Maier, D. E., Channaiah, L. H., Martinez-Kawas, A., Lawrence, J. S., Chaves, E. V., Coradi, P. C. and me, G. A. 2010. Monitoring carbon dioxide concentration for early detection of spoilage in stored grain. *Julius-Kühn-Archiv* 425, 505.

Maity, J. P., Kar, S., Banerjee, S., Chakraborty, A. and Santra, S. C. 2009. Effects of gamma irradiation on long-storage seeds of *Oryza sativa* (cv. 2233) and their surface infecting fungal diversity. *Radiation Physics and Chemistry* 78(11), 1006–10. doi:10.1016/j.radphyschem.2009.06.002.

Marín, S., Hodžić, I., Ramos, A. J. and Sanchis, V. 2008. Predicting the growth/no-growth boundary and ochratoxin A production by *Aspergillus carbonarius* in pistachio nuts. *Food Microbiology* 25(5), 683–9. doi:10.1016/j.fm.2008.03.006.

Marín, S., Ramos, A. J. and Sanchis, V. 2012. Modelling *Aspergillus flavus* growth and aflatoxins production in pistachio nuts. *Food Microbiology* 32(2), 378–88. doi:10.1016/j.fm.2012.07.018.

Martín Castaño, S., Medina, A. and Magan, N. 2017a. Comparison of dry matter losses and aflatoxin $B_1$ contamination of paddy and brown rice stored naturally or after inoculation with *Aspergillus flavus* at different environmental conditions. *Journal of Stored Products Research* 73, 47–53. doi:10.1016/j.jspr.2017.06.004.

Martin Castaño, S. M., Medina, A. and Magan, N. 2017b. Impact of storage environment on respiration, dry matter losses and fumonisin $B_1$ contamination of stored paddy and brown rice. *World Mycotoxin Journal* 10(4), 319-26. doi:10.3920/WMJ2017.2237.

Medina, A., Gilbert, M. K., Mack, B. M., Obrian, G. R., Rodriguez, A., Bhatnagar, D., Payne, G. and Magan, N. 2017. Interactions between water activity and temperature on the *Aspergillus flavus* transcriptome and aflatoxin b1 production. *International Journal of Food Microbiology* 256, 36–44. doi:10.1016/j.ijfoodmicro.2017.05.020.

Mokiou, S. and Magan, N. 2008. Physiological manipulation and formulation of the biocontrol yeast *Pichia anomala* for control of *Penicillium verrucosum* and ochratoxin contamination of moist grain. *Biocontrol Science and Technology* 18(10), 1063–73. doi:10.1080/09583150802585769.

Mutasa, E. S. and Magan, N. 1990. Utilization of potassium sorbate by tobacco spoilage fungi. *Mycological Research* 94(7), 965–70. doi:10.1016/S0953-7562(09)81313-0.

Mylona, K. and Magan, N. 2011. *Fusarium langsethiae*: storage environment influences dry matter losses and T2 and HT-2 toxin contamination of oats. *Journal of Stored Products Research* 47(4), 321–7. doi:10.1016/j.jspr.2011.05.002.

Mylona, K., Sulyok, M. and Magan, N. 2012. Relationship between environmental factors, dry matter loss and mycotoxin levels in stored wheat and maize infected with *Fusarium* species. *Food Additives and Contaminants. Part A, Chemistry, Analysis, Control, Exposure and Risk Assessment* 29(7), 1118–28. doi:10.1080/19440049.2012.672340.

Mylona, K., Kogkaki, E., Sulyok, M. and Magan, N. 2014. Efficacy of gaseous ozone treatment on spore germination, growth and fumonisin production by *Fusarium verticillioides in vitro* and *in situ*. *Journal of Stored Products Research* 59, 178–84. doi:10.1016/j.jspr.2014.08.001.

Mylona, K., Garcia-Cela, E., Sulyok, M., Medina, A. and Magan, N. 2019. Influence of two garlic-derived compounds, propyl propane thiosulfonate (PTS) and propyl propane thiosulfinate (PTSO) on growth and mycotoxin production by *Fusarium* species *in vitro* and in stored cereals. *Toxins* 11(9), 495. doi:10.3390/toxins11090495.

Nazzaro, F., Fratianni, F., Coppola, R. and De Feo, V. 2017. Essential oils and antifungal activity. *Pharmaceuticals* 10(4), 86; doi:10.3390/ph10040086.

Needham, R., Williams, J., Beales, N., Voysey, P. and Magan, N. 2005. Early detection and differentiation of spoilage of bakery products. *Sensors and Actuators B* 106(1), 20–3. doi:10.1016/j.snb.2004.05.032

Njoroge, A. W., Affognon, H. D., Mutungi, C. M., Manono, J., Lamuka, P. O. and Murdock, L. L. 2014. Triple bag hermetic storgae delivers a lethal punch to Prostephanus truncatus (Horn) (Coleoptera: Bostrichidae) in stored maize. *Journal of Stored Products Research* 58, 12–19.

Olsson, J., Börjesson, T., Lundstedt, T. and Schnürer, J. 2002. Detection and quantification of ochratoxin A and deoxynivalenol in barley grains by GC-MS and electronic nose. *International Journal of Food Microbiology* 72(3), 203–14. doi:10.1016/s0168-1605(01)00685-7.

Otniel, F. S., Hector, M. V. and Armando, V. 2013. Potential of aqueous ozone to control aflatoxigenic fungi in brazil nuts. *ISRN Biotechnology* 6, 859830.

Patil, S., Torres, B., Tiwari, B. K., Wijngaard, H. H., Bourke, P., Cullen, P. J. and Valdramidis, V. P. 2010. Safety and quality assessment during the ozonation of cloudy apple juice. *Journal of Food Science* 75(7), 437–43.

Prakash, B., Kedia, A., Mishra, P. K. and Dubey, N. K. 2015. Plant essential oils as food preservatives to control loulds and mycotoxin contamination and oxidative deterioration of agri-food commodities – potentials and Challenges. *Food Control* 47, 381–91.

Ramirez, M. L., Chulze, S. and Magan, N. 2006. Temperature and water activity effects on growth and temporal deoxynivalenol production by two Argentinean strains of *Fusarium graminearum* on irradiated wheat grain. *International Journal of Food Microbiology* 106(3), 291–6. doi:10.1016/j.ijfoodmicro.2005.09.004.

Ribes, S., Fuentes, A., Talens, P. and Barat, J. M. 2018. Prevention of fungal spoilage in food products using natural compounds: a review. *Critical Reviews in Food Science and Nutrition* 58(12), 2002–16, doi:10.1080/10408398.2017.1295017.

Sadhasivam, S., Britzi, M., Zakin, V., Kostyukovsky, M., Trostanetsky, A., Quinn, E. and Sionov, E. 2017. Rapid detection and identification of mycotoxigenic fungi and mycotoxins in stored wheat grain. *Toxins* 9(10), 302; doi:10.3390/toxins9100302.

Sahab, A., Hassanien, F., El-Nemr, S., Abdel-Alim, H. and Abdel-Wahhab, M. 2013. Effect of ozone gaseous on aflatoxin degradation and fat and protein content in peanut seeds. *Journal of Applied Sciences Research* 9, 2170–5.

Sanchis, V. and Magan, N. 2004. In Mycotoxins in food: detection and control. In: Magan, N. and Olsen, M. (Eds), *Environmental Profiles for Growth and Mycotoxin Production* . Woodhead Publishing Ltd., Abingdon, UK, pp. 174–89.

Schmidt-Heydt, M., Magan, N. and Geisen, R. 2008. Stress induction of mycotoxin biosynthesis genes in relation to abiotic factors. *FEMS Microbiology Letters* 284(2), 142–9. doi:10.1111/j.1574-6968.2008.01182.x.

Sudini, H., Ranga Rao, G. V., Gowda, C. L. L., Chandrika, R., Margam, V., Rathore, A. and Murdock, L. L. 2015. Purdue Improved Crop Storage (PICS) bags for safe storage of groundnuts. *Journal of Stored Products Research* 64, 133–8. doi:10.1016/j.jspr.2014.09.002.

Suhr, K. I. and Nielsen, P. V. 2004. Effect of weak acid preservatives on growth of bakery product spoilage fungi at different water activities and pH values. *International Journal of Food Microbiology* 95(1), 67–78. doi:10.1016/j.ijfoodmicro.2004.02.004.

Sultan, Y. 2010. Biodiversity of mycotoxigenic *Aspergillus* species in Egyptian peanuts and strategies for minimizing aflatoxin contamination (PhD thesis). Cranfield University, Cranfield, UK.

Suslow, T. V. 2004. *Ozone Applications for Postharvest Disinfection of Edible Horticultural Crops* (vol. 8133). ANS Publication, pp. 1-8.

Tassou, C. C., Natskoulis, P. I., Magan, N. and Panagou, E. Z. 2009. Effect of temperature and water activity on growth and ochratoxin A production boundaries of two *Aspergillus carbonarius* isolates on a simulated grape juice medium. *Journal of Applied Microbiology* 107(1), 257–68. doi:10.1111/j.1365-2672.2009.04203.x.

Tothill, I. E. and Magan, N. 2003. Rapid detection methods for microbial contamination. In: Tothill, I. E. (Ed.), *Rapid and On-Line Instrumentation for Food Quality Assurance*. Woodhead Publishing, pp. 136–60. Chapter 8. Woodhead Publishing Ltd, Cambridge CB22 3HJ, U.K.

Tubbs, T., Baribusta, D. and Woloshuk, C. 2016. Impact of opening hermetic storage bags on grain quality, fungal growth and aflatoxin accumulation. *Journal of Stored Products Research* 69, 276–81. doi:10.1016/j.jspr.2016.10.003.

Uddin, M. S., Armstrong, P. R. and Zhang, N. 2006. Accuracy of grain moisture content prediction using temperature and relative humidity sensors. *Applied Engineering in Agriculture* 22(2), 267–73. doi:10.13031/2013.20273.

Udomkun, P., Wiredo, A. N., Nagle, M., Muller, J., Vanlauwe, B. and Bandyopadhyay, R. 2017. Innovative technologies to manage aflatoxins in foods and feeds and the profitability of application: a review. *Food Control* 76, 127–38. doi:10.1016/j.foodcont.2017.01.008.

Wallace, H. A. H. 1973. Fungi and other organisms associated with stored grain. In: Sinha, R. N. and Muir, W. E. (Eds), *Grain Storage: Part of a System*. AVI Publishing Co., Westport, CT, pp. 71-98.

Wang, J. and Yu, Y. 2010. Inactivation of mildew in rough rice and wheat by gamma irradiation. *Radiation Physics and Chemistry* 79(6), 740-3. doi:10.1016/j.radphyschem.2009.12.001.

Wang, J. H., Zhang, J. B., Chen, F. F., Li, H. P., Ndoye, M. and Liao, Y. C. 2012. Amultiplex PCR assay for genetic chemotyping of toxigenic *Fusarium graminearum* and wheat grains for 3-acetyldeoxynivalenol, 15-acetyldeoxynivalenol and nivalenol mycotoxin. *Journal of Food, Agriculture and Environment* 10, 505–11.

Williams, S. B., Baribusta, D. and Woloshuk, C. 2014. Assessing Purdue Improved Crop Storage (PICS) bags to mitigate fungal growth and aflatoxin contamination. *Journal of Stored Products Research* 59, 190–6. doi:10.1016/j.jspr.2014.08.003.

Williams, S. B., Murdock, L. L. and Baribusta, D. 2017. Storage of maize in Purdue Improved Crop Storage (PICS) bags. *PLoS ONE* 12, e01686224. doi:101371/journal.pome.0168624.

# Chapter 12

## Advances in techniques for monitoring the quality of stored cereal grains

*Roger G. Aby and Dirk E. Maier, Iowa State University, USA*

1 Introduction
2 Methods and techniques used to monitor stored grain
3 Conclusion
4 Future trends
5 Where to look for further information
6 References

## 1 Introduction

Grains are seasonally harvested in large quantities while their consumption is continuous throughout the year. Thus, grain has to be stored at multiple points in the supply chain and preserved until its usage. Grain storage periods vary from a few weeks to a few months and sometimes years. Grain can deteriorate throughout any of these periods (Metz, 2007). Stored grain deterioration occurs because of factors such as grain temperature, moisture content, maturity, birds, dockage, insects, microorganisms, geographical location, storage time, and environmental conditions.

Temperature, relative humidity, and moisture content of grain are the main factors affecting grain quality (Gonzales et al., 2009). Insect activity, mold growth, and the subsequent production of mycotoxins occur at a relatively high temperature and relative humidity. For instance, mold growth increases at a relative humidity of over 70% and at temperatures ranging from 25°C to 35°C, and insect activity begins at temperatures above 15°C (Mills, 1989). Therefore, whenever an increase in temperature and relative humidity is observed while monitoring stored cereal grains, an increased rate of insect activity, mold growth, and fungal spoilage is likely the cause (Lacey, 1988; Magan et al., 2003).

Similarly, increased levels of carbon dioxide are also an indicator of deterioration within the grain mass and point to the presence of insects and fungi. Biological organisms generate carbon dioxide and other odor volatiles that indicate spoilage of stored cereal grains (Magan and Evans, 2000; Maier

http://dx.doi.org/10.19103/AS.2020.0072.15

et al., 2006). The intergranular air in the grain mass and headspace air above the grain mass can be analyzed for the presence of carbon dioxide and volatile organic compounds to rapidly identify spoilage by fungi (Borjesson et al., 1989).

Consequently, it is important to monitor the quality of grain regularly during storage to prevent devastating losses before it is too late. It is recommended that, at a minimum, grain be monitored for temperature, relative humidity, moisture content, carbon dioxide, insects, odor, and signs of mold spoilage every two weeks when stored above 10°C, or at least monthly when stored at cooler temperatures. Data points should be taken from multiple locations at several depths with emphasis on the grain mass surface, core, and any hot spots that may be developing (Hodgson and Holscher, 2009). Monitoring stored grain for these indicators requires special sensors and measurement equipment that are the subject of this chapter. Data collected and analyzed provides information on whether problems exist or are developing within the grain mass, such as insects and molds. Awareness of changes alerts stored grain managers and allows them to implement steps to mitigate potential risks and damage to the quality of stored cereal grains.

State-of-the-art monitoring systems have sensors that can early detect the onset of spoilage before 'hot spots' develop and alert the user of their location through real-time data that can be delivered directly to cloud-enabled devices (TSGC, n.d.). The function and utility of sensors and monitoring equipment for temperature, relative humidity and moisture content, carbon dioxide and oxygen, phosphine and other gasses, and volatile organic compounds will be reviewed in-depth in this chapter.

## 2 Methods and techniques used to monitor stored grain

### *2.1 Manual grain inspection*

Various methods have been used for monitoring stored grain quality. The most basic one is to manually check the temperature and moisture content of a grain sample and to visually inspect its condition. This requires someone to sample a stream of grain that is being conveyed in or out of a storage structure or the grain mass at rest with a probe. The sample is then inspected for relevant quality parameters in an on- or off-site grain grading room or grain quality laboratory. Stored grain managers routinely check stored grain for smell (odors) and look (discoloration, surface crusting), by sniffing exhaust air from aeration fans or peeking into the tops of silos to check for visible signs of insect infestation or grain spoilage and moisture condensation on the underside of roofs or on the grain surface (Maier et al., 2010). This method is labor-intensive, time-consuming, and ineffective in large silos and storage structures and is subject to human error. Furthermore, removing a sample from a stored grain mass or inspecting stored grain from the top of a silo or storage structure would require someone to enter the headspace and step onto the grain surface, which is dangerous and has resulted in way too many life-threatening entrapments and tragic engulfment deaths (Freeman et al., 1998). Entering the headspace of a storage structure should be avoided, requires special safety gear and procedures, is prohibited in many commercial grain facilities, and given today's state-of-the-art technology, is entirely unnecessary for monitoring stored grain quality.

## 2.2 Temperature cable sensors

The use of temperature cables for stored grain monitoring has been available and utilized for more than 60 years (Jones, 1957). Temperature cables are either made with thermocouple or thermistor sensors that can be checked at any time to obtain a grain temperature reading. A thermocouple consists of two different types of metal, connected at one end (Fig. 1). When the junction of the two metals is heated or cooled, a voltage is created that can be correlated back to the temperature. Thermocouples were accidentally discovered by Thomas Johann Seebeck in 1821. He found that when the ends of two conductors are at different temperatures, a voltage occurs between them. Seebeck's experiment demonstrated that this voltage is proportional to the temperature difference. His discovery soon became the basis for the

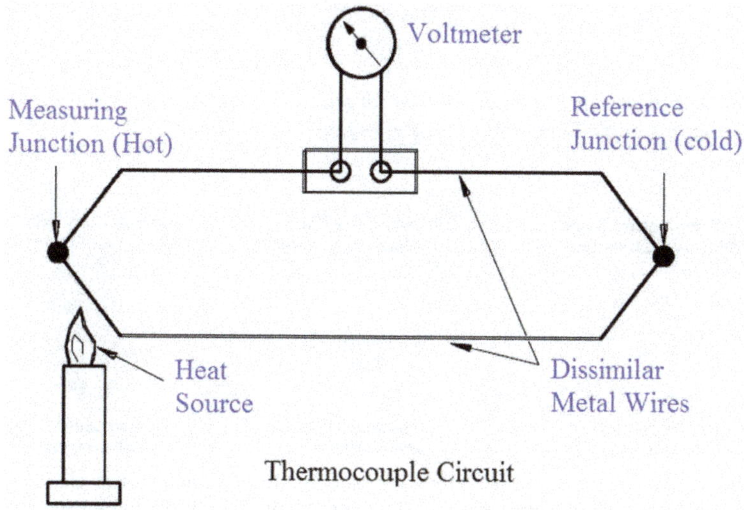


**Figure 1** Thermocouple circuit to establish a voltage between two dissimilar metal wires as a function of the temperature difference measured at the junctions. Figure Credit: https://instrumentationtools.com/thermocouple-construction/.

thermocouple, which today remains the most popular temperature cable sensor (Fig. 2) (Anon, 2003).

In comparison, thermistors function based on resistance. Thermistors are made up of metallic oxides, binders, and stabilizers; pressed into wafers; and then cut to chip size. The precise ratio of the composite materials governs their resistance versus the temperature curve. As the temperature increases (or decreases), the thermistor resistance also increases (or decreases) in a nonlinear fashion, following its resistance versus temperature curve.

Typically, temperature cables are hung in a bin, silo, tank, or building from the roof rafters in a uniformly distributed pattern within the grain mass (Fig. 3). Temperature cables remain an important tool to check grain temperature. They last long after being installed in storage structures. However, they have several limitations. For example, temperature cables are restricted by the number of

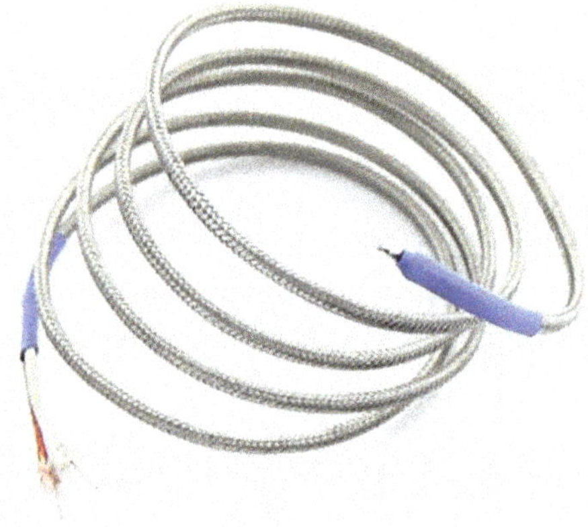

**Figure 2** Thermocouple temperature sensor and wire. Figure Credit: www.seeedstudio.com/Thermocouple-Temperature-Sensor-K-Type-1M-p-3132.html.

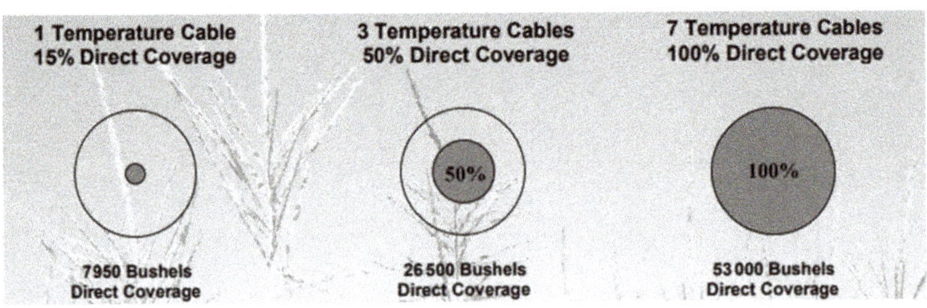


**Figure 3** Three recommended temperature cable configurations and claimed grain mass coverage for a 14.6 m diameter, 10.1 m eave height, and 14.6 m peak height grain silo rated to hold 1346 MT of peaked maize. The 15% direct coverage with one cable supposedly represents 202 MT, the 30% coverage with three cables represents 673 MT, and the 100% coverage with seven cables represent 1346 MT.

sensors that can be mounted on them and by the number of cables that can be hung from the roof without jeopardizing load limits. Moreover, significant loads are added to the roof structure as a result of friction occurring between the grain and the cables during loading, settling, and unloading (Thompson et al., 1991). These loads can potentially damage the roof of a grain silo unless the roof is reinforced. Strengthening the roof structure is costly and must be part of the original silo design and construction. This is also the reason why temperature cables should never be fixed to the floor of a storage structure. During the filling process, grain flow would bow out of the cables and place excessive force on the roof and floor tie-off points, and thus the floor and especially the roof to the point that the weaker of the two - the roof - will give way and collapse. In order to prevent the bottom of cables from being pushed out when the filling process begins, stored grain managers will tie the bottom of cables to the floor with a fishing line that will break once the first amount of grain engulfs the tie-off point. That is enough to prevent cables from floating on the sloped surface of the filling grain mass and being placed at an angle. This management practice keeps cables vertically in place without causing excessive force on the roof.

Temperature cables monitor the grain mass only at a few locations preventing stored grain managers from knowing what happens at other locations. Moreover, they are not commonly placed near the sidewalls of the silo, limiting the ability to reliably monitor the periphery of the grain mass where external solar radiation substantially influences grain temperature. These extreme fluctuations in temperature can cause grain spoilage and wall caking when moisture content is too high or condensation occurs, or result in excessive moisture shrink due to evaporative cooling during repeated aeration.

## 2.3 Moisture content and relative humidity

Moisture content is the most important quality characteristic of cereal grains that influences all stages of handling and processing, starting from harvesting, threshing, marketing, storage, milling, and distribution. Moisture content determines grain quality, price, grade, and storability. For instance, price discounts are applied when selling grain is delivered to grain facilities with moisture contents greater than those specified for marketing and safe storage. The dry matter of grain has more value than the water it contains, and costs of drying the grain to have a safe storage moisture content must be taken into account. Moreover, higher grain moisture content causes mold spoilage and insect infestation that affect the quality of grain during storage.

There are several methods to measure grain moisture content. They fall into two main categories: direct measurement and indirect measurement.

Direct methods measure weight loss of water due to evaporation from air oven drying, infrared radiation, microwave heating, and distillation. Indirect methods measure properties correlated to moisture using dielectric capacitance or conductance, near-infrared (NIR) reflectance or transmittance, and relative humidity – all compensating for temperature. Four of these methods are most widely used in the grain industry and the research community, that is, air oven drying, which is the standard laboratory method, dielectric capacitance, or resistance, which is the basis of moisture meters used at grain handling facilities, NIR machines, which are utilized in grain processing facilities, and relative humidity (and temperature) sensors, which are utilized to measure the interstitial air within a stored grain mass.

**Air oven drying method**: This method is the basis for all official laboratory standards and remains the most accurate moisture measurement method. Unfortunately, it is labor-intensive and time-consuming. All meters that measure moisture content are calibrated against the ovendry standard. In this method, a sample of grain is weighed on a scale, then dried in an air oven at a grain-specific temperature and for a predetermined time, during which all water evaporates, and then the dried sample is weighed. The following equation is used to calculate the dry basis moisture content of a sample:

$$MC = \frac{\text{Initial weight} - \text{Ovendry weight}}{\text{Ovendry weight}} * 100\%$$

**Dielectric properties of grain**: The use of electrical measurements to determine grain moisture content has been around for many decades (Nelson, 1977). Grains have two important electrical properties important for moisture measurement, that is, capacitance and conductance. Measurement of these electrical properties allows for estimating moisture content based on their high correlation to the presence of water molecules inside the grain kernel matrix. Several modern electronic moisture meters use the capacitance principle in which moisture is calculated from electrical measurements on a capacitor having a sample of seeds packed between its plates, or a combination of capacitance and conductance, which additionally considers the electrical resistance of the seeds between the plates (Bhattacharya, 2013). This method has been widely used because it is fast, reliable, relatively low cost for use in instruments, and has good measurement accuracy for all types of grain. Accuracy can be checked and achieved with proper calibration against the standard ovendry method. For instance, when farmers deliver grain to a handling facility, the operator needs to measure the moisture content of a representative sample from each load as quickly as possible because, during harvest season, farmers deliver many loads and do not want to wait long before returning to their fields. The combination of capacitance and conductance method is used by several moisture meter manufacturers such as DICKEY-john, Ag-MAC, Farmex MT-PRO, and so on (Fig. 4).

However, it is important to keep in mind that grain moisture measurement based on these dielectric properties can be erroneous unless other factors such as grain temperature, bulk density, kernel size, composition, and sorption-desorption are accounted for in the calibration based on grain type (Nelson, 1982). Therefore, moisture meters using this method need to be regularly checked and calibrated against the reference ovendry standard method; otherwise, results will be meaningless. The conductivity effect of grain is sensitive to the distribution of moisture within kernels and to kernel morphology. This makes moisture meter calibration unstable across different grain types (Funk et al., 2007). As a result, research continues to improve the reliability and accuracy of electrical and electronic instruments for grain moisture measurement. One of the most well-known outcomes of these efforts is the Unified Grain Moisture Algorithm (UGMA), where the range for the measurement frequency for dielectric moisture properties was adjusted from 1–20 MHz to about 1–149 MHz. The UGMA technique was led in the United States by the Federal Grain Inspection Service (FGIS) and helped reduce the conductivity effects on moisture measurements (Funk et al., 2007). It also resulted in a de facto standard that moisture meter manufacturers must comply with if they want their meters used by FGIS-licensed grain inspection service providers to add moisture content to official grain inspection certificates. A limitation of dielectric sensors is their size, which makes them unsuitable for deployment inside a grain mass (Neethirajan and Jayas, 2007).

**Near-infrared transmission and transmittance of grain**: Near-infrared analysis remains one of the most valuable tools for proximate analysis of grains.

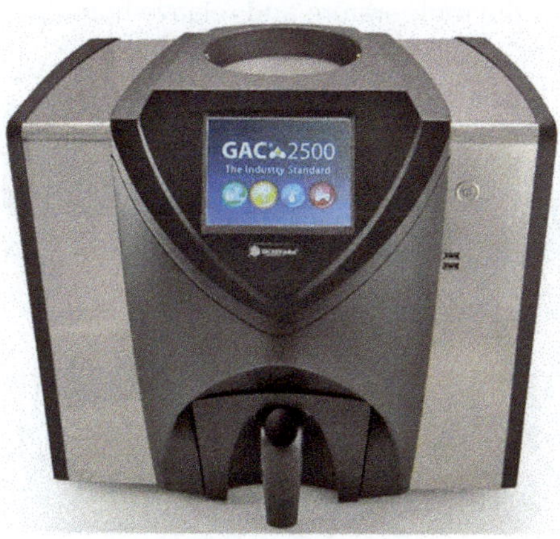


**Figure 4** GAC-2500 moisture meter utilizing the Unified Grain Moisture Algorithm. Figure Credit: https://seedburo.com/.

Near-infrared reflectance (NIR) is commonly used for the analysis of ground grain samples (Lamb and Hurburgh, 1991). However, since grinding grain samples before analysis is a tedious and time-consuming process, the use of near-infrared transmittance (NIT) for rapid estimation of whole-grain moisture content was investigated (Stermer et al., 1977) and has become increasingly popular (Williams et al., 1985). Determining whole-grain quality parameters (moisture, protein, etc.) by the NIT method is advantageous for several reasons. For instance, sample preparation is simple and takes less time. Moreover, the method is nondestructive and thus provides the possibility for replication, additional tests, and storage of the sample for future analysis.

The principle of NIT is based on the emission of wavelengths within a certain frequency region from near-infrared light-emitting diodes. Light at discrete wavelengths passes through the sample and is received in the form of log (1/T) (apparent transmission) by a silicon detection system from which it is amplified and transposed into the second derivative of log (1/T) for computation of protein and moisture by a microprocessor (Williams et al., 1985).

**Temperature and relative humidity of the air surrounding grain**: Temperature and relative humidity can indirectly indicate the moisture content of grain based on the equilibrium moisture content (EMC) relationship (Neethirajan and Jayas, 2007). Grain moisture content in the form of internal water vapor pressure equilibrates with the water vapor pressure of the air surrounding kernels at a given combination of temperature and relative humidity. Relative humidity sensors can predict grain moisture content with a high level of accuracy (Neethirajan and Jayas, 2007). There are numerous types of relative humidity sensors commercially available such as SHT31, HMP77, H1H900, etc. (Fig. 5). When these sensors are placed within the stored grain mass, they measure the temperature and relative humidity of the interstitial air

**Figure 5** SHT31-D relative humidity and temperature sensor. Figure Credit: www.adafruit.com/product.

within the grain mass, and with the help of EMC equations, those readings are converted to grain moisture content at the measured temperature.

Relative humidity sensors usually contain a humidity sensing element along with a thermistor to measure temperature. For a capacitive humidity sensor, the sensing element is a capacitor. Here the change in electrical permittivity of the dielectric material is measured to calculate the relative humidity values. Low resistivity materials are used for the construction of a resistive humidity sensor. The change in the resistivity value of this material is used to measure the change in humidity. Once the sensor signals are converted into temperature and relative humidity value representing a location of the interstitial air within the grain mass, a mathematical relationship such as the modified Henderson equation is used to calculate the EMC of the grain (in dry basis). This calculated EMC value is converted into a wet basis and assumed to represent grain moisture content having equilibrated with the surrounding environment at that location in the grain mass.

$$MC_{db} = \left[ \frac{\ln(1-RH)}{-K(T+C)} \right]^{1/N}$$

$MC_{db}$ = EMC (decimal), dry basis
K, N, and C = coefficients defined for various grain types
RH = relative humidity (decimal)
T = temperature (°C)

It should be noted that during the aeration process, forced convection causes grain temperature and moisture content to change as a function of the ambient air conditions introduced into the grain mass and the changing air conditions as the air either cools or warms grain. Grain is a good insulator and takes time to change in temperature. However, the temperature change is an order of magnitude faster than moisture change. Thus, during aeration, temperature and relative humidity sensors in the grain mass indicate real-time interstitial air conditions that the grain is equilibrating over time. The actual grain moisture content cannot be determined until after the cooling (or warming) front has moved through the grain mass, and the aeration fans have been turned off.

## 2.4 Moisture cable sensors

Moisture sensing cables were commercialized in the early 2000s. They consist of a thermistor and relative humidity sensors that measure interstitial air at fixed locations and calculate the corresponding EMC of the grain (Fig. 6). The cable sensors send data to a gateway typically installed on the roof of the storage structure and from there via cellphone modem to a cloud-based server. The

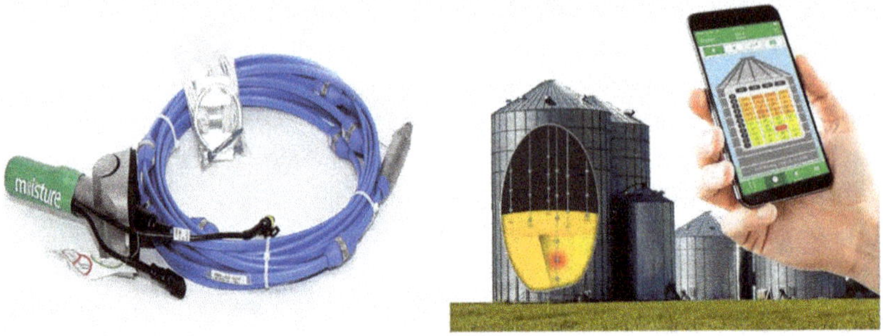


**Figure 6** Moisture cable that includes multiple temperatures and relative humidity sensors spaced along its length (left) and graphic (right) showing multiple cables installed in a grain silo with readout values displayed on a cloud-enabled handheld device. Figure Credit: http://grainequipment.com/.

actual sensors are visible as peanut-size knots on the cables, and as a result, cause greater frictional loads on roofs than temperature cables.

Thus, they have even greater constraints than temperature cables in terms of the number of locations that can be monitored in a stored grain mass. Moreover, the sensors have a limited range and represent grain conditions only within a small volume of 30–60 cm (Mills, 1989). The sensors fastened to each cable are typically vertically separated by 1.2 m for smaller on-farm storage facilities and up to 2.4 m for commercial storage facilities. The resulting cross-sectional spacing on a horizontal plane is one moisture cable per 40 $m^2$ (Singh and Fielke, 2017). No research has been done that provides a scientific basis for optimally spacing temperature, relative humidity, and thus moisture sensors in a stored grain mass.

Consequently, detecting hot spots in grain, which is a poor heat conductor, due to biological self-heating and spoilage, not too far away from a sensor's range, occurs only when the hotspot has grown sufficiently large and thus caused substantial damage (Muir, 2000). These inherent constraints limit the effectiveness of temperature and moisture cables for use with automated stored grain monitoring and aeration fan control systems. They are an important tool to help stored grain managers but should not be solely relied on for making informed decisions. Instead, $CO_2$ sensors should be added to automated systems for improved decision making.

### 2.5 Wireless temperature and moisture sensors

A wireless sensor network (WSN) is composed of several low-power and multifunction wireless sensor nodes equipped with sensing wireless communication and computation capability (Singh et al., 2010). In most WSN applications, once the sensor nodes are deployed, they autonomously organize themselves into a wireless communication network (Gupta et al., 2015). The wireless sensor nodes communicate over a relatively short distance via wireless antennae (Fig. 7). WSN technology has only recently become available for application in stored grain management. The current generation of wireless sensors is capable of measuring temperature and relative humidity (Fig. 8). All nodes communicate with each other before sending data to a base station (also known as a sink, gateway, or hub) that transmits it via cellphone modem to a cloud-based server. Data is sent to the cloud at certain intervals or whenever a measurement changes by more than a preset limit at any time during the storage season.

Wireless sensing has great potential to improve stored grain quality. For instance, it will reduce structural roof loads as cables would no longer be

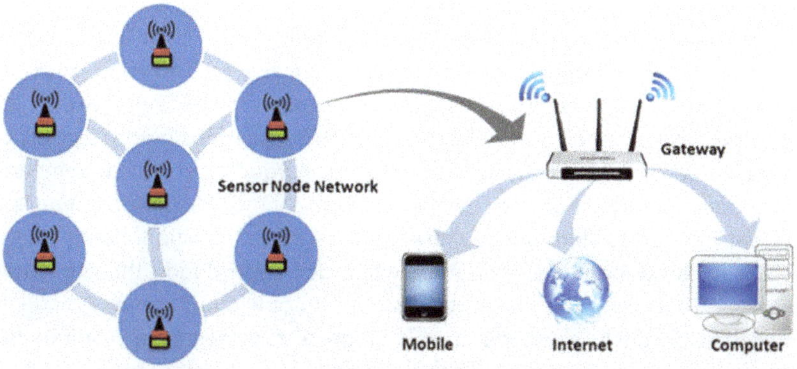


**Figure 7** Components of the wireless sensor network with sensor nodes communicating with each other and a gateway that makes data available for cloud-based access. Figure Credit: http://zomko.facach.org/wireless-sensor-network/.

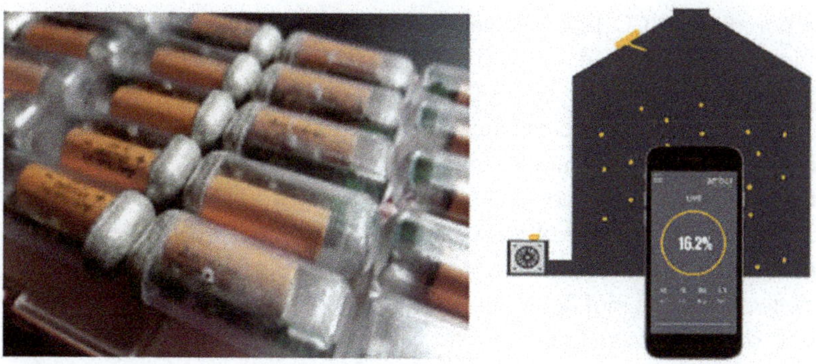


**Figure 8** Set of wireless temperature and relative humidity sensors (left) and graphic (right), illustrating the potential distribution of wireless sensors in a stored grain mass and readout of values on a cloud-enabled handheld device. Figure Credit: www.amber.ag/.

connected to roofs. Moreover, grain quality can be monitored during storage and transport, for example, in rail cars, barges, and ship holds. Eventually, wireless sensors will be capable of measuring $CO_2$ in the grain mass or headspace above the grain surface, which has proven to be an effective early warning indicator of biological activity in stored grain caused by grain respiration, mold spores, and insects. Wireless sensing of the dew point temperature of the headspace air and the dry bulb temperature of the roof will also provide an easier means for determining condensation conditions and activating roof exhausters. This will prevent grain surface crusting with all of its negative consequences for grain quality, such as plugging floor outlets during unloading and entrapment or engulfment of people who should have absolutely no reason to enter a silo and try to fix a grain spoilage problem. Thus, wireless sensors have the additional potential benefit of making grain handling facilities a safer place to work.

While wireless sensing systems are a potential tool for improved stored grain quality monitoring, they have several drawbacks. The most obvious is battery life. In a stored grain mass, wireless sensors cannot be accessed after deployment, so batteries must last for at least 9–12 months before sensors can be captured during unloading and can be recharged or replaced. Sometimes grains are stored more than one season at a time. Moreover, there is no current method to determine where the sensors self-position, horizontally or vertically, throughout the grain mass when deployed during the silo loading process. Controlling the horizontal distribution of wireless sensors is critically important because, given the limited range of representative readings from any one sensor, stored grain managers need to know whether grain conditions measured are representative of grain mass core and perimeter.

## 2.6 Carbon dioxide ($CO_2$) sensors

Carbon dioxide ($CO_2$) is an indicator gas of biological activity due to germ, fungi or insect respiration in a stored grain mass. In 2018, $CO_2$ in ambient air was 407 parts per million (ppm) (NOAA, 2020). Interstitial air in a grain mass with 450–600 ppm $CO_2$ indicates a safe range to continue the storage of a stable grain mass (Maier, 2016). Elevated $CO_2$ levels are a definite sign of the onset of deterioration in stored grain. Increasing levels of $CO_2$ from week to week correlate to higher levels of grain deterioration from self-heating, mold spoilage, or increase in insect infestation (Marlow, 2011). Measuring $CO_2$ in stored grain is an effective tool to monitor changes in quality and to help stored grain managers make decisions to mitigate the escalation of problems caused by biological activity. $CO_2$ concentrations around 600–1 500 ppm indicate the onset of mold, insects, or moisture infiltration that a stored grain manager should check into, and 1 500–4 000 ppm is a severe condition that requires immediate mitigation.

Ileleji et al. (2006) used handheld commercial $CO_2$ sensors to measure $CO_2$ levels in exhaust air from silos, tanks, warehouses, and piles (Fig. 9). They also measured $CO_2$ in the headspace of silos, tanks, and warehouses with sensors suspended from roof rafters or placed in the outlet of roof vents. The researchers concluded that $CO_2$ concentration could reliably detect the onset of spoilage and the formation of hot spots earlier than temperature cable sensors that only provide data at respective physical locations. Thus, $CO_2$ sensors can provide a quicker and more accurate way of detecting grain spoilage than temperature cables because $CO_2$ changes the composition of air while heat from warming grain changes the temperature of air (Marlow, 2011).

Researchers have also measured intergranular $CO_2$ by placing sensors in the grain mass and showed that they could pinpoint specific locations where spoilage may be occurring (Olsen et al., 2013). In comparison, $CO_2$ sensors are larger and costlier than temperature and relative humidity sensors. They also consume more power and thus cannot be easily deployed as cable-based sensors in the grain mass.

Sensors have become commercially available to monitor $CO_2$ concentrations by mounting them in the headspace of storage structures, or ducts of aeration or roof exhaust fans depending on the direction of airflow through the grain mass. It is difficult to interpret $CO_2$ readings from sensors without understanding how air circulates in and through a grain mass. $CO_2$ generated in the grain mass is carried by slow-moving natural convection currents upward, constantly being mixed with interstitial air at higher or lower $CO_2$ concentrations, and eventually reaching the headspace where $CO_2$ concentration is further diluted. As headspace temperature increases during the daytime, its air volume increases and causes $CO_2$ to be pushed past sensors and out of the structure. At nighttime, the reverse occurs, and $CO_2$ accumulating

**Figure 9** Handheld $CO_2$ sensor commercially available and used to measure $CO_2$ in the exhaust air from a storage structure. Figure Credit: www.tempcon.co.uk/shop/telaire-7001-co2-sensor.

in the headspace is diluted with ambient air being drawn in. Therefore, $CO_2$ readings in a headspace typically show a 12-h cyclical pattern unless aeration fans, roof exhausters, or strong winds force air from the headspace. Aside from cooling grain, aeration fans can be turned on to sweep $CO_2$ accumulating in the grain mass up (or down) and out of the storage structure past a handheld $CO_2$ sensor or $CO_2$ sensors installed in the roof vents or fan ducts. At a typical aeration airflow rate of 0.11 $m^3$/min per MT, it only takes 5 min to exchange the entire volume of interstitial air in a stored grain mass. During an air sweep, a $CO_2$ sensor will show an increase in concentration to a maximum level and then a decrease to the background level. The area under the curve can be integrated to calculate the total amount of $CO_2$. Given that storage structures are leaky, sweeping interstitial air past fixed or handheld $CO_2$ sensors by turning on aeration fans for a few minutes every week or two - ideally at a time between midnight and sunrise - is considered by the authors a more reliable indicator of biological activity than solely relying on cyclical $CO_2$ readings from the same sensors during non-aerated periods. Key to interpreting $CO_2$ readings is the change in concentration over time and not to overreact, because $CO_2$ is highly sensitive and can detect the onset of spoilage represented by less than 0.1% of a stored grain mass (Ileleji et al., 2006).

$CO_2$ levels in the grain are also measured for achieving and monitoring gas concentrations during modified atmosphere storage in hermetic bags and silos. In the United States, monitoring $CO_2$ in stored grain using handheld sensors has been widely adopted by commercial stored grain managers. At last count, at least four companies in the United States offer automated grain monitoring and aeration control systems with the option to incorporate $CO_2$ sensors. In Argentina, $CO_2$ monitoring was commercially introduced in the early 2000s when silobags were rapidly adopted as an alternative storage technology (Bartosik et al., 2008). There, handheld $CO_2$ units were equipped with GPS and barcode reading technology (Fig. 10). Service providers check $CO_2$ readings at several locations along the length of a silobag and upload readings to the cloud where they become instantly accessible to the stored grain owner. Color coding of green, yellow, and red provide indication to what extent readings have changed over time and which silobags may need to be unloaded and which ones can continue to be stored with confidence.

## 2.7 Phosphine and other gases

Measuring concentration of phosphine and other gases (e.g., nitrogen, sulfuryl fluoride) is critical for successfully monitoring fumigation of grain for the purpose of eradicating an insect infestation. Australia's Grains Research and Development Corporation (GRDC) recommends that a minimum concentration of 300 ppm of the fumigant gas be maintained for 7 days, or 200 ppm for

**Figure 10** Handheld $CO_2$ sensor and reader commercially available in Argentina and used to measure $CO_2$ in silobags that store different grain types. Figure Credit: D. E. Maier.

10 days, to achieve effective fumigation treatment (GRDC, 2013). Monitoring phosphine concentration in a grain mass allows for calculating the gas leakage rate in real time, and as needed, fixing gas leaks and topping off the structure under fumigation with additional phosphine gas. Tracking phosphine concentrations allows the fumigator to make sure the correct concentration over time product (ppm-h) is achieved to assure the death of all insects at all life stages.

Brabec et al. (2019) described the typical approach used for monitoring phosphine gas concentration in grain silos during fumigation. It involves the use of monitoring lines or tubing placed at different depths of the grain mass and in the headspace, drawing air samples periodically through these tubes via a control manifold into a calibrated gas sensor that quantifies phosphine concentration. Sampling airlines can be permanently or temporarily installed in fumigated spaces and connected to a gas sensor. Manually monitoring gas concentrations every few hours is time-consuming and labor-intensive given the length of typical fumigation is three days (72 h). Sampling at multiple locations is limited due to the difficulty of placing too many lines into the grain mass and the number of people needed to take readings at regular time intervals, especially during the night.

A simple but less accurate method for detecting whether phosphine has penetrated the grain mass is to place a piece of copper wire at the opposite end of where the phosphine was introduced and observe whether the copper has corroded as a result of phosphine exposure (GRDC, 2013).

Recently, a system has been commercialized that includes phosphine sensors that communicate wirelessly and allow for real-time, cloud-based data collection and analysis (Kaloudis et al., 2018). Brabec et al. (2019) tested the system extensively in a 120 MT silo. The sensors measured concentrations reliably every two hours for six days during three trials. Concentrations were compared to data collected with monitoring lines that were distributed at several positions and depths in the silo. The study concluded that although the fumigation data were significantly different in terms of gas concentration patterns over time, the two sampling methods had similar trend lines. The automated sampling system provided a more detailed picture of the fumigation process because of the frequency of data collection. More detailed information should help fumigators better evaluate real-time fumigation success and thus assure insect control. Wireless gas sensors provide fumigators a quality assurance tool with which they can document the success of fumigation to their customers and assuring them that fumigation was successful (Brabec et al., 2019).

## 2.8 Volatile organic compounds (VOC)

Grain spoilage due to insects, fungi, and mycotoxins produces odors in the form of volatile organic compounds (VOCs) that are detectable in the air of a silo. VOCs are organic chemicals that are numerous, varied, and ubiquitous. They include naturally occurring chemical compounds such as scents or odors. The majority of VOCs are produced by plants including grains and others by animals and microbes such as fungi and mycotoxins. VOCs can be detected based on different principles and interactions between the organic compounds and the sensor components. In the case of spoiling grain, VOCs are detectable to the human nose. Compared to $CO_2$, microbial VOCs are more difficult to measure. Electronic devices (i.e., e-nose) can detect ppm concentrations and others can predict with reasonable accuracy the molecular structure of a VOC. For most accurate quantification, an air sample has to be collected and injected into a mass spectrometer. VOC types and levels can provide an indication of the spoilage agent and allow differentiation between the presence of insects, fungi, or mycotoxins. Currently, there are no commercial sensors available to measure VOCs or odors in real time inside a grain silo.

Odor, or more correctly off-odor, is an important factor in grain grading in the United States (Seitz and Sauer, 1992). If a sample submitted for grading has a slight off-odor and/or cannot be easily classified, then the sample will be graded as 'off-odor' and assigned as sample grade. That is the lowest of the US grain grades and reduces the economic value of the entire grain lot from which the sample was drawn. Thus, maintaining stored grain quality is a critically important responsibility for every stored grain manager.

Assessment of odor is done by experienced grain inspectors who are remarkably accurate and consistent. It is difficult, if not impossible, to be done by inexperienced individuals because of background odors and subjectivity (Neethirajan and Jayas, 2007). Spoilage is likely at an advanced stage by the time these organic compounds are detectable, which may not leave much time to take action to mitigate further grain deterioration (Neethirajan and Jayas, 2007).

Costello et al. (2003) fabricated a prototype of metal oxide sensors to detect odors caused by fungi and tested the sensors in stored wheat. The results of the study showed that the system could detect fungal odors but could not measure the concentration or identify the types of VOCs present in the grain mass (Neethirajan and Jayas, 2007). Not much research has taken place since to advance VOC sensor development for deployment in a stored grain mass and detecting the onset of spoilage and the amount of quality deterioration.

## 2.9 Electromagnetic imaging

Electromagnetic imaging is widely used in medical applications and for nondestructive evaluation of materials. It allows for detecting differences in both conducting and dielectric materials by generating 2D or 3D images based on electromagnetic principles. As a result of electromagnetic frequency waves penetrating into grain, temperature, and moisture differences can be quantified and displayed as images. Temperature and moisture changes in grain affect grain permittivity and thus the ability to store electrical potential energy under the influence of the electric field generated by the electromagnetic waves (Nelson et al., 2000). Differences in permittivity are displayed as color differences in images (Nelson and Trabelsi, 2006). The 3D imaging technology has only recently been commercialized by one company for monitoring stored grain in steel silos (Fig. 11). The technology consists of plate sensors installed at several

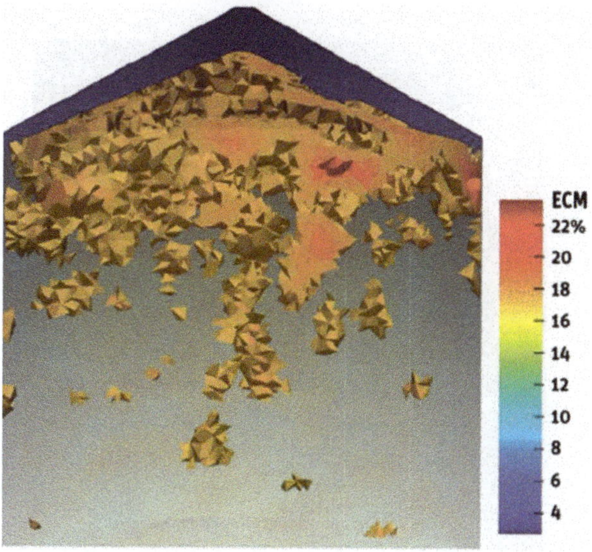


**Figure 11** 2D image of moisture content distribution in a stored grain mass generated by a novel electromagnetic imaging sensor technology installed at several locations of the silo wall. Figure Credit: http://grainviz.com/product.html.

locations on the inside of the silo. It can be used to create comprehensive 3D maps of the stored grain mass displaying, for example, moisture content, one 2D horizontal or vertical slice at a time. The 3D image of a stored grain mass can pinpoint a specific color (such as deep red) at the location where grain spoilage occurs. The associated permittivity can be used to calculate the corresponding grain moisture content (Gilmore et al., 2017). Once fully developed, the system is supposed to be capable of detecting insect infestation, quantifying shrink as a function of changes in weight and volume, and control the aeration fan automatically. This new technology seems to be effective but complex and expensive. The overall system costs about $8 000 per silo (http://grainviz.com/product.html). It remains to be seen how widely it will be adopted by stored grain managers in the coming years.

### 2.10 Insect probe traps

Insect probe traps can be used to collect live insects within the stored grain mass as an indication of level and type of infestation. Insect probe traps are hollow 'plastic' tubes with a series of downward sloping holes along the wall (Fig. 12). These traps are simple to use and can be inserted into the grain mass where insects crawl through holes in the trap and fall to the bottom where they are unable to escape. Traps should be connected to a nylon line to allow easy retrieval. Once retrieved, the contents can be carefully analyzed and provide an indication of the type and level of infestation (Johnson and Townsend, 2009).

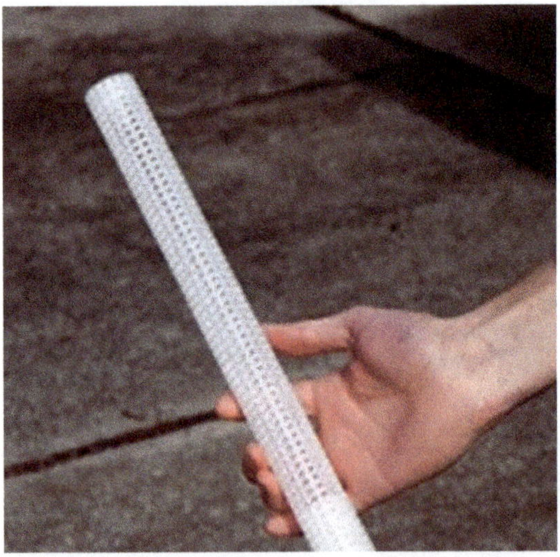

**Figure 12** Insect pitfall probe trap to be deployed in the upper layer of a stored grain mass. Figure Credit: www.graintec.com.au/products/grain-testing/insect-management/insect-probe-trap/description/.

Remedial action such as cooling, fumigation, or partial unloading can then be carried out as required.

The key problem with these traps is that they require physical placement in the top surface of the grain mass after filling a silo and physical recovery to remove the sample of trapped insects every week or two during the warmer period of storage. Entering a storage structure and walking on a grain surface to remove insect probe traps is dangerous and should be avoided. Most commercial grain handling companies either require a permitting process each time an employee enters a structure filled with grain or categorically no longer allow employees to enter structures for grain quality monitoring reasons. One company addressed this challenge by developing an electronic probe that is tied in with its automated monitoring system (Fig. 13). The probe is equipped with a set of LED lights that allow for the counting of insects that fall into the bottom of the trap. As a result of the length of time the light beam is interrupted, the size of the insect can be determined, and thus the stored product insect category can be identified. Category determines whether the insect is deemed a primary or secondary feeder. Primary feeders are the most destructive, and depending on count, the stored grain manager will decide whether to fumigate and mitigate insect infestation.

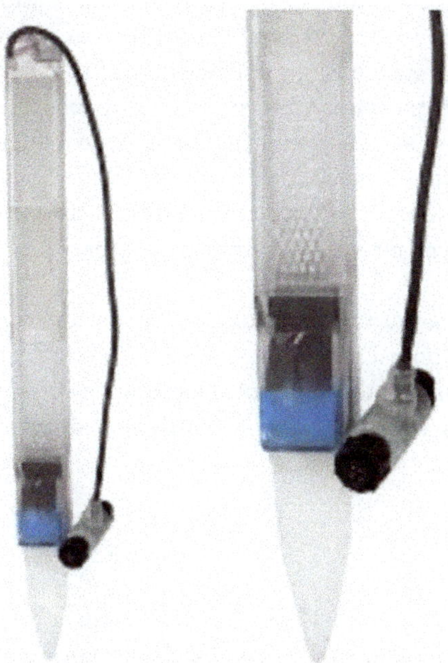

**Figure 13** Insect pitfall probe trap with electronic count and identification mechanism. Figure Credit: www.grainstoragesolutions.ca/opi-systems/.

### 2.11 Acoustical sensors

Acoustical sensors have been developed to monitor insect activity in stored grain. They have shown to be effective in detecting stored grain insect populations eating and moving within a stored grain mass (Hagstrum et al., 1991). The activity of insects within a grain bulk produces noises in the audible range of wavelengths, which can be detected by high-performance acoustic sensors (Hagstrum et al., 1996). When an acoustic wave (input) travels through a certain material or along the surface of a material, it is influenced by the different material properties and obstacles it travels through. Any changes to the characteristics of this travelling path, such as insects chewing inside grain kernels, will affect the velocity and/or amplitude of the wave. These characteristics are translated into a digital signal (output) using transducers. These changes can be monitored by measuring the frequency within a stored grain mass.

However, acoustical sensors cannot monitor all insect species nor differentiate between insect species based on sound. For instance, the sensors cannot detect insects feeding on mold or moths. Moreover, other background noise (such as wind, aeration fans, roof exhausters, conveying equipment) can be captured by the sensors which might affect the sound recording unless the sensors are properly calibrated and sufficiently advanced to filter out the sound from the noise. Acoustical sensors are not feasible for placement in grain stored in large silos or other structures such as warehouses and outdoor piles because they require wiring for power and sound transmission and are limited in terms of signal strength as to how many sensors would be needed to cover the grain mass in a sufficiently representative manner. After all, the main goal is not to use sound detection to monitor insect population development but rather to early detect the occurrence of insects and make management decisions to prevent insects from multiplying and causing damage to the stored grain.

Therefore, a method used to detect the presence of insects within the grain is useful but only if it can be applied on a large scale for making pest-management decisions that control insects such as weevils. Unfortunately, acoustical sensors have significant drawbacks that prevent their use to effectively monitor stored grain.

## 3 Conclusion

It is critically important to monitor stored grain regularly to prevent devastating quality and quantity losses. For instance, in Africa alone, it is estimated that 25% of grains harvested are subject to postharvest losses valued at over $2 billion annually (Anon., 2011).

Monitoring stored grain requires sensing technologies and monitoring systems that can be deployed manually using a single sensor to measure, for example, $CO_2$ of air exhausted from a storage structure, or installed permanently such as cables equipped with temperature and relative humidity sensors that are tied together into an automated system that is accessed with a cloud-enabled device. The key to successfully preserving grain is to maintain its temperature and moisture content at safe storage conditions. Grain needs to be as cool as feasible and dry as necessary for the geographic location and weather conditions during the storage period. Most storage problems are caused by moisture infiltration and insect penetration into the storage structure which cause biological respiration from germ, fungi, and insects to increase. Thus, monitoring $CO_2$ with a handheld sensor is the most effective tool for a stored grain manager to utilize. Measuring temperature in the stored grain mass is the second most important tool but requires cables to be installed throughout the storage structure, or wireless sensors deployed during filling and recovered during unloading. Combined with $CO_2$ sensing, aeration cooling using ambient air or grain chillers can be tracked and the storage period can be monitored for any increases in biological activity. The addition of relative humidity sensors allows for calculation of moisture content which is important to minimize shrink loss but adds a level of complexity and cost that needs to be evaluated judiciously. Other useful technologies for monitoring the presence of insects using electronic insect probe traps and gases such as phosphine during fumigation are important for mitigating stored grain pests once they occur. Other technologies remain futuristic and may require additional research and economic analysis such as deploying electromagnetic 3D imaging and measuring volatile organic compounds to discern the types of mycotoxins.

## 4 Future trends

In terms of future trends in research and development on advanced stored grain monitoring techniques, wireless sensors and electromagnetic imaging are top the list.

Wireless sensors have shown their capability to automatically and accurately monitor stored grain temperature and relative humidity and calculate EMC (Guy, 2020). However, several questions still need to be investigated further for successful commercial deployment. The most obvious is the battery life, which needs to last for 9-12 months, or even 14 months. Moreover, there is no current method to determine where the sensors self-position horizontally or vertically throughout the grain mass during the filling process. Knowing the horizontal distribution tendencies of the wireless sensors is critically important in evaluating whether they can monitor grain conditions in a representative manner (including grain mass core and perimeter). Last but not least, the minimum number of sensors needs to be determined to achieve sufficient accuracy for stored grain quality monitoring as a function of different silo sizes. This factor is important to find an appropriate balance between placing many sensors to achieve maximum coverage versus a few sensors to minimize the system cost.

Electromagnetic imaging is a powerful technology to visualize and check temperatures and moisture content of the overall stored grain bulk. Research has shown that a comprehensive 3D moisture map can be created by using radio signals from a set of antenna plates placed on the periphery wall of a grain silo displaying in real time changes in moisture content caused by shrink due to aeration, condensation on the grain surface, water dripping from downspouts, and so on. However, the technology is still rather expensive and, therefore, not affordable for most commercial grain handlers and farmers. One way to reduce the cost of the system is to further simplify system installation.

## 5 Where to look for further information

- www.gipsa.usda.gov/fgis/publication/ref/primer.pdf.
- www.cmciel.com/applications/grain-temperature-monitoring.html.
- www.advancedgrainmanagement.com/products/.
- www.feedandgrain.com/product/category/temperature-monitors.
- www.tsgcinc.com/portable-grain-monitoring/bin-temp.
- https://intellifarms.com/solutions/bin-check.
- www.topcropmanager.com/new-tools-for-grain-bin-monitoring-21071/.
- www.world-grain.com/articles/10238-monitoring-co2-in-stored-grain.
- www.feedandgrain.com/magazine/co2-monitoring-a-new-grain-management-tool.
- https://grdc.com.au/news-and-media/news-and-media-releases/south/2015/01/correct-fumigation-of-stored-grain-a-priority.
- https://grdc.com.au/resources-and-publications/grdc-update-papers/tab-content/grdc-update-papers/2018/12/grain-storage-updates-for-fumigation-in-large-silos-and-grain-protectants.
- https://dl.acm.org/doi/10.1016/j.compag.2015.10.016.

## 6 References

Anon. 2003. Introduction to thermocouples. *Control Engineering* 50(11), 70. Available at: https://www.dataforth.com/catalog/pdf/an106.pdf.

Anon. 2011. Effective grain storage for better livelihoods of African farmers project. *The Swiss Agency for Development and Cooperation (SDC)*, 52(June 2008). Available at: https://www.shareweb.ch/site/Agriculture-and-Food-Security/focusareas/Documents/phm_egsp_2008_2011.pdf.

Bartosik, R., Cardoso, L. and Rodriguez, J. 2008. Early detection of spoiled grain stored in hermetic plastic bags (silo-bags) using $CO_2$ monitoring. *Proceedings of the*

*8th International Conference Controlled Atmospheres and Fumigation of Stored Products*, pp. 550–4.

Bhattacharya, K. R. 2013. Analysis of rice quality. *Rice Quality*, 431–530. Available at: https://www.sciencedirect.com/science/article/pii/B9781845694852500131?via%3Dihub.

Borjesson, T., Stollman, U., Adamek, P. and Kaspersson, A. 1989. Analysis of volatile compounds for detection of molds in stored cereals. *Cereal Chemistry* 4(66), 300–4.

Brabec, D., Campbell, J., Arthur, F., Casada, M., Tilley, D. and Bantas, S. 2019. Evaluation of wireless phosphine sensors for monitoring fumigation gas in wheat stored in farm bins. *Insects* 10(5). doi:10.3390/insects10050121.

Climate Change: Atmospheric Carbon Dioxide | NOAA Climate.gov. 2020. Available at: https://www.climate.gov/news-features/understanding-climate/climate-change-atmospheric-carbon-dioxide (Accessed on March 14, 2020).

Costello, B. P. D. L., Ewen, R. J., Gunson, H., Ratcliffe, N. M., Sivanand, P. S. and Spencer-Phillips, P. T. N. 2003. A prototype sensor system for the early detection of microbially linked spoilage in stored wheat grain. *Measurement Science and Technology* 14(4), 397–409. doi:10.1088/0957-0233/14/4/301.

Freeman, S. A., Kelley, K. W., Maier, D. E. and Field, W. E. 1998. Review of entrapments in bulk agricultural materials at commercial grain facilities. *Journal of Safety Research* 29(2), 123–34. doi:10.1016/S0022-4375(98)00008-5.

Funk, D. B., Gillay, Z. and Meszaros, P. 2007. Unified moisture algorithm for improved RF dielectric grain moisture measurement. *Measurement Science and Technology* 18(4), 1004–15. doi:10.1088/0957-0233/18/4/007.

Gilmore, C., Asefi, M., Paliwal, J. and LoVetri, J. 2017. Industrial scale electromagnetic grain bin monitoring. *Computers and Electronics in Agriculture* 136, 210–20. doi:10.1016/j.compag.2017.03.005.

Gonzales, H. B., Armstrong, P. R. and Maghirang, R. G. 2009. Simulataneous monitoring of stored grain with relative humidity, temperature, and carbon dioxide sensors. *Applied Engineering in Agriculture*, 25(Mc), 595–604.

Grain Research & Development Corporation 2013. *Grain Storage Fact Sheet: Western Region Grain Storage Pest Control Guide*. Available at : https://grdc.com.au/__data/assets/pdf_file/0017/142622/gsfs-1_gspestcontrol_wr_2013_lr_final-pdf.pdf

Gupta, R., Science, C. and Noida, G. 2015. A survey on routing protocols of MANET Wireless Sensor Network . *International Journal of Computer Science and Mobile Computing*, 4(4), 812–22.

Guy. 2020. *Wireless Sensors for Quality Monitoring and Management of Stored Grain Inventories*. MS Thesis, Iowa State University, Ames, IA.

Hagstrum, D. W., Flinn, P. W. and Shuman, D. 1996. Automated monitoring using acoustical sensors for insects in farm-stored wheat. *Journal of Economic Entomology* 89(1), 211–7. doi:10.1093/jee/89.1.211.

Hodgson, E. and Holscher, K. 2009. *Think about Stored Grain Pests Before Harvest*. Ames, IA. Iowa State University Digital Repository. Available at: https://lib.dr.iastate.edu/cgi/viewcontent.cgi?article=1554&context=cropnews.

Ileleji, K. E., Maier, D. E., Bhat, C. and Woloshuk, C. P. 2006. Detection of a developing hot spot in stored corn with a $CO_2$ sensors. *Applied Engineering in Agriculture* 22(2), 44–5.

Johnson, D. W. and Townsend, L. H. 2009. *Controlling Insects in Stored Grain*. University of Kentucky - College of Agriculture (Cooperative Extension Service). Available at: https://entomology.ca.uky.edu/files/efpdf1/ef145.pdf

Jones, C. E. 1957. Thermocouple unit for grain temperature measuring equipment. United States Patent Office.

Kaloudis, E., Analytics, C., Athanassiou, C. and Agrafioti, P. 2018. Modeling the distribution of phosphine in cylindrical grain silos with CFD methods for precision fumigation (October). doi:10.5073/jka.2018.463.153.

Lacey, J. 1988. *Biodeterioration*. Elsevier Applied Science, New York, NY.

Lamb, D. T. and Hurburgh, C. R. 1991. Moisture determination in single soybean seeds by near-infrared transmittance. *Transactions of the American Society of Agricultural Engineers*, 2123–9. doi:10.13031/2013.31848.

Magan, N. and Evans, P. 2000. Volatiles as an indicator of fungal activity and differentiation between species, and the potential use of electronic nose technology for early detection of grain spoilage. *Journal of Stored Products Research* 36(4), 319–40. doi:10.1016/S0022-474X(99)00057-0.

Magan, N., Hope, R., Cairns, V. and Aldred, D. 2003. Post-harvest fungal ecology: impact of fungal growth and mycotoxin accumulation in stored grain. *European Journal of Plant Pathology* 109(7), 723–30. doi:10.1023/A:1026082425177.

Maier, D. E. 2016. World grain Maier July 2016 - $CO_2$ Monitoring.pdf. *World Grain*.

Maier, D. E., Hulasare, R., Qian, B. and Armstrong, P. 2010. Monitoring carbon dioxide concentration for early detection of spoilage in stored grain. *Julius-Kühn-Archiv* 425, 505–9. doi:10.5073/jka.2010.425.332.

Maier, D. E., Hulasare, R., Qian, B. and Armstrong, P. 2006. Monitoring carbon dioxide levels for early detection of spoilage and pests in stored grain. In: *9th International Working Conference on Stored Product Protection*, Sao Paulo, Brazil.

Marlow, B. 2011. Carbon dioxide levels in grain offer a quick and accurate measure of quality. *GEAPS Exchange*. Portland, OR.

Metz, N. 2007. The WA guide to high moisture harvest management, grain storage and handling. CBH Group. CBH Group and South East Premium Wheat Growers Association with Assistance from GRDC and DAFF.

Mills, J. D. 1989. Spoilage and heating of stored agricultural products: prevention, detection and control. *Minister of Supply and Services*.

Muir, W. E. 2000. *Grain Preservation Biosystems*. Spoilage and Heating of Stored Agricultural Poructs: prevention, detection and control.

Neethirajan, S. and Jayas, D. S. 2007. Sensors for grain storage. *2007 ASABE Annual International Meeting, Technical Papers 12 Book (February)*. doi:10.13031/2013.23510.

Nelson, S. O. 1977. Use of electrical properties for grain-moisture measurement. *Journal of Microwave Power* 12(1), 67–72. doi:10.1080/16070658.1977.11689030.

Nelson, S. O. 1982. Factors affecting the dielectric properties of grain. *Transactions of the American Society of Agricultural Engineers* 24(4), 1045–9, 1056. doi:10.13031/2013.33665.

Nelson, S. O. and Trabelsi, S. 2006. Dielectric spectroscopy of wheat from 10 MHz to 1.8 GHz. *Measurement Science and Technology* 17(8), 2294–8. doi:10.1088/0957-0233/17/8/034.

Nelson, S. O., Kraszewski, A. W., Trabelsi, S. and Lawrence, K. C. 2000. Using cereal grain permittivity for sensing moisture content. *IEEE Transactions on Instrumentation and Measurement* 49(3), 470–5. doi:10.1109/19.850378.

Olsen, J. W. W., Wilhelmi, C. R., Zandonadi, R. S., Danao,M. G. C. and Gates, R. S. 2013. Monitoring grain bed conditions during truck transport of soybeans in Brazil. *Paper*

*Number 131619395, 2013 Kansas City, Missouri, July 21–July 24, 2013*. doi:10.13031/aim.20131619395.

Seitz, L. M. and Sauer, D. B. 1992. Off-odors in grains. *Developments in Food Science* 28, 17–35. doi:10.1016/B978-0-444-88558-6.50007-6.

Singh, C. B. and Fielke, J. M. 2017. Recent developments in stored grain sensors, monitoring and management technology. *IEEE Instrumentation and Measurement Magazine* 20(3), 32–55. doi:10.1109/MIM.2017.7951690.

Singh, S. K., Singh, M. P. and Singh, D. K. 2010. A survey of energy-efficient hierarchical cluster-based routing in wireless sensor networks. *International Journal of Advanced Networking and Applications* 570(02), 2–570.

Stermer, R. A., Pomeranz, Y., and McGinty, R. J. 1977. Infrared reflectance spectroscopy for estimation of moisture of whole grain. *American Association of Cereal Chemists* 54(2), 345.

Thompson, S. A., Schwab, C. V. and Ross, I. J. 1991. Forces on temperature cables in a model bin under restrained conditions. *Transactions of the American Society of Agricultural Engineers*, 34(5), 2187–92.

TSGC. n.d. *Grain Bin Temperature & Moisture Monitoring System*. Available at: https://www.tsgcinc.com/

Williams, P. C., Norris, K. H. and Sobering, D. C. 1985. Determination of protein and moisture in wheat and barley by near-infrared transmission. *Journal of Agricultural and Food Chemistry* 33(2), 239–44. doi:10.1021/jf00062a021.

# Chapter 13

## Supporting smallholder farmers in developing countries to improve postharvest management of staple grains: the role of loss reduction technologies

*Brighton M. Mvumi, University of Zimbabwe, Zimbabwe; and Tanya Stathers, Natural Resources Institute (NRI) – University of Greenwich, UK*

1 Introduction
2 Improvement strategies
3 Institutional, policy and market-related issues for PHM transformation
4 Creating sustainable markets: case study models
5 Conclusion and future trends
6 Where to look for further information
7 References

### 1 Introduction

The postharvest challenges faced by smallholder farmers in developing countries have not changed much from those identified by Best et al. (2005) in the mid-2000s. They include:

- Low-value goods. The majority of smallholders produce basic staples for their own subsistence, with a little surplus for sale to cater for other livelihood needs.
- Market pressure. Structural adjustment and market liberalisation in the mid-1990s have meant smallholder farm families in developing countries have faced ever-increasing pressures to fit into market economies to generate enough income to cater for their basic needs and improve their livelihoods.
- Undermining of existing markets. Relief agencies supply free or highly subsidised imported food to many developing countries for the

http://dx.doi.org/10.19103/AS.2020.0072.16

food-insecure. This has often led to bypassing or erosion of local marketing structures and systems.

- Lack of market information. Most farmers receive limited information about grain prices, whereas traders can easily access such information. This skew in access to basic market information impedes farmers' ability to negotiate better prices and thus build market confidence and trade links. Provision of such services is of low cost, yet few governments invest in them.
- Poor farmer organisation. Smallholder farmers in many developing countries are poorly organised, which prevents them from doing business at scale. This typically means trade suffers from poor quality, low volumes, weak links and is consequently often uncompetitive because of the high transaction costs. The collapse of government-led cooperative systems, due to poor management and marketing skills, is often blamed for the lack of farmer organisation in many developing countries. In many countries, this has resulted in long-term distrust of working together by farmers.
- Deteriorating infrastructure. Farmer access to national and regional markets is hindered by poor infrastructure, such as roads, storage facilities, market sheds, power and transportation. With increasing poverty in some parts of rural Africa, these facilities are decreasing, and therefore smallholder farming communities are less able to engage in more distant markets than before.
- Shocks. Natural and civil shocks limit faster market growth in many developing countries. These severe impediments are caused by numerous factors including droughts, floods, war, political and religious disputes; the devastating effects of diseases such as HIV/AIDS, malaria, ebola virus disease and more recently Corona Virus Disease (COVID-19); and attacks by crop pests and diseases including locusts, African army worm, fall army worm, maize streak virus, larger grain borer (LGB) and so on.
- Market consolidation. Consolidation and merging of buyers has created monopolies and shifted power from the hands of producers to small- and medium-sized Enterprises (SMEs) and corporations who can bulk-buy directly from smallholder farmers and position them against each other.
- Complexity of policies. There is evidence suggesting that attempts to alleviate poverty and hunger through improving staple cash-crop production are flawed. Where agroclimatic conditions are favourable, it has been demonstrated that the production of staples such as maize and beans can be increased through government-assisted programmes providing subsidised agricultural inputs (seed, fertilisers, pesticides, credit). However, these interventions are often and incomplete, with no medium- to long-term vision or support for the development of sustainable local and regional markets, or adequate and transparent management of strategic grain reserves, resulting in the reversal of seemingly positive productivity

outcomes into situations where crop prices decline, and farmers lose money and are unable to repay loans, while countries continue to require food aid.

Addressing these complex challenges thus requires a multi-dimensional and multi-stakeholder approach.

Cereal grains such as sorghum, millets, maize and rice must be harvested, dried, shelled/threshed, cleaned, treated, packaged, stored and transported using methods that preserve their quality as seed, foodstuff or raw material. Some grains are consumed shortly after harvest and require little handling other than cleaning of the grains. Storage can be for varying lengths of time, ranging from short-term storage on-farm (less than 3 months) to medium storage (3-6 months) to long-term storage (6–12 months). Farmers usually store grains medium- to long-term, not only for household food security, but in speculation of advantageous market conditions later in the storage season. As more erratic rainfall leads to an increasing frequency of crop failure many smallholder farmers are now compelled to store part of their harvest as strategic reserves and only dispose of the grain after assessing the situation regarding the new harvest (Stathers et al., 2013).

Smallholders, as defined by the FAO (2017), constitute the majority of the farming community in developing countries in contrast to industrialised countries, where a small fraction of the population dominate farming and typically, use highly mechanised farming operations. For example, in sub-Saharan Africa (SSA), 60–70% of the food is produced by smallholders. Most of these farmers rely on grain staples such as maize, sorghum, millet, beans or rice for sustenance and income generation. Hence, it is imperative that these smallholder farmers, who supply most of the total national food production, be adequately supported in producing these staples. This includes ensuring that sound postharvest management (PHM) practices are followed to avoid food losses which might result in food shortages and food insecurity.

It is projected that by 2050, the global human population will be 9.8 billion, up from the current 7.6 billion (UNDESA, 2017). Most of the population increase is going to be in developing countries. This means that there will be an increased demand for food in such areas. This is particularly the case in SSA, where the population is expected to double from about 1 billion to 2.2 billion (UNDESA, 2017), yet 307 million people are already suffering from severe food insecurity (FAO et al., 2017). To meet the required food demand, a 60–70% increase in production would be needed (FAO et al., 2017). Food insecurity is going to be exacerbated by global warming and SSA is expected to be severely affected. The rise in temperature is expected to affect storage pest distribution, the intensity of the attack and multiplication (Stathers et al., 2013). It is anticipated that storage pesticides may degrade faster due to the rising temperature (Mlambo et al., 2017). There is evidence suggesting that natural enemies of field insect pests could be more affected by the rising temperature, which will result in unchecked multiplication of pest populations (Thomson et al., 2010). The majority of smallholder farmers in developing countries are highly vulnerable to climate change because they mostly depend on rain-fed agriculture, cultivate in areas of marginal rainfall and soil fertility and lack access to the technical or financial support required to help them invest in more climate-resilient agriculture (Morton, 2007; Stathers et al., 2013; Holland et al., 2017; Donatti et al., 2019).

Postharvest loss (PHL) can describe a quantitative and/or qualitative reduction in food (FAO, 2019). Developing countries are already burdened by malnutrition and hunger, often necessitating diversion of developmental resources towards food. For example, in SSA alone, each year US$4.1 billion worth of cereal grains are lost after production, which constitutes about 13.5% loss in annual value of African cereal (World Bank et al., 2011). This is equivalent to the annual caloric requirement of at least 48 million people and is substituted by food imports (World Bank et al., 2011).

The objectives underpinning investments in reducing PHL in developing countries and regions such as SSA have been discussed by Sheahan and Barrett (2017) and include: (1) improve food security via all four internationally recognised pillars: availability, access, utilisation and stability; (2) improve food safety, as distinct from food security, and avoid food loss due to perceptible and imperceptible spoilage or contamination, leading to adverse health effects when the food is consumed including mycotoxicosis, in the form of fumonisin and aflatoxin, which are linked to the development of oesophageal and liver cancers, respectively, and are growth retarding and immunosuppressive; (3) reduce unnecessary resource use of water, chemical fertilisers, pesticides, labour, finance, energy and land which may pose sustainability challenges. Anticipated PHL by farmers may mean that more of these resources are used than is necessary to meet production or consumption targets. Reducing PHL could, therefore, in principle, reduce fuel costs, transport-related pollution, energy consumption in processing and so on. Thus limiting input use has environmental and/or human health benefits, in addition to reducing costs for value chain players (farmers, traders, processors), resulting in more profits and possible reduced consumer food prices; (4) increase profits for food value chain actors, elaborating objective (3), aforementioned. The private sector, including smallholder farmers, plays a critically central role in making food available to consumers. In SSA, PHL reduction could mean improving the livelihoods of both smallholder farmers and large-scale agribusinesses. The latter will be supplied with larger quantity of higher quality produce. Recognising this natural profit motive to reduce PHL is essential for a clear understanding of the likely benefits of direct interventions that reduce PHL.

Meeting these PHL reduction objectives obliges technocrats and policymakers to review the 'fitness' of the currently available hard and soft PHM technologies, including the existing institutional arrangements or 'organowares' which should deliver PHM services to smallholders, in order to achieve food and nutrition security in transforming landscapes, agri-food systems and consumer demands. This provides an opportunity to re-examine the nature of support that smallholder farmers require to ensure adequate food and nutrition security.

While researchers have identified many potentially useful postharvest technologies for use in developing countries, there is a lack of information regarding the costs and financial benefits of these postharvest technologies, since costs are rarely documented during research studies (Kitinoja et al., 2011). Generally, the adaptive research step between gathering laboratory findings and extension of the results is missing, or local costs are simply not considered when investigating the technology and its field applications. Technically useful practices, therefore, tend to be lost because the practices are promoted without adequate consideration of cost and benefit factors in different farmer circumstances. Although extension services in developing countries are increasingly involved in providing educational programmes and training activities on postharvest topics, often there is a lack of follow-through and support after the training (Kitinoja et al., 2011). Those participating in postharvest training may be convinced and willing to implement improved practices but cannot do so when needed supplies, tools or equipment are not locally available. Kitinoja et al. (2011) observed that even when users are initially provided with these tools and supplies as part of a training programme or development project, there tends to be a lack of local support, supply chain development, expertise and services once the programmes end.

The purpose of this chapter is to review available literature on existing and future postharvest interventions, associated or potential adoption challenges in the smallholder context, and identify means to tackle such constraints in support of the end-users of the technologies/interventions - the smallholder farmers. The chapter unpacks the multifaceted complexities in the post-production food grain pipeline and proposes support mechanisms to unblock the constrictions along the food grain supply system. The chapter also interrogates and attempts to illuminate the paradox that so many postharvest technologies have been developed to address food grain postharvest challenges faced by smallholder farmers, yet most remain largely unused by the target users and instead are gathering dust on the shelves or workshops of research institutions. The chapter proposes how postharvest practice and policy can be done differently to ensure wider socio-economic use of developed technologies. The chapter attempts to provide scope for research and development (R&D) practitioners to identify new and appropriate directions in designing, development, testing and rolling-out of future policies, strategies, research, extension and marketing programmes related to PHM in support of smallholder farmers in developing countries.

The chapter draws heavily from the authors' own and colleagues' experiences in postharvest R&D as well as from published and unpublished literature on the subject. This chapter will benefit practitioners and scholars of postharvest R&D in developing countries including national agricultural research and extension system personnel, international agricultural research centres, academia, private sector, NGOs and agencies based in the North who are involved in postharvest R&D in the South.

# 2 Improvement strategies

## *2.1 PHL reduction technologies: merits, limitations, adoption and uptake challenges*

Several PHL reduction technologies have been developed for smallholder farmers, but many of these innovations are either not used or not fully utilised for various reasons (Table 1). These problems of limited or lack of uptake are complex in nature and are not just due to technical reasons but due to many other socio-economic and institutional issues. The problems are diverse, and some of the more technical ones relate to:

- unavailability of adequate technologies;
- inappropriate technologies;
- unfamiliarity with available technologies;
- inadequate use of or adaptation difficulties of effective available technologies;
- disadoption or refusal to use effective available technologies, due to various reasons; and
- inadequate technical support.

Technology uptake and adoption are influenced by efficacy, culture, socio-economics, cost, awareness raising, political stability and the way the technology is introduced. Many PHL technology projects do not sufficiently involve key stakeholders, often resulting in lack of technical backstopping, iterative participatory design of technologies, or acceptance by the target group, or other services such as informed marketing and distribution networks.

Postharvest R&D partners face a number of challenges including:

- Lack of data from properly designed loss assessment studies for various crops along the whole postharvest system based on credible and comparable methodologies to support estimates (for further discussion, see Stathers and Mvumi, 2020).
- Inadequate involvement of the private sector in PHL reduction activities.

**Table 1** Grain postharvest loss reduction technologies across sub-Saharan Africa: brief analysis of their merits and limitations

| PHL reduction technology | Merits | Limitations |
|---|---|---|
| Synthetic pesticides | Fast, effective and convenient. | Concerns with environmental, human health and resistance challenges. In some countries pesticide access, cost and adulteration are hampering the use of this technology. |
| Diatomaceous earths (DEs) | Extremely effective and acceptable to farmers; local DE deposits available in many parts of Africa. | Initial private sector investment required for wider-scale availability. Needs to be combined with other materials to control full pest spectrum and to help reduce the application rate and therefore reduce dustiness. |
| Breeding for resistance to storage pest attack | Host-plant resistance could be a sustainable and practical grain protection strategy as famers do not have to do anything except acquiring the right varieties; some local varieties it is not always a priority of breeding programmes have some degree of natural resistance. | Farmer's access to the few varieties with improved resistance still limited and will cost them more. High susceptibility to bird attack (small grains) and storage insect pest attack (cereals) still discouraging farmers from growing some otherwise improved and/or high-yielding varieties. |
| Botanicals | A lot of R&D done. Low chances of resistance development; locally available; farmers familiar with technology; easy to use. | Limited botanical products on the market. Technology bogged down by key areas requiring strengthening which include cultivation, propagation, sustainable harvesting, safety, variable performance under different environmental conditions and private sector engagement for sustainable supply of products. Most of the work has been laboratory-based. |
| Biocontrol of LGB using the predatory beetle, *Teretrius nigrescens* | Comes at low cost to the farmers as centrally delivered; self-sustaining; may reduce field infestation and forest-to-store infestation. | Limited evidence of effectiveness in stores; not applicable in stores where pesticide is used; slow population growth rate of *T. nigrescens* compared to its prey; poor establishment of the predator in some areas leading to possible extinction, so may need regular release by national agricultural research service-providers. Only works at community or geographical scale rather than individual farm level and thus needs to be facilitated. |

(*Continued*)

**Table 1** (*Continued*)

| PHL reduction technology | Merits | Limitations |
|---|---|---|
| Hermetic storage systems | Huge potential for hermetic bags which are extremely effective when used correctly; farmers already familiar with bag storage and the concept of airtightness from traditional practice; bags can be reused for two or three seasons if not perforated. Perforations can be sealed by packaging tape; prices can be reasonable if locally manufactured; bags facilitate flexibility regards volumes of storage, are easy to manage and farmers can retain the hermetic bags at marketing. | More evidence needed that hermetic bags work in LGB- and rodent-infested areas. Some farmers do not trust that the bags work with no pesticide and so still apply pesticide to manage the risk of grain loss. Market access still a major challenge in many countries especially where the bags have to be imported. Private sector not willing to take the manufacturing risk where the market is not well developed and guaranteed. Environmental issues with disposal of non-biodegradable plastic. |
| | Plastic silos have great potential in non-LGB infested areas; can be used indoors and therefore secure; variable sizes to cater for different farmer requirements and different grain types. | Can be bored through by LGB. Need careful transportation to avoid damage; security issues when installed outdoor. |
| | Metal silos are highly effective; can be pre-abricated and installed indoors; durable (15–20 year lifespan) and cheaper than most options in the long run. Variable sizes to cater for different farmer requirements and different grain types. | Metal silos ar3e initially costly to low-income farmers; challenges exit in facilitating trained local artisans to take over manufacture and supply; workmanship to ensure metal silos are airtight coupled with farmer maintenance of hermetic conditions needs further strengthening. |
| | Airtight 'cocoons' have potential for commercial or local entrepreneurs purchasing grain for later sale or community grain banks for later sale or withdrawal during the lean period; can be used to tap on the growing niche organic markets. | Airtight 'cocoons' require the zipping mechanism to be well managed. Cocoon handling, installation and management require expertise; multiple 'reuse' still needs to be verified; LGB has potential to penetrate the cocoon fabric. Mostly used together with bag storage which is an extra cost. |
| Solarisation/thermal disinfestation | Uses concepts familiar to farmers, and can be repeated as needed. Most stored product insects can be killed at 50–60°C after up to 1 h exposure. Works well for small quantities of grain and especially against bruchids in pulses. Bruchids are highly susceptible to heat. | Low adoption; the use of an artificial method of grain disinfestation by heat of solarisation, such as a solar dryer, is not cost-effective to most of smallholder farmers, who do not have sufficient capital to invest in dryer construction. Some pests such as LGB are particularly tolerant to elevated temperatures hence higher temperatures (>50°C) or a longer exposure period is required. Insect reinfestation possible if grain is not stored in an insect-proof facility. Local manufacture of solarising (greenhouse effect) plastic/pouches may be a challenge in some countries; heat may affect seed viability where recycling of seed is practised (e.g., in pulses and small grains such as sorghum and pearl millet). |

| | | |
|---|---|---|
| Improved brick or earth granaries | Huge potential for uptake since it is an adaptation of the typical traditional granaries already familiar to farmers; durable. | Affordability issues at installation. Indoor polypropylene bag storage coupled with insecticide treatment is increasing in Eastern and Southern Africa and replacing the solid-walled granaries. This is mainly for security, volumetric flexibility and ease of marketing reasons. The bags are affordable and easy to manage, especially for young couples who either lack the granary construction skills or have other competing investment priorities. |
| Grain banks and warehouse receipt systems | Have the potential to reduce PHLs, increase food security and increase income generation within farming communities. | Require functional institutional arrangements, strong management and external injection of resources to kick off the process; market linkages and group dynamics are key drivers. Also requires farmers' willingness to 'aggregate' into a common pool and losing identity of 'their' grain bags in the lot. |
| Mechanised threshers/ shellers, cleaners, dehullers | Reduce processing time, labour and food losses with significant impact on women's drudgery; grain quality is greatly enhanced (no grit/stones; low mycotoxin contamination). | The appropriate machine may not be easily available or affordable. May work best under group arrangements or 'custom' service-provision such as that provided by local millers (e.g. through mobile machinery as an agribusiness). Some areas may not be easily accessible because of poor road infrastructure. Mechanisation of these processes can result in labour displacement problems of some of the most vulnerable people in the community. |
| PHM training of farmers and service-providers (skills/ knowledge) | Provides sustainable access to contextualised information within or close to the communities. | Essential to optimise the use of, adaptation of and scaling out of PHL technologies. However, very limited funding for PH training/ capacity building exists compared to the support for tangible technologies. ToT approaches need to be followed-up to determine efficacy of the cascading effect and to determine what has changed because of the training. |

Adapted from: Mvumi and Stathers (2015).

- Lack of national policy support, which would enable budgetary allocation.
- Heavy bias by policymakers, development partners and research expenditure towards the crop production phase.
- Insufficient emphasis being given to the postharvest-relevant food safety aspects of food security, such as pesticides or aflatoxins.
- Effectiveness of the 'Training of Trainers' (ToT) approach in cascading postharvest (PH) knowledge and skills is not certain. Follow-ups for quality control monitoring and lesson learning are often missing in the programmes.
- Challenges of taking pilots to scale; many innovations have remained small and localised. Such 'islands' of success need to be scaled up and out.
- Many funders do not provide funding to support upscaling of projects or post-project impact assessment.
- Accessible and appropriate training materials for different circumstances, commodities and levels. Good quality training manuals have been developed but are they sufficiently available and easily adapted to the specific contexts of different countries?
- Lack of coordination among postharvest value chain actors; normally, this should be a central government role. This has often resulted in duplication of efforts, inefficient use of resources, lack of cross-learning and neglect of some parts of the postharvest value chain.
- There are no readily accessible platforms to reflect, interrogate, exchange ideas, learn and plan interventions, especially in Africa.

## 2.2 PHL reduction technologies: Dissemination strategies and approaches

Lack of stakeholder buy-in is a major hindrance to the uptake of research findings. To ensure the utilisation of the results, it is paramount that key stakeholders and end-users (farmers, farmer representatives, policymakers, donors, agricultural extension staff, private sector players such as small and medium enterprises and agrochemical companies, NGOs and other researchers) are involved in the needs assessment, design, testing, dissemination and marketing of technologies and solutions. This reduces chances of developing inappropriate technologies and increases chances of acceptability, adoption and continued use in the future. R&D partners need to work together to facilitate adoption, adaptation and dissemination of successful technologies and solutions developed in one country or subregion to another, thus creating a multiplier effect and avoiding 'reinventing the wheel', a common drawback of research for development projects.

Stories of successful change need to be shared widely to inform and attract additional support for further work to keep expanding, improving and adapting interventions. The new knowledge needs sharing with different stakeholders through platforms, including field days, workshops/seminars at community, district, national, regional and international levels alongside electronic pathways. Sharing of R&D results at different innovation platforms enhances knowledge gain, consolidates understanding of technologies and facilitates the adoption of proven effective mitigation strategies to reduce postharvest losses by non-research participants. Functional multi-stakeholder innovation platforms involving grain producers, trader associations, women and youth organisations, farmer associations and private and public sectors should result in continuous generation of innovations, iterative learning, demand for services and goods and the accompanied social and economic change.

One of the R&D approaches used in Zimbabwe is the setting up of multi-locational storage trials, involving a wide range of stakeholders. The trials allow experiential learning by farmers (and other stakeholders), who learn best by 'doing and seeing things' in their own circumstances. This is the 'learning alliance approach' as articulated by Lundy et al. (2005), Morris et al. (2005) and Mvumi et al. (2008). Using this approach, grain storage technologies at host farmers' homes are evaluated from different perspectives by fellow farmers and key stakeholders, including researchers and extension agents, who evaluate the comparative quantitative and qualitative (e.g. nutritional) losses of the samples. These centres become the nuclei for learning and dissemination of new effective technologies by the surrounding communities and stakeholders when these groupings periodically come together to observe, share and learn lessons on-site through direct participation and by asking pertinent questions (Mubayiwa, 2019). The learning centre approach, as described by Mashavave et al. (2011), can also be used to evaluate and promote the uptake of a range of postharvest technologies including drying technologies and facilities set up at farmers' homesteads as learning centres instead of the conventional fields as is normally the case with agronomic trials. The approach provides an opportunity for farmers to incorporate their own suggestions into the trials from one season to another, and together the consortium can co-learn and co-innovate what works and what does not in the context of smallholder farmer circumstances. Hence, the experiential and collaborative experience of working in learning centres is convincing and can be used to accelerate the adoption of effective and appropriate postharvest technologies. The approach can help create a demand for goods and services by stakeholders and especially the smallholder farmers themselves. However, this only works if the private sector then takes over the distribution of the technologies verified and demanded (i.e. passed) by the stakeholders. This process can be enhanced by including private sector players in the learning centre activities. Mashavave et al. (2013) showed that smallholder farmers involved in learning alliances had a denser network structure of social interactions and higher communication efficiency in comparison to non-participants. The study found that field-based learning alliances shortened the innovation dissemination cycle and hence increased the likelihood of technology adoption.

However, participatory research processes are time-consuming and require a good understanding of the context by all players before smallholder farmers and other stakeholders buy-in the research ideas. Many a time, farmers participate in trials, hoping to get some tangible freebies in one way or another, without really fully engaging in gaining knowledge for their longer-term benefit. In addition, development partners, especially funding agencies, do not usually have a budgetary provision for these participatory processes because they are under pressure to produce 'number of farmers or product uptake' type results in the shortest possible periods.

Lundy et al. (2005) forewarned that a learning alliance approach is not suitable for all agricultural research projects but can be appropriate for researchers and projects working directly with or hoping to effect locally-owned and sustained change on a given rural innovation system. These authors highlighted the following challenges:

- *Partner and participant selection*: The selection of partner agencies and appropriate individuals within those agencies is critical to the success of a learning alliance. Both agencies and individuals should be open to critical reflection and learning about their own practices. In addition, partner and participant turnover should be avoided, or at least minimised, since this disrupts the learning process.
- *Process facilitation and coordination*: Establishing a learning alliance, while it takes time, is relatively straightforward once appropriate partners and individuals have been identified. The maintenance of the learning alliance and its ongoing facilitation, however, is a separate challenge. To stay vibrant, a learning alliance must adapt and change as learning occurs and new questions arise. While the largest share of development project budgets will be allocated to project implementation, financial resources are also needed to support time for personnel to engage in learning alliance activities.
- *Funding*: Finding development partners interested in funding an open-ended learning process is unlikely. It may be easier to get funding for specific R&D projects that use learning alliance as an implementation mechanism. Funding for learning alliances can also be secured by riding on large development initiatives since this makes efficient use of researchers' time, and alliance activities can complement dissemination and training budgets.
- *Linking learning across levels*: Documenting, analysing and sharing learning from diverse partner agencies at the micro, meso and macroscales are very demanding for all participants, while drawing out key livelihood

and policy implications from such a wide range of experiences takes a good deal of thought and time. Equally, the creative application of diverse tools and methods to promote network building among partners and the use of decentralised processes of reflection and learning will also help (e.g. regional learning fairs with local partners, topic-driven short workshops and virtual discussions).

While scientists engaged in on-farm adaptive research must observe strict protocols to secure reliable, evidence-based findings (e.g. pesticides research for registration purposes), which typically casts farmers in supporting roles, such work invariably provides wider opportunities for sharing and building trust (Morris et al. 2005). Researchers and technical staff must, however, recognise that farmers are experts in their domain and will ultimately determine the uptake. Researchers or development practitioners must humble themselves to learn from the farmers as well.

Learning alliance approaches allow and encourage the active engagement of key organisations and individuals within the innovation system but outside the research realm. Early contact and information sharing with local stakeholders during the research phase are strategic in securing buy-in. Trust is a key ingredient in building learning alliances, but this takes time and is a long-term commitment by individuals to establishing and maintaining good relations.

Learning alliances, which are inevitably interdisciplinary and multi-agency, can be expected to outlive conventional project time frames (Morris et al., 2005). In the long term, the learning alliance approach has the potential to deliver on co-learning, co-innovation and dissemination of technologies and information. However, budgetary issues need to be considered and discussed early on with all parties at the project design stage to avoid funding problems during implementation.

## 2.3 Capacity development (training)

### 2.3.1 Extensionists and technicians

Efforts have been made to train various stakeholders in improved PHM. However, many of these training programmes have often been one-off with no further follow-ups to determine the impact of the training. Refresher training is often required to consolidate understanding and practice and share experiences and possible new dimensions from the field - positive and negative - as part of the learning and internalising process. It must be recognised that many extension agents and farmers have a lot of diverse field experience based on what they hear, encounter, observe and practice in the field as they conduct their daily business, but this needs to be harnessed through training workshops and consolidated.

The information also needs to be 'winnowed' for accuracy and validated through interaction with other stakeholders, including researchers. Misinformation and disinformation also need to be addressed as part of the learning process.

The content and duration of training sessions are often inadequate and not standardised in terms of the minimum body of knowledge in the curricula. Critical components such as gender, agribusiness and socio-economics are often missing, yet PHM is not just about technical issues. Cultural issues are key to technology adoption and hence need to be incorporated in training programmes. Some farmers or extension agents only get one or two days of training, while other comprehensive training programmes for extension staff may last up to two weeks. One training strategy that could help standardise training programmes at the national level is to convene training workshops for core teams of trainers at various levels, who will, in turn, continue to train their peers upon return to their workstation. These core teams could be composed of extension agents (including NGOs), traders or lead farmers. A major challenge to this strategy is that these cascading training programmes are often not well thought out, planned and budgeted for, and therefore the objectives are not achieved.

Another strategy is to include grain PHM in tertiary curricula (agricultural colleges or universities) to ensure that graduates from these institutions are already equipped with the necessary knowledge and skills when they are recruited to serve the agricultural community. In many countries, there are very few postharvest technocrats or practitioners; hence there is minimal lobbying and advocacy for meaningful inclusion of PHM issues in the national agricultural development plans, resulting in limited budgetary allocations.

There are not many postharvest scientists/technocrats in developing countries, yet that is where postharvest losses, food deficits and malnutrition remain high. University postgraduate training is critical so that context-specific and relevant evidence can be generated to convince industry and policymakers to invest in the development of the postharvest sector, which will also benefit smallholder farmers.

In the mid-1980s to late-1990s, there used to be very strong in-service training programmes in agriculture, run annually at national levels, for all new staff in the national extension service of many SSA countries. The training courses were designed to run for between 1 and 2 weeks, and PHM was a key topic. Some of the courses were obligatory for any staff member in the national extension department. These courses were required for promotion to senior grade in government. However, these trainings have since stopped because of funding issues from the central government, and most of those trained at that time have since moved on. Such cources are now being conducted on an *ad hoc* basis when funding becomes available, usually through externally funded projects. The original model could be resuscitated in a modified form and in partnership with development agencies and the private sector.

Training needs to be ongoing due to:

1 New extension staff joining the service.
2 New generations of stakeholders (farmers, policymakers, researchers) coming onboard.
3 Changes in the biophysical environment (e.g. new pesticides, new crop varieties, new pests, climate change).
4 New awareness of global challenges (e.g. health and safety issues, nutrition).
5 Incorporation of new knowledge, interventions and approaches into the services provided.
6 Changes in population demographics and consumer preferences.

Similarly, in the mid-1980s to the late 1990s, there were many international institutions offering postharvest training programmes at all levels - from short courses to PhD - but after 2000, these shrunk drastically for various reasons including drastic downsizing of the number of scientists and changes in foci by donor agents. It is therefore imperative that strong continental or subcontinental programmes be mounted to support PHM training in developing countries. Appropriate training curricula and materials are required to support this endeavour. There are a number of quality manuals that have been developed (e.g. World Food Program and Natural Resources Institute, 2012). However, these training materials may not be easily and widely accessible to the trainers or the materials may not be in a form that is readily understood by the targeted level of trainers.

## 2.4 Integrated PHM and implementation options

Preharvest and PHM practices need to be integrated at each stage as opposed to focusing on each stage. In Benin, Hell et al. (2000) found a positive correlation between insect damage in the field and aflatoxin levels in harvested and stored maize. This highlights the need for effective preharvest and postharvest insect pest management to mitigate fungal and mycotoxin contamination of maize. Effective preharvest pest management involves the implementation of a number of strategies that include selecting varieties with complete and tight husk cover to minimise pest entry and moisture ingress; selecting inherently resistant varieties; and good general husbandry to minimise cross-infestation and cross-infection by insects and fungi, respectively. Other strategies are presented in Table 2. Field studies in Kenya and Zimbabwe demonstrated that increases in aflatoxin and fumonisin levels can be prevented by the use of storage of well-dried grain in hermetic containers such as hermetic bags and metal silos (Murashiki et al., 2018; Walker et al., 2018; Dembedza et al., 2019) but mycotoxin levels present during field stages remain.

**Table 2** Farming practices associated with higher and lower aflatoxin levels in stored maize

| Lower aflatoxin levels | Higher aflatoxin levels |
|---|---|
| *Production practices* | |
| Crop rotation | Maize mono cropping |
| Use of resistant cultivars | Use of susceptible cultivars |
| Maize in a mixed cropping system | Cowpea, peanut or cassava intercrop |
| Fertiliser application | No fertiliser |
| Use of Aflasafe | No application of Alfasafe |
| Complete and tight husk cover | Incomplete and loose cob husk cover allow field infection and infestation |
| *Harvest and pre-storage practices* | |
| Harvest at crop maturity | Delayed harvest |
| Harvest of maize with husk | Harvest maize in heaps; cobs shelled later |
| Sun-drying on platform or plastic sheeting | 'Field' drying on the plant |
| Drying of maize without the husk | Delayed drying |
| Immediate removal of damaged or infected cobs | No sorting of harvest |
| Shell as soon as moisture content is ≤13% | Repetition |
| Shelling methods that reduce grain damage | Kernel damaging shelling methods |
| *Storage practices* | |
| Cleaning of the storage structure or disinfestation of bags | No adequate preparation of the storage structure |
| Maize stored for just 3–5 months | Maize stored for long term (8–10 months) |
| Use of smoke or effective insecticides or hermetic containers | No effective insect control |
| Maize stored in well-aerated stores | Maize stored in poorly aerated stores |

Adapted from: Hell et al. (2008).

The development of locally adapted solutions to reduce PHL requires relatively modest investments that can result in higher returns compared to agronomic-based efforts alone. Gitonga et al. (2013) and Ndegwa et al. (2016) provided evidence of locally made metal silos as examples of cost-effective technologies for maize storage in Kenya. However, wider-scale adoption of the technology is still limited.

While major PHM upscaling efforts have yielded positive results ( Baributsa et al., 2014; AgResults, 2019; www.rockefellerfoundation.org/YieldWise), more still needs to be done to ensure wider coverage across other developing countries and to penetrate remote areas. Maize storage challenges that still need to be addressed include:

- While residual pesticides may remain the most common and effective control method for maize storage pests such as *Prostephanus truncatus*

(Coleoptera: Bostrichidae), farmers face challenges of pesticide: cost, abuse, adulteration and efficacy; exposure during treatment; and chemical residue during consumption. It behoves technocrats and farmers alike to adopt an integrated approach to reduce pesticide use and pest control costs through, inter alia, timely harvesting (Borgemeister et al., 1998), use of resistant varieties (Meikle et al., 1998), regular inspection/scouting for pests (Meikle et al., 2002), shelling the cobs for easy treatment, use of storage structures with no timber or timber that is not in direct contact with grain, various combinations including sorting the cobs prior to shelling (Adda et al., 2002). These methods, when applied carefully, will also help to manage other postharvest pests.

- There is a widespread (mis)use of fumigants in many SSA countries (Brice and Golob, 1999). Fumigants are not recommended for smallholder use because of a plethora of challenges including concerns over the safety of the farmer and those in the vicinity and high risk to the environment due to high toxicity; lack of airtightness of smallholder storage facilities; lack of personal protective equipment during application; and lack of knowledge as to how phosphine works (Brice and Golob, 1999; Golob and Tran, 2001; Singano et al., 2019). The misuse of phosphine in short exposures at high temperatures in poorly sealed enclosures has led to the development of phosphine-resistant insect pest strains (Bell, 2000).
- The most promising practical natural enemy that can be used within LGP IPM frameworks is the predatory beetle *Teretrius nigrescens* (Lewis) (Coleoptera: Histeridae), which is strongly attracted to and feeds on *P. truncatus* larvae. Field-releases of the predator have been done in several countries in SSA including Togo, Benin, Ghana, Guinea, Kenya, Mozambique, Malawi and Zambia. However, the optimism has not been fully realised as later studies found that sufficient control of *P. truncatus* may not be feasible in stored infested grain due to the slow population growth rate of *T. nigrescens* compared to its prey (Holst and Meikle, 2003; Muatinte and Cugala, 2015). The predator is known to be highly sensitive to synthetic pesticides (Stathers et al., 2002), which are commonly used in grain stores.
- While the *P. truncatus* pest status is alleged to have declined in West Africa (Bandyopadhyay et al., 2019), this cannot be corroborated by any scientific data, and a clear explanation of the causes of the decline is yet to be derived and published. The interactions of the *P. truncatus* and the *T. nigrescens* in the forest and those in the store are not clearly understood. There is also evidence of the poor establishment of *T. nigrescens* in some parts of Africa, which has been ascribed to the influence of environmental factors and the mismatch between the original biotypes from Mesoamerica and the local ecology (Omondi et al., 2014). The successful establishment

of the predator reported in West Africa was ascribed to the use of more genetically diverse strains of *T. nigrescens* in large numbers and at more sites compared to Kenya (Omondi et al., 2014). Ecotypes of *P. truncatus* have been reported (Guntrip et al., 1996), and hence corresponding genetically diverse *T. nigrescens* strains could be more effective in controlling *P. truncatus*.

Mutiga et al. (2019) proposed a conceptual model in which the establishment of a local grain production and banking system (LGPBS) could enhance food security and safety in East Africa (Fig. 1). Conceptually, LGPBS refers to centres that provide farm inputs, management practice advisory services, grain aggregation, grain storage, grain drying, grain safety assessment and credit facilities to farmers within their communities (Fig. 1). The model, set up by a collaborative consortium of investors, local and/or national governments and primarily implemented by the national and local authorities, provides a suite of extension/advisory, technical and financial support within warehouse receipt systems to enhance optimal grain production and post-production management in smallholder farming communities. While the proposed conceptual model is focused on maize production and post-production management, emphasising food safety issues, a similar approach could be applied to any other grain, and even concurrently under the same framework.

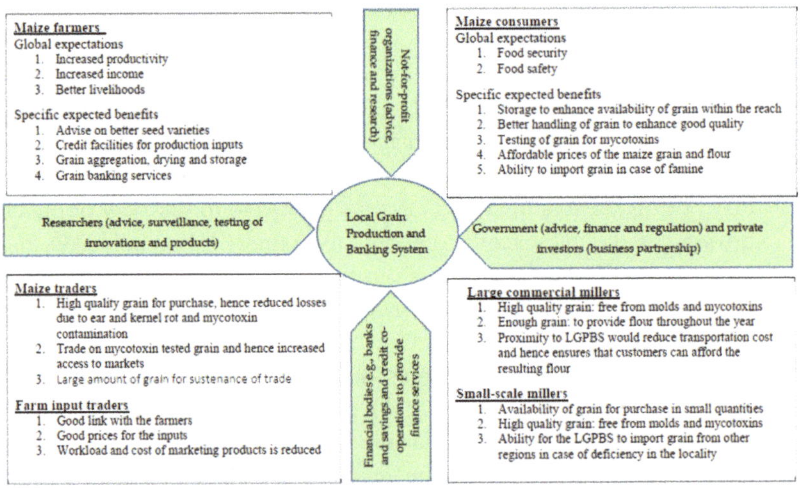


**Figure 1** Conceptual model for the establishment of local grain production and banking system (LGPBS) to enhance food security, safety and trade in East Africa. The potential support (for example, advisory, credit, donations, research, and so on) from different stakeholders to the LGPBS are shown in light green arrows. The potential benefits of the LGPBS to millers, traders, farmers and consumers are contained in white rectangular boxes. Source: Mutiga et al., 2019.

Major stakeholders of the LGPBS would comprise the farmers, ministries of agriculture, agricultural research institutions, health and water organisations from local and national governments, private investors, agro-dealers, seed companies, national food safety organisations, microfinance organisations and insurance companies. Each of these entities would contribute synergistically at different stages in the grain value chain. Although the government agencies would have a stake, a preferred model is to have LGPBS operate as independent businesses, with autonomous management to enhance efficiency and eliminate bureaucracy and political interference, hence, better service delivery to the farmers. However, the management would need to be certified as that and corruption are the big challenges with this kind of system. Together the consortium is expected to ensure that all major services are provided efficiently to the farmers, with an overall goal of enhancing increased maize productivity and safety within the respective production schemes. The model is worth testing in different developing countries under different circumstance to determine if this model could be used to support smallholder farmers in reducing grain PHLs on a sustainable basis. Shepherd (2018) presented another example of strategies for improving aggregation and coordination in smallholder paddy rice milling (Table 3). Again such strategies need to be validated under different circumstances.

Mycotoxins are known to be a major food safety challenge in developing countries and are associated with perpetual hidden human health problems caused by chronic exposure (Sun et al., 1999; Shirima et al., 2013; Misihairabgwi et al., 2017) and lethal poisoning emanating from acute exposure to aflatoxin (Azziz-Baumgartner et al., 2005; Buguzi, 2016). There is an urgent need to support smallholder farmers to establish effective and sustainable mycotoxin mitigation strategies because the toxins cause many health and socio-economic

**Table 3** Ways of improving aggregation and coordination: the case for paddy rice milling

| Commodity purpose | Methods to improve aggregation and coordination | Commodity or value chain characteristics |
|---|---|---|
| For small local mills | • Group marketing | • Requires centralised milling<br>• Dominance of small mills |
| For sale to larger mills | • Group marketing<br>• Alliances between producer organisations (POs) and buyers<br>• Coordination by buyers | • Consolidation and modernisation of milling can enhance efficiency and quality but implies greater distance from production |
| For paddy with clearly defined characteristics | • Contract farming<br>• Alliances between POs and buyers | • Some but limited opportunities for differentiation, except in the case of highly distinct varieties |

Adapted from: Shepherd (2018).

impacts including from international trade bans when mycotoxins are found to be above the acceptable specified levels tolerated for products ingested by humans or livestock (Marasas et al., 2008).

Ideally, strategies to effectively manage aflatoxin and fumonisin contamination include preventing crop stress during the field production stage, ensuring proper handling during harvesting and drying, and storing the grain under conditions that do not favour proliferation and contamination by the toxigenic moulds (Mutiga et al., 2019). Unfortunately, achievement of such ideal conditions by all stakeholders in the maize value chain is not an easy task due to many reasons including resource constraints, lack of technical advice and lack of capacity to address the challenges that cause preharvest and postharvest contamination.

The majority of small-scale grain traders are known to sort or blend grain in an effort to reduce the concentration of apparent mouldiness in the purchased grain, although this practice has worked only in significantly reducing fumonisin and not aflatoxin (Mutiga et al., 2014). A strong correlation is reported to occur between visually apparent mouldiness and fumonisin levels, and thus visual assessment to remove mouldy grain can significantly lower the fumonisin contamination in maize (Mutiga et al., 2014). *Fusarium verticillioides*, which is one of the main fumonisin-producing fungi causing kernel rot, is restricted to the pericarp of the grain (Duncan and Howard, 2010), while *Aspergillus flavus*, an aflatoxin-producing fungus, can exist internally in a maize kernel, therefore hindering the effectiveness of visual inspection (Smart et al., 1990).

If grain is harvested with a high moisture content and sold, the risk of mould contamination can be exacerbated during temporary storage, when there are long delays that occur while selling the grain to grain processors and the grain reserve. In Kenya and Tanzania, Mutiga et al. (2019) report that it is common for farmers and traders to spend weeks in queues of trucks waiting for their grain to be off-loaded, during which significant deterioration and mycotoxin contamination of the grain could occur, especially in cases where the grain has a moisture content exceeding 13%. Thus, government policies that facilitate the efficient marketing of grain are urgently required to minimise the presale waiting period, which frustrates both farmers and traders and allows the development of fungi and possible build-up of mycotoxin contaminants.

In developing countries, and especially SSA, where resources are limited and sophisticated technologies are lacking, relevant, culturally acceptable, simple, economically feasible, practical and sustainable methods of mycotoxin reduction are required in rural subsistence farming communities (Misihairabgwi et al., 2017). These authors recommend a range of mitigation strategies, including: awareness creation, use of improved varieties, and dietary diversity that limits maize and groundnut consumption.

Drying maize on a mat/raised platform, hand-sorting (damaged, discoloured and moulded grains) and applying reduced doses of residual insecticides during storage are strategies associated with less contamination of maize with aflatoxin and fumonisin, and therefore reduced mycotoxin exposure by subsistence farming communities and other consumers of it (Matumba et al., 2015; Kamala et al., 2016). However, the issue of using residual insecticides, though very practical, is raising serious concerns among consumers. In the long run, synthetic pesticides should be replaced by non-chemical methods including hermetic containers (see Table 1). Direct contact of maize with soil increases the chance of exposure to fungal spores in the soil that can then infect the drying maize. Preharvest and postharvest insect infestations of maize grain are also closely associated with fungal infection, which increases the risk of mycotoxin contamination.

Strategies for reducing aflatoxin levels in grain production and post-production and those that facilitate their occurrence are known (e.g. see Hell et al., 2008; Mutiga et al., 2019; Table 4). Institutional support is required to enable farmers to implement the recommended strategies as elaborated in Table 4.

## 2.5 Information and communication technologies (ICTs) in PHM

Modern information and communication technologies (ICTs) are playing an increasingly important role in agricultural value chains. At the simplest level, the fact that most farmers now have access to mobile phones means that they can call buyers to check on the price before making a decision to harvest or to transport the produce to a buyer (Shepherd, 2018). This means producers can get the best available prices and avoid transporting produce to the market where there is no buyer or where prices are unfavourable. Marketing transaction costs, such as travelling to the marketplace, are significantly reduced. Research in Bangladesh, China, India and Vietnam found that 80% of farmers had mobile phones and half of them had used those phones to reach deals with buyers (Shepherd, 2018).

Agricultural production and market information services that utilise ICTs, have been provided by government and some companies for decades now. Examples include EcoFarmer in Zimbabwe, M-Pesa in Kenya and E-Soko in Ghana. In West Africa, a French NGO has set up N'kalô, which offers both farmers and traders a 'prospective analysis': an impartial analysis and forecast of supply, orders and demand (Shepherd, 2018). Sharing this with both sides means they can both make optimal decisions based on the best available information. However, it should be noted that ventures of this type have yet to develop viable business models. ICTs can reduce the costs of coordination and transport and increase transparency, thereby reducing the potential for mistrust between partners (Shepherd, 2018). For instance, smallholder

**Table 4** Support services for mycotoxin reduction in maize: a case of East Africa

| Point | Activity/support | Problem tree issue | Sections of local grain production and banking system (LGPBS) |
|---|---|---|---|
| Production | Provision of certified seed of cultivars with desirable traits | Less susceptible maize genotype | • Agronomy advisory team - this would provide farmers with the appropriate information on the best cultivars to grow in each agroecology<br>• Finance credit - this would provide information on monetary support to enhance acquisition of the appropriate seed |
| | Input for control of soil-borne pests/weeds/other pests and diseases | Improved soil/ plant health | • Crop protection section - to provide information on what pesticides /herbicides and the appropriate timing and application rates<br>• Finance credit - to enhance acquisition of the appropriate input |
| | Provision of information on plant spacing | Reduced competition and enhanced plant vigour | • Agronomy advisory team - this would provide advice to the farmers based on the type of cultivar they grow in their fields |
| | Irrigation | Management of water stress | • Water harvesting section - this would support farmers to ensure that the crop gets optimal amount of water<br>• Finance credit - to support water harvesting initiative for farmers |
| Preharvest | Information on proper harvesting equipment | Good harvesting practices | • Agricultural mechanisation team - to provide appropriate information on the most sustainable harvesting methods. Could adopt harvesting using special equipment which is provided by the LGPBS or hired *via* 'fee-for-service' |
| | Information on shelling devoid of kernel breakages | Good harvesting practices | • Agricultural mechanisation team - to provide appropriate information on the most sustainable harvesting methods |
| Postharvest | Information on appropriate grain packaging | Proper grain handling after harvest | • Postharvest loss prevention team - can advise the farmers on how to package the grain. If possible, the LGPBS should take the responsibility of packaging in bags that can allow enough grain drying prior to storage |
| | Collection of grain for delivery to LGPBS | Proper grain handling after harvest | • Postharvest loss prevention and transport sections of the LGPBS to facilitate delivery of the grain to the local aggregation centre or warehouse. This should avoid exposure to additional moisture |
| | Provision of grain drying services by the LGPBS | Drying to attain optimal grain storage moisture | • Postharvest loss prevention team - the LGPBS should have sustainable/inexpensive grain drying methods. To reduce the cost of running the system, modern solar powered driers could be acquired and utilised. They would provide information to farmers about maize cultivars with fast kernel drydown |
| | Prevention of damage by storage pests | Control of storage insect pests and other storage pests | • Postharvest loss prevention team - grain could be stored at conditions that do not favour fungal infection or infestation by storage insects and rodents. The section could promote use of hermetic containers or apply recommended pesticides to keep the grain free from insect or rodent damage |

Adapted from: Mutiga et al. (2019).

farmers may accuse companies or middlemen of exploiting them when the national grain marketing boards or miller/processor price is increased without their knowledge. With mobile phones, they can easily cross-check this on the internet or with relatives in urban areas. Some service-providers circulate regular mobile phone text messages containing market price information to their farmer groups.

Smartphone apps are increasingly being used to supply information to smallholders on weather, transport, pests and diseases and farming practices and techniques which can provide a unique opportunity to facilitate agriculture-related technological adoption and access (Harris and Achora, 2018). The use of videos that farmers can watch on their phones is a growing practice. For instance, in India, Digital Green makes available 6000 videos for download (Shepherd, 2018). Thus smartphones are beginning to replace traditional extension services. However, the commercial profitability of such services remains unclear, and they may not be available to poorer farmers, who cannot afford phones or the data bundles required to access the services or videos (Shepherd, 2018; Kamau et al., 2018; Harris and Achora, 2018).

ICT improves access to financial services, for example, the use of mobile money through M-Pesa in Kenya and E-Soko in Ghana. Furthermore, the focus on the development and spread, especially through the development of broadband technologies, in order to ensure affordable and reliable ICT access to almost all parts of each and every country as most of the farmers are based in rural areas. This will not only provide new opportunities for rural farmers to have access to information on agricultural technologies, but will also enable them to use ICTs for better access to agricultural services such as those offered by national agricultural extension staff (Harris and Achora, 2018), rather than rely on physical visits by, or meetings with, the extension staff.

There have been a few local success stories, including the use of information technology kiosks in India, cell phone text messages for providing market information in Rwanda and formation of smallholder groups for direct marketing (Kitinoja et al., 2011). A study of existing internet communication technologies and a wide variety of commercial services in Uganda, Indonesia and India indicate that farmers are willing to pay for IT-based advisory services, but their access to text-based messages is limited by high costs and illiteracy rates (Kitinoja et al., 2011).

Past research on household food grain sales behaviour in developing countries has tended to focus on the role of market access and price-related factors to explain why many rural households do not sell staple crops such as maize. However, a key concern raised in recent literature is that low household asset endowments may constrain the ability of many smallholders to take advantage of public goods that reduce the cost of market access. Mather et al. (2013) used econometric analysis of nationally representative smallholder panel data sets from Kenya, Mozambique and Zambia to examine the question of how to achieve increases in marketed surplus of maize, the most widely marketed cereal food staple in Eastern and Southern Africa. Their analyses showed that, *inter alia*:

1 Trader presence even in 'remote' villages has greatly improved compared with the situation a decade ago, perhaps a result of increased investments in road construction as well as the recent proliferation of cell phones in rural areas. Thus, while there still may be significant transport costs to the nearest relevant market for farmers in 'remote' villages (which would cause traders to push down maize buying prices), even these farmers now face considerably lower search costs for price information and access to traders than they did a decade ago.
2 In Mozambique, household receipt of market price information results in large increases in the probability of maize sale and sale quantities. In Kenya and Zambia, radio and cell phone ownership were a proxy for household access to market price information. There were also significant positive effects of such assets on quantities sold. These findings suggest that funding to increase the spatial coverage and frequency of radio broadcasts in these countries could potentially lead to large increases in both quantities of maize sold as well as the numbers of households selling maize. Increasing spatial coverage of mobile phones would be expected to have a similar effect.

Today a new paradigm of agricultural development is fast emerging in both developing and developed countries, and the traditional ways of delivering services to rural communities is being challenged, and traditional tacit knowledge societies are being transformed into modern explicit knowledge societies (Kamau et al., 2018; Anon., 2018a), although it can be argued that in practice, the two forms of knowledge actually co-exist (Anon., 2018a). At the postharvest stage, lack of sufficient information makes it difficult for farmers to get fair prices for their crops or to control postharvest pests. A lot of postharvest practices require farmers to be in constant touch with local agricultural extension officers and other agricultural specialists; including getting markets for produce, market price analysis, best storage practices to avoid damage of grain by devastating pests such as *P. truncatus*, value addition technologies for small grains for example, packaging and accounting. In developing countries, however, these agricultural specialists are few and are not able to be physically available to offer their expertise. Even if available, they may not have the capacity to advise farmers adequately, which could lead to farmers being discouraged in their attempts to invest more in inputs and technologies.

ICT solutions are being used to communicate relevant knowledge to rural farmers and provide a cost-effective way to facilitate access to markets. ICT solutions for postharvest management can be divided into several areas, as outlined by Kamau et al. (2018).

### 2.5.1 Knowledge management and training

Farmer organisations in most African countries are faced by the challenge of lacking the knowledge and proper tools to identify, assess and communicate postharvest knowledge and advisory services to the rural farming population. Extension service agents also face similar problems and often lack the means and tools, if they possess the knowledge, to disseminate information to these farmer organisations in a timely way. This gap in information flow and tools limits effective decision-making in harvesting, shelling/threshing and cleaning, treatment and storage or marketing and general business management, resulting in crop loss, low productivity and subsequent low sales and incomes. ICT can assist in generating, capturing and disseminating the valuable knowledge and information that could help smallholder farmers increase the productivity and quality of their crops and the ability to sell them at a good price.

### 2.5.2 Market price information and marketing

One of the most prevalent uses of ICTs by African farmers is for marketing. They use ICTs to determine current market prices and gain access to markets, to bargain and conduct transactions and to send and receive money. Having access to up-to-date market information on commodity and input prices, as well as demand trends, boosts the farmers negotiating positions and informs decisions about when and where to buy and sell, what to produce and the quantity and quality of future production. African smallholder farmers often sell their farm produce at low farm-gate prices to middlemen, even when prevailing market prices are higher, due to lack of information. For many smallholder farmers, the only source of information about the market rate for crops comes from middlemen. The lack of pricing transparency means that farmers do not always get the best deals. These low prices mean that many farmers are unable to afford farm inputs, thus resulting in low yields.

To counter this trend, ICT tools such as mobile applications, webpages, television and radio stations have been developed to provide market price information to farmers. These applications centralise market information from a number of complementary sources, including access to expert analysis of the market and comparisons of prices in different regions, and provision of historical market price trends. This helps farmers to improve marketing logistics and reduce transportation costs. It increases profitability by facilitating farmers' capacity to organise themselves into groups to leverage economies of scale. Aggregation of produce by farmer groups results in the marketing of larger volumes, which can improve access to bulk buyers and increase collective and individual profits. These ICT tools help farmers to gain greater control over their product sales by finding new sources of demand; improving product quality to meet market conditions; and cutting out intermediaries by selling directly to large-scale buyers, traders and processors.

### 2.5.3 Community groups/news sharing

For farmers to adopt new agricultural methods or other external information, it involves a process of integrated learning, participation, evaluation and validation. By being transferred horizontally, generic knowledge can be made more location-specific and can encourage self-directed, sustained learning among rural communities. ICT solutions facilitate space for community groups to share news and knowledge. In some cases, a representative in a community is elected who can share information they receive from their sources/intermediaries to the rest of a group, or else farmers can form their own individual groups where they share information among themselves. Examples include: (1) Wefarm – the largest farmer-to-farmer social media platform where farmers can learn from each other regardless of where they are located; (2) YouTube – Farmers can access postharvest information through dedicated YouTube channels for farming such as Shamba Shape Up and Farmers' TV Kenya. The video format provides a comprehensive pool of expert advice that provides viewers with an up-close, real-life perspective of various farming methods; and (3) WhatsApp and other social media – WhatsApp is a widely used ICT tool in Africa for communication between farmers and agricultural extension officers. Farmers form or join relevant WhatsApp groups and share information on agricultural practices that can improve PHM and quality of their produce. As the penetration of mobile internet in developing countries is increasing, other social media platforms such as Facebook, Twitter and YouTube are gaining popularity, especially amongst the younger generation who are starting agricultural ventures due to influence from these social media tools.

### 2.5.4 Financial services

Beyond disseminating knowledge, some ICT solutions are also helping to provide farmers access to financial services including credit, insurance, savings, as well as non-financial services like training and market access support. Farmers use financial services to purchase inputs and expand the commercialisation of their produce. Only 22% of rural populations in low-income countries have formal bank accounts. The reason for the gap in financial access is attributed to the difficulty of financial service-providers to penetrate and reach out widely to smallholder farmers in rural areas. ICT solutions close the gap and increase financial inclusion because digital finance providers can reach large numbers of people in remote areas with services that are secure, fast, cost efficient and transparent. There is immense potential growth in digital financial services to accommodate the needs of a large number of small entrepreneurs. Financial service-providers often offer incentives or conditions for farmer access to their services by providing information on good agricultural practices for preharvest, harvest and postharvest of their crops. The interaction benefits both the farmer and the financial service-provider because it increases the chances of a good return on investment.

### *2.5.5 Logistics, aggregation and supply chain*

Smallholder farmers generally own small and often dispersed pieces of land making it challenging to transport their produce to market and find sellers interested in purchasing their products. With the use of mobile technologies to manage multiple contracts, payments and receipts, agribusinesses are now able to cut down on time and transportation costs, making them more willing to work with remote farmers. In addition, ICT interventions can allow farmers to form virtual groups and help synchronise logistics of collection, aggregation and delivery by setting out organised collection routes, assembling produce of multiple smallholder farmers and delivering them to large traders.

In summary, ICT solutions are being increasingly integrated at the postharvest stage to address the challenges of stock, marketing, sale and delivery of farmers produce. ICT holds many potential benefits to smallholder farmers that include increasing productivity and quality of produce because of access to information and record keeping; access to daily market prices leading to competitive prices for crops; access to finance, microloans and crop insurance via mobile banking and digital financing; and more direct contact with buyers and procurement streams. The strategic application of ICT to the agricultural industry, the largest economic sector in most African countries, opens vast untapped opportunities for farmers, investors and entrepreneurs to improve the efficiency of food PHM and consumption. There are still many challenges that they face in scaling up and operating sustainably, including poor regional infrastructure, low literacy rates and limited internet access (Harris and Achora, 2018; Kamau et al., 2018). While some applications use basic phones with short message service (SMS), others require user-training to make the jump to more complicated smartphone applications. Smartphones are more expensive and need to be charged more frequently, requiring access to electricity, which is another major continental challenge. Development and promotion of solar charging systems could be a solution. Mobile internet also remains too expensive for low-income populations (Harris and Achora, 2018; Kamau et al., 2018).

To get a sense of the relative cost of internet connectivity, the Alliance for Affordable Internet (A4AI) - an initiative of the Web Foundation - assessed the cost of purchasing 1 GB of mobile broadband data across 59 low- and middle-income countries. They calculated affordability of 1 GB price as a percentage of income and found that while mobile broadband affordability continues to improve across the board, the cost to connect remains out-of-reach for many, especially in Africa (Kamau et al., 2018). The study revealed that across all 59 countries, the 3-year average broadband cost to connect is 5.2% of average monthly income while it is most affordable in the Asia-Pacific region, with 1 GB of mobile data available for 2.5% of average monthly income and most expensive in Africa, where 1 GB costs 9.4% of average income (Fig. 2). However, Africa also experienced the most significant cost reductions compared to any region - an average drop of 4.7% across the continent (Fig. 2).

## 3 Institutional, policy and market-related issues for PHM transformation

Agricultural investments have largely been focused on production aspects with very little attention being paid to PHM, which is equally important as a complementary strategy to sustainable crop production in increasing food and nutrition security (see the introductory section), poverty alleviation and overall economic growth of households. In developing countries, most households spend almost all their income on food, and when food is available,

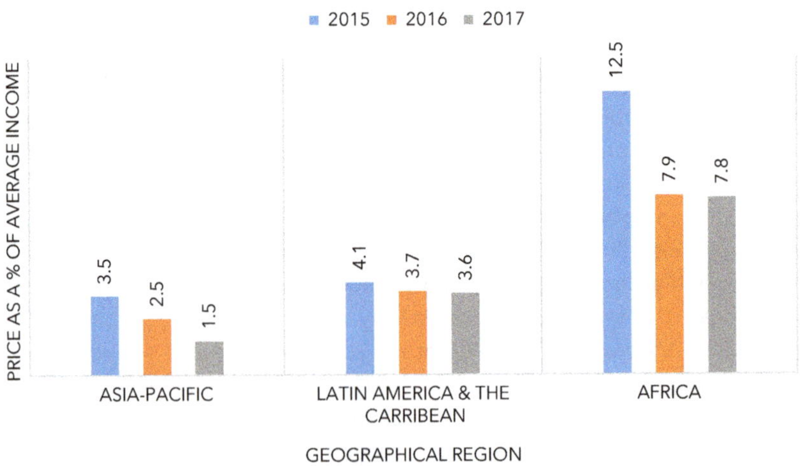


**Figure 2** Affordability of 1 GB price as % of income in developing countries, 2015–2017 (recalculated based on data from: https://a4ai.org/mobile-broadband-pricing-data/).

households can invest in other non-agricultural economic sectors (Chauvin et al., 2012). Surplus food can generate income for the household when sold while simultaneously supplying the agro-processing subsector. However, greater value and ownership of outcomes could be realised if farmers or farmer groups were directly involved in the agro-processing.

Policymakers need to review current national agricultural policies and prioritise investment in areas where they have comparative advantage including PHM, agro-processing and the emerging sectors such as organic agriculture. Some issues that require attention by policymakers are summarised as follows:

1 Policy changes to transform agriculture in developing countries should highlight and integrate PHM and agro-processing issues in design of food loss reduction programmes in the context of national/regional conditions. This has also been emphasised globally through the Bellagio Statement on PHM (Anon., 2018b) and in Africa through the Africa Union Commission Communique (AUC, 2019).
2 Detailed analyses of factors causing food losses and how addressing them can contribute towards agricultural productivity need studying. Understanding constraints that limit technology uptake is paramount if the benefits of investments are to be realised. Some of the constraints include:
   i Impact of climate change (increasing temperature and declining rainfall) on PHM.
   ii Lack of appropriate technology for farmers and entrepreneurs.
   iii Lack of appropriate extension advice on PHM.
   iv Lack of investment prioritisation by farmers on PH investments, despite the increasingly erratic rains, resulting in more frequent crop failure and need to rely on the previous harvest.
   v Unavailability of affordable credit and loan terms for farmers/entrepreneurs to invest in improved technologies and reap the benefits of grain sales after the harvest period when prices have risen substantially.
3 Demand factors: A number of issues need to be interrogated and put in place if farmers are to be able to sell their grain. These include:
   i Creation/development of markets close to producers.
   ii Having a clear pricing policy which allows farmers to plan in time.
   iii Transport bottlenecks which hinder farmers from selling their produce.
   iv Farmers need to be paid on time for their produce and preferably on the spot. Cases of delayed payments when farmers have delivered their grain to marketing agencies are common in Africa.

4 Supply factors: Technological packaging, research and extension

Innovations in technology, adoption and application are critical in driving agricultural transformation. Many brilliant PH technologies have been developed and are available in developing countries, but somehow they have not reached the end-user or farmers may be aware of them but are not using them for one reason or another. Some of the causes of this disconnect are that the technologies:

i Have not been properly domesticated or adapted to local circumstances.
ii Have not been introduced and extended in the correct way and hence have not been widely accepted.
iii Are too costly and therefore unsustainable for use in some remote areas without access to affordable finances.
iv Are too sophisticated to operate without locally available and affordable backup technical service.

A starting point for policymakers would be to conduct an inventory of locally/regionally available technologies and examine how they were introduced to learn of any adoption challenges or benefits and develop realistic and appropriate approaches to address the constraints where necessary. Some factors that constrain the adoption of new PH technologies include:

i Sociological factors such as natural inertia change to new technologies or mistrust of new technologies, for example, metal silos. Innovative approaches are required to facilitate agricultural extension agents, researchers, farmers, entrepreneurs and community leadership to work together.
ii Gender considerations are important in the design and application of technologies. An understanding of the gender roles in PHM is crucial for targeting the right users and for appropriately packaging the training interventions.
iii Where PH technologies have failed to be adopted, studies have to be conducted to determine why new technologies have failed.
iv Existing institutions which provide services to farmers (public, private and/or farmer organisations) and manage PH programmes need to be re-examined to explore opportunities for combining forces and to develop or enhance synergies.

5 Input and output markets, agricultural processing and access to finance

Farmers need the right incentives for them to invest in PHM practices that reduce PHLs. Price factors such as input and output prices and non-price factors such as access to markets, credit, technical support and so on need to be favourable. Some questions that policymakers need to answer pertaining to existing commodity/sector value chains in each country and to identify critical entry points for strengthening the value chains are provided in Box 1.

**Box 1**: Main questions for policymakers to answer (Adapted from: Chauvin et al. 2012)

1. Which existing profitable supply and value chains that have established direct links with farmers (especially smallholders and local industries)?
2. For small grains, for which value chains are weak or non-existent, what is lacking? How can linkages be established or strengthened, for example, with primary and secondary processors?
3. Where there are no direct links between producers and industries, are there marketing organisations for the crops? How has that lack of 'guaranteed markets' affected production, productivity and PHM of these crops?
4. What are the main constraints in setting up credible and effective professional organisations that can contribute to the creation of postharvest supply and value chains? Are there possibilities for public-private partnerships in PHM?
5. How available are modern postharvest inputs/technologies (i.e. synthetic or botanic pesticides, bags, bag stitches, pesticide-incorporated bags (polypropylene or plastic), hermetic bags, hermetic metal silos, 'cocoons', pheromones, (for detecting presence and monitoring of pests))? What supply and distribution channels exist for these? Is there competition between government-driven input programmes and private ones, for example, provision of metal silos for free?

The USAID's Food Security Bureau under the Feed the Future Programme, and in collaboration with country missions, has developed the Agricultural Scalability Assessment Toolkit (ASAT) for assessing the scalability of agricultural innovations with the greatest potential for both successful scaling and significantly improving food security and reducing malnutrition (Kohl and Foy, 2018; Box 2). The toolkit can be adapted for application in developing countries for identifying PHM innovations for scaling up.

The ASAT is designed to provide a qualitative appraisal of an innovation's scalability. While innovations do have intrinsic features that may make them more or less scalable in general, most of the factors affecting scaling potential can only be assessed relative to a specific socio-economic context and the characteristics of target adopters. The ASAT provides information on the strengths and weaknesses of the innovation relative to scalability, the most promising scaling-up pathways (i.e. commercial, public or public-private partnerships (PPPs)), and information on the extent to which target contexts – locations and populations – and their market and public sector capacity currently facilitate scaling.

The ASAT is not meant to be 'the' decision-making toolkit. Instead, the toolkit is intended to identify constraints to, and opportunities for, scaling. These will serve to inform decisions about whether and where to invest in the upscaling of specific innovations, or for further investment in R&D. The ASAT can also inform design efforts to improve the scalability of innovation, improve and strengthen market and public sector systems to facilitate scaling or both. The ASAT's greatest value is not the scoring per se, but the assessments behind the scoring and the conversations provoked in making and justifying an assessment. It is strongly recommended that the ASAT be applied by a team comprised of at least three people: a researcher with knowledge of the relevant sector, a markets expert and someone with considerable knowledge of the country(ies) or region(s) targeted for scaling. The ASAT approach resonates very well with the AgResults Pay-for-Results (PfR) prize model implemented in Kenya to develop, strengthen and scale market systems for postharvest storage technologies (see Section 4.2 for details).

### **Box 2**: The agricultural scalability assessment toolkit (Adapted from: Kohl and Foy, 2018)

The ASAT consists of two components, or tools: an Agriculture Scaling Decision Tree (ASDT) and an Agricultural Scalability Assessment Matrix (ASAM). A dashboard also summarises the results of the tools and provides recommendations based on that analysis. The ASDT is used to help select the appropriate upscaling pathway for an intervention (i.e., private, public or donor driven). It should be applied first, since scoring via the ASAM is contingent on the choice of pathway.

When referring to potential upscaling pathways, it is important to keep in mind that in almost all cases, the upscaling of agricultural innovations is a multi-stakeholder partnership. Even when one sector plays the predominant role in scaling, other actors (e.g., donors, private and public sectors, NGOs) are almost always involved even if minimally, such as in sanctioning and legitimising scaling. A second caveat here is that the relative importance of different sectors may change over time. In the early stages of creating demand and strengthening market institutions, a donor may play the predominant role. Once the market is established and risk is reduced, the private or public sector may then drive the scaling.

The ASDT prioritises commercial scaling as the preferred choice, along with PPPs. In other words, it prioritises any pathway that is likely to achieve something approaching the potential scale, retains impact, and is financially and institutionally sustainable. This favours pathways that primarily leverage domestic 'resources' in the broadest sense of the word (i.e. pathways that are profitable for adopters and value chain actors, and/or where there are public fiscal support and institutional capacity already available). For a largely commercially driven scaling pathway to be viable, one or more upstream or downstream private sector actor must have the means (e.g. financial resources, organisational infrastructure technical knowledge) and incentives (e.g. profitability or 'the business case', taking into account risk) to scale the innovation. Upstream actors are importers, producers or distributors of the innovation, who assess that they could make enough money off the innovation at scale to justify assuming the costs of scale. Downstream actors are usually processors, buyers and exporters.

The preference for private sector scaling is based on two assumptions. First, in many developing countries - especially low-income and the least developed countries - the public sector lacks the resources, incentives and ability to scale, let alone to do so sustainably. Second, donors have neither the mandate nor resources to support scaling beyond five to ten years, let alone sustainably on an ongoing basis. If scaling is profitable for private sector actors, in contrast, it is likely to be sustainable and to eventually achieve something approaching maximum potential scale.

The ASDT considers four primary tasks involved in scaling to determine the preferred pathway. These are the resources, incentives and ability to (1) produce and distribute the innovation (upstream activities); (2) drive scaling overall; (3) create demand among target or potential adopters; and (4) provide training and technical support to adopters to be able to implement and apply the innovation correctly.

The ASAT can inform decisions regarding upscaling innovation at each stage of research, testing, piloting, planning and implementation. The tool can be used to integrate upscaling considerations; assess progress; decide whether upscaling makes sense; and modify current approaches to innovative design, test and scale, based on new evidence. A partial list of other uses includes (1) identifying areas where scalability of innovation could be improved so they can be included in the objectives of phase II or phase III research; (2) creating the foundation for a scaling strategy for a new procurement; (3) designing a scaling strategy prior to implementation; and (4) monitoring the scaling process during implementation.

---

Verma et al. (2019) provided a conceptual framework to identify macro/structural factors responsible for food loss at various supply chain stages. They found that poor institutional infrastructure appears to be as prevalent a reason behind food loss as bad physical and financial infrastructures. This suggests that while technology-based solutions to food loss are important, the role of institutions deserves equal importance if not more attention. The framework further brings to the fore the complex nature of the interconnected reasons underlying food loss in supply chains and emphasises the need to see food loss as a systemic outcome.

The main advantages of the Verma et al. (2019) approach are that:

- It can quickly help to identify problem areas in the supply chain without the need to gather data on the whole supply chain of a product in any region. Once identified, the points of possible intervention should be explored further to devise the right solution.
- This manner of looking for solutions using a systems view lowers the risk of ineffective solutions and unintended negative consequences of the proposed intervention.
- While food loss is hard to measure, data on indicators listed earlier are more easily available and comparable across regions.
- A solution towards bettering a given indicator affects not one but multiple causes associated with that indicator, whether in the same or in different stages of the supply chain; therefore, this approach can be used to identify solutions with most potential. Similarly, a given cause can be perpetuated due to bad performance on multiple indicators. Therefore, a single intervention aiming a single indicator/category would often fail to achieve its full potential in the presence of other bottlenecks in the system.

The authors concluded that their framework can be a useful tool to identify socio-economic indicators that can signal the presence of food loss in supply chain stages. Furthermore, the framework can be used for linking practices in a given supply chain stage with the respective underlying causes that appear across multiple supply chain stages. However, examples of the application of this approach are still scarce.

In the face of numerous challenges faced by farmers in increasing crop productivity, for example, climate change, gender disparity in agriculture (societal barriers that hamper activities of women farmers), declining soil fertility and land degradation, rural-urban migration by the youths, HIV/AIDS, resistant pests and so on, Gaffney et al. (2019) contend that the best - and possibly the only - way to bridge that gap is to expand global investment in agricultural R&D, especially through collaborative efforts such as public-private partnerships (PPPs). They argue that solutions conceived from public-private collaborations could provide effective advances in the pursuit of answers to global agricultural challenges and can result in systemic change. PHM is not an exception in these partnership arrangements. Greater involvement in PPPs by agribusinesses could help bridge the innovation gap in developing countries where public investment in agricultural research is a low priority and helps drive more agricultural research where it has been historically strong. The PPP model has been successful for agriculture, helping to deliver productivity- and sustainability-improving technologies developed by agribusiness, which in turn helps to improve their revenue-generating products into the hands of small-scale farmers in low-income countries. More still needs to be done, though, to ensure affordability of, and access to, modern postharvest technologies by smallholder farmers. Greater focus is needed on PHM improvements for relatively marginalised crops like pearl millet, finger millet, sorghum, teff and many legume crops such as cowpeas.

Gaffney et al. (2019) called for greater resilience, perseverance and patience from development partners to support research through to the end, in order to deliver valuable products to farmers and consumers. Private sector are a critical player since delivering products is the core of maintaining a viable business. Science-based, objective and harmonised regulatory policies are required to expedite access to improved postharvest technologies by farmers and consumers.

In the Rift Valley of Ethiopia, Abebe et al. (2016) collected data from 345 farmers to investigate the determinants of the choice of farmers for trading through middlemen and to analyse the effect of choosing a middleman or a wholesaler as their trading partner on the welfare of farmers. Although focused on the potato value chain, the concepts may apply to other agricultural crops in developing countries. Farmers selling directly to wholesalers realised a gross profit of 225% more than those selling through middlemen. Such farmers had access to better quality inputs, better contract specifications and received higher prices for their products. What was puzzling was that despite receiving lower prices, the majority of farmers continued to trade via middlemen. The researchers postulated four reasons to explain these findings: (1) wholesalers seem to prefer to work with middlemen to guarantee critical quantity; (2) the presence of any form of quality uncertainty in trade may encourage both wholesalers and farmers to use middlemen to reduce the cost of measuring quality. Middlemen, living in the same community with farmers, incur lower cost in monitoring farmers and/or assessing quality than wholesalers who are external to the community. The wholesalers would need to spend a significant amount of time inspecting the produce delivered by each farmer; (3) trading via middlemen can enhance smallholder commercialisation by linking low resource-endowed farmers to traders and final markets. Normally, these farmers are more likely to be excluded from the relatively high value market channels as transaction cost per farmer would be too high for wholesalers to buy directly from such farmers; (4) personalised relationships might attract smallholders into trading through middlemen regardless of income losses. Social ties are reportedly weak between farmers and wholesalers, who directly buy from these farmers. However, it was observed that the majority of farmers using the middlemen route had strong ethnic and/or religious ties with those middlemen, and these ties are essential in building trust in business in rural economies. What might then be required is to capacitate and organise these farmers so that they are more empowered to bargain with the middlemen since the latter play a vital role in the agricultural value chains. An even better strategy would be for farmers to form their own cooperatives to aggregate their grain and negotiate with buyers - whether middlemen or wholesalers - for fairer prices.

# 4 Creating sustainable markets: case study models

## 4.1 Development of the hermetic bag market in Central and West Africa

Upscaling extension efforts is a prerequisite for building a strong customer base for developing markets for new agricultural technologies. Extension services play a major role in building markets for agricultural products and are therefore a critical integral component in an intervention such as PHM for loss reduction.

Market development requires awareness building among potential users. The scope and intensity of awareness building are important because they determine the size of the market. Increasing the number of trained farmers will likely increase the adoption of technology. Large-scale extension efforts can help to create demand for the technology and provide incentives to the private sector to tap into a market opportunity.

In a major initiative to develop a sustainable market for the Purdue Improved Crop Storage (PICS) hermetic bag in Central and West Africa for farmers between 2007 and 2012, a double-pronged approach was adopted: (i) large-scale extension activities to build awareness by training farmers how to use the technology, thereby creating demand for it (Table 5) and (ii) supply chain development efforts to support local manufacturing and distribution of the PICS bags to make the technology available to farmers in rural areas (Baributsa et al., 2014). Village activities were implemented by field technicians and included (i) awareness building just before or during cowpea harvest, (ii) demonstrations right after harvest, (iii) follow-up visits during the storage period and (iv) open-the-bag ceremonies at the end of the storage period to witness the results of the 'trial'. The large-scale initiative reached about 2 million people through the demonstrations, and more than 2.4 million triple bags were sold in the first 5 years of the project.

Baributsa et al. (2014) derived the following key lessons from implementing the project which could apply to any other PHM technology in developing countries:

**Table 5** Participation in different village activities during the implementation of the PICS project in Central and West Africa from 2007 to 2012

| Country | Number of trained field technicians | Number of participants to different village activities | | |
|---|---|---|---|---|
| | | Awareness building | Demonstrations | Open-the-bag ceremonies |
| Niger | 348 | 217 748 | 259 203 | 208 792 |
| Burkina Faso | 178 | 52 933 | 127 116 | 212 873 |
| Nigeria | 1095 | 604 215 | 717 082 | 881 306 |
| Benin | 136 | 46 251 | 33 741 | 32 233 |
| Togo | 131 | 52 450 | 40 370 | 43 183 |
| Mali | 184 | 104 361 | 95 087 | 117 713 |
| Cameroon | 138 | 70 099 | 61 105 | 143 636 |
| Ghana | 187 | 132 626 | 113 054 | 120 033 |
| Senegal | 168 | 136 079 | 113 957 | 139 402 |
| Chad | 61 | 83 103 | 62 593 | 78 850 |
| Total | 2626[a] | 1 499 865 | 1 623 308 | 1 978 021 |

[a]Total field technicians has 26% women: includes extension agents, social workers, teachers and so on.
Source: Data provided by partners who implemented extension activities in West and Central Africa; Baributsa et al. (2014).

1 Financial support from public or private sector is needed to build the market for a new agricultural input or technology. Market building is an expensive endeavour and requires financial resources. Initial support is essential to demonstrate to farmers and other value chain participants that the technology indeed is effective. Providing incentives during the introduction of the technology (e.g. financing initial manufacturing of the technology or guaranteeing purchases during the first year) is very important. Private entrepreneurs are risk-averse and will not absorb the whole risk for a new input or technology if there is no existing positive track record.
2 Direct contact and communication with farmers is essential by means of demonstrations and rigorous media campaigns (including print, audio and audiovisual messages via radio and television messages, print posters and flyers and cell phone videos). Diversifying awareness building strategies is important for reaching more potential clientele. Awareness building is needed at all levels of the value chain including policymakers and implementers, manufacturers, vendors, traders, farmers, processors and consumers.
3 Market building activities for a storage technology should focus on geographical areas where grain production is significant and important such that farmers are likely to store. Market building activities need to

be large enough to stimulate a substantial demand for the technology to incentivise private sector investment in the business. Small-scale extension activities do not build large enough demand to sustain market development of a new product on their own.

4 Supply chain development is key for increasing the use of a new technology by farmers. Increasing the density of retail networks and diversifying sales strategies would improve the availability of the technology to farmers in rural markets. Facilitating meetings and exchanges among the supply chain participants helps build trust and collaboration. In countries that lack well-developed agricultural input retail networks, it takes longer (more than 2 years) to develop and strengthen the retail distribution system.

## *4.2 AgResults initiative*

AgResults, a US$147 million multilateral collaborative initiative between the governments of Australia, Canada, the United Kingdom, the United States, the Bill and Melinda Gates Foundation and the World Bank Group, used prize competitions to incentivise the private sector to overcome deep-rooted market failures or barriers in agricultural systems and develop appropriate solutions to enhance food security. Under the model, these competitions encouraged grantees to achieve predetermined result thresholds. Different projects were implemented in different countries including Kenya, Nigeria, Tanzania, Uganda, Vietnam and Zambia. One of the projects was the Kenya On-Farm Grain Storage Challenge Project, a 4-year US$6.25 million Pay-for-Results (PfR) project, launched in May 2015. The project was designed to incentivise private sector companies to develop, market and sell new or redesigned on-farm storage devices to smallholder farmers in Kenya.

The objectives of AgResults were to:

1 Overcome agricultural market failures by offering results-based monetary prizes to competing private actors for the development and/or upscaling of new agricultural technologies.
2 Test the effectiveness and efficiency of PfR financing in comparison with traditional approaches to the promotion and adoption of innovative agricultural technologies.

The AgResults' theory of change was premised on the idea that, if appropriately incentivised, the private sector can respond by creating and/or scaling new technologies to benefit smallholder farmers. The project's prize incentive encouraged private sector actors to develop distribution networks and marketing strategies to promote their improved on-farm storage products to farmers, who through increased familiarity, then invested in improved storage. Increased use of on-farm storage devices would then lead farmers to store more grain with fewer losses, reducing purchases of grain and improving household incomes and food security (Anon. n.d.). Theory-based evaluations of Abt Associates (2017) hypothesised that 'pull' mechanisms can have inherent trade-offs between development impact - such as inclusion of smallholders and marginalised groups - and creation of a sustainable market for the technology.

AgResults' key innovation is the 'market-pull mechanism', which uses cash prizes to encourage private sector investment in creating and marketing technologies that benefit smallholder farmers. The AgResults approach contrasts with the traditional 'technology-push' efforts - such as demonstrations, training and subsidised distribution - which have not had much impact at scale.

According to Anon. (n.d.), PfR prizes or 'pull' mechanisms have several advantages over traditional 'push' funding, including:

- Heightened awareness about global problems. Prizes can educate, inspire and even mobilise the public, focusing attention on otherwise minimised issues.
- Results-based funding. Prizes are paid only if and after the predefined results are achieved, which minimises the risk of loss of that funding.
- Process agnostic. Funders of the initiative do not need to predict which actions will yield the desired result, as prizes are awarded based solely on the achievement of and not the processes or steps to obtain results.

'The prize is a tool to strategically partner with commercial actors already in the (market) system and facilitates strengthening of relationships (with smallholder farmers)' (Ms Aviva Kutnick, Private Sector Engagement, Division Chief in USAID's Bureau for Food Security and AgResults Steering Committee Chair).

There are indications that in Kenya, AgResults boosted sales of on-farm storage devices, which reduced smallholder farmers' postharvest grain losses (AgResults, 2019). The pilot offered private companies cash prizes for successfully developing a product attractive to smallholders and getting the farmers to buy it. Prizes were paid based on actual sales of improved on-farm storage devices to smallholders. During the sales period from 1 May 2015 to 1 May 2018, competitors sold 1 390 777 improved storage devices, generating 413 265 MT of improved storage capacity for smallholder households. The successful companies received a total of US$6.25 million in incentives. Details of the competitors and the respective storage devices that they distributed to smallholder farmers are presented in Table 6. The competitors used a variety of distribution channels to reach out to as many smallholder farmers as possible (see Table 7).

**Table 6** AgResults competitor profiles and on-farm grain storage devices distributed to smallholder farmers in Kenya

| Competitors | Profile | Storage devices |
|---|---|---|
| African Farms and Markets Ltd. (AFMA)/ GrainPro | AFMA is an agribusiness company that works with smallholder farmers in Kenya to increase income by improving their capacity to trade. To participate in the project, AFMA entered into a consortium agreement with GrainPro where AFMA would promote, sell and distribute GrainPro's storage solutions in the targeted regions. | GrainPro SuperGrainBag FarmTM 90 kg<br>GrainPro SuperGrainBag IV-RTM 90 kg<br>GrainPro GrainSafe MiniTM 250 kg |
| A to Z Textile Mills Ltd. | A to Z Textile Mills, a Tanzanian company established in 1977, manufactures and sells plastic products (hermetic bags, PP woven bags, plastics, bed nets, garments, mineral water, etc.). A to Z joined the competition to expand on-farm storage sales into Kenya, selling two types of their AgroZ Bag, a multi-layered hermetic bag. | AgroZ Bag 100 kg<br>AgroZ Bag 50 kg |
| Bell Industries | Bell Industries is an agribusiness, health products and solutions company. Since 1992, the company has sold agriculture inputs, fertilisers, herbicides and farming machinery to help smallholder farmers in the eastern region increase yields and profits. Bell expanded operations to the Rift Valley to participate in the project. | PICS Bag 100 kg<br>PICS Bag 50 kg |
| Forms Ltd. (CBF)/ Vestergaard | CBF Limited is a Kenyan supplier and distributor of Vestergaard Products. These two companies partnered to participate in the project and sell the ZeroFly Storage Bag, an insecticide-incorporated woven polypropylene bag developed for postharvest storage for smallholder farmers. | ZeroFly Hermetic Storage Bag 100 kg |
| Ekima Engineering Works | Ekima Engineering Works, founded in 2004 and based in Kenya's eastern region, trains local artisans to design, fabricate and market metal silos in Ethiopia, Kenya, Rwanda, Somalia, South Sudan and Uganda. Ekima produces its custom-sized metal silos for smallholder farmers at a specialised workshop in Machakos. | Metal Silo (Nine sizes ranging from 50–540 kg) |
| Elite Innovations Ltd. | Elite Innovations was founded in 2013 in Eldoret, Kenya, as a social enterprise to promote improved grain storage and mitigate postharvest losses. In 2015, Elite was purchased and became a commercial business. Elite joined the project to sell its hermetic Elite Bags to smallholder farmers. | Elite Bag Heavy Duty 100 kg<br>Elite Bag Light Duty 90 kg<br>Elite Bag Light Duty Double Layer 90 kg |

**Table 6** (*Continued*)

| Competitors | Profile | Storage devices |
|---|---|---|
| Kentainers Ltd. | Kentainers is a leading manufacturer of polyethylene roto moulded products in Eastern and Central Africa to improve water management, food production and storage for rural and urban consumers. To participate in the competition, they adapted their plastic water tanks and developed a plastic GrainSilo. | GrainSilo 380 kg |
| Kenya Promotion and Marketing Company (KPMC) | KPMC is a value chain service-provider that collaborates with private and public sector players to address smallholder farmer needs. The company sells their Grain Storage Bag (GSB), a multilayer polyethylene bag, across East Africa. Prior to the project, KPMC sold approximately 80 000 bags through Ministry of Agriculture programmes. | KPMC GSBL 90 kg<br>KPMC GSBL 60 kg |
| Post Harvest Africa | Post Harvest Africa is a Kenyan private enterprise that targets smallholder farmers across Africa. They sell Grainz bags, hermetic grain storage devices that require no pesticides, so food is free of chemicals and insects. | Grainz bag 100 kg |

*Source*: AgResults (2019).

Although these competitors used different channels (see Table 7), one common denominator was that they were willing to dedicate skills and resources to adapt their business models; investing in distribution networks and targeted marketing to increase sales to the target smallholder farmer population. This willingness to make drastic operational changes and ability to manage risk set the winning participants (Table 8) a cut above the rest and led them to successfully outcompete others in the prize competition. The tactics used by each of these three competitors provide important lessons to those companies which may wish to participate in similar projects in the future (Table 8). Other key lessons learnt from the project design of AgResults through to implementation can be very useful in designing and implementing similar interventions elsewhere in developing countries (see Fig. 3) and can cut the learning curve. One of the major highlights is that only competitors with access to finance were able to reach the threshold within the stipulated period.

The AgResults prize competition model demonstrated that stimulating private sector investment can contribute immensely towards availing of appropriate postharvest technologies to smallholder farmers in developing countries. The big challenge now is to implement similar models elsewhere

**Table 7** AgResults competitor distribution channels for grain storage devices in Kenya

| Distribution channel | Channel overview |
|---|---|
| Distributors | Large aggregators of various agro-inputs in the big urban centres sell devices to downstream smaller distributors (agro-dealers or agro-vets) in rural areas, and sometimes to the farmer. Distributors often require high-profit margins from implementers, up to 25% |
| Direct sales to farmers | This entails bypassing all the distribution chain intermediaries and selling directly to the end-user. Direct selling is used by start-ups and smaller companies like Elite; however, for bigger companies like Bell, this selling method is rarely cost-effective or feasible |
| Agro-dealers | Agro-dealers, or agro-vets, are smaller distributors spread across the countries. Agro-dealers are the last point of sale as they are closest to the farmer. There are over 10 000 agro-dealers in Kenya |
| Sales representatives/field staff | These are implementer-hired staff who act as proxy 'distributors'. They mainly liaise with agro-dealers to assess demand and stock levels in the field and ensure agro-dealers are amply supplied with devices. In addition, they carry out farmer and agro-dealer sensitisation and trainings |
| Commercial villages | One of the competing companies, African Farms and Markets Ltd. (AFMA) uses a 'commercial villages' model developed and prescribed by the parent company, Farm Concern International |

*Source*: AgResults (2019).

in developing countries with different contexts to test whether the models would still work. Additionally, it would be key to revisit the countries where the AgResults model was implemented to evaluate the model again say 5 years down the line to determine the sustainability and expansion of the intervention without further external input. Did the companies involved continue to invest and scale the storage innovations beyond the project areas and are smallholder farmers responding positively and continuing to purchase the products? This would be the 'litmus' test of the quality of the model!

## 5 Conclusion and future trends

Moving smallholder farmers from a subsistence or household food security goal to a greater level of market orientation requires major changes in terms of how farmers and their communities perceive global developments, interact with each other and with other actors beyond their communities. These changes mean that smallholder farmers must take on new skills to:

**Table 8** Tactics used by successful competitors to reach threshold sales of grain storage devices in Kenya

| Competitor | Tactics used to reach sales threshold |
|---|---|
| A to Z Textile Mills Ltd. | Investments: Purchased a new hermetic bag production line in Arusha to compete in the project. Heavily invested in marketing and distribution (launch, radio ads, vehicle purchases for agronomists, etc.)<br>Distribution: Initially focused on direct sales to smallholder farmers but then started using agro-dealer networks and sales agronomists to expand reach.<br>Marketing and sales: Established over 400 agro-dealer linkages in 36 counties since October 2015. Employs 11 sales agronomists equipped with vehicles. |
| Bell Industries | Investments: Expanded manufacturing capacity by opening a new production facility to increase their annual production<br>Distribution: Initially used big distributors but expanded to smaller distributors to boost sales, increasing distribution outlets from 30 to 144 in 2016.<br>Marketing and sales: Used mass media, mainly radio/TV ads, and expanded coverage from three to eight sales representatives<br>Other: Was an early market entrant (pre-AgResults) and relied on brand name recognition associated with PICS bags. |
| Elite Innovations Ltd. | Investments: Attracted start-up investments to expand their business and participate in the project.<br>Distribution: Initially sold directly to farmers and hired three sales representatives and sales agents on commission who sold directly to farmers. Incorporated agro-dealers to work with sales representatives, growing from 3 agro-dealers in 2015 to more than 1000 in 2018<br>Marketing and sales: Attended trade fairs and organised group meetings to reach smallholder farmers directly.<br>Other: Initially operated as a community-based empowerment group in North Rift Valley and became a commercial business in 2015. |

*Source*: AgResults (2019)

- Identify the right product for the right buyer at the right quality, price and time.
- Establish commodity production systems that make efficient and sustainable use of existing financial, human and natural resources.
- Incorporate postharvest handling and processing techniques so that farmers' products meet buyers' requirements while farmers get more value for their investments.
- Improve business and marketing organisational schemes that increase competitiveness by reducing costs and expanding marketable volumes.
- Strengthen relations and market links among market chain actors.
- Improve access to flows of both market-based information and new production and postharvest handling technologies and financial services.

**Figure 3** Lessons learnt in implementing the AgResults project in Kenya (Source: https://agresults.org/projects/kenya).

This transition for the majority of farmers, especially those located in more marginal/remote areas and distant from markets, requires reorientation and mentoring from development organisations, access to a range of public and private services (inputs, land, transport, information, finance, etc.) and a set of government policies that provide an enabling environment that supports business in remote locations.

The International Center for Tropical Agriculture (CIAT) proposed a territorial or geographical approach (Lundy et al., 2007) that seeks to contribute towards the development of local capacities to facilitate rural enterprise development in a flexible, dynamic and coordinated manner. The approach includes the following components:

1 The identification and strengthening of an interest or working group composed of diverse local organisations with common goal and strategies for rural enterprise development.
2 Identification, management and development of market opportunities available to the region.
3 Participatory production to market chain analysis, consensus building with diverse actors along the chain and design of a shared strategy to increase chain competitiveness.
4 The implementation of selected options that were identified and negotiated by the actors in the previous step.
5 Identification and promotion of appropriate and sustainable business development services and markets for these services for the region.

The entry point for this approach is the identification and consolidation of the local interest group. The other intervention components are then subsequently developed in collaboration with that group. During the process, a number of policy - or enabling environment-related issues - will be identified, and the local interest group needs to progressively incorporate these into their strategy and plan of action for their geographical location.

Heavy emphasis should be placed on the process of enterprise development. The process is important for two reasons:

1 It helps to define the situation as it is now and the desired situation in the future.
2 It makes it possible to identify possible steps that need to be taken to achieve the desired situation.

An example of this process is the territorial approach to rural agro-enterprise development, which involves a sequence of steps including:

- Interest group formation and characterisation of a geographical area territory.
- Organisation of farmers.
- Identification of markets and chain analysis.
- Development of a strategy and business plan with chain actors.
- Strengthening of support services for selected chains.
- Advocating for policy changes.

Trust-building processes are complex; they take time and require facilitation, championing and front-end financing (Mvumi et al., 2008). This has a number of implications. Development practitioners - both government and NGOs - need to more clearly define their roles: Are they facilitators of development, or are they service-providers? The natural tendency has been for them to be service-providers, who have, perhaps wittingly or unwittingly, distorted market mechanisms by substituting themselves for potential private service-providers. For most institutions, this paradigm shift is not easy and requires buy-in and support from senior management. It demands changes in attitude and behaviour and learning of new skills.

Moving towards a market and enterprise orientation means that the skills of front-line facilitators have to be upgraded and complemented. An increasing minority have the required skills but to have real meaningful impact on the magnitude and plethora of the challenges in developing countries, massive investment is required in capacity development. Understanding market and business principles are critical. Skills are required for identifying stakeholders, understanding their motivations and facilitating interactions among them. In many cases, development institutions will need to build in-house capacity in these areas, either through retraining, retooling or hiring personnel with appropriate backgrounds and hands-on skills to mentor and support high-level facilitators. Capacity building is not a one-off event but rather a process that requires iterations of learning; field-level practice; analysis of what has been addressed; and further learning, practice and analysis cycles (Remington et al., 2005).

Champions of supply chains are likely to be private sector actors from within the supply chains themselves or from associated services. Their emergence may be fortuitous and spontaneous, or it may be that in existing market situations, champions may be 'latent' and require external facilitators to activate them or catalyse them to play the role for which they were destined. There are a number of questions that come to mind:

1 How can potential champions be identified early in the agro-enterprise process and the champions be facilitated and supported in their role(s)?
2 Is the existence of a supply chain champion a precondition for the sustainability of an intervention once project funds run out?
3 How to determine whether levels of linkage and trust are at a sufficiently high level such that external facilitation is no longer required?

Further research is required to provide guidance to field facilitators on how to mobilise different actors in support of improved PH system performance.

There is a natural tendency to confine linkages and trust to actors within a supply chain. However, it is evident that the concept of linkages and trust also applies to relations among different stakeholders at different stages of the agro-enterprise development process. Highlights of some of the key linkages include:

- Among facilitators and service-providers at a geographical location.
- Within and among farmers' groups.

- Between the public and private service-providers and the farming community.
- Between R&D institutions and the private sector.
- Between civil society organisations, government policymakers and donors, leading to the connection between micro- and meso-level interventions and the macro- or policy-related issues that are likely to favour or hinder success at grass root level.

The discourse in the R&D community is that many improved technologies and work practices are effective on a small scale but have had limited systemic uptake. In Kenya, the AgResults project demonstrated that a prize competition model can be used to leverage private sector innovation and investment and thus provide a sustainable solution to scaling innovative postharvest technologies. The AgResults work showed how a prize competition creates a potential road map for scale using short-term incentives and opportunities for replicability. The model is worth testing in different geographical areas, commodity value chains and contexts.

## 6 Where to look for further information

### *6.1 Further reading*

AFAAS - African Forum for Agricultural Advisory Services - https://afaas-africa.org/postharvest-management-in-sub-saharan-africa-phm-ssa-project/.

APHLIS - https://aphlis.net/en#/.

FANRPAN - Food Agricultural & Natural Resources Policy Analysis Network - www.fanrpan.org/archive/projects/postharvest/sdc-phm/.

FAO - http://fao.org/ethiopia/news/detail-events/en/c/1171248/.

FAO - Community of Practice on Food Loss Reduction www.fao.org/food-loss-reduction/news/detail/en/c/1043703/.

FAO - The State of Food and Agriculture 2019. Moving forward on food loss and waste reduction. http://www.fao.org/3/ca6030en/ca6030en.pdf (2019b).

FAO - The State of Food Security and Nutrition in the World 2019. Safeguarding against economic slowdowns and downturns. - http://www.fao.org/3/ca5162en/ca5162en.pdf.

FAO - SAVE FOOD - http://fao.org/food-loss-reduction/resources/.

GIZ/GTZ - https://giz.de/projektdaten/index.action?request_locale=en_GB.

Ceres2030 - https://ceres2030.org/reducing-food-loss/.

HELVETAS Swiss Intercooperation - Bellagio Statement on Postharvest Management. https://www.helvetas.org/Publications-PDFs/bellagio_statement_postharvest_management_1.pdf.

Africa Union Commission (AUC) http://postharvest.org/Communique_of_the_2nd_All_Africa_Postharvest_Congress_and_Exhibition.pdf.
The Postharvest Education Foundation. www.postharvest.org.
International Fund for Agricultural Development (IFAD). The Food Loss Reduction Advantage: Building sustainable food systems https://www.ifad.org/documents/38714170/41335091/food_loss_advantage.pdf/373e787f-fc6f-0426-eea0-d730f79d38c3.
International Development Research Centre (IDRC) – https://idrc.ca/en/resources/publications.
Rockefeller Foundation – https://rockefellerfoundation.org/YieldWise.
WEB FOUNDATION – https://webfoundation.org/2017/11/what-does-it-cost-to-purchase-broadband-data-in-developing-countries/.
WFP – Training Manual for Improving Grain Postharvest Handling and Storage – https://wfp.org/content/p4p-training-manual-improving-grain-postharvest-handling-and-storage.

## 6.2 Key journals and conferences

### 6.2.1 Journals

*Journal of Stored Products and Postharvest Research.*
*Journal of Stored Products Research.*
*Crop Protection.*
*Journal of Agricultural Education and Extension.*
*Journal of Applied Entomology.*
*Journal of Pest Science.*
*International Journal of Tropical Insect Science.*
*International Journal of Pest Management.*
*African Crop Science Journal.*
*Journal of Economic Entomology.*
*Pest Management Science.*
*Insects.*
*African Entomology.*
*Food Security.*
*Food Policy.*
*World Mycotoxin Journal.*
*Sustainability.*

### 6.2.2 Conferences

Agricultural and Biosystems Engineering Conferences.
African Crop Science Conference.

Conference on Agricultural Extension & Food Security in Africa.
African Forum for Agricultural Advisory Services (AFAAS) Africa-wide Agricultural Extension Week.
International Congress of Entomology.
International Working Conference on Stored Product Protection.
IOBC-WPRS Working Group on the Integrated Protection of Stored Products.
International Conference on Controlled Atmosphere and Fumigation in Stored Products.
All Africa Postharvest Congress and Exhibition.

## 7 References

Abebe, G. K., Bijman, J. and Royer, A. 2016. Are middlemen facilitators or barriers to improve smallholders' welfare in rural economies? Empirical evidence from Ethiopia. *Journal of Rural Studies* 43, 203–13. doi:10.1016/j.jrurstud.2015.12.004.

Abt associates. 2017. Evaluations: money motivates private sector to adopt best practices - if used appropriately. Available at: https://www.abtassociates.com/who-we-are/news/feature-stories/evaluations-money-motivates-private-sector-to-adopt-best-practices (Accessed on 20 July 2019).

Adda, C., Borgemeister, C., Biliwa, A., Meikle, W. G., Markham, R. H. and Poehling, H. -M. 2002. Integrated pest management in post-harvest maize: a case study from the Republic of Togo (West Africa). *Agriculture, Ecosystems and Environment* 93(1–3), 305–21. doi:10.1016/S0167-8809(01)00344-9.

AgResults. 2019. AgResults Kenya On-Farm Storage Challenge Project: final report. Available at: https://agresults.org/learning/45-final-report-kenya-on-farm-storage-challenge-project/file (Accessed on 16 November 2019).

Anon. n.d. Using prize competitions to create sustainable agricultural markets. Available at: https://agresults.org/our-approach (Accessed on 30 July 2019).

Anon. 2018a. The different types of knowledge. Available at: http://www.knowledge-management-tools.net/different-types-of-knowledge.html (Accessed on 26/05/2020).

Anon. 2018b. Bellagio statement on postharvest management. HELVETAS Swiss intercooperation. Available at: https://www.helvetas.org/Publications-PDFs/bellagio_statement_postharvest_management_1.pdf (Accessed on 10/05/2020).

AUC 2019. *Communique of the 2nd All Africa Postharvest Congress and Exhibition. African Union Commission, Addis Ababa, Ethiopia. 20th September, 2019.* (Accessed 10/05/2020).

Azziz-Baumgartner, E., Lindblade, K., Gieseker, K., Rogers, H. S., Kieszak, S., Njapau, H., Schleicher, R., McCoy, L. F., Misore, A., DeCock, K., Rubin, C., Laurence Slutsker, L. and the Aflatoxin Investigative Group 2005. Case-control study of an acute aflatoxicosis outbreak, Kenya, 2004. *Environmental Health Perspectives* 113(12), 1779–83. doi:10.1289/ehp.8384.

Bandyopadhyay, R., Orteg-Beltran, A., Schulthess, F., Meikle, W., Setamou, M. and Cotty, P. J. 2019. Identifying and managing plant health risks for key African crops: maize. In: Neuenshwander, P. and Tamò, M. (Eds), *Critical Issues in Plant Health: 50 Years of Research in African Agriculture*. Burleigh Dodds Science Publishing Limited, Cambridge, UK, pp. 173–212.

Baributsa, D., Abdoulaye, T., Lowenberg-DeBoer, J., Dabiré, C., Bokar Moussa, B., Coulibaly, O. and Baoua, I. 2014. Market building for post-harvest technology through large-scale extension efforts. *Journal of Stored Products Research* 58, 59–66. doi:10.1016/j.jspr.2014.02.012.

Bell, C. H. 2000. Fumigation in the 21st century. *Crop Protection* 19(8–10), 563-9. doi:10.1016/S0261-2194(00)00073-9.

Best, R., Ferris, S. and Schiavoe, A. 2005. Building linkages and enhancing trust between small-scale rural producers, buyers in growing markets and suppliers of critical inputs. In: Almond, P. R. and Hainsworth, S. D. (Eds), *Beyond Agriculture-Making Markets Work for the Poor*. Proceedings of the International Seminar 28 February - 01 March 2005, Westminster, London, UK. Crop Post-Harvest Programme (CPHP), Natural Resources International Limited Aylesford and Practical Action, Bourton on Dunsmore, UK, 176p.

Borgemeister, C., Adda, C., Sètamou, M., Hell, K., Djomanou, B., Markham, R. H. and Cardwell, K. F. 1998. Timing of harvest in maize: effects on post harvest losses due to insects and fungi in central Benin, with particular reference to *Prostephanus truncatus* (Horn) (Coleoptera: Bostrichidae). *Agriculture, Ecosystems and Environment* 69(3), 233–42. doi:10.1016/S0167-8809(98)00109-1.

Brice, J. and Golob, P. 1999. Fumigation of grain in farmer-level mud stores and metal grain tanks using phosphine. In: Jin, Z., Liang, Q., Liang, Y. and Guan, L. (Eds). Proceedings of the 7th International Working Conference on Stored -Product Protection, 14–19 October 1998. Sichuan Publishing House of Science and Technology, Beijing, China, Chengdu, China.

Buguzi, S. 2016. Tanzania: food poisoning linked to 14 deaths in two regions. *The Citizen Newspaper*. Nation Media Group, Dar es Salaam, Tanzania.

Chauvin, N. P., Mulangu, F. and Porto, G. 2012. Food production and consumption trends in Sub-Saharan Africa: prospects for the transformation of the agricultural sector. United Nations Development Programme Working Paper 2012-2011: February 2012. Regional Bureau for Africa, New York, NY. Available at: https://www.undp.org/content/dam/rba/docs/Working%20Papers/Food%20Production%20and%20Consumption.pdf (Accessed on 20 April 2019).

Dembedza, M. P., Chidewe, C., Benhura, M. A., Mvumi, B. M., Manema, L. R. and Nyanga, L. K. 2019. Effectiveness of hermetic technology in limiting aflatoxin exposure in women and children from smallholder farming areas, Zimbabwe. *World Mycotoxin Journal* 12(3), 233–43. doi:10.3920/WMJ2019.2435.

Donatti, C. I., Harvey, C. A., Martinez-Rodriguez, M. R., Vignola, R. and Rodriguez, C. M. 2019. Vulnerability of smallholder farmers to climate change in Central America and Mexico: current knowledge and research gaps 11, 264–86. *Climate and Development*. doi:10.1080/17565529.2018.1442796.

Duncan, K. E. and Howard, R. J. 2010. Biology of maize kernel infection by *Fusarium verticillioides*. *Molecular Plant-Microbe Interactions* 23(1), 6–16. doi:10.1094/MPMI-23-1-0006.

FAO, 2017. Defining small scale food producers to monitor target 2.3 of the 2030 Agenda for Sustainable Development. *FAO statistics division working paper series ESS/17-12*, 51pp. Available at: http://www.fao.org/3/a-i6858e.pdf (Accessed on 10 May 2020).

FAO, IFAD, UNICEF, WFP and WHO, 2017. *The State of Food Security and Nutrition in the World 2017. Building Resilience for Peace and Food Security*, 132pp. Available at: http://www.fao.org/3/a-I7695e.pdf (Accessed 10 December 2019).

FAO, 2019. The state of food and agriculture 2019. Moving forward on food loss and waste reduction. Food and Agriculture Organization, Rome.

Gaffney, J., Challendera, M., Califfa, K. and Harden, K. 2019. Building bridges between agribusiness innovation and smallholder farmers: a review. *Global Food Security* 20, 60–65. doi:10.1016/j.gfs.2018.12.008.

Gitonga, Z. M., De Groote, H., Kassie, M. and Tefera, T. 2013. 'Impact of metal silos on households' maize storage, storage losses and food security: an application of a propensity score matching. *Food Policy* 43, 44–55. doi:10.1016/j.foodpol.2013.08.005.

Golob, P. and Tran, B. 2001. Developing countries hazards: dangerous phosphine practices in West Africa. *Pesticide News* 53, 4.

Guntrip, J., Silby, R. M. and Smith, R. H. 1996. A phenotypic and genetic comparison of egg to adult 1068 life history traits between and within two strains of the Larger grain Borer *Prostephanus* 1069 *truncatus* (Horn) (Coloeptera: Bostrichidae). *Journal of Stored Products Research* 32(3), 213–23. doi:10.1016/S0022-474X(96)00019-7.

Harris, C. and Achora, J. C. 2018. Designing ICT for agriculture (ICT4A) innovations for smallholder farmers: the case of Uganda. *Conference Paper Interacción 2018*, 12–14 September 2018, Palma, Spain. doi:10.1145/3233824.3233830.

Hell, K., Cardwell, K. F., Sètamou, M. and Poehling, H. -M. 2000. The influence of storage on aflatoxin contamination in stored grains in four agroecological zones in Benin, West Africa. *Journal of Stored Products Research* 36(4), 365–82. doi:10.1016/s0022-474x(99)00056-9.

Hell, K., Fandohan, P., Kiewnick, S., Sikora, R. and Cotty, P. J. 2008. Pre- and postharvest management of aflatoxin in maize: an African perspective. In: Leslie, J. F., Bandyopadhyay, R. and Visconti, A. (Eds), *Mycotoxins: Detection Methods, Management, Public Health and Agricultural Trade*. CABI Publishing, Wallingford, UK, pp. 219–29.

Holland, M. B., Shamer, S. Z., Imbach, P., Zamora, J. C., Medellín, C., Leguía, E., Donatti, C. I., Martínez-Rodríguez, M. R. and Harvey, C. A. 2017. Mapping agriculture and adaptive capacity: applying expert knowledge at the landscape scale. *Climatic Change*. doi:10.1007/s10584-016-1810-2.

Holst, N. and Meikle, W. G. 2003. *Teretrius nigrescens* against larger grain borer *Prostephanus truncatus* in African maize stores: biological control at work? *Journal of Applied Ecology* 40(2), 307–19. doi:10.1046/j.1365-2664.2003.00805.x.

Kamala, A., Kimanya, M., Haesaert, G., Tiisekwa, B., Madege, R., Degraeve, S., Cyprian, C. and De Meulenaer, B. 2016. Local post-harvest practices associated with aflatoxin and fumonisin contamination of maize in three agro ecological zones of Tanzania. *Food Additives and Contaminants: Part A, Chemistry, Analysis, Control, Exposure and Risk Assessment* 33(3), 551–9. doi:10.1080/19440049.2016.1138546.

Kamau, T., Vitswamba, E. and Vyas, H. 2018. Trend analysis of post-harvest ICT solutions in East Africa. *Engineering for Change*. Available at: https://www.engineeringforchange.org/research/trend-analysis-post-harvest-ict-solutions-east-africa/ (Accessed on 15 March 2020).

Kitinoja, L., Saran, S., Roy, S. K. and Kader, A. A. 2011. Postharvest technology for developing countries: challenges and opportunities in research, outreach and advocacy. *Journal of the Science of Food and Agriculture* 91(4), 597–603. doi:10.1002/jsfa.4295.

Kohl, R. and Foy, C. 2018. Guideline to the Agricultural Scalability Assessment Toolkit for assessing and improving the scaling potential of agricultural technologies. United States Agency for International Development. Available at: https://www.agrilinks.org/post/guide-agricultural-scalability-assessment-tool (Accessed on 20 April 2019).

Lundy, M., Gottret, M. V. and Ashby, J. 2005. Learning alliances: an approach for building multi-stakeholder innovation systems. The Institutional Learning and Change (ILAC) Brief No. 8, August 2005. International Center for Tropical Agriculture (CIAT), pp. 1–4.

Lundy, M., Gottret, M. V., Ostertag, C., Best, R. and Ferris, S. 2007. *Participatory Market Chain Analysis for Smallholder Producers*. International Centre for Tropical Agriculture (CIAT), Cali, Colombia.

Marasas, W. F. O., Gelderblom, W. C. A., Shephard, G. S. and Vismer, H. F. 2008. Mycotoxins: a global problem. In: Leslie, J. F., Bandyopadhyay, R. and Visconti, A. (Eds), *Mycotoxins: Detection Methods, Management, Public Health and Agricultural Trade*. CABI Publishing, Wallingford, UK, pp. 29–39.

Mashavave, T., Mapfumo, P., Mtambanengwe, F., Chikowo, R., Gwandu, T., Nezomba, H. and Siziba, S. 2011. Factors influencing participation of smallholder farmers in knowledge sharing alliances around SOFECSA field-based learning centres. *African Crop Science Conference Proceedings* 10, 335–8.

Mashavave, T., Mapfumo, P., Mtambanengwe, R., Gwandu, T. and Siziba, S. 2013. Interaction patterns determining improved information and knowledge sharing among smallholder farmers. *African Journal of Agricultural and Resource Economics* 8, 1–12.

Mather, D., Duncan Boughton, D. and Jayne, T. S. 2013. Explaining smallholder maize marketing in southern and eastern Africa: the roles of market access, technology and household resource endowments. *Food Policy* 43, 248–66. doi:10.1016/j.foodpol.2013.09.008.

Matumba, L., Van Poucke, C., Njumbe Ediage, E., Jacobs, B. and De Saeger, S. 2015. Effectiveness of hand sorting, flotation/washing, dehulling and combinations thereof on the decontamination of mycotoxin-contaminated white maize. *Food Additives and Contaminants. Part A, Chemistry, Analysis, Control, Exposure and Risk Assessment* 32(6), 960–9. doi:10.1080/19440049.2015.1029535.

Meikle, W. G., Adda, C., Azoma, K., Borgemeister, C., Degbey, P., Djomamou, B. and Markham, R. H. 1998. Varietal effects on the density of *Prostephanus truncatus* (Horn) (Col.: Bostrichidae) and *Sitophilus zeamais* Motsch. (Col.: Curculionidae) in grain stores in Benin Republic. *Journal of Stored Products Research* 34(1), 45–58. doi:10.1016/S0022-474X(97)00020-9.

Meikle, W. G., Markham, R. H., Nansen, C., Holst, N., Degbey, P., Azoma, K. and Korie, S. 2002. Pest management in traditional maize stores in West Africa: a farmer's perspective. *Journal of Economic Entomology* 95(5), 1079–88. doi:10.1093/jee/95.5.1079.

Misihairabgwi, J. M., Ezekiel, C. N., Sulyok, M., Shephard, G. S. and Krska, R. 2017. Mycotoxin contamination of foods in Southern Africa: a 10-year review (2007–2016). *Critical Reviews in Food Science and Nutrition*. doi:10.1080/10408398.2017.1357003.

Mlambo, S., Mvumi, B. M., Stathers, T., Mubayiwa, M. and Nyabako, T. 2017. Field efficacy of hermetic and other maize grain storage options under smallholder farmer management. *Crop Protection* 98, 198–210. doi:10.1016/j.cropro.2017.04.001.

Mlambo, S., Mvumi, B. M., Stathers, T., Mubayiwa, M. and Nyabako, T. 2018. Field efficacy and persistence of synthetic pesticidal dusts on stored maize grain under contrasting agro-climatic conditions. *Journal of Stored Products Research* 76, 129 – 39. doi:10.1016/j.jspr.2018.01.009.

Morris, M. J., Mvumi, B. M., Stathers, T. E. and Riwa, W. H. 2005. Post-harvest innovation to improve food security in Tanzania and Zimbabwe: learning alliance lessons. In: Smits, S., Fonseca, C. and Pels, J. (Eds). Proceedings of the Symposium on Learning Alliances for Scaling up Innovative Approaches in the Water and Sanitation Sector. Delft. IRC International Water and Sanitation Centre, 7–9 June 2005, pp. 99–109.

Morton, J. F. 2007. The impact of climate change on smallholder and subsistence agriculture. *Proceedings of the National Academy of Sciences of the United States of America* 104(50), 19680–5. doi:10.1073/pnas.0701855104.

Muatinte, B. L. and Cugala, D. R. 2015. Monitoring the establishment and dispersal of *Teretrius nigrescens* Lewis (Coleoptera: Histeridae), a predator of *Prostephanus truncatus* Horn (Coleoptera: Bostrichidae) in Manica Province, Mozambique. *African Entomology* 23(1), 250–4. doi:10.4001/003.023.0123.

Mubayiwa 2019. Smallholder postharvest technologies to manage climate-related risks in selected sorghum production systems. MPhil Thesis. Department of Soil Science and Agricultural Engineering, Harare, Zimbabwe, xiii+160pp.

Murashiki, T. C., Chidewe, C., Benhura, M. A., Manema, L. R., Mvumi, B. M. and Nyanga, L. K. 2018. Effectiveness of hermetic technologies in limiting aflatoxin B1 and fumonisin B1 contamination in stored maize grain under smallholder conditions in Zimbabwe. *World Mycotoxin Journal* 11(3), 459–69. doi:10.3920/WMJ2017.2288.

Mutiga, S. K., Were, V., Hoffmann, V., Harvey, J. W., Milgroom, M. G. and Nelson, R. J. 2014. Extent and drivers of mycotoxin contamination: inferences from a survey of Kenyan maize mills. *Phytopathology* 104(11), 1221–31. doi:10.1094/PHYTO-01-14-0006-R.

Mutiga, S. K., Mushongi, A. A. and Kangéthe, E. K. 2019. Enhancing food safety through adoption of term technical advisory, financial, and storage support services in maize growing areas of East Africa. *Sustainability* 11(10), 2827. doi:10.3390/su11102827.

Mvumi, B. M. and Stathers, T. 2015. The food security challenges in sub-Saharan Africa: the potential contribution of postharvest skills and technology in closing the gap. In: Arthur, F. H., Kengkanpanich, R., Chayaprasert, W. and Suthisut, D. (Eds), 11th International Working Conference on Stored Product Protection, Chiang Mai, Thailand, 24–28 November 2014, pp. 32–43.

Mvumi, B. M., Morris, M., Stathers, T. E. and Riwa, W. 2008. Doing things differently: post-harvest Innovation Learning Alliances in Tanzania and Zimbabwe. In: Sanginga, P. C., Bayer, W., Kaaria, A., Njuki, S. J. and Wettasinha, C. (Eds), *Innovation Africa: Enriching Farmers' Livelihoods*. Earthscan Publications, London, UK, pp. 186–202.

Ndegwa, M. K., De Groote, H., Gitonga, Z. M. and Bruce, A. Y. 2016. Effectiveness and economics of hermetic bags for maize storage: results of a randomized controlled trial in Kenya. *Crop Protection* 90, 17–26. doi:10.1016/j.cropro.2016.08.007.

Omondi, B. A., van den Berg, J., Masiga, D. and Schulthess, F. 2014. Molecular markers reveal narrow genetic base and culturing-associated genetic drift in *Teretrius nigrescens* Lewis populations released for the biological control of the larger grain borer in Africa. *Bulletin of Entomological Research* 104(2), 143–54. doi:10.1017/S0007485313000552.

Remington, T., Best, R., Lundy, M. and Ferris, S. 2005. *CRS and Its Learning Alliance with CIAT and FOODNET: Building Sustainable R&D Partnerships for More Efficient Generation and Delivery of New Knowledge to Improve the Livelihoods of the Rural Poor*. Catholic Relief Services (CRS) and Centro Internacional de Agricultura Tropical (CIAT), Nairobi, Kenya.

Sheahan, M. and Barrett, C. B. 2017. Food loss and waste in Sub-Saharan Africa: a critical review. *Food Policy* 70, 1–12. doi:10.1016/j.foodpol.2017.03.012.

Shepherd, A. W. 2018. *Addressing the Aggregation and Coordination Problems in Smallholder-Based Value Chains*. World Bank, Washington, DC.

Shirima, C. P., Kimanya, M. E., Kinabo, J. L., Routledge, M. N., Srey, C., Wild, C. P. and Gong, Y. Y. 2013. Dietary exposure to aflatoxin and fumonisin among Tanzanian children as determined using biomarkers of exposure. *Molecular Nutrition and Food Research* 57(10), 1874–81. doi:10.1002/mnfr.201300116.

Singano, C. D., Mvumi, B. M. and Stathers, T. E. 2019. Effectiveness of grain storage facilities and protectants in controlling stored-maize insect pests in a climate-risk prone area of Shire Valley, southern Malawi. *Journal of Stored Products Research* 83, 130–47. doi:10.1016/j.jspr.2019.06.007.

Smart, M. G., Shotwell, O. L. and Caldwell, R. W. 1990. Pathogenesis in *Aspergillus* ear rot of maize–aflatoxin b1 levels in grains around wound inoculation sites. *Phytopathology* 80(12), 1283–6. doi:10.1094/Phyto-80-1283.

Stathers, T. and Mvumi, B.M. 2020. Challenges and initiatives in reducing postharvest food losses and food waste: sub-Saharan Africa. Chapter 27. In: *Preventing food losses and waste to achieve food security and sustainability*. E. M. Yahia (Ed.) Burleigh Dodds Science Publishing. ISBN: 978-1-78676-300-6.

Stathers, T., Lamboll, R. and Mvumi, B. M. 2013. Postharvest Agriculture in Changing Climates: its importance to African smallholder farmers. *Food Security* 5(3), 361–92. doi:10.1007/s12571-013-0262-z.

Stathers, T., Mvumi, B. M. and Golob, P. 2002. Can diatomaceous earths have an integrated role in small-scale tropical grain storage? *International Organisation for Biological Control (IOBC) Bulletin* 25(3), 231–7.

Sun, Z. T., Lu, P. X., Gail, M. H., Pee, D., Zhang, Q. N., Ming, L. H., Wang, J. B., Wu, Y., Liu, G. T., Wu, Y. Y. and Zhyu, Y. 1999. Increased risk of hepatocellular carcinoma in male hepatitis B surface antigen carriers with chronic hepatitis who have detectable urinary aflatoxin metabolite M1. *Hepatology* 30(2), 379–83. doi:10.1002/hep.510300204.

Thomson, L. J., Macfadyen, S. Hoffmann, A. A. 2010. Predicting the effects of climate change on natural enemies of agricultural pests. *Biological Control* 52, 296–306.

UNDESA 2017. World population prospects 2017. United Nations Department of Economic and Social Affairs. Available at: https://www.un.org/development/desa/publications/world-population-prospects-the-2017-revision.html (Accessed on 10 May 2020).

Verma, M., Christine Plaisier, C., van Wagenberg, C. P. A. and Achterbosch, T. 2019. A systems approach to food loss and solutions: understanding practices, causes, and indicators. *Sustainability* 11(3), 579. doi:10.3390/su11030579.

Walker, S., Jaime, R., Kagot, V., Probst, C. 2018. Comparative effects of hermetic and traditional storage devices on maize grain: Mycotoxin development, insect infestation and grain quality. *Journal of Stored Products Research* 77, 34–44.

World Bank, NRI, FAO. 2011. *Missing* Food: The Case *of* Postharvest Grain Losses *in* Sub-*Saharan Africa*. The World Bank, Washington DC, 116p. Available at: http://siteresources.worldbank.org/INTARD/Resources/MissingFoods10_web_final1.pdf (Accessed on 23 April 2019).

World Food Programme and Natural Resources Institute 2012. Training manual for improving grain postharvest handling and storage. Developed by Hodges, R. J. and Stathers, T. E. for the World Food Programme, Rome, 246pp. Available at: https://www.wfp.org/content/p4p-training-manual-improving-grain-postharvest-handling-and-storage (Accessed on 10 May 2020).

# Index

Note: Page numbers in italics denotes figures and tables.


Above-floor ducts 153
*Absidia* spp. 38, 41
Absolute estimates 200
*Acanthoscelides obtectus* 279, 288
Acoustical sensors 382
Actellic Super Dust 112
Acute rodenticides 108
Adzuki bean weevil. *see Callosobruchus chinensis*
Aeration 133-135, 142, 149-150, 245-246
  with ambient air 152-154
  chilled, grain preservation by 252-253
  with chilled air 154
  objectives of 247-251
  simulation models, in cooling grain 253-255
Aérial-Technology Resource Centre (France) 327
Aerial traps 207-208-
AFB1 160
Aflatoxins 38, 41, 51, 101, 163, 348-349, *404*
A4AI. *see* Alliance for Affordable Internet (A4AI)
African Farms and Markets Ltd/ GrainPro 179
AGRA. *see* Alliance for Green Revolution in Africa (AGRA)
AgResults initiative 179, 426-430, 435
Agricultural Scalability Assessment Toolkit (ASAT) 419-421
AgriSense-BCS Ltd 212
AgroZ bag 172-174, 178
AgroZ bag Plus 174
Airflow rate 153
Air oven drying method 368
*Akodon azarae* 104
Alliance for Affordable Internet (A4AI) 416
Alliance for Green Revolution in Africa (AGRA) 179
*Alloeocranum biannulipes* 286
*Alphitobius diaperinus* 283
*Alternaria* spp. 37, 351
Aluminum phosphide 155
Ambient aeration 134
*Amphibolus venator* 286
*Anisopteromalus calandrae* 287
Anobiidae 213
Anticoagulants 109
  first-generation 108
  second-generation 108
*Antrocephalus mitys* 287
Applied Energy Devices (Albuquerque, NM) 326
*Araecerus fasciculatus* 279
Argentina 52, *96*, 101, 141, 158, 243, 376
Arsenic 82
ASAT. *see* Agricultural Scalability Assessment Toolkit (ASAT)
*Aspergillus* spp. 36-38, 41, 44-46, 50, 51, 53-55, 341, 351
  *candidus* 38, 40, 45
  *carbonarius* 45, 353
  *clavatus* 45
  *flavus* 37, 38, 40, 41, 45, 46, 48, 50, 51, 54, 341, 348, 350, 353, 408
  *glaucus* 36-40, 45, 46
  *niger* 37, 45
  *ochraceus* 38, 40, 45, 53, 55
  *parasiticus* 45
  *restrictus* 36, 38, 40, 45, 46
  *versicolor* 40, 45
  *westerdijkiae* 349, 353
A to Z Textile Mills Ltd (Tanzania) 174, 179
Attractants 213
Augmentative biological control 285

Australia 52, 77, 78, 98, 138, 144, 151, 158, 200, 243, 244, 320, 333
Australian bunkers 144, 147
Australian grain industry 74

B3 film 330
*Bacillus* spp.
  *subtilis* 325
  *thuringiensis* 272
Bait bags 213
Bambara nuts. *see Vigna subterranea*
*Bandicota* spp.
  *bengalensis* 99, 100, 102, 105
  *indica* 100, 102
Bangladesh 48, *95*, 101, 409
BCR. *see* Benefit-cost ratio (BCR)
Bean/Cowpea Collaborative Research Support Program 21
Bell Industries 179
Benefit-cost ratio (BCR) 12
Berlese funnel 201, *202*
Bethylid wasps 288
Bill & Melinda Gates Foundation 21, 178
Bioactive compounds 293
Bioassay 74, 77, 200
Biochemical pesticides 271
Biologically based control strategies 267-270
  biopesticides and 270-275
  of stored-product pests 285-296
  *see also* Semiochemicals
Black flour beetle (*Cynaeus angustus*) 48
Boerner-type divider 202, *204*
Bolivia 243
Bolted steel silos 238-241
Botanical active substance 273-274
*Bracon hebetor* 281
Braconid wasp. *see Habrobracon hebetor*
Bran 52
Brazil 7, 43, *96*, 243, 253
Bruchidae 213
Bruchids 201
Buckwheat 33
Bulk storage 125-127
  best management practices 161-163
  horizontal building structures and 137, 142-144
    disadvantages of 137-138
    construction materials 138-140
    structural style 140-142
  quality and quantity and preventative strategies and 150-154
    sealing of structures and exclusion of pests 151-152
  quality and quantity maintenance and preservation 154-161
  structure types and upright silo structures 127-128, 133-137
    construction materials 128-132
    structural style 132-133
  *see also* On-the-ground piles
Bunkers and piles 144-145
Bunker storage 243
Burkina Faso 10, 188

Cabinet X-ray system 326-327, *326*
*Cadra cautella* 278, 282, 287
*Cajanus cajanus* 176
*Callosobruchus* spp.
  *chinensis* 279, 288
  *maculatus* 173, 288
*Calomys* spp.
  *callosus* 104
  *laucha* 104
  *musculinus* 104
Canada 52, 200, 243, 244
Canada Grain Act 200
Canadian Grain Commission 204
CAP. *see* Common Agricultural Policy (CAP), Europe
CAPS. *see* Cleaved amplified polymorphic sequence
Carbonation 136
Carbon dioxide 84
Carbon dioxide sensors 374-376
Carbonyl sulfide 84
Carter Dockage Tester 201
Cats, for rodents 110
CCPs. *see* Critical control points (CCPs)
Cellulose baits 109
Central America, metal silos in 19-20
*Cephalonomia* spp.
  *tarsalis* 288, 294
  *waterstoni* 288, 294
Chalcid wasp. *see Antrocephalus mitys*
CHCs. *see* Cuticular hydrocarbons (CHCs)
Chile 243
China 43, 48, 126, 138, 139, 144, 158, 200, 212, 409
CIAT. *see* International Center for Tropical Agriculture (CIAT)
Cigarette beetle. *see Lasioderma serricorne*
*Cladosporium* spp. 37, 39, 48, 50
Classical biological control 285

Cleaved amplified polymorphic sequence (CAPS) 77
Codex General Standard for Irradiated Foods 319
Coffee bean weevil. *see* *Araecerus fasciculatus*
COFO. *see* Commercially objectionable foreign odors (COFO)
Cold plasma treatment 160-161, 351-352
  ozone fumigation of stored cereals for spoilage control 352-354
Combustible gases 234
Commercially objectionable foreign odors (COFO) 251
Common Agricultural Policy (CAP), Europe 268
Common beans. *see* *Phaseolus vulgaris*
Community groups/news sharing 414
Concrete floor and sidewalls 140-142
Concrete grain elevators 241-242
Concrete silos 128
  benefits and limitations of 129
Condensation, risk of 137
Conservation biological control 285
Controlled atmosphere. *see* Modified atmosphere (MA)
*Corcyra cephalonica* 287
Coring 260
Corporate Business Forms Ltd/ Vestergaard 179
Corrugated bolted silos 129-130
Corrugated cardboard traps 210, *211*
*Cotesia glomerata* 295
Cowpeas. *see* *Vigna inguiculata*
Critical control points (CCPs) 339
Cross-flow aeration 260
Cryotolerants 35
*Cryptolestes* spp. 201, 214
  *ferrugineus* 69, 70, 78, 85, 207, 213, 218, 219, 236, 239, 241, 243, 253, 288
  *pusillus* 288
  *turcicus* 288
*Curvularia* spp. 37
Cuticular hydrocarbons (CHCs) 294
*Cynaeus angustus* 48
Cyprus 158, 241, 243
Cytochrome-b5-related protein (*Cyt-b5-r*) 80-81

DBD. *see* Dielectric barrier discharge (DBD)
Delta traps 208-209, *208*
Deming, W. E. 13
Deoxyribonucleic acid (DNA) 42
  assay 75
Depressed flour beetle. *see* *Palorus subdrepressus*
Dermestidae 213
Devitalisation 84
Diamond traps. *see* Delta traps
DICE. *see* Dry, inspect, clean and examine (DICE)
Dielectric barrier discharge (DBD) 160
Dielectric properties, of grain 368-369
Digital Green app (India) 411
Dihydrolipoamide dehydrogenase (DLD) 81
*Dinoderus porcellus* 286
DLD. *see* Dihydrolipoamide dehydrogenase (DLD)
DMLs. *see* Dry matter losses (DMLs)
DNA. *see* Deoxyribonucleic acid
Dome traps *210*, 213
Double hermetic bags *174*
Double sleeve grain triers, for sampling bags *207*
Downward aeration 153
*Dozynki* 33
Dried bean beetle. *see* *Acanthoscelides obtectus*
Drucker, P. 13
Dry, inspect, clean and examine (DICE) 126-127
  as best management practice 162-163
Dry grain biological heating prevention 250
Dry matter losses (DMLs) 345, *346*
Duct spacing 149

East African Grain Council 185
Ecologically based rodent management (EBRM) 112
Economic loss 5
Egypt *95*
Ekima Engineering Workers 179
Electromagnetic imaging 379-380
Electromagnetic imaging 384
Electron beams 325
*Eligmodontia* spp. 103
Elite bag 173, 178
Elite Bag Light Duty Double Liner 174
Elite Innovations (K) Ltd (Kenya) 174, 179
El Salvador 20
EMC. *see* Equilibrium moisture content (EMC)
Endosperm 43, 52

Enhancing the Quality of U.S. Grain for International Trade' 252
Environmental loss 5
Environmental Protection Agency (EPA) 233, 271
*Ephestia* spp. 208
  *elutella* 278
  *kuehniella* 278, 282, 283, 287
*Epicoccum* spp. 37
EPPO Standard PM 6/3 274, 285
EPSON V-600 perfection scanner 330
Equilibrium moisture content (EMC) 370, 371
Equilibrium relative humidity (ERH) 36, 39, 238
Equity Bank 180
ERH. *see* Equilibrium relative humidity
*Escherichia coli* 325
E-Soko (Ghana) 411
Ethanedinitrile 84
Ethiopia 95, 183, 188, 423
Ethyl formate 84
Europe 138
*Eurotium* spp. 36, 37, 47
  *amstelodami* 47
  *chevalieri* 47
  *rubrum* 47

Facebook 414
Family of curves 254
FDA. *see* U.S. Food and Drug Administration (FDA)
Federal Grain Inspection Service (FGIS) 369
Federal Insecticide, Fungicide, and Rodenticide Act (FIFRA) 272-273
Feed the Future Programme 419
FEERIX 327
FGIS. *see* Federal Grain Inspection Service (FGIS)
FHB. *see Fusarium* Head Blight' (FHB)
Field mycoflora *36*, 37
FIFRA. *see* Federal Insecticide, Fungicide, and Rodenticide Act (FIFRA)
Financial services and ICTs 414-415
Flat-bottomed grain storage 132-133
Flat-bottom warehouses 140, *141*
Flotation 201-202
Flour beetle. *see Tribolium confusum*
FLW. *see* Food losses and wastes
Food and Agriculture Organization (FAO) 274, 319
Food irradiation 319
Food losses and wastes (FLW) 31
Food Security Bureau (USAID) 419
Forced aeration 246
France 200
Free fatty acids (FFAs) 232, 238
Fully perforated floors 153
Fumigant residues and odors, removal of 251
Fumigants, qualitative comparison of *68*
Fungal contamination in cereals 31-32
  cereal grains and 34
    ecological succession of molds in storage 41-42
    ecophysiology 34-36
    fungal invasion mechanisms 40-41
    gas composition 40
    genera of storage fungi 37-38
    temperature 39-40
    water activity 39
  humankind and 32-33
  *see also* Maize; Rice; Wheat
*Fusarium* Head Blight' (FHB) 53
*Fusarium* spp. 37, 39, 47, 48, 50-51, 53, 54, 101, 341, 349-351
  *asiaticum* 53
  *boothii* 53
  *culmorum* 53
  *graminearum* 53, 341, *342*,349
  *proliferatum* 47, 53
  *semitectum* 47
  *verticillioides* 47, 349, 353, *354*, 408

*Galleria mellonella* 281
Gas chromatography-mass spectrometry (GC-MS) 342
Gas fumigation 154-156
Gas supply, from pressurized cylinders 233-234
GC-MS. *see* Gas chromatography-mass spectrometry (GC-MS)
Genetics, of resistance inheritance 72-73, *73*
*Geosmithia* spp. 38
Ghana *95*, 188
Grain bulk cooling, for pest suppression 247-248
Grain chilling 252-253
Grain Probe Trap 212
Grainpro Inc. 172, 178
GrainPro Ultra Hermetic Bunker™ 159
Grains Research and Development Corporation (GRDC) 153-156, 376-377

Grain Storage Pest Control Guide 150
GRDC. *see* Grains Research and Development Corporation (GRDC)
Greater wax moth. *see Galleria mellonella*
Ground piles 144
Guatemala 19

Habitat management 110
*Habrobracon hebetor* 287, 296
Hazard Analysis Critical Control Points (HACCP) 332, 339
HDPE. *see* High-density polyethylene (HDPE)
Headspace and downspout condensation prevention 250
*Helminthosporium* spp. 37
Hermetic bags
  market development, in Central and West Africa 424-426
  types of *174*
Hermetic storage 44, 157-159, 242-243
  of grain, and use of natural compounds 348-349
Hermetic storage bags 171-172
  case study 185-190
  challenges 181
    affordability and access 182-183
    availability 181-182
    need for drying technology and moisture assessment 184
    need for standards and environmental impact 184-185
    usability and durability 183-184
  development of 172-174
  dissemination, to reduce postharvest loss 178-181
  efficacy of 176-178
  future trends 191-192
  working of 175-176
*Heteromys* spp. 103
Heterozygotes 75, 80
Hexaploid wheat 52
High-density polyethylene (HDPE) 173, 174
High-energy X-ray treatment 332
  experimental design 332-333
  results and discussion 333-334
High pressure carbon dioxide treatment 157
Historical use, for stored product disinfestation 327-328
*Holepyris sylvanidis* 288, *292*, 293, 294, 295
Honduras 19
Hopper-bottomed silos 132
Hygrophiles 35
Hypercarbia 234
Hyperspectral imaging 213-214
Hypoxia 176, 234

Ichneumonid wasp. *see Venturia canescens*
ICT. *see* Information and Communication Technology (ICT)
Idiobiont ectoparasitoid 287-288
Idiobiont wasp 287
India 14, 15, 48, 77, 78, 95, 126, 144, 205, 409
Indianmeal moth. *see Plodia interpunctella*
Indonesia 48, *95*, 105
Information and Communication Technology (ICT) 191
In-ground storage 238
Insect flight 79-80
Insector system 200-201
Insect pest management, controlled atmosphere and temperature control 231-232
  future trends 258-259
  temperature manipulation 245-255, 258-260
    high temperatures 255-256
    recent developments 257
  *see also* Modified atmosphere (MA)
Insect pest management, detection and monitoring 199-200
  detection
    using chemistry 215
    using imaging 213
    using sound 214
  future trends 221
  sampling theory
    representative sample 219-221
    sample unit and size 215-218
    sequential sampling 218-219
  sampling 200-201
    methods for grain storage types 201-205
  *see also* Traps
Insect pest management, novel techniques 319-320
  case studies 328-334
  future trends 334-335
  ionizing radiation sources and 324-328
  irradiation for stored product pests control 320-321
    desired response 321
    generic and specific doses 321-322

public perception of irradiated food 323-324
resistance management 322-323
Insect pests. *see Callosobruchus maculatus*
Insect probe traps 380-381
Insects Limited 213
Institutional void 18
Integrated pest management (IPM) 199
*Integrated pest management* (Shankar and Abrol) 100
Intermediate mycoflora *36*
Internal rate of return (IRR) 12, 21
Internal transcribed spacer (ITS) 42
International Center for Tropical Agriculture (CIAT) 432
International Plant Protection Convention (IPPC) 320
International Rice Research Institute (IRRI) 173
International Working Conference on Stored Product Protection (IWCSPP2018) 289
Ionizing radiation 320
IPM. *see* Integrated pest management (IPM)
IPPC. *see* International Plant Protection Convention (IPPC)
IRR. *see* Internal rate of return (IRR)
IRRI. *see* International Rice Research Institute (IRRI)
Iso-concentration lines *240*
Israel 243
Italy 243
ITS. *see* Internal transcribed spacer
IWCSPP 2018. *see* International Working Conference on Stored Product Protection (IWCSPP2018)

Japan 50, 333
Jet plasma 160

Kairomones 213
KAVES. *see* Kenya Agricultural Value Chains Enterprises Project (KAVES)
Kazakhstan 243
Kentainers Ltd 179
Kenya 5, 101, *174*, 178, 180, 232, 403, 404, 408, 412, 427, 435
Kenya Agricultural Value Chains Enterprises Project (KAVES) 179
Kenya Bureau of Standards 132, 185
Kenya On-Farm Grain Storage Challenge Project 426
Kenya Promotions and Marketing Company Ltd 179
Khapra beetle. *see Trogoderma granarium*
Kingdom Fungi (Mycota) 34
Kissinger, H. 4
Knockdown' test 74
Knowledge management and training 413
Koinobiont parasitoids 287

Laos *95, 99*
Large accelerator irradiator machine *327*
Larger grain borer. *see Prostephanus truncatus*
*Lariophagus distinguendus* 288, 296
Larval ectoparasitoid. *see Bracon hebetor*
*Lasioderma serricorne* 209, 213, 282, 284, 287-288, 328
Latin America *96*, 101, 104
LDPE. *see* Low-density polyethylene (LDPE)
Learning alliance approach 399-401
Lesser grain borer. *see Rhyzopertha dominica*
LGPBS. *see* Local grain production and banking system (LGPBS)
Limited grain drying, by aeration 251
Lindgren multiple funnel trap *209*
LMICs. *see* Lower middle-income countries (LMICs)
Local grain production and banking system (LGPBS) 406-407
Low-density polyethylene (LDPE) 159
Low-energy electron beam 325-326, 328-329
experimental design 329-330
results and discussion 330-332
Lower middle-income countries (LMICs) 340

MA. *see* Modified atmosphere
Macrobials/invertebrates 275
Macroorganisms 274
Maize. *see Zea mays*
Maize, fungal contamination in 42-43, 54
factors affecting mold growth 47-48
molds and associated damage 44-47
storage 44
Maize weevil. *see Sitophilus zeamais*
Malawi 95, 188
Malaysia 95, 99
Manual grain inspection 364-365
Market price information and marketing 413-414

Mass trapping 282
*Mastomys* spp. 108
*natalensis* 102, 105
Mating disruption (MD) 282–283
Mealworm. *see Alphitobius diaperinus*
Mesophiles 35
Metallocene Linear Low-Density Polyethylene (MLLDPE) 174
Metal silo users 11–12, 19–20
Methyl bromide 83, 231, 320
Mexico *96*, 243
Microbial pesticides 272
Microbials 274
Microflora growth suppression 248
Microorganisms, definition of 273
Milling 50
MLLDPE. *see* Metallocene Linear Low-Density Polyethylene (MLLDPE)
Modified atmosphere (MA) 232–233, 257–259
controlled atmosphere effects 243–245
combination of high carbon dioxide and low oxygen levels 237
commercial use 242–245
high carbon dioxide level 236
low oxygen level 235–236
under normal atmospheric pressure 233–235
product quality preservation 238
recent developments 245
seed germination 237–238
structure types 238–241
temperature and relative humidity effects on fumigation 237
Modified atmosphere treatment 156–159
categories of 157
Moist grain spontaneous heating prevention 250–251
Moisture cable sensors 371–372
Moisture content and relative humidity 367–371
Moisture migration prevention in grain bulk 249–250
Molecular assay 77
Molecular methods, in stored grain 347–348
*Moniliforme* spp. 48
Monitoring, of stored cereal grain quality 363–364
future trends 383–384
methods and techniques 364–383
Montreal Protocol 83
Mozambique 412
M-Pesa (Kenya) 411
Mucorales group 38
*Mucor* spp. 38, 39, 41
Mung beans. *see Vigna radiata*
Murdock, L. 21
*Mus* spp. 103
*musculus* 99, 100, 102, *104*
Myanmar 48, *95*, 101, 105
community village-level rodent management in 110–112
Mycotoxin 50–51, 53, 54, 160, 176
*Myrothecium* spp. 50

Natural biological control 285
Natural preservative compounds 349–350
Near-infrared 214
Near-infrared reflectance (NIR) 370
Near-infrared transmittance (NIT) 369–370
*Neotoma* spp. 103
New Zealand 320, 324
Nicaragua 20
Niger 188, 189
Nigeria 188, 232
*Nigrospora* spp. 37
NIR. *see* Near-infrared reflectance (NIR)
NIT. *see* Near-infrared transmittance (NIT)
Nitrogen 84
Nobbe grain triers, for sampling bags *207*
Non-thermal' plasma. *see* Cold plasma

OECD. *see* Organization for Economic Co-operation and Development (OECD)
*Oligoryzomys longicaudatus* 104
On-site $N_2$ generators 234–235
On-the-ground piles 126, 144–145
construction materials 145–147
structural style 147–150
OPIsystems Inc. *201*, 212
Organization for Economic Co-operation and Development (OECD) 271
Program on Biological Pesticides 274–275
*Oryzaephilus* spp.
*mercator* 288
*surinamensis* 80, 236, 241, 242, 288
*Oryza* spp.
*barthii* (Barth's rice) 49
*glaberrima* (African rice) 49
*indica* 49
*japonica* 49
*latifolia* (broadleaf rice) 49

*longistaminata* (longstamen rice) 49
*punctata* (red rice) 49
*rufipogon* (brownbeard rice) 49
*sativa* 49, 176
*Ototylomys* spp. 103
Ozonation and cold plasma treatment 159-161

P45S/P49S gene 78
Pad 145
Paddy rice 49, 176
*Paecilomyces* spp. 38
Pakistan 99
Para-benzoquinones 284
Paraguay 243
Parasitoids 270, 286-287
application of 289-291
research, future trends in 294-296
and semiochemicals 291-294
of stored-product beetles 287-289
of stored-product moths 287
Passover (Jewish) 33
Pay-for-Results (PfR) prize model 420, 426, 427
PCR. *see* Polymerase chain reaction (PCR)
PC Trap 212
*Penicillium* spp. 36-39, 41, 46-47, 50, 51, 53-55, 248, 341, 351
*aurantiogriseum* 37, 46, 51
*citreonigrum* 51
*citrinum* 46
*commune* 47, 51
*digitatum* 47
*glabrum* 47
*islandicum* 51
*oxalicum* 46
*palitans* 40
*roqueforti* 40, 350
*rugulosum* 51
*variabile* 47
*verrucosum* 37, 46-47, 53, 55, 341, 342, 347, 349, 350, 352
*viridicatum* 40, 46
Perforated bulkheads 150
*Peromyscus* spp. 103
Pest control organisms 272-273
PfR. *see* Pay-for-Results (PfR) prize model
*Phaseolus vulgaris* 176
Pheromones 213, 269-272, 275, 282
of stored-product pests 276-277
aggregation and sex pheromones in species with long-lived adults 278-280
sex pheromones in species with short-lived adults 277-278
Philippines 95, 99, 173
PHL, Post-harvest loss 3
*Phoma* spp. 50
Phosphine 232, 320, 322-323, 376-378
Phosphine resistance 67-68
alternative fumigants 82-83
biochemical and genetic basis of 80-81
toxicity models 81-82
breakers 83-85
fitness costs associated with 79
fecundity 80
flight 79-80
future trends 85
genes and 75
genetic assays for 75-76
complementary resistance assays 77
strategic management of resistance 77
genetic basis understanding and 74
genetic testing and 77-78
impacting management 72-73
meaning of 69-70
practical conditions for pest management 70-72
resistance gene variations and 78-79
significance of 69
*Phyllocnistis citrella* 283
Phytosanitary irradiation doses 321, 322
*Pichia anomala* 350
PICS. *see* Purdue Improved Cowpea Storage (PICS)
PICS Global 189
PICS™ bags 172
*Pieris* spp. 295
Pigeon peas. *see Cajanus cajanus*
Pile cover 146
PIPs. *see* Plant-incorporated protectants (PIPs)
Pitfall traps 210-213, *211*, *380*, *381*
Plant extracts/botanicals 275
Plant-incorporated protectants (PIPs) 272, 273
Plant protection products 268
Plant Protection Regulation 1107/2009 (EU) 273
Plasma 159-160
Plastic aeration tubes *142*
Plastic silos 131
portable 132
*Plodia interpunctella* 160, 208, 210, 278, 282, 283, 287, 322

Pneumatic delivery system 203, *205*
Pollination 33
Polymerase chain reaction (PCR) 42, 77, 347–348
Polypropylene (PP) 44
Polypropylene bag systems 348
Poly-vinyl chloride (PVC) 159
Postcosecha Program (Central America) 11, *13*, 19–20
Post-frame and construction materials 138–139
Post Harvest Africa 179
Post-harvest loss (PHL) 3–5
economic interpretation of goal to eliminate 8–9, *8*
  economically sustainable reduction of 9–10, *9*
  excessive, 'hidden tax' from 10–13
  future progress of 6–7
  intervention and 16–17, *17*
  as multidimensional 5–6
  paradox, overcoming 15–18
  reduction and management 13–15
  sustainable reduction of 19–23
  *see also individual entries*
PP. *see* Polypropylene
PPPs. *see* Public-private partnerships (PPPs)
Precision, significance of 218
Predatory beetle. *see Teretrius nigrescens*
Pressure aeration 134, 135
Pressure-swing absorption systems 234–235, 244
*Proechimys* spp. 103
*Prostephanus truncatus* 174, 280, 286, 404–406, 412
Pseudocereals 33
Psocids 69, 214
Public-private partnerships (PPPs) 422–423
Purdue Improved Cowpea Storage (PICS) 11, *12*, 20–21, 172, 173, 178, 185–190, 348, 424
  approach to commercialization *188*
  countries using *187*
  implementation in Central and West Africa *425*
private sector and commercialization of *189*
PVC. *see* Poly-vinyl chloride (PVC)

Qualitative and quantitative loss 5

Radiotolerance, in insects 320
Radura 323
*Rattus* spp. 103, 106
  *argentiventer* 106
  *exulans* 102
  *norvegicus* 99, 102, *104*, 108
  *rattus* 99–102, *104*, 105, 108
Reactive oxygen species (ROS) 81
Red flour beetle. *see Tribolium castaneum*
Reduviid. *see Amphibolus venator*
Refrigerated aeration 252
Refuge traps 210
Relative estimates 200
Relative humidity 36–37
Relative humidity sensors 370–371
*Rhizopus* spp. 38, 41, 351
Rhodotron® (IBA, Belgium) 327
*Rhyzopertha dominica* 60, 76, 78, 80, 158, 207, 209, 213, 235, 239, 241, 242, 256, 280, 283, 287, 288, 329, 331
Rice, fungal contamination in 48–49, 54
  infection in kernels during storage 50
  mycotoxin producing fungi in stored rice 50–51
  storage and harvest 49–50
Rice moth. *see Corcyra cephalonica*
Rice weevil. *see Sitophilus oryzae*
RISO HDRL 330
Riso Scan 1.03 software 330
Rockefeller Foundation 179
Rodent pests 93–98
  and biology 102–105
  case studies 110–113
  characteristics *104*
  damage, types of 105–106
  damage/losses to grains in storage facilities *95–98*, *103*
future trends in 114
  impact on stored grain 99–102
  strategies to reduce damage of
    physical and management practices 109–110
    rodenticides 107–109
    rodent management principles in grain stores 106–107
Roof and headspace ventilation 135, 143
Roofs, of horizontal warehouse structures 140
ROS. *see* Reactive oxygen species (ROS)
Round silos 127
*rph1* gene 73, 75–77, 80–82
*rph2* gene 73, 75–81
Rural Development Initiative (RUDI) 180
Russia 52, 243

Rust-red flour beetles 153
Rusty grain beetle (*Cryptolestes ferrugineus*) 69
Rwanda 33, 188, 411

Sample divider, in grain elevator *207*
Sanitation, Loading, Aeration And Monitoring (SLAM) 126, 161
  aeration in 162
  loading in 161-162
  monitoring in 162
  sanitation in 161
*Scleroderma guani* 294
Scoop 215-216
*Scopulariopsis* spp. 38
SDC. *see* Swiss Agency for Development and Cooperation
Sealed steel silos 241
Sealed versus sealable silos 136-137, 143-144
Seed and grain quality maintenance 248-249
Semiochemicals 269, 273, 275, 285-286
  application of 281-283
  effects on stored-product environment 280-281
  future trends in 283-284
  and parasitoids 291-294
  use in stored-product protection 275-280
Serricornin 282
SHFs. *see* Small-holder farmers (SHFs)
Sidewall support 146
Sierra Leone *95*
*Sigmodon hispidus* 104
Silobags, Argentine 148-149
Silos
  aeration diagram *133*
  significance of 127-128
  *see also individual entries*
Single hermetic bag *174*
*Sitophilus* spp. 201, 214, 215, 287, 288
  *granarius* 236, 242, 279, 321
  *oryzae* 69, 207, 235, 236, 239, 321-323, 329, 331-334
  *zeamais* 48, 80, 158, 209, 247, 321
*Sitotroga cerealella* 278, 288, 322
SLAM. *see* Sanitation, Loading, Aeration And Monitoring (SLAM)
Small-holder farmers (SHFs) 3, 6, 10-11, 15, 20
Smallholder farmers, supporting 389-394
  future trends 430-435
  improvement strategies
    extensionists and technicians 401-403
    information and communication technologies (ICTs) 409, 411-416
    integrated PHM and implementation options 403-409
    PHL reduction technologies 394-401
  institutional, policy and market-related issues for PHM transformation and 416-424
  sustainable market creation and 424-430
Smooth-wall silos 130
Societal loss 5
South Africa 243
South America 243
South Korea 50, *95*
Soviet Union 328
Spoilage moulds and mycotoxin contamination
  $CO_2$ monitoring for real-time identification of risk of 344-345
  control strategies 345-354
    irradiation as control method 350-351
  factors affecting 341-342
Spontaneous heating 250-251, 260
SPP. *see* Stored-product protection (SPP)
Sri Lanka *95*
Steel framing 138-139
Steel silos 129-130, 135
  benefits of 130
  limitations of 130-131
*Stemphylium* spp. 37
Sticky traps 209, 213
Storage mycoflora *36*, 37
Stored-product protection (SPP) 270
Strip auger samplers *206*
Strong resistance 73
Sub-acute rodenticides 108
Sub-Saharan Africa (SSA) 4, 172, *174*, *177*, 178, *190*, 391, 392, 405
Suction aeration 134-35
Sudan 243
Sulfuryl fluoride 83-84
SuperGrainbags™ bags 172, 173, 178
Supply chain 17-18, *17*
importance, in development project 20-21

Surface traps 210, 213
Sustainable Use of Pesticides Directive (EU) 290
Swaziland 232
Sweep conveyors 133
Swiss Agency for Development and Cooperation (SDC) 11
Syngenta 112

Tanzania *95*, 100, 179, 188, 408
  maize in 21–23
  rural community rodent management in 112–113
Tarp 145–147
Temperature and relative humidity of air, surrounding grain 370–371
Temperature cable sensors 365–367
Temperature equalization throughout grain bulk 249
Temperature sensors 344
*Tenebrio molitor* 280, 294
*Teretrius nigrescens* 280, 286, 405–406
Tetraploid durum wheat 52
Thailand 48, *95*
Thanksgiving Day 33
*Theocolax elegans* 288
Thermal plasma 160
Thermistors 366
Thermocouples 365
Thermotolerants and thermophiles 35
Thinline traps 209
3D imaging technology 379–380
Three-step approach 14–15
*Toxoplasma gondii* 110
Traps 110, 205–213, 380–381
Trécé Inc. 209, 210, 212
*Tribolium* spp. 201, 210, 213, 214, 235, 280, 288, 294
  *castaneum* 48, 69, 76, 78, 80, 158, 207, 209, 235, 236, 239, 283, 352
  *confusum* 48, 236, 241, 279, 288, *292*, 293, 294
*Trichogramma* spp. 280, 287
  *evanescens* 281
  *pretoisum* 295
Triple hermetic bags *174*
Triple layer storage bags, in Africa 20–21
*Triticum* spp. 52
  *monococcum* 52
  *turgidum* 52
*Trogoderma* spp. 279
  *granarium* 70, 200, 210, 279, 283, 322
Turkey 77
Twitter 414

Uganda 10, *11*, 188
UGMA. *see* Unified Grain Moisture Algorithm (UGMA)
Ukraine 52, 243, 328
*Umuganura'* day 33
Uncovered piles 145
Underground pits 172
Under-roof condensation 250
Unified Grain Moisture Algorithm (UGMA) 369
United States Department of Agriculture (USDA) 320
  Agricultural Research Services 135
Upward aeration 153
Uruguay 243
US Code of Federal Regulations 272
USDA. *see* United States Department of Agriculture (USDA)
US Food and Drug Administration (FDA) 46, 51, 320, 324
US grain pile 147–148

Vacuum treatment 157
Value Added Tax (VAT) 183
Vapormate 84
Variance 216
VAT. *see* Value Added Tax (VAT)
V-bottom buildings 140–141
*Venturia canescens* 287
Vietnam 48, 409
*Vigna* spp.
  *inguiculata* 173
  *radiata* 176
  *subterranea* 176
Volatile organic compounds (VOCs) 283, 284, 378–379
Volatiles, as early indicator of stored cereals spoilage 342–343

Warehouse pirate bug. *see Xylocoris flavipes*
Warehouses 126, 137, 138. *see also* Bulk storage
Weak resistance 73, 80
Web Foundation 416
Wefarm 414
West Africa 180

WFP. *see* World Food Program (WFP)
WhatsApp 414
Wheat, fungal contamination in 52, 54
  infection in wheat kernels during storage 52-53
Wireless monitoring 245, 257
Wireless sensor network (WSN) 372-374
Wire netting, for rodents 110
Wood framing 139
World Food Program (WFP) 132, 179-180
WSN. *see* Wireless sensor network (WSN)

Xerophiles 35-36
X-ray radiation 325-327
*Xylocoris flavipes* 286

Yellow mealworm beetle. *see Tenebrio molitor*
YieldWise initiative 21-23, 179
YouTube 414

Zaire *96*
Zambia *96*, 412
*Zea mays* 176
ZeroFly® hermetic storage bag 173, 174, 178
Zero loss, significance of 7-10
Zimbabwe 399, 403

Lightning Source UK Ltd.
Milton Keynes UK
UKHW050106220820
368601UK00001B/31